THE INFORMATION PARADOX

THE INFORMATION PARADOX

Robert Wiles

MB BS FRCGP FRACGP FACRRM

Information Press

First Published in 2014
By Information Press
PO Box 159
Cooma. NSW. 2630
Australia
www.infopress.com.au

ISBN-13: 978-0-9875622-0-3

National Library of Australia
Cataloguing-in-Publication data:
Information systems / Bioinformatics / Medical informatics / Australian

See http://catalogue.nla.gov.au

Design: Derrick I. Stone, Melbourne
Cover photo: Holly Kuchera/iStock/Thinkstock

Printed by Lightning Source

Dedication: To those who thirst for wisdom.

Foreword

How wonderful that we have met with a paradox. Now we have some hope of making progress.[1]

NIELS BOHR

For better or for worse, we live in the Age of Information. Information is ubiquitous and unavoidable for citizens of the 21st-Century. We use it every day to prepare our meals or to pay our bills, not to mention in the running of our businesses or governments. Telephones are a common example of how we have become increasingly dependent on information. A generation ago, a telephone did not need an instruction manual. However, mastering a new mobile device today with its message bank, text applications and cloud capability can be daunting without instructions to expound the myriad functions. Banks are yet more information-dependant. If their computers malfunction, such as during a power failure, the banks close their doors as they can no longer trade. And the rate at which new, information-hungry devices are invading our lives does not seem to be slowing. If anything, the speed of innovation appears to be accelerating: in the home, for work, at leisure and in the sciences. Moore's Law[2] offers a typical example, stating that the processing power of new computers doubles every two years (which equates to exponential growth). In fact, the only way that we will be able to manage the stresses on our planet: the growing population pressure on land, the dwindling food supplies in the seas and the increasing resistance of pathogens to antibiotics, is through the rigorous study and efficient management of information.

I began investigating information eight years ago in an effort to better explain to my children the evolution debate, as our eldest child was studying biology in high school. I had become frustrated with the seemingly endless dispute. Initially I sought for an entity or principle to which all scientists would agree. Just as a spike hammered into the rock can act as an anchor point for mountaineers, I was searching for a common idea from which further scientific accord could be reached. The entity I discovered that could be acknowledged by all scientists was *information*. When I started I intended to limit my exploration to biology and in particular the appearance of the first cell. At school and university I had always accepted Darwinian evolution unequivocally. However, using information to analyse what I was reading, yielded results that were not as I had anticipated. These unexpected findings led me to wonder what information

1. Jones, A.Z.: About.com Education: Physics – Niels Bohr Quotes http://physics.about.com/od/nielsbohr/tp/Niels-Bohr-Quotes.htm

2. Gordon Moore, founder of Intel Corporation made this observation in 1965. Finkel A.: Moore is less with Physics. *Cosmos* 54:31 (2013).

analysis might reveal about the evolution of a multi-celled organism. The results were just as unforseen, so I probed further to examine the appearance of species, the formation of the biosphere and finally, the existence of the Universe. At each stage I was prompted to delve deeper because of the surprising results that emerged from using information to assess the preceding topic. The research rapidly progressed to the point where I could not organise all the information in my head, especially the Law of Coding (see Chapter 3). So I started writing the findings down to clarify the ideas. I was intrigued that relatively little is known about the entity of information itself, even though most people acknowledge that we are living in the 'Information Age'. This is the first paradox. However, since information proved such a useful tool for the extraction of understanding, which was otherwise obscured, setting out the findings in a book seemed the most appropriate way to share these discoveries.

Prior to 1980, the World conformed to a different paradigm. Since the beginning of the Industrial Revolution, the cost and immobility of manufacturing plants had exerted an attractive force on workers who moved to where the infrastructure was located. The result in Western Europe was the growth of the large, industrial cities such as Birmingham and Cologne. The invention of the microchip is reversing this process, with the trend toward centralisation continuing to slow for the last two decades. The reason is that the power and the portability of microchips when connected to the internet, have allowed many businesses to mobilise and literally operate from anywhere in the World, with far fewer workers than in the past. At the same time, the pressure on land and services and hence the cost of operating a business is often higher in the centre of a big city, not to mention the attendant problems of traffic, pollution and crime. Consequently many business owners find it cheaper and more attractive to move away from the city centre, as they can still access most of the benefits that were once restricted to an inner-urban location but without the associated costs.[3] In Australia, this process is called the 'Sea Change' or 'Tree Change' phenomenon, where significant numbers of people are moving out of the cities to live in coastal or rural areas. This change in the forces shaping our societies is not only relocating people but it has reshaped whole industries, sometimes extremely rapidly. Under the previous paradigm big was beautiful. Nothing epitomised this more than IBM which, in the 1970s, was known as 'Big Blue' and controlled eighty per cent of the World computer market. Within a decade, however, the advent of cheaper, more powerful microchips in the 1980s caught IBM unawares. Before the company could change to meet the downsizing trends created under the new paradigm (from mainframes to desktop computers), IBM had lost this market dominance, which it has never regained. Manufacturing the hardware was also supplanted by software production as the highest value product and Microsoft replaced IBM at the top of the tree. This information/microchip/internet revolution appears to be progressing unabated. Therefore, if we are to avoid being overtaken by the information tsunami, like IBM, we must radically re-evaluate some of our most cherished ideas to incorporate an information perspective. As we explore this 'brave new world', we will discover that an *Information Model* of the Universe emerges from this new paradigm that will stretch and extend our minds in novel directions and provide explanatory power that has been lacking in previous models.

This Model uses information both qualitatively as a powerful investigative tool and quantitatively to provide a benchmarking function to measure and hence

3. Davidson, J.D. & Rees-Mogg, W.: *The Great Reckoning: How the World will change before the Year 2000*. Pan Books, London (1994) pp.83–85.

appraise the many vastly different entities in the Universe. Information provides a common currency by which we can compare the contrasting fields of biology, chemistry, geology, physics and cosmology, thus giving a new understanding to things we may have previously taken for granted. Yet as the common currency, it also acts as the 'gold standard' to keep us intellectually honest as we use science and philosophy to explore our Universe. In addition, information can be audited and information leakage or injection can be tracked to prevent erroneous conclusions.

This book is a journey of discovery, visiting places few have ventured before and I invite you to join me as we investigate the length and the breadth, the height and the wonder, the antiquity and the paradoxes of information and its relevance to our lives. As information is integral to all the sciences, this book addresses a wide range of disciplines. However, as no one person can be an expert in every area of scientific endeavour and very few books have been specifically written about information per se, I have quoted a wide range of authors to glean the understanding required. I expect that no one will agree with the views of all of the authors cited but I ask you to judge this book on the strength and veracity of the arguments (the information), rather than on personalities; in other words, please debate the message, don't shoot the messengers! I have experienced a very steep but utterly enjoyable learning curve in researching this book; I sincerely hope that as you read it, you will too.

Robert Wiles
February 2014

Contents

Acknowledgements

This book has proven to be a much bigger undertaking than I ever envisaged at the outset and I am very grateful to all those who have listened to the developing arguments and offered advice to put me back on the path when I have strayed off it. In particular, I thank Dr Mark Glasgow for his prescient choice of books and discussion to provide inspiration, Professor John Lennox for advice on refining the ideas and terminology, Dr Alastair Noble for assistance with identifying new sources, Dr Desmond Ford and Matthew Hiorns for proof reading and feedback and Bernard d'Abrera for offering new avenues and refining the arguments and presentation, not to mention the robust debate. Also, I would like to acknowledge the considerable assistance of Derrick Stone in designing this book and assisting with bringing it to publication.

This book contains a vast array of information and every effort has been made to quote accurately and appropriately and to correctly reference the material and ideas presented, however, if errors have occurred, I apologize in advance.

Finally, I would like to acknowledge my wife, Dawn and our four children who have listened patiently during umpteen retellings of the concepts, acting as sounding boards for the development of new ideas and for their assistance in chasing references and sources and in particular, Amy, for proof reading.

1

The Information Age

To follow knowledge like a sinking star,
Beyond the utmost bound of human thought.[1]

Alfred, Lord Tennyson

The Explosion of Information

We live in the Age of Information, a time of accelerating change that is fuelled by a rapidly growing deluge of facts and figures, observations and calculations which we collectively call: 'information'. My school physics textbook stated that at the beginning of the 20th-Century, all the knowledge in the Western World was known by just five men (yes, in those days, women like my grandmother were usually denied a university education) but by the close of the 20th-Century, five people would not have been able to know even one branch of science or the arts. This has especially been so in biology and medicine where our understanding of the functioning of the cell and living organisms has progressed further than could ever have been dreamt of 100 years ago.

The existence of information is not a new discovery, though for most of man's history, it has not been recognised as a discrete entity worth studying for its own sake. Stretching back into antiquity, previous civilisations *used* information but in general, they had other more pressing priorities than the study of information. The ancient Sumerians and Babylonians had a great appreciation of the stars and left many clay tablets, while the Ancient Egyptians built the pyramids but their technology was limited to the abacus and the fulcrum. As time advanced, the Chinese invented paper and discovered gun powder, the ancient Greeks learnt to smelt metals to make bronze and the Romans forged iron. Despite these achievements, however, all the civilisations of the ancient world primarily depended on agriculture. The manipulation of materials, such as timber and metals, was generally limited to articles required for the survival of that civilisation, principally food. In other words, they were agricultural economies. Information was a tool to derive their livelihoods from the land or to protect themselves from invasion. Information was primarily a means to meld matter for sustenance rather than an end in itself; hence the term: the Agricultural Age.

In Europe, the Agricultural Age lasted until about 1750, when the invention of the steam engine changed the World forever. The ability of the steam engine to increase productivity through the harnessing of stored energy (coal) allowed a larger proportion of the population to be relieved of agricultural

1. Alfred Lord Tennyson: *Ulysses*. (1842).

duties in favour of higher value activities and the pace of information acquisition increased. In Western Europe, and Britain in particular, the Agricultural Age gave way to the Industrial Revolution. However, information generally served a similar purpose to the previous agricultural era, as a means to feed and house the population. The generation and/or application of energy now became the high value pursuit and the efficient application of energy determined the winners under this new paradigm.

The Industrial Revolution expanded around the World and lasted well into the 20th-Century. In the USA, in 1837, however, Samuel Morse (1791–1872) was granted patents for the Wire Telegraph Machine and the seed was sown for a new paradigm, although it would take more than a century for that new paradigm to supersede the existing industrial one.

Although Morse code could rapidly traverse great distances and had a significant role to play in changing the prevailing paradigm, the origins of information exchange are much older and are lost in the mists of time. *Personal* human communication began before recorded history and had already progressed from the spoken word, to the written word and then into print with the invention of Gutenberg's movable-type in 1440.[2] In parallel to these local forms of communication, remote information transfer had also been achieved for millennia by using light and smoke signals. A Roman lighthouse still stands in Dover Castle, England and it is well known that the Native Americans used smoke signals to communicate. However, the volume of data that was able to be communicated was severely restricted. Morse code could deliver significantly more information than light or smoke signals but with time, even this quantity of data transmission became a limiting factor for Morse code too. Then in 1876, Alexander Graham Bell (1847–1922) was granted the first US patent for a telephone. The advent of the telephone radically increased the volume of data that was able to be communicated and heralded the Communications Era.

The next significant step came when Italian-born inventor, Guglielmo Marconi (1874–1937) developed the first practical radio transmitter in 1895, which freed communications from the constraints of copper wires. Sound wave broadcasting became moving-picture transmission just 30 years later when, a Scotsman, John Logie Baird (1888–1946) demonstrated the first television in 1926. Initially, the signals were in waveform (analogue) but improving technology provided greater precision for the transmission of information with digital-signal broadcasts, the system we have today. After World War II, the industrial paradigm transitioned into a communications one. The World began to shrink as communications companies grew and the transmission of information commanded a higher value than the utilisation of energy, which had defined the Industrial Age.

After the Communications Era came the Information Age, the origins of which can be traced back at least as far as the abacus. The abacus was in use for commercial purposes in ancient Babylon, as early as 2400 BC but increasing technology and knowledge eventually produced *mechanical* calculators. The Antikythera mechanism from ancient Greece dates to 100 BC and is the earliest discovered mechanical calculator.[3] Calculators continued to develop over the centuries but they were limited in their power and scope until the invention of the radio valve (also called a vacuum tube), shortly after the turn of the 20th-Century. This electronic method of manipulating information dramatically improved performance. The radio valve was followed by the transistor in the early 1950's and

2. Lloyd, S.: *The Computational Universe. in* Davies, P. & Gregersen, N.H.: *Information and the Nature of Reality: from Physics to Metaphysics.* Cambridge University Press (2010) pp.92–3.

3. Freeth, T., Jones, A.R., Steele, J.M., & Bitsakis, Y.: Calendars with Olympiad display and eclipse prediction on the Antikythera Mechanism. *Nature* 454: 614–617 (2008).

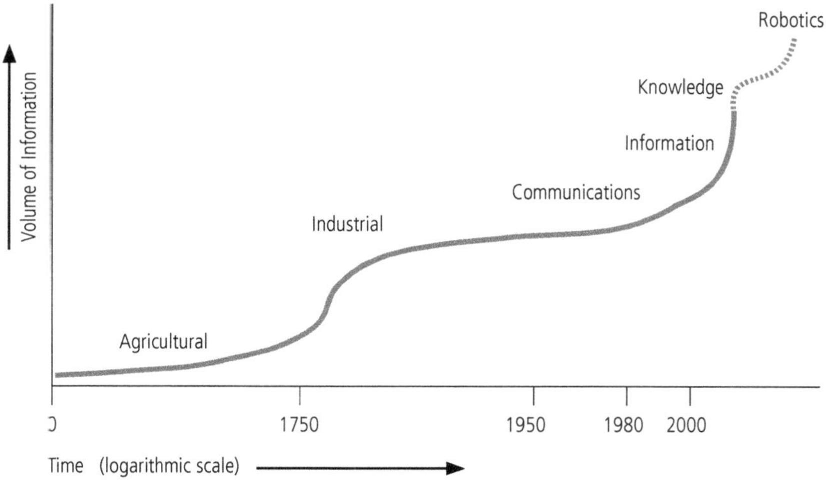

Figure 1.1: The Historical Ages of Europe illustrating Information Growth
Successive paradigms have lasted shorter and shorter periods before being replaced, which reflects the accelerating pace in the growth of information.

then the microchip in the 1960's, which allowed increasingly powerful computers. In 1981, with the launch of the IBM personal computer, computers became small enough, cheap enough and practical enough for business and personal ownership, ushering in the Information Age. The highest value commodity now became the collection, storage and retrieval of information and the pre-eminence of the communications companies began to wane as their dominance was challenged by the computing industry. Apple Computer enthusiasts are certain to dispute this date as the Apple II predated the IBM Personal Computer by a number of years but the lack of a standardised operating system (developed by Microsoft at the behest of IBM) hindered the uptake of computing technology by the average business and home user.

In the same way that knowledge of the information sciences has exploded in the latter half of the 20th-Century, biology has also experienced dramatic growth in its understanding of how biological systems work. The biological revolution began with James Watson and Francis Crick (1916–2004) when they deciphered the x-ray crystallographic pictures of DNA and deduced its structure in Cambridge, England in 1953. Crick, who had been a cryptographer during World War II, introduced the notion that DNA was an *information carrier* and not just a chemical. This understanding instigated the modern era of biological information and the concept that the cell is also a repository and processor of information.[4]

However, the history does not end there. The rapid accumulation of massive volumes of data has necessitated interpretive and analytical programmes that are able to recognise and process raw information, such as that found in the financial and healthcare systems. This foreshadows the dawn of the Knowledge Revolution. The Knowledge Revolution is likely to spawn even more sophisticated computer programmes, allowing increasingly intelligent machines to be developed over the next 50 years. The connections between the millions of computers that make up the hardware component of the Internet and the software programmes, dubbed the 'World Wide Web', are products of the Communications and Information ages which present vast amounts of information in a form that is readily usable for further study and analysis. The automated analysis of this data by ever more efficient search engines, such as 'Google', marks the emergence of the Knowledge Revolution.

4. Missler, C. & Meyer, S.: *In the beginning there was... information.* (CD) Interview with Dr Meyer. Koinonia House (2000).

The Age of Robotics is yet to come, where new and novel information will be able to be generated and analysed without direct human input. Machines with complex integrated management systems, exploiting vast amounts of information (such as on the Internet), will operate with increasing independence of their human designers as our skills in programming and manufacture increase. Such self-learning systems are complicated to design and are at present in their infancy. Along with this continuous increase in information has been a massive escalation in complexity—the sophistication in the way the information is linked.

This brief history of the modern development of the human species has been summarised in Table 1.1. The table shows that food (matter) and energy production have been primary requirements for the early development of modern civilisations but that matter and energy are not enough; our modern society uses a third entity: vast quantities of relevant information as well.[5] So I am inviting you to put on the cloak of a 21st-Century Sherlock Holmes and rather than calling yourself a detective, think of yourself as an 'Informationist' and join me for a grand tour of the information that has swamped us in an information tsunami. It is only through the careful study of information that we will be able to fully understand both who we are and the Universe in which we live.

Historical Ages of Europe	Principal informational activity	Major activity
Agricultural Age	Information Generation/Recognition	Food production
Industrial Revolution	Information Application (value adding)	Energy utilisation
Communications Age	Information transfer	Information exchange
Information Age	Information accumulation & interpretation	Information collection
Knowledge Revolution	Information complexing & understanding	Information manipulation
Robotics Era	Information auto-analysis & integration	Information auto-generation

Table 1.1: Summary of the Dominant Activity in each Historical Age
It can be seen that information has been integral in every age but in an increasingly powerful way.

The competitive advantage that successive civilisations gained through the application of information led to its increasing usage. This has now developed to the extent that the study and analysis of information has been recognised by governments as an essential requirement for the efficient functioning of our modern societies and my own children's school curriculum now requires the study of *mandatory* subjects in Information Science.

Although information was initially used for practical purposes, such as to manufacture more efficient ploughs and to produce better steel from which to craft swords, in time, the recognition of the usefulness of studying the world around us led the ancient Greeks and Romans to seek knowledge for its own sake. When that knowledge, which is applied information, became regularised and systematised it became known as 'philosophy'. Philosophy is defined by the *Concise Oxford Dictionary* as the *"love of wisdom or knowledge, especially that which deals with ultimate reality, or with the most general causes and principles of things"*.[6] It is the study of the causes and laws of all things, both concrete and abstract (physical or metaphysical). Philosophy establishes the fundamental rules for the accountability of all human endeavours, through empiricism and logic and has as its aim: the pursuit of understanding. Philosophy involves the study of information or ideas that cannot readily be tested or verified by repeated experimentation, such as origins. In contrast, 'science' is the study of information that has been submitted to repeated experimentation or observation with consistent results. Aristotle is considered to be a philosopher but also the father of science and as science

5. Thaxton, C.B., Bradley, W.L., & Olsen, R.L.: *The Mystery of Life's Origin: Reassessing Current Theories.* Philosophical Library, New York (1984) *in* Woodward, T.: Darwin Strikes Back. Baker Books, Grand Rapids (2006) p.117.

6. *Concise Oxford Dictionary.* (5th Ed.) Oxford University Press, London (1973).

involves the study and verification of information, science comprises much of this book. Therefore, science is a useful place to begin our journey into information.

Defining Science

So, what is science? Double Nobel laureate, Linus Pauling (1901–1994) stated: *"Science...is the search for the truth."* [7] Which leads to the next question—what is truth? From the above paragraph it can be seen that science developed out of philosophy, with which it needs to be consistent. 'Scientific method' derives from the empiricism and logic required by 'philosophical method'. These principles keep science objective and accountable. Science works well with tangible objects and concepts such as solids, liquids and gases, because experiments can be conducted, results can be demonstrated and the process can be repeated to confirm outcomes. Conclusions can be rigorous with tangible entities but science struggles with intangible concepts such as thoughts and feelings. We know that thoughts and feelings exist because we experience them every day. However it is one thing to be able to experience something and it is quite another to be able to understand and explain it. One current concept of science maintains that there is a *physical* cause for every effect; this is known as 'Materialism'. So where does that leave thoughts and feelings? Science is seen by many as the final arbiter in distinguishing truth from error, yet it struggles with intangible and non-material entities. These concerns will be discussed in detail in later chapters.

Information is another intangible entity. In Western Society we deal with it and use the term every day yet it is only vaguely explained by scientific method. The idea that there are things that we cannot see, nor perceive with our five senses, is not new. Radio waves and television signals are not perceivable by our immediate senses and were not known to Sir Isaac Newton yet we know they exist, that is, they are real. Radio waves became demonstrable when the appropriate technology was invented and we may find other entities in the future which, although they are currently imperceptible, may become discernible in time. We know that when we think, our brains generate minute electrical signals. These electrical signals can be recorded on very sensitive recording equipment called an electroencephalogram (or EEG) to create a 'brain wave' pattern. Our hearts beat due to tiny electrical impulses too, which can be replaced by an electrical pacemaker if our hearts become damaged and our own biological pacemakers cease to operate efficiently. Artificial pacemakers require information for their manufacture and operation, so is there also information in the natural pacemakers, (the sino-atrial and atrio-ventricular nodes) in our heart? In later chapters we will discover that there is, therefore we will not be able to fully understand ourselves if we do not study the information that allows our bodies to function. Thus, in both the physical sphere (such as radio waves) and biological sphere (such as thought), intangible yet verifiable phenomena occur. As information is integral to both of them, the relationship of science, not only to tangible but also to intangible and non-material entities needs further investigation and clarification.

Defining Information and Knowledge

One aphorism I heard recently on the radio said: *"We are drowning in information but are starved of knowledge; hundreds of newspapers are produced daily and*

7. Pauling, L.: *No More War!* Dodd, Mead & Co., New York (1958) p.209 *in* Strobel, L.: *The Case for a Creator.* Zondervan, Grand Rapids, Michigan (2004) p.28.

thousands of books are published weekly so that information is doubling every four years." [8] How are we to make sense of this accelerating growth of information? What does it all mean? To dig deeper, we need to define some specific terms such as: 'data', 'information', 'knowledge' and 'understanding'. By consulting the Oxford English Dictionary I found the following definitions:

Datum (plural data): *"1646. A thing given or granted; something known or assumed as fact, and made the basis of reasoning or calculation."* [9] So a datum is something which is known to be true that can be relied upon as a starting point; which is why good science starts with accurate research data to establish a solid foundation for a line of reasoning.

Information is a term derived from the Latin word *informationem* meaning 'to inform'. It was in use in Britain by 1450 and appears to have come into Middle English via Old French although the term was undoubtedly used during the Roman occupation of Britain until the fall of the Roman Empire in the 5th-Century. The Shorter Oxford English Dictionary defines it as: *"The action of informing; training, instruction; being told, items of knowledge, news."* [9] This definition is good as far as it goes, but our modern usage of information would appear to be more than this. Professor Emeritus of Law at Berkeley, Philip Johnson writes: *"Information...[is] a message which conveys meaning."* [10] American Associate-Professor of Philosophy, Stephen Meyer defines information as: *"The number of instructions required to specify a structure."* [11] British anthropologist and linguist, Gregory Bateson (1904–1980) stated in 1972 that information is: *"a difference that makes a difference."* [12]

Information is also integrated data, or data that produces a useful outcome. Compare this to 'static' or 'noise' which conveys no meaning nor confers any productive outcomes. Information could be defined as a "useful entity" but in a living cell, a protein which has no use today may prove to be useful tomorrow when environmental conditions change. Therefore the definition of information must include complexity which has the potential to aid survival now *or in the future*. Thus the term 'information' is used in a number of different settings with different meanings and *all* of them are relevant to what we will be studying.

Knowledge is a term which dates from early Middle English in the 14th-Century. The Oxford Dictionary defines it as: *"The fact of knowing a thing, state or person. Familiarity; intellectual acquaintance with, or perception of fact or truth; the fact, state, or condition of understanding. The sum of what is known; [A] person's range of information (in philosophy) certain understanding, [as] distinct from opinion."* [6]

Understanding is: *"1. To comprehend; to apprehend the meaning or import of;* 2. *grasp mentally, perceiving the significance or explanation or cause or nature of, know how to deal with* [something]; 3. *infer, especially from information received.*[6,9] But bare definitions do not always give us the best appreciation of a topic. Therefore, looking at the characteristics of the issue being investigated can often help with understanding the subject. Scientists now recognise that information is not matter. It requires matter for storage and transmission but information and matter are entirely different entities. Information can be both quantitative (it can be counted) and qualitative (it can be compared). Information may also be digital, thus it can be manipulated by mathematics and computers. The term 'digital' can have a

8. Vision Radio, Australia (2010).

9. *Shorter Oxford English Dictionary.* Oxford University Press, London (1986).

10. Johnson, P.: *Wedge of Truth.* InterVarsity Press, Downers Grove, IL (2000) p.42.

11. Meyer, S.C.: The Explanatory Power of Design – DNA and the Origin of Information, *in Mere Creation – Science, Faith & Intelligent Design.* Dembski, W.A. (Ed.): InterVarsity Press, Downers Grove, IL (1998) p.142.

12. Bateson, G.: *Steps to an Ecology of Mind: Collected Essays in Anthropology, Psychiatry, Evolution, and Epistemology.* University of Chicago Press, Chicago, IL (1972) *in* Davies, P. & Gregersen, N.H.: *Information and the Nature of Reality: from Physics to Metaphysics.* Cambridge University Press (2010) p.163.

number of meanings including: representing data in a numerical format but also, a signal that is apportioned in whole discrete units rather than fractions. The last concept to be appreciated at this stage is summed up by astrophysicist, Paul Davies who said: *"No physical laws will define information."* [13] The reason is because information is not physical but intangible. So even at the beginning of our journey, we have been able to discern a number of significant characteristics of information. But there is more...

The Multi-layering of Information

Information has been increasing rapidly over the 20th-Century, to the extent that we are now drowning in it. Scientists have known for hundreds of years that information is not restricted to human activity alone but that it is also found in biology and the microscopic world which we cannot see with the naked eye. Modern technology has been helping us to recognise the full extent of nature's reliance on information. A typical cell is a good example. In 1998, Bruce Alberts, President of the US National Academy of Sciences said: *"The entire cell can be viewed as a factory that contains an elaborate network of interlocking assembly lines, each of which is composed of a set of large protein machines ... why do we call [them]* machines? *Precisely because, like machines invented by humans to deal efficiently with the macroscopic world, these protein assemblies contain highly coordinated moving parts."* [14] Thus a cell is packed full of information-containing structures but the factory analogy equally applies to a house or even the complexity of a mediaeval town surrounded by a city wall. They all provide a contained environment protected from would-be invaders with areas for all essential functions: (See Table 1.2)

Table 1.2: Examples of Multi-layered Information
It can be seen that information is ubiquitous, densely packed and multilayered.

	Function	Factory	House	Town	Living Cell
1	**Nutrition**	Loading dock	Kitchen	Cafes / restaurants	Vacuoles
2	**Power**	Electricity supply	Meter Board	Power station	Mitochondria
3	**Waste disposal**	Waste bins / sewer	Toilet/bathroom/ laundry	Sewerage works	Golgi apparatus
4	**Repair / manufacture**	Maintenance department	Workshop / garage	Industrial estate	Ribosomes
5	**Information storage**	Manager's office	Bookshelf/ computer	Town Hall / library	Nucleus/ DNA
6	**Transport links**	Conveyor belts	Hallways	Roads	Cytoplasmic reticulum
7	**Ports**	Entrance	Doors /cat flaps	Gate house	Portals
8	**Defence**	Walls	Walls and roof	City wall	Membrane
9	**Management**	Manager	Occupant/ owner	Town council	Epigenome

It is worth noting that cells in an organism have great variability and connectivity, just as do houses, factories and towns. Also note:

- Houses, factories and towns are connected to essential services in their immediate environment and the region.
- Life can be simple like an isolated farm in the country. To survive in this situation, the farmhouse needs to be self-contained with its own water supply and septic system (just as an amoeba is self-reliant).
- Some people live in suburbia but they are dependent on being connected to the centre via transport arteries, such as most mammalian muscle cells.
- Inner-city tenements/ high-rise (back-to back) have similarities to liver cells.

13. Dawkins, R. & Lennox, J.C.: *The Great Debate, Oxford.* (DVD) Fixed Point Foundation, USA (2008).

14. Alberts, B.: The Cell as a Collection of Protein Machines. *Cell* 92:291 (1998) *in* Strobel, L.: *The Case for a Creator.* Zondervan, Grand Rapids (2004) p.193.

Thus, it can be readily seen that there are multiple layers of information all around us, in the macrocosm of our society and the microcosm of our bodies.

Investigating Information – the Basic Paradoxes

The title of this book implies that there are paradoxes within information systems and we met the first one in the Foreword. There are a number of information paradoxes and it is time to examine two more. When investigating information, it can be seen that information is not matter but matter can convey or store information, if meaning is contained within the physical arrangement of the matter. That is, there must be a degree of organisation to allow the storage of information, such as on a compact disc. In the same way, *energy* can also convey or store information such as in the flashing beam of light from a lighthouse. Not only does the beam of light reveal the presence of the lighthouse but the frequency of the flashes within the beam reveals which lighthouse is sending the signal. The greater the organisation within the medium, the higher is the density of information that can be stored and hence, the larger the potential for meaning and utility. The meaning comes from order and thus, we infer *no meaning* from disorder (which is often called 'static'). The natural tendency of all systems is to move from order to disorder which is in accordance with the Second Law of Thermodynamics (Law of Entropy). It is a common observation that information does not arrange itself spontaneously either. (When was the last time you saw a lighthouse or a compact disc create themselves?) Therefore, in summary, there are two paradoxes evident here:

Paradox 2: *Information is neither matter nor energy;* yet matter and energy can convey or store information as long as the physical arrangement within the matter or energy has meaning;

Paradox 3: *Matter is not meaning;* yet meaning can be conveyed or stored by matter or energy, as long as the physical arrangement within the matter or energy contains information.[15]

These counter-intuitive facts need to be borne in mind as we analyse information.

Information Transmission

Thus matter and energy can store and transmit information and these physical entities are quite well understood but what about the information itself? John Lennox, Professor of Mathematics at Oxford University in England writes: '... *information is invisible. The carriers of information may well be visible —like paper and writing, smoke signals, television screens or DNA—but the information itself is invisible*.'[16] Thus, to detect the information we need to make the mode of communication of the information visible, so that it can be identified (like the radio waves) but the *information itself* still remains invisible, which is why much information has gone unrecognised for so long. Another way to approach this problem is to appreciate that information needs a language for its expression. However, learning to speak English or French does not provide specific information. It simply provides a vector to deliver information. Further, differing delivery systems can convey the same information. One can receive information

15. d'Abrera, B.: personal communications. (2010–11).

16. Lennox, J.C.: *God's Undertaker. Has Science Buried God?* Lion Books, Oxford (2007) p.168.

that: 'a knife is sharp' by hearing someone describe it, reading a report on a piece of paper or feeling it with one's fingers, yet the information (about the sharpness of the knife) is independent of the way that one finds out about it. To complicate matters further, a language is also independent of *its* medium.[17] The language is 'software' that depends on 'hardware' for its delivery. This creates some challenges for the trainee informationist. As we cannot pick up a ruler and physically measure information, to better understand it we can consider specific examples to make this clearer. The message (information) might be Hamlet's soliloquy: *"To be or not to be..."* A thought in Shakespeare's mind generates the thought: *'Hamlet is contemplating death'*, which becomes the message. The message requires a language and William Shakespeare (1564–1616) wrote in Middle English, but that message may be translated into Spanish, Russian or Chinese without losing its essential meaning. However, as the thought, the message, the language and our comprehension of the message are all intangible, we cannot see or interpret them until a vehicle delivers them to us. (How many times have you heard someone say: *'If you don't tell me I do not know. I cannot read your mind'?*) In a stage performance of *Hamlet*, the vehicle might be the sound waves from the actor's voice or the soliloquy might be printed in a book to be studied by high school students. There may or may not, then, be an effect or result of the message. The audience may applaud or even throw rotten tomatoes as in Shakespeare's day. (See Figure 1.2)

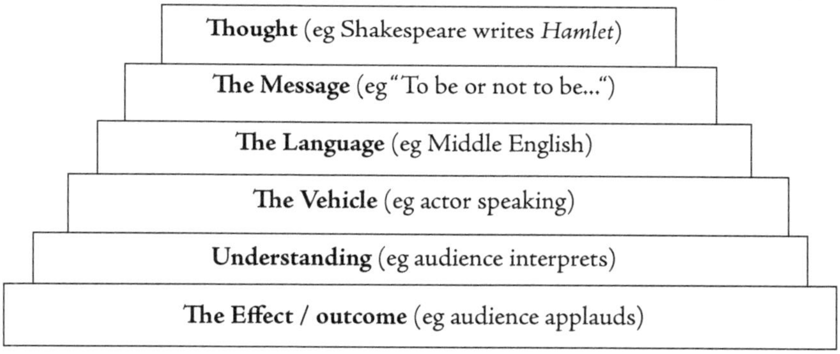

Figure 1.2: Steps in Information Generation, Transmission and Communication.

It can be seen from this example that information lacks any matter or substance of its own and as such, it is a *mass-less* entity. Therefore, material explanations alone will not be sufficient to explain information because information is non-material and the normal rules of matter do not apply. (To test this, consider whether gravity affects information?) In fact, it will be observed in Figure 1.2 that only the vehicle and the effect can be measured in physical terms. Yet each step is merely another part of a sophisticated information delivery system for a packet of information which began as an idea in Shakespeare's mind.

This example of a performance of *Hamlet* at the Globe Theatre, London in 1602 can illustrate another interesting property of information. *Hamlet* has become one of the best-known plays of all time but if it had been a complete flop and no one had turned up to hear Hamlet's second performance of his soliloquy, would the information still exist? The answer of course is yes, as it persists in written form. Four hundred years later we still have the soliloquy to enjoy, so information can exist without someone to hear or interpret it, at least under certain circumstances. Although the message (information) is intangible, the information can persist if a suitable vehicle exists to preserve it. Another example is engraved on the information plates of the Voyager Space Probes sent into outer

17. Missler, C.: *Cosmic Codes.* Koinonia House, Coeur d'Alene, Idaho (2004) p.305.

space in 1977. If Voyager 1 and 2 are never found or the plates are never read, the plates still contain information that was coded into them before they left Earth.

The first person to systematically study information was Claude Shannon (1916-2001) who published his research in 1948. Shannon looked at information quantitatively. Andrew Gosler, Chair of the Institute of Human Sciences at Oxford University wrote: *"Shannon* [was] *a mathematician and engineer working for the Bell Telephone laboratories in the 1940s. Shannon recognised that information (the signal) is the antithesis of entropy (noise), and that entropy declines as information increases."* [18] But in biological systems the content (the quality of the information) is just as important as the amount of information, so a qualitative measure is also required.[19] To illustrate this, imagine a string of DNA with an endlessly repeating segment and compare that with the genome of an amoeba. Although the endless sequence has more letters, it creates nothing, whereas the amoeba is an independent organism. As we look at the world around us, the number of entities that contain information, both quantitative and qualitative, appear endless. However, before we look further into qualitative measures of information, one common misconception needs to be addressed.

Just a Concept

To summarise our findings to date: we have established that information is contained in many different entities, we use it every day and it is intangible, which means that we cannot measure it with a ruler. But what is information? We have begun to discover what information is *not* but have not yet been able to say exactly what it is. Many people comment that information is *"just a concept"*. They consider that it is a useful way of thinking about the way we see our world but that is all. However, I discovered that information was a discrete entity after I purchased my first computer in 1983. When I first switched on my beautiful new device, I was annoyed to find that it was useless as an information processing machine until I had purchased extra programs and installed them. I had to purchase software in the form of an operating system (MS DOS at that time) and applications programs, such as a word-processing program as well. Once these were installed onto the new computer it became a very useful machine. But after the software was installed, did the computer become any wider, deeper, longer or heavier than the hardware alone? There is no doubt that something of value was transferred from the floppy discs and installed onto the computer (try using a computer without the software) and most software packages involve many hours of intellectual pursuit to write. This is why our society recognises the value of information/knowledge by allowing it to be copyright protected and the special term 'intellectual property' has been coined to describe it. Without the software packages the computer is useless as an information processing machine, although it could be argued that it makes an attractive ornament or effective paperweight. It can be seen therefore, that a discrete entity, which is contained within the product that we call 'software', that is more than the disk that carried the information, was installed onto the computer. If the information was *"just a concept"* then the information alone could not have affected the *physical* outcome, which was to allow the computer to function purposefully. This is the 4th Paradox: *Information has no mass and is independent of energy, but it is real* as it changes the function and outcomes of using the hardware that was purchased. Information is a discrete (yet non-material) entity in its own right and must be considered separately from

18. Gosler, A.G.: Surprise and the Value of Faith, *in Real Scientists Real Faith.* Berry, R.J. (Ed.): Monarch Books, Oxford (2009) p.183.

19. Gitt, W.: *In the Beginning was Information.* (3rd Ed.) CLV Neuhausen, Germany (2003) p.45.

matter and energy. Thus, in any investigation of the Universe, there are now three fundamental entities that must be considered: matter, energy and *information*.

Measuring Information and Complexity

Every decade has seen the uncovering of more and more of the vast number of intricacies and complexities within living systems. If we are to properly understand the world around us, we need to learn firstly to describe and then to specify such complexity. This specification requires analysis of the information within it.[20] As mentioned already, complexity is structured information.[21] Structured information requires a relationship to be specified, which in turn requires activity to occur to link those pieces of information in that particular way. If effort is expended to define information, then it is an active, not a passive process and from the Second Law of Thermodynamics it can be deduced that information in an isolated system cannot build itself spontaneously.

To measure the complexity of a structure, an arbitrary value may be assigned to each link or relationship which I have named an 'infobit'. Living organisms are made up mostly of proteins which are folded strings of component parts called amino acids. A typical protein composed of 250 amino acids has a *minimum* of 250 infobits specifying its sequence of amino acids and *at the least* another 250 infobits specifying its folded three-dimensional structure which equals a minimum of 500 infobits. If the information in the human genome with its 3 billion (10^9) base-pairs or as Stephen Hawking, Professor of Mathematics at Cambridge and Caltech physicist, Leonard Mlodinow express it: *"...the thousand trillion trillion molecules in the human body"*,[22] there are 10^{20} molecules. Expressed in infobits there are, therefore, potentially between 10^9 and 10^{20} infobits in each human being, an exceedingly vast amount of information. There is an information potential stored in each meaningfully-related piece of information, which would appear to be lost if that relationship between adjacent, specified entities is disrupted. Increasing entropy, which is equivalent to the disintegration of links specified by information, is a natural progression in our universe which derives from the Second Law of Thermodynamics. Thus, it can be seen that not only is activity and/or energy required to build complexity but activity and/or energy is also required to *maintain* entities linked by information so that they do not break apart or disintegrate over time. Therefore, complexity comes at a cost (both an initial cost to build it and an ongoing cost to maintain it).

If information is incorporated into an event, then meaningful or structured complexity can result if the information is materially preserved (for example: a landscape painting). This may be recognised as a work of art, containing qualities lacking in the blank canvas and paints before they were combined. But if no meaningful information is involved, as for example, if a can of paint was to accidentally fall and burst open on the dirt floor of the shed, then we do not recognise that a significant amount of new information was incorporated into the event, but we call it a 'paint spill'. On the other hand, if the artist signs their landscape painting, then further meaningful information has been placed on the canvas.

Thus complexity depends upon (or can be measured by):

1 The storage mechanism: The number of steps or bits of information required to describe the preservation method. This is a measure of the amount of effort required to *store* the information.

20. Behe, M.J.: *The Edge of Evolution.* Simon & Schuster, New York (2007) p.192.

21. For discussion on the types of complexity, three are given by Lennox, see Lennox, J.C..: *God's Undertaker. Has Science Buried God?* Lion Books, Oxford (2007) pp.145–6.

22. Hawking, S. & Mlodinow, L.: *The Grand Design.* Bantam Press, London (2010) p.32.

2 The information stored: The amount of information embedded within the storage mechanism. This also reflects the effort required to *create* the information.

3 The functional outcomes: The number of useful functions the information can perform.

As an example, think of the landscape painting and the paint spill described previously. There is no doubt that the autographed painting is more complex and contains more information than the random paint-spill in the dirt. Information was required to manufacture the canvas and paints. More information was required to create both the picture (for example: a rendition of the Tower of London) and any aesthetic value derived from the degree of skill with which the painting was executed. In contrast, the paint spill in the dirt has random information in the shape that the paint made as it splattered on the ground (which some may consider to be 'modern art'). There was information that went into manufacturing the paint stored in the can, but there is no canvas and no intelligible meaning to the arrangement of the paint molecules apart from the inference that: 'a spill occurred'. In addition, the painting of the Tower of London may also be used as a decoration or traded as an item of value but the paint spill has no useful purpose. This example demonstrates the various ways that information can be multilayered but also that there is often some detective work required to identify all the information encoded.

Thus, the linkage of entities involving information implies meaningful complexity, because if the linkage is broken then there is a loss of information or function (such as when the hyperlink on a Web page is disrupted). Such a simple linkage is a primary level of complexity. If three pieces of information become linked then there is a second degree of complexity etc. A number value can thus be ascribed to the complexity of an object to the value of (n−1). Thus: (n−1) complexity = n connections.

But knowing the number of linkages alone is not enough. Knowing the linkage of each paint molecule in the painting above cannot explain everything about the picture of the Tower of London. However, the painting is far too complex to analyse easily, so we shall begin by examining something much simpler, such as the amount of information or complexity that is required to successfully set a mousetrap.

At its most basic level, the simple mousetrap with its five parts has a complexity of n-1 which equals a fourth degree of complexity. We must, of course, choose the correct five parts to make a functional mousetrap. This is *constitutional complexity* but there is also information in the relationship of each part to every other part. Having the five parts loose in a box and disassembled does not catch mice either; so there is a further (n-1) complexity describing the relationship of each piece to its neighbour. This is *relational complexity*.

However, just building the mousetrap with the correct components in their correct relationships does not allow us to catch mice. Further information needs to be specified to activate the mousetrap. Cheese needs to be placed on the trigger and it needs to be set. This is *functional complexity*. One more piece of information needs to be provided - locating the mousetrap where it can catch mice. It is obvious that putting a correctly loaded mousetrap in a sealed refrigerator will catch no mice. We, therefore, need a further piece of information relating to *geographic or locational complexity*. In the simplest situation there would, therefore, be at least one piece of information specifying location. Unfortunately, the complexity does not finish there because each of the physical components that make up the mousetrap had information specified within them as well. These include the materials from which the components were made. This is *compositional complexity*. It is a gross oversimplification, but, for the sake of simplicity, let us assume that there is one piece of information specifying the composition of each part. This is significantly under-representing the true figure but it illustrates the principle. Thus, to measure the total complexity involved (in setting a mousetrap to successfully catch a mouse), one requires a combination of compositional, constitutional, relational, functional and locational complexity. As each component varies independently of every other one, the total complexity can be expressed as:

Total minimal complexity (TC_{min})
= Compositional x Constitutional x Relational x Functional x Locational complexity

In the case of the mousetrap this would be:

$$
\begin{aligned}
TC_{min} &= n \times (n-1) \times (n-1) \times 2 \times 1 \\
&= 5 \times 4 \times 4 \times 2 \times 1 \\
&= 160 \text{ infobits}
\end{aligned}
$$

Notice that this is the *minimum* amount of information that is required to specify how to set a mousetrap to catch a mouse. In reality, the complexity of most situations makes the calculation of the actual degree of complexity, as measured in infobits, almost impossible. However, this methodology does provide a way to measure and compare degrees of complexity. Note that infobits measure the number of steps or effort required but not the significance of the information or the amount of energy expended to create that information. The multi-layering of information is, once again, evident.

Information in Proteins

Protein molecules are both the building blocks and the workhorses of life. So to fully appreciate the amount of information that is involved in our everyday existence, we need to study proteins in detail. The average protein contains between 100 and 500 amino acids. There are over 150 naturally occurring amino acids but only 20 of them occur in living organisms on Earth. These can be arranged in any order. If a protein molecule was 100 amino acid units in length, there would be 10^{130} alternative ways to arrange the amino acids for that DNA string.[23] This figure is so large it is difficult to comprehend but to put it into perspective, physicists estimate that there are about 10^{80} particles in the Universe so 10^{130} is an extremely large number. Usually, there is only one permutation of this vast number of alternatives that will provide a biologically active molecule. To identify this one active combination of amino acids by chance is unimaginable unless there is a guiding mechanism by which to select that one active combination. To complicate the situation further, this 100-amino-acid protein must be precisely folded to form a three dimensional shape to make the protein molecule functional.

Hulbert Yockey, Cambridge nuclear physicist and bio-informatician, calculated that if a supply of suitable amino acids had been available for the estimated period that life has been on the Earth of 1 billion years, the largest

23. Lennox, J.C.: *God's Undertaker. Has Science Buried God?* Lion Hudson, Oxford (2007) p.147.

protein that would have been produced by chance would be just 49 amino acids long.[24] The likelihood of this protein being biologically useful would be extremely remote. Thus, in real life situations, there is a vast amount of information but it does not form complex relationships very readily by chance. Further, in life it is only *meaningful* information that can produce a useful outcome.

1	Human cell
20	Amino acid types[25]
23	Chromosomal pairs[26]
25,000	Genes[27]
100,000	Proteins[28]
3.5 billion	Characters[28]
10^{10}	Molecules[29]

Table 1.3: The Multi-layering of Information in Biology

From the preceding discussion it can be seen that accurately calculating the number of infobits in even one human cell is beyond our ability at this stage, let alone a complete human being.

Apart from calculating such a large number of variables, there is yet another problem. Some information is linked because it just happened by chance (*incidental information* creating *incidental complexity*) such as the 'modern art' on the shed floor when the paint tin fell off the shelf. Whereas the painting of the Tower of London has meaningful information that delineates the relationship between the paint particles which were arranged by the artist and contains *organised information*. However, it is possible for someone to misunderstand the meaning that is implicit within the arrangement of the information and attribute the situation entirely to chance. In the case of an investigator discovering the paint spill, there is no proof that the paint spill was not a deliberate act by an artist to see what might be created by intentionally pushing the paint tin off the shelf. Conversely though, it is impossible to infer that the painting of the Tower of London was an accidental paint-spill because of its organisation and structure and its visual similarity to another recognisable entity (in this instance, a building beside the Thames in London). In a different example, a murder may be thought to have been an accident because of certain circumstances. It is also possible to mistake the chance discovery of a corpse at the bottom of the staircase and read malevolent meaning into a situation where there was none (the death really was the result of an accidental fall down the stairs). Thus, the existence of information does not necessarily explain nor define causation.

Complexity is not the only consideration when assessing the significance or value of information within an entity, as the complexity measure (information content) of a rubbish dump would be far higher than that of our mousetrap. However, the dump has a low measure of functionality: it is simply a place to leave rubbish. In contrast, a Swiss pocketknife which has multiple blades and tools in one neat, small unit has a high level of versatility, and hence, functionality. Instead of versatility, another way to look at functionality is to look at the efficiency with which an object or tool performs its functions. Electronic devices such as a camera or a DVD player are very efficient at performing the particular function for which they were designed. Most enzyme systems in the human body are similarly highly specific and efficient and also fall into this category. We will explore enzymes further in later chapters. If we wish to compare the complexity of objects, however, and especially if we need to measure the *increase* in complexity (which is an identifying characteristic of living things), we need to analyse not

24. Yockey, H.P.: A calculation of the Probability of Spontaneous Biogenesis by Information Theory. *J. Theor. Biol.* 67:377–398 (1977) *in* Sarfati, J.: *By Design.* Creation Book Publishers (2008) p.158.

25. Dawkins, R.: *The Greatest Show on Earth.* Bantam Press, London (2009) p.236.

26. Aubisson, P. et al.: Biology in Context – *The Spectrum of Life.* (2nd Ed.) Oxford Univ. Press, Melbourne (2004) p.195.

27. Sarfati, J.: *By Design.* Creation Book Publishers, Brisbane, Australia (2008) p.169.

28. Lennox, J.C.: *God's Undertaker. Has Science Buried God?* Lion Books, Oxford (2007) p.129.

29. Hawking, S. & Mlodinow, L.: *The Grand Design.* Bantam Press, London (2010) p.32.

only the information inherent within a situation or object but also the degree of *functionality*. Functionality can be assessed as zero, low, medium or high. The problem is that this assessment involves a degree of subjectivity. So to measure a *true* increase in information (not just the addition of meaningless data, such as refuse in a rubbish dump), we need both an increase in the complexity score *and* an increase in the functionality score. If you are now lost, I am not surprised as there are so many variables to consider. Infobits are a useful tool when comparing simple situations but unfortunately, when infobits are applied to the real world, the whole process becomes very messy. Attempting to build up a complete understanding from infobits, although providing an objective measure, has not clarified the situation to date, so is there some other way to refine and simplify our quest? Fortunately, there is.

Information, Evidence and Science

It is here that we turn to science to help us out. Scientific method is an analytical tool which we can use to help clarify complex situations. Science is based on the best information we have at the time. We derive our information from the evidence that we can collect: but what is evidence? Evidence, as defined by the Oxford Dictionary is: *"Clearness, obviousness, conspicuous testimony, facts, making for a conclusion; information tending to establish fact."* [6] Evidence is a good indicator of how something really is. Another way of saying this is that: evidence identifies what is 'true'. Often, however, the circumstances do not allow us to be 100% certain that something is true. So, we often work with the best evidence that we have available. The evidence, then, needs to be assessed as to its degree of reliability, which will determine the confidence that can be placed in the conclusion. In medicine, the term 'evidence-based medicine' has come into the language over the last two decades. Evidence-based medicine is a term which is synonymous with the best way to practice medicine at this time. Logic says that this should also be so in science. In the *Concise Oxford Dictionary*, Science is defined as: *"Systematic and formulated knowledge; pursuit of this or principles regulating such pursuit"* [6] and is based on 'scientific method'. The biggest problem in 'evidence-based medicine' is to find enough reliable evidence to gain an accurate picture, so that a useful course of action can be devised. In most branches of medicine, the complete evidence is not yet available. In fact, evidence usually covers much less than 50% of the decisions required to be made every day in medical practice. Where there is insufficient evidence, an educated guess or an interpretation based on previous experience is used instead. Where there is evidence, however, the general consensus is that all relevant evidence should be applied, with the expectation that this will achieve the best results.

We use these principles every day in many ways. A typical example would be trying to locate an address in an unfamiliar city. We could look for a map to provide accurate directions, but if the map is poor, or we have no map at all (or a GPS navigation aid), then we would have to work out our route by some other method, such as by asking people along the way. Sometimes though, through lack of sufficient information, we get lost. This is also the case with science. Even so, the principle of seeking the best evidence should apply not just to medicine but to all the other sciences too. Therefore, this book will focus on the best evidence available. Alistair McGrath, molecular biophysicist and Professor of Historical Theology at Oxford University uses the term: *"evidence-based thinking"*.[30] Evidence-based

30. McGrath, A. & J.: *The Dawkins Delusion.* (2007) p.3.

thinking is the approach that will be taken here, but in addition, we will look at the evidence from an *information perspective*, with the aim of sharpening our focus, to better understand the reliability and hence, the significance of the evidence (information) available. Unfortunately due to our lack of a full understanding of the reliability and completeness of the entities we are observing, much of the evidence is open to differing interpretations, depending on any initial assumptions (presuppositions) and the method of assessment. What might be 'evidence' and 'important' to one scientist may be deemed 'chance' or 'assumption' by another. Steven Rose, Professor of Biology and Neurobiology at the University of London observes: *"Despite our ignorance of the overwhelming majority of life-forms which exist on Earth today (indeed, most biochemical and genetic generalisations are still derived from just three organisms: the rat, the fruit fly, and the common gut bug, Escherichia coli), and our inability to do more than offer informed speculations about the processes that have given rise to them over the past 4 billion years, we biologists are beginning to lay claims to universal knowledge of what life is, how it emerged and how it works... Biology also makes claims as to who we are, about the forces that shape the deepest aspects of our personalities, and even about our purposes here on Earth."* [31] From Professor Rose's comment, it is clear that when we are assessing far-reaching claims, it is necessary to ask whether the evidence supports the extent of the claims being made, or are we extrapolating further than the evidence (information) allows?

Once there is agreement on what constitutes valid evidence, the next problem comes when the evidence is interpreted. There is an even greater problem with the meaning or *significance* of the evidence. A current example is the much discussed topic of global warming or climate change. There is a lot of evidence but it is very difficult getting consensus on which evidence is significant and the meaning of that evidence. Alistair McGrath wrote: *"The entire point of the scientific method is to reduce, and where possible eliminate...bias, to strive to give as objective and fair account as possible."* [32] However, it is usually easier to ask the right question than to discern the right answer and the history of science is littered with erroneous conclusions to valid questions. The classic example is the cause of microbiological disease. For hundreds of years scientists were asking the right question: *"What causes it?"* but until the 19th-Century, the usual conclusion was miasma or *"bad air"*.[33] This book, as much as possible, will consider all the possible interpretations of the evidence presented by science and address each one of them in turn, analysing them through an informational frame of reference. Philip Johnson writes: *"A real science does not employ propaganda and legal barriers to prevent relevant questions from being asked, nor does it rely on enforcing rules of reasoning that allow no alternative to the official story."* [34] Sometimes asking the right questions is harder than following the status quo but discerning the right questions, selecting the appropriate information and then deducing logical answers, is of course, ultimately going to be the most rewarding course to take when it can be achieved.

Information and the Universe

So, as we begin our journey into information, what are the right questions? Let's start by looking at the big picture. Information is embedded in our Universe. Professor Emeritus at Colorado State University, Holmes Rolston III writes: *"Four fundamental forces hold the world together: the strong nuclear force, the weak*

31. Fowler, T.B. & Kuebler, D.: *The Evolution Controversy.* Baker Academic, Grand Rapids (2007) pp.36–7.

32. McGrath, A &J: *The Dawkins Delusion.* (2007) p.35.

33. Powell, K.: Enteric Diseases, contagion and miasms in the 19th-Century. *Canberra Doctor* (Feb. 2007).

34. Johnson, P.: *Wedge of Truth.* InterVarsity Press, Downers Grove. Illinois (2000) p.141.

force, electromagnetism, and gravitation. These range over 40 orders of magnitude; some involve repulsion as well as attraction—the push as well as the pull is used to hold things together. The mix of forces is both remarkable and complex." [35] The weak nuclear force and electromagnetic force are now thought to be two different facets of the same entity called the 'electroweak' force.[36]

All of these forces act in predictable and precisely defined ways. We use them to make predictions, such as: the motion of the planets that cause eclipses and the behaviour of a laser beam. They are extremely reliable, even over thousands of years (and as far as we can tell), across the entire Universe. But why are they so consistent and what defines them? What compels these forces to unswervingly behave the way they do? Why is mathematics able to describe the behaviour of so many objects within the Universe at all? The premise is that they have information embedded within them that determines their behaviour. The behaviour of water molecules is very predictable because the nuclear and electromagnetic forces act the same way each time an experiment involving steam, water or ice is performed. This is because the information defining the nature of these fundamental forces is fixed and does not change. This is an effect alluded to by English biologist, T.H. Huxley (1825–1895) who wrote: *"there is a wider teleology which is not touched by the doctrine of Evolution. This proposition is that the world...is the result of the mutual interaction, according to definite laws, of the forces possessed by the molecules of which...the universe was composed. If this be true, it is no less certain that the existing world lay potentially in the cosmic vapour, and that a sufficient intelligence could, from a knowledge of the properties of the molecules of that vapour, have predicted say, the state of the fauna of Britain in 1869, with as much certainty as one can say what will happen to the vapour of the breath on a cold winter's day."* [37] As we will see later, it is not possible to predict the future as accurately as Huxley imagined might be possible, based on our knowledge of the laws of Quantum Physics, but his basic premise that the laws of physics are consistent and outcomes are predictable is true. It is because these forces are consistent that we can reliably predict not only that the Sun will rise tomorrow but also the time and the position in which the Sun will rise. This premise that information underpins our Universe will be tested throughout this book.

Information and Life

From the preceding discussion, it has been seen that information underlies the four basic forces of the physical Universe but what about closer to home — on planet Earth — where life is our most distinguishing feature? It was previously noted that there are large numbers of infobits in proteins which are the building blocks of life but is information significantly involved with life when we look at organisms, species and bio-systems too? The first point to note is that life is very varied, ranging from the relatively simple amoeba to very complex mammals such as ourselves. We are made up of literally trillions of cells, each of which is made up of trillions of atoms. As already observed, if the formula for complexity was able to be calculated for a human being, it would result in an astronomically large number of infobits, each of which specifies a piece of information. Further, most scientists argue that Homo sapiens is the most successful species in the history of life on Earth, placing human beings at or near the top of the functionality scale. The mousetrap's complexity and functionality is derived from information which allowed the manufacturer to make and assemble the parts. As life also appears to

35. Rolston, H.III: Care on Earth: generating informed concern. *in* Davies, P. & Gregersen, N.H.: *Information and the Nature of Reality: from Physics to Metaphysics.* Cambridge University Press (2010) p.208.

36. Andrews, E.: *Who Made God? Searching for a Theory of Everything.* EP Books, Darlington, England (2009) p.46.

37. Huxley, T.H.: *The Academy* 1:13–14 (1869) *in* Lennox, J.C.: *God's Undertaker. Has Science Buried God?* Lion Books, Oxford (2007) p.90.

contain significant amounts of information, which specifies its complexity, where did the information come from? Can it just have happened by chance or is there some other process? Certainly it is a valid scientific question to enquire how we acquired the information of which we are composed.

However, let's start with the mousetrap which is much simpler. As already established, the Universe is made up of matter, energy and information. If we apply this observation to the construction of the mousetrap, it is clear that its assembly involved matter in the substance of the five parts from which it was fabricated and energy was expended to achieve the finished product. The manufacturer also had a plan which contained information that allowed the pieces to be assembled in the correct order to produce a functional mousetrap. So, if information is necessary to account for the construction of a mousetrap, what about the construction of a human being? The modern theory of evolution asserts that the complexity we observe in human beings has evolved through chance mutation associated with natural selection over billions of years. The question then arises: 'Is *information* produced by this process too?' As human beings require trillions of infobits to describe them, if evolution is to explain the growth in complexity from single-celled organisms to human beings, then it must be able to generate information, which should be able to be audited. This will be investigated further in Chapter 3.

Thus, information can be seen to be an integral part of our human make-up and for the same reason, it is an essential part of all life. Richard Dawkins commented: "*The machine code of the genes is uncannily computer-like.*" [38] How then might biological information be defined? Is it the same as in a computer? A definition for biological information could be: "*An attribute or stored resource which enhances survival of a living entity.*" To qualify as biological information, rather than noise, information must provide an increase in survival value either now or in the future. But how can this be assessed? This may be difficult to describe but it is important not to make our definitions too narrow and exclude valid alternatives, such as, a mutation lying dormant in a genome that would be advantageous should there be a propitious change in the environment.

From this line of reasoning, Darwinian evolution would suggest that the information generation process to make a mousetrap is different from the information generation process to make a man. One is intentionally manufactured with information, matter and energy from a plan and the other, according to the majority of biologists today, evolved by inducting information, matter and energy by chance. Is this right? Although the details of the information that specify a mousetrap and a man are different, there only appears to be one type of information. However, intention and chance are the antithesis of each other. Thus there appears to be an incongruity here. This observation is all the more surprising because the random process produces a man who is so much more complex (when measured in infobits) than a mousetrap, which cannot occur except by the intentional introduction of information. This is the 5th paradox. This is also the first major incongruity to surface on our journey. One of the driving themes of this book is the resolution of significant incongruities.

Incongruities and Contradictions

The ideas put forward by Charles Darwin 150 years ago have resulted in the mechanism of random mutation with natural selection being overwhelmingly

38. Dawkins, R.: *River out of Eden* Basic Books, New York (1995) p.17 *in* Meyer, S.C. et al.: *Explore Evolution* Hill House Publishers, London (2009) p.94.

accepted by the scientific community as the explanation for the presence of life on the Earth. So why question or seek to change what is a very popular concept? Isn't it just rocking the boat for no good reason? Well, if Darwinian Theory explained all that was observed in life then this attitude may be justified, although some would say that science should never stop asking questions. But there are still significant observations that are not explained well or are even contradictory to Darwinian Theory. The first and most basic question here is: how did life begin? How was the information for the first cell generated?

As a mental exercise, I ask my students to imagine how much information there is in the Universe and then to estimate the percentage of that information that is currently known and understood. Some estimate less than 10%, others about 30%. Few proffer figures over 50%. It is clear that we lack the information necessary to answer this question but with the vast amount of new information appearing in journals each week, it is equally obvious that we still have a lot to learn. Some of the most difficult incongruities that are yet to be understood include: the intersection of Pasteur's Law with Darwinian evolution, the wave/particle duality of light and Einsteinian Relativity versus Quantum Mechanics. Each of these conundrums will be dealt with in turn as we search for our information origins.

Science, Truth and Post Modernism

Clearly we have a long way to go to solve the enigmas listed above and we have only just commenced our journey into information, still trying to understand science and how it can help us on our quest, but what is science? The English word derives from the Latin word: 'scientia' meaning knowledge, but it is more than that.[10] Science is easier to understand as a method or a tool rather than an entity in its own right. Better expressed as the *'discipline of Science',* it acknowledges those facts or ideas that can be explained by *'scientific method'.* Thomas Fowler, Senior Principal Engineer at the Centre for Information Technology and Telecommunications at Noblis and Daniel Kuebler, Assistant Professor of Biology at the Franciscan University in Steubenville, USA, wrote in their book, *The Evolution Controversy*: *"...it is useful to distinguish two types or classes of explanations, rational and scientific.* A rational explanation *is one that uses some type of plausible reasoning from premises to explain an observed phenomenon or state of affairs."* [39] Much of philosophy is in this category, as it does not lend itself to experimental testing, such as, questions regarding origins or of an afterlife. Nearly every traditional society describes an afterlife and most religions are based on and proffer evidence for an afterlife. Millions of people over thousands of years, throughout recorded history, have believed in an afterlife to the extent that they have sacrificed their lives for this belief. It could, therefore, be argued that with this many people over that much time, with all the evidence which the religions present, that it is *rational* to believe in an afterlife. In opposing this view, a scientist might respond that it is not a common experience for people to come back from the dead to give a first-hand account of an afterlife and science has not demonstrated a physical mechanism for this to happen. Nor does science present concrete evidence that it has happened, so, although some might argue that it is *rational* to conclude that there is an afterlife, it is not *scientific*. Fowler and Kuebler note however that: *"This does not mean that* [rational arguments] *are therefore inferior or incorrect, of course, nor does it mean that they are exempt from scrutiny."* [40]

39. Fowler, T.B. & Kuebler, D.: *The Evolution Controversy.* Baker Academic, Grand Rapids (2007) pp.124–5.

40. *Ibid.* p.128.

Although *science* has no evidence of an afterlife, the lack of scientific evidence does not prove that an afterlife is impossible either. By the same line of reasoning, Galileo might have 'proved' that radio waves did not exist because the technology and mathematics had not been developed in his day. Fowler and Kuebler continue: "Scientific explanations *are a specific subset of rational explanations that are based upon the scientific method…Such explanations may take many forms, but in general they require a hypothesis that is advanced to explain or generalize observed phenomena and which in turn is used to make predictions about what additional phenomena will be observed, and under what conditions. If these phenomena are indeed observed, the hypothesis is considered to be confirmed, but the hypothesis (which now may be referred to as a theory or a law) is subjected to further tests of this type on an ongoing basis. Both the ability to make testable predictions and the possibility of being refuted are considered essential to scientific explanation.*" [40]

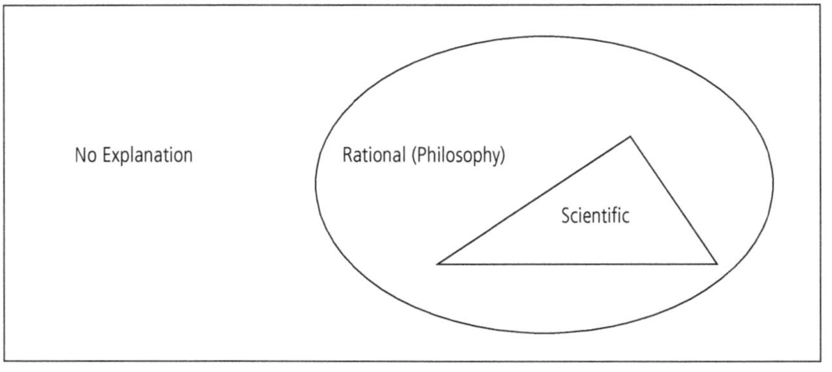

Figure 1.3: Types of Explanation
Philosophy involves the study of rational arguments which are consistent with observation. Science is a subset of philosophy, where the explanations are not only consistent with the evidence but repeatable and testable.

If a theory cannot be tested, then it ceases to come under the umbrella of science and is considered to be philosophy. Fowler and Kuebler reiterate: "*One central criterion of a scientific hypothesis is that the predictions it makes must be* falsifiable; *unfalsifiable predictions cannot be meaningfully verified by experimental tests, as no test can disprove them. In the case of evolution, for instance, Darwin told investigators to gather more fossils and they would find that most of the fossil record would consist of transitional forms. This is a perfectly valid type of experiment, even though it refers to events that happened millions of years earlier. The key point is that investigators* are told to look somewhere they have not looked before, for something they have not seen before. *It is this ongoing experimental verification and feedback that distinguishes science from philosophy and other types of knowledge.*" [40]

It is interesting to note that this key prediction of Darwinian Theory continues to be vigorously debated, with Neo-Darwinians contending that it has been well satisfied by new fossil finds and their opponents contending that it has not. This debate will be further examined in Chapter 5. Thus, even well supported scientific theories are not exempt from controversy. As Fowler and Kuebler explain, neither philosophy nor science is above scrutiny. Fowler and Kuebler go on to list 10 essential characteristics of any genuine scientific theory; [41]

1. Compactness: well defined with tight boundaries.
2. Simplicity: simple and elegant.
3. Falsifiability: capable of falsification.
4. Verifiability: verifiable by experiment and observation.
5. Retrodiction: ability to explain previously known facts and observations.

41. Fowler, T.B. & Kuebler, D.: *The Evolution Controversy.* Baker Academic, Grand Rapids (2007) pp.129–131.

6. Prediction: ability to direct researchers to yet unknown facts.
7. Exploration: suggest new areas of fruitful research.
8. Repeatability: able to be repeated or observed by independent researchers.
9. Clarity: general agreement about what the theory says and means.
10. Intuitiveness: transparent and intrinsically satisfying.

Science depends upon the premise that some events happen reproducibly and predictably and that some do not. That is, some conclusions are 'right' and some are 'wrong'. To test this, look at our physical Universe. On Earth, water always runs downhill due to the effect of gravity unless there is a pump to push it back up the hill again. This is a very straightforward observation but unfortunately, not all situations are as clearly defined. When the conclusions are diametrically opposed, it is not so easy, as we will see when Darwin meets Pasteur later in this chapter. Either Pasteur *or* Darwin was right but both cannot be correct simultaneously.

However, this is not a simplistic debate. Fowler and Kuebler summarised it succinctly when they wrote: "*The debate over the scope of naturalistic explanations in science is, in a sense, a battle for the soul of science: its contents, its place in the context of all human knowledge, and its limits.*" [42] The risk is that the evidence can be contorted to make science support an individual theory, rather than designing the theory to explain the evidence assembled through observation and experimentation. To complicate matters even further, human beings are not automatons, lacking feelings or preconceived beliefs and so, even the most devout scientist will have a bias due to their worldview, no matter how careful they are to try to exclude it. The other problem is that the stronger our feelings or preconceived beliefs, the harder it is to exclude personal bias and there is almost no topic in science that is more hotly contested than origins and evolution. So defining the discipline of science under these conditions is fraught with danger, like walking through a minefield. When the debate leaves scientific circles and enters the public domain, the waters become even muddier. Post-modernism defines truth as a personal choice rather than a requirement to adhere to a rigorous set of criteria or external standards such as Fowler and Kuebler's ten characteristics listed above. The result is that most people believe that Darwin and Pasteur were both right, even though their theories are mutually contradictory.

So, in summary, science can be recognised as a tool, a system of analysis, where the accuracy of evidence and the rigour of the analytical process are equally as important as the result. It, therefore, should logically lead us to precise and enduring outcomes that are shown to be correct over the passage of time. The best measure of a true fact is one that is still predicting accurate outcomes and is, hence, considered 'correct', even after the passage of hundreds or even thousands of years.

Some More Definitions

Fowler and Kuebler emphasise that it is vitally important to distinguish between facts, the explanation of facts and between a theory and a fact.[43] So how is each defined?

Fact: The Oxford Dictionary defines a fact as: "*Something that has really occurred or is the case; hence, a datum of experience, as distinct from conclusions.*" [9] Fowler &

42. *Ibid.* p.123.
43. *Ibid.* pp.343–355.

Kuebler write: *"In science, an observation that has been repeatedly confirmed and for all practical purposes is accepted as 'true'. Truth in science, however, is never final, and what is accepted as a fact today may be modified or even discarded tomorrow."*[41] By this statement, Fowler and Kuebler acknowledge that our knowledge and understanding of the Universe is incomplete and science can only assess the *known* facts. New discoveries may falsify previously accepted 'wisdom' which was based upon a paucity of evidence and /or understanding.

Postulate: The Oxford Dictionary definition states: *"A thing claimed or assumed as a basis of reasoning, a fundamental condition; prerequisite."* [9] It is a claim that is used as a starting point for a line of reasoning and is often used in geometry such as: *"The shortest distance between two points is a straight line."* [6] A postulate which is true may assist the observer to discover new information but a false postulate may direct an investigator away from fruitful areas of investigation and result in the incorrect interpretation of observations.

Hypothesis: The Oxford Dictionary states: *"A supposition or conjecture put forth to account for known facts; especially in the sciences, a provisional supposition which accounts for known facts, and serves as a starting-point for further investigation by which it may be proved or disproved-1646."* [9] Fowler & Kuebler write: *"A tentative statement about the natural world leading to deductions that can be tested. If...verified, the hypothesis is provisionally corroborated. If the deductions are incorrect, the original hypothesis...must be abandoned or modified."* [41]

Theory: The Oxford Dictionary defines a theory as: *"A scheme or system of ideas or statements held as an explanation or account of a group of facts or phenomena; a hypothesis which has been confirmed or established by observation or experiment, and is propounded or accepted as accounting for the known facts-1638."* [9] Fowler & Kuebler have restated this as: *"In science, a well-substantiated explanation of some aspect of the natural world that can incorporate facts, laws, inferences and tested hypotheses."* [42] *"The true measure of a scientific theory is its ability to make predictions as well as explain convincingly all known facts."* [44] This does not prove that it is necessarily correct, only that it is widely accepted as a useful tool for investigating the subject being studied.

Law: This term appears to have come into Old English from Old Norse meaning: *"something laid or fixed"*.[9] Fowler & Kuebler adapt this to science as: *"A descriptive generalisation about how some aspect of the natural world behaves under stated circumstances."* [41] Hawking and Mlodinow write: *"...a law of nature is a rule that is based upon an observed regularity and provides predictions that go beyond the immediate situations upon which it is based."* [45] To qualify as a law, there must be no exceptions, therefore, to falsify a law requires only one exception.

Fowler and Kuebler note: *"Science operates on the basis of theories and evidence and constant criticism. Theories either stand on their own merits and can be defended on the basis of empirical evidence, or they are discarded."* [46] This does not mean that every frivolous question must be taken seriously but legitimate questions need to be acknowledged and answered in scientific terms and not dismissed out of hand. Figure 1.4 is a summary of the scientific method.

44. *Ibid.* p.199.

45. Hawking, S. & Mlodinow, L.: *The Grand Design.* Bantam Press, London (2010) p.27.

46. Fowler, T.B. & Kuebler, D.: (2007) pp.338–9.

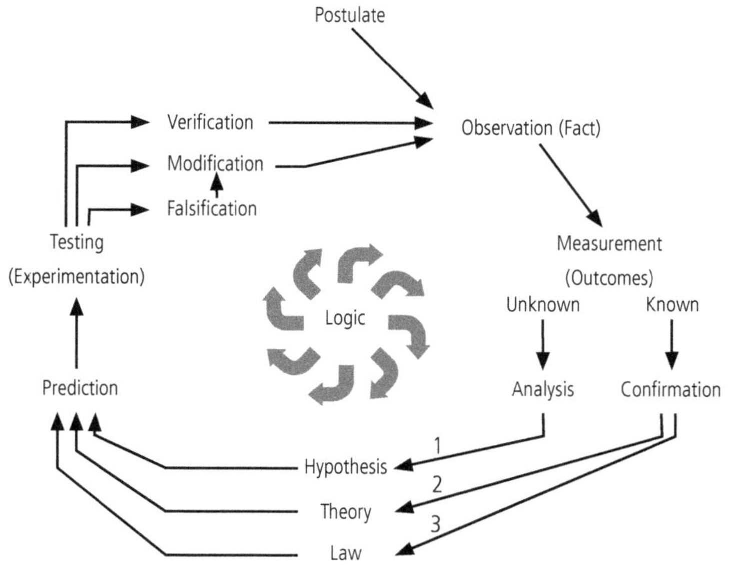

Figure 1.4: Summary of Scientific Method

Each time the information rotates around the model, the degree of certainty increases. The idea passes from an hypothesis to a theory and if it continues to survive the rigours of the scientific process, it has the potential to become a law.[47]

Looking for Biological Laws

Discovering the basic 'laws of nature' is a challenging process and biology is definitely the poor cousin when compared to physics and chemistry. Logic says that a biological law must have the ten characteristics that identify a scientific theory, as elucidated by Fowler and Kuebler above, but it must satisfy the standards to a higher degree of certainty. In particular, as stated previously, there must be no exceptions.

If it is difficult to judge between competing theories, it may be constructive to try to determine which hypothesis or theory can account for more of the known facts, while at the same time, having no glaring deficiencies that disqualify that model. It is no good having a theory which accounts for many facts but has been shown to be impossible (such as Lamarck's theory of inheritance). A little-known theory which yet has few details but no obvious deficiencies (for example: Mendelian inheritance in the 19th-Century) may eventually be more useful in directing future research and, hence, prove more fruitful for further discoveries. If a theory is likened to a sturdy, wooden tower from which to survey the World and Universe around us, then an hypothesis might be envisaged as a raised wooden platform and a law might be considered to be like a brick tower which can now be used as a foundation upon which to construct further wooden platforms or towers to give an even better view. From this analogy, it can be seen that hypotheses, theories and laws are not finished entities but works in progress. They are tools to aid further discovery and, hence, the nature of the tower may aid or hinder our quest. Therefore, not all hypotheses, theories or laws will be of equal usefulness.

47. Andrews, E.: (2009) p.55.

The Law of Cells (Pasteur's Hypothesis)

In the early Victorian era, the concept of *vitality* or the spontaneous generation of life was a widely held belief because, as microbes could not be seen, it appeared that maladies just spontaneously arose and people sickened and died. Maggots appeared in dead meat while hay spontaneously grew mould and rotted, so simple life forms such as maggots and mould appeared to be able to spontaneously form literally 'out of thin air'. The explanation given for these observations was that it was caused by *'miasma'* (literally 'bad air'). Perhaps it was this widely-held concept that led Charles Darwin (1809–1882) to write to his friend, Joseph Hooker (1817–1911), introducing the idea of life spontaneously forming in a *"warm little pond"* in 1871. Joseph Hooker was a botanist and explorer who was one of the founders of geographical botany and the Director of the Royal Botanic Gardens at Kew, London for 20 years. Darwin may not have been aware that only a few years earlier, in the early 1860's, the French chemist and microbiologist, Louis Pasteur (1822–1895) showed that spontaneous generation did not happen.

Although Pasteur's work in 1861was ground-breaking, it was not all together new. Two centuries earlier, similar experiments had been conducted by Francesco Redi (1626–1697), who was an Italian physician, naturalist and poet. In 1668, Redi published a series of experiments where he had placed fish and meat in glass jars, some of which were covered with gauze and others which were open. Further experiments showed that maggots only appeared if live flies had access to the meat.[48] Richard Dawkins writes: *"Louis Pasteur had said, in a lecture at the Sorbonne, 'Never will the doctrine of spontaneous generation recover from the mortal blow struck by this simple experiment.' The simple experiment was the one in which Pasteur showed, contrary to popular expectation at the time, that broth sealed off from access by micro-organisms would not spoil."* [49] Through his research and reasoning, Pasteur was able to solve the problem faced by the French wine industry, where the wine sometimes inexplicably turned into vinegar. We now know that this occurs because of fermentation caused by microorganisms but at that time microorganisms were still unknown. Later, the English physician, John Snow (1813–1858), the Prussian physician, Robert Koch (1843–1910), the US Army physician, Walter Reed (1851–1902) and many others confirmed Pasteur's results. As a result of this work, Pasteur perfected 'pasteurisation' and the British surgeon Joseph Lister (1827–1912), using Pasteur's rationale, developed antisepsis, processes that are still in use today.

Pasteur's findings have subsequently been variously formalised as the 'Theory or Law of Biogenesis' which states: *"Life comes from life"* which became the first Law of Biology.[50] Only a few years later on the other side of the English Channel, Thomas Henry Huxley, a colleague and supporter of Charles Darwin and a champion of the Theory of Evolution, coined the term *abiogenesis* in 1870. In an essay he wrote: *"Carbon, hydrogen, oxygen, and nitrogen are all lifeless bodies. Of these, carbon and oxygen unite in certain proportions and under certain conditions, to give rise to carbonic acid; hydrogen and oxygen produce water; nitrogen and other elements give rise to nitrogenous salts. These new compounds, like the elementary bodies of which they are composed, are lifeless but when they are brought together, under certain conditions, they give rise to the still more complex body, protoplasm, and this protoplasm exhibits the phenomena of life. I see no break in this series of steps in molecular complication..."* [51]

48. *The World Book Encyclopaedia.* World Book Inc., Chicago (1984) Vol. 18, p.628.

49. Dawkins, R.: *The Greatest Show on Earth.* Bantam Press, London (2009) p.418.

50. Whether you consider this a theory or a law will depend on whether you believe Darwin or Pasteur was correct! Compare *The World Book Encyclopaedia.* World Book Inc., Chicago (1984) Vol. 2, p.242 with Seiglie, M.: 10 Ways Darwin Got It Wrong. *The Good News* 14:6 (2009) p.8.

51. Huxley, T.H.: *Autobiography and Selected Essays.* (Ed. Snell, A.L.F.) Riverside College Classics (2009) *in Journal of Creation* 23:3 (2009) p.125. [www.scribd.com/.../Autobiography-and-Selected-Essays-by-Huxley-Thomas-Henry-1825–1895]

The following year, Darwin penned his now famous letter to Hooker: *"It is often said that all the conditions for the first production of a living organism are now present, which could ever have been present. But if (and oh! what a big if!) we could conceive in some warm little pond, with all sorts of ammonia and phosphoric salts, light, heat, electricity, etc, present, that a proteine compound was chemically formed ready to undergo still more complex changes, at the present day such matter would be instantly devoured or absorbed, which would not have been the case before living creatures were formed."* [52]

Commenting on this apparent contradiction between Darwin and Pasteur, Richard Dawkins maintains, as Darwin suggests, that life could not form again, as it might have once in the (sterile) warm little pond because bacteria would now consume the proto-organism before it had a chance to establish itself. Dawkins writes: *"Evidently, the spontaneous generation of life is a very rare event, but it must have happened once, and this is true whether you think the original spontaneous generation was a natural or a supernatural event."* [53] There are some problems with this line of argument. The Oxford English Dictionary defines the word spontaneous as: *"occurring without external cause"*,[6] which is not the case if the appearance of life, as Dawkins writes was *"a supernatural event"*. The other difficulty is that the presence of bacteria consuming our proto-organism is not the biggest problem for the *'spontaneous'* formation of life, which requires *information generation*. This will be explored in detail in Chapter 3.

Huxley was partially correct, in that simple elements are lifeless and protoplasm supports life, but he grossly underestimated the complexity of the process by which simple elements link together to form into living protoplasm, which we now call *cytoplasm*. The on-going study of cells has revealed amazing complexity in even the simplest cells. Huxley and Darwin lacked the electron microscopes we have today which have revealed the microscopic machinery necessary for a cell to function. Instead, from what could be seen in Darwin's day with the light microscopes that were available, the cell appeared to be a little bag of simple chemicals, so abiogenesis would have seemed to be a reasonable conclusion at that time. This concept of a simple 'blob' of chemicals is another example of a rational explanation which has foundered on the rocks of new observations and has now been shown to be unscientific, as previously explained by Fowler and Kuebler.

The complex biomechanics of the flagellum has been extensively described by other authors (see Chapter 4) but that aside, even the simplest one-celled organism has thousands of proteins, each of which is composed of a minimum of 100 to 300 amino acids (and some significantly more). These create complex cell membranes and molecular machines such as ion channels, ribosomes, mitochondria and many others needed for the maintenance and replication of the cell. Michael Behe, Professor of Biological Science at Lehigh University in the USA, has described *irreducible complexity* and the problems associated with the development of living organisms below a minimum level of sophistication, where some of the essential molecular machines or elements may be missing. An analogy would be the development of the internal combustion engine for your car. It is a complex piece of machinery that needs many parts which must be precisely arranged and they all need to be present for the engine to work. Anyone who has suffered a malfunction in their car knows that even a minor detail, such as: a loose lead to the battery or a cracked distributor cap, can result in the engine failing to work altogether. If life is to begin from inanimate components then

52. Darwin, C. *The Life and letters of Charles Darwin.* Vol 3. John Murray, London. (1887) *in* Dawkins, R.: *The Greatest Show on Earth.* Bantam Press, London (2009) p.417.

53. *Ibid.* p.218.

all the essential components need to be assembled in the same place at the same time and then commence functioning almost simultaneously. This has not been achieved even artificially in a laboratory and the chance of it happening in an uncontrolled environment is even less probable to the point of being impossible. American biologist, Craig Venter has artificially copied a Mycoplasma genome and then inserted it into an already living cell cytoplasm but this is vastly different from designing, as well as constructing, the bacterial genome, the cell membrane, mitochondria and all of the cytoplasmic elements from scratch and then finding that the cell was now alive.[54] The Neo-Darwinist will reply that such complexity need not have existed at the start and that, as Darwin hypothesised, a simple model can grow through the process of evolution to achieve a very complex result. In Chapters 3, 4 and 5 this point will be discussed in detail. However, Redi and Pasteur demonstrated that without a pre-existing working model, a new cell does not begin. So in plain English, the First Law of Biology can be restated as:

> *"A living cell is required to produce another living cell."*

This does not explain from whence the first cell came, but it is a fixed point from which further research may proceed as we investigate this question. The corollary of Pasteur's Law is that once a cell dies it cannot be reanimated unless it begins again from both its *component DNA* and another *living cell*. To test this requirement, try growing a cell without its nucleus (which contains the DNA). Alternatively, observe that a dead cell which still contains the correct DNA will not grow either. It is not just any DNA that will do, however. The DNA must be exquisitely delineated in both quality and quantity. The precision required for the DNA suggests that the information required to produce another living cell needs to be just as specifically defined.

The sequencing of the human genome in 2000 has revealed the vast amount of information required to 'describe' a human being but it has also shown the efficiency with which this information is packaged. Not only does the human DNA sequence specify the materials out of which we are constructed but it provides the construction process so that the correct procedure can be followed. The microscopic machines that carry out the construction process are inherited from the cytoplasm of the living parent cell from which the new organism has grown. This is called cytoplasmic inheritance and the DNA of the cytoplasmic nano-machines is *not* coded in the nucleus. Cytoplasmic inheritance is responsible for the origin of much of the cytoplasmic machinery, such as: the golgi bodies, mitochondria, centromeres and endoplasmic reticulum, all of which are essential for the proper functioning of cells. (See Chapter 4) Most surprising is that some sequences of both DNA and proteins can be sliced up in different ways to perform alternative functions at different parts of the construction process or as end-user proteins. A typical example is the hormone/catalyst, insulin (a protein), which can be cleaved to produce C peptides which are also active proteins.

DNA and Information

It has been noted that the complexity within living cells is much greater than Darwin or Huxley imagined but the intricacy and sophistication of the information systems within a cell is something few of us have fully appreciated. So in an effort to understand this better, we can compare it to a modern computer.

54. Venter, C. et al.: Creation of a Bacterial Cell Controlled by a Chemically Synthesized Genome. *Science* 329 (5987):52–56 (2010).

When studying computers, we know that looking at the external casing or even opening it up and surveying all the physical components does not tell us the most important details about the capability and speed of the microchips as well as the information stored within the computer's architecture, which includes its memory and programming. Thus, we speak of computers possessing 'hardware' and 'software'. In a similar way, living cells and hence organisms have hardware and software too. The 'hardware' in living organisms is that part which can be seen and felt and has been studied for as long as man has been able to feel his arms and study his own reflection in a pool of water. As already mentioned, the information that specifies the hardware is stored in both the DNA in the nucleus and the cytoplasmic organelles, such as, the mitochondria. The software is the information that runs the electrical pacing of the heart as well as the other autonomic (automatic) systems of the body. These processes are hidden within all living organisms and are largely invisible until you have a fairly sophisticated understanding of how each living system works. It is vitally important to recognise that despite software being intangible or non-material, such that it cannot be physically picked up, this does not make it unreal. It is just as real as the hardware but it is different. This cellular software (information) which is stored in the DNA or RNA has the ability to grow our bodies from a single cell and then run all our functions. In addition, it defines our abilities to communicate and to live the way we do. It may be the loss of certain essential software components at cell death that leaves the hardware as a corpse, which, without its corresponding software, cannot come back to life. It is significant at this point to note that the information that defines the hardware and software is different from a third category of information which accounts for human thought and probably memory which run on the hardware and software. Thought and memory are discussed further in Chapter 10.

We take DNA and RNA for granted today as the ubiquitous store of information in all living things but it is both a very powerful yet very simple language that spells out the web of life. It is in fact the repository of our genetic language. It has an alphabet (A, C, G, T & U) and grammar (rules of engagement: A = T; C = G). These letters are symbols for the five nucleobases: adenine (A), cytosine (C), guanine (G), thymine (T) and uracil (U). Its power resides in its ability to store immense meaning. It was previously ascertained that meaning in a spoken or written language does not happen by accident. However, the meaning encoded in the genetic language of DNA could not have been placed there by human beings, so where did it come from? For millennia, people had believed the explanation given in the Bible which states that God created life but Charles Darwin (among others), noticed similarities between living organisms and fossil specimens and proposed the Theory of Evolution.

Defining Evolution

Charles Darwin was not the first person to describe the concept of evolution but he was the first to conceive of a scientific explanation as to how it might have occurred. There are various definitions of the word: 'evolution'. The term has been used by different people in different ways and they are not all exactly the same. To avoid confusion the process needs to be precisely defined. The Oxford English Dictionary gives two different definitions which are both relevant. The first is a broad definition that dates from 1647 which states: *"The appearance in orderly succession of a long train of events."* [9] Fowler and Kuebler comment: *"This is what one could call the basic or 'wide' meaning of evolution; it corresponds to what is least controversial about evolution, the simple notion of change over time."* [55] In biological terminology it is often expressed as: *'Change in gene frequency over time.'* The other narrower definition is expressed as: *"Origination of species of animals and plants by a naturalistic process of development from earlier forms, and not by a process of 'special creation'"*.[9] Darwin used the term: *"Descent with modification"*.[56]

Since Darwin's time, 'Chemical or Molecular Evolution' has emerged to address the process described by Huxley, whereby it is proposed that inorganic molecules combine to create life. More recently, the terms 'Macro' and 'Microevolution' have been coined. Macroevolution describes the process where Darwin hypothesized that the information in one biological species can increase to the extent that it *changes into another biological species* and Microevolution describes the fluctuations of information *within* a species. A species is a group of organisms that can interbreed to produce fertile offspring.[57]

If evolution is defined as: *"random variation with natural selection"* as Darwin did, then any change, no matter how small, can be seen as 'evolution'. But from an information perspective, it can be seen that change, without the addition of information, will not build a more complex organism.[58] Change alone (Microevolution), will not be enough to explain the vast diversity of life on Earth. To change a single-celled amoeba into a mammal or a man (which is Macroevolution) requires the addition of *new* information which involves both qualitative as well as quantitative enhancement.[15]

Once we delve into the realm of information relating to evolution, there is a need to distinguish between the appearance of *new traits* and the formation of *new information* because sometimes the rearrangement of existing information code will create a new trait *without* any new information. A typical example is the development of bacterial resistance to drugs by the shutting down of cell pumps. This is important because rearranging deck chairs on a cruise ship doesn't build a bigger, faster ocean liner.

No one was there to observe and record Macroevolution happening in real time, so there is inevitably a lot of inference. Thus, Macroevolution (from amoeba to man) cannot be proven by the first method of science, which is direct observation, to verify the progressive change that has been hypothesised. It has not been repeated either, so a second method of scientific endeavour to investigate the process is also denied us. Therefore, scientists must resort to indirect means to establish a case, but as the evidence is incomplete (such as the fossil record) and evolution involves excessively slow processes over exceedingly long time periods, the result is not surprisingly controversial, as evidenced by the multiple disputes that have sometimes spilled over into court, such as the Scopes Trial in 1925 and the US District Court case in Pennsylvania regarding the teaching of Darwinian Evolution and Intelligent Design in US schools, in 2005.

So in summary, in its most basic form, evolution can be viewed as the progression from simplicity to complexity. This was applied to biology by Darwin in the 19th-Century, however it has now been used to describe anything that follows this step-by-step path, such as the 'evolution' of a language, the motor car or computers. Sometimes this wide usage of the term makes the situation in biology more confusing, as biological and social constructs do not behave according to the same rules (the Laws of Nature). However, we will discover that the analysis of the information that is intrinsic to life and the Universe can assist

55. Fowler, T.B. & Kuebler, D.: *The Evolution Controversy.* Baker Academic, Grand Rapids (2007) p.26.

56. Darwin, C.: *The Origin of Species.* (1859) reprinted by Penguin Classics (1985) p.433.

57. Lennox, J.C.: *God's Undertaker. Has Science Buried God?* Lion Book, Oxford (2007) pp.100–1.

58. Johnson, P.: *Wedge of Truth.* InterVarsity Press, Downers Grove. Illinois (2000) p.48.

in teasing out what is occurring, even though the information is often embedded within the entity and is not immediately obvious.

The Influence of the Ancient Greeks

So if Charles Darwin was not the first to describe the process of evolution, where *did* it begin? Well, much of Western scientific thought can be traced back to the ancient Greeks.[59] The Greeks lacked the technology of the modern era and depended on their unaided senses to interpret the Universe. What could not be observed directly was inferred through theoretical models (as we do today), but we have far more powerful tools to observe our Universe and hence, more data to make more accurate models. Hawking and Mlodinow, writing about the ancient Greeks in *The Grand Design* note that: "*...the connection of cause and effect in nature was invisible to their eyes...*" and cite Thales of Miletus (c.624–c.546 BC) as the first person to see patterns in nature. They continue: "*The idea arose that nature follows consistent principles that could be deciphered.*" [60] Others followed. Aristotle, the most famous of the ancient Greek philosophers, credits Empedocles (c.495-435 BC) with the first description of the Universe as being created from the four material elements of fire, air, water and earth, controlled by the opposing forces of 'love' and 'strife'. Love was the combining force that built things up and strife was a destructive influence that broke things apart. Empedocles was a native of the city of Acragas in Sicily, where he was a politician, physician, philosopher and poet. At that time, Sicily was part of the Greek Empire. With time, the unobservable and non-material influences of love and strife were discarded but the tangible and materialist parts of his theory persisted.[61]

Aristotle (384–322 BC) was born in Chalcidice, Macedonia in northern Greece, 50 years after Empedocles died. Although he lived more than 2300 years ago, he has been one of the most influential thinkers the World has ever known. He developed the concept of perfection in the celestial (heavenly) bodies, possibly from observing the full moon on a clear night! But, due to lack of information, he incorrectly inferred that all celestial orbits were perfectly circular (a perfect geometric form) and that every object in the heavens was a perfect sphere, that rotated around the Earth (which is what our senses intuitively tell us.) Despite this error, Aristotle's greatest strength was that he used a logical cause and effect approach to explain physical events which he observed in the World and the Universe and hence, he was able to reason that the World was round rather than flat. Unfortunately he also reasoned that if the World was spinning, then we would feel a constant east wind as the Earth turned.[62] Similarly, archers would have had to adjust their trajectory for the spin of the Earth. (Everyone knows that an arrow fired straight up in the air on a still day comes straight down, not somewhere to the West, as would be anticipated if the World really rotated! Unfortunately, our intuition is not always correct.) However, Aristotle did view the Universe as purposeful.[63] If there was not enough information to understand an observed phenomenon, then a theory was devised to supply the missing details, which was much of the time. Aristotle is referred to as a philosopher, because although most of his theory is based on observation and logic he could not conduct the scientific experiments (such as measuring the orbits of the planets as Kepler would do), due to a lack of technology to supply the evidence. As the Sun and the other planets appear to revolve around us, Aristotle placed the Earth at the centre of the Solar System, which is known as the 'Geocentric model'. Aristotle's ideas were pervasive

59. Gonzalez, G. & Richards, J.W.: *The Privileged Planet*. Regnery Publishing, Washington DC (2004) pp.225–245.

60. Hawking, S. & Mlodinow, L.: *The Grand Design*. Bantam Press, London (2010) p.17.

61. Stanford Encyclopaedia of Philosophy *at* http://plato.stanford.edu/

62. *Andriessen, M. et al: Preliminary course physics one*. (2nd Edition) Jacaranda, Milton, Qld (2004) p.251.

63. Gonzalez, G. & Richards, J.W.: *The Privileged Planet*. Regnery Publishing, Washington DC (2004) p.230.

and they lasted for more than 1800 years until Copernicus and Galileo, with the aid of telescopes, supplied new information to question the established theories (assumptions). About 240 BC, another Greek, Aristarchus (c.310–c.230 BC), did propose an 'Heliocentric Model' of the Solar System[64] but although Aristarchus' model was very similar to what we observe today, *"it did not gain favour because it was not sufficiently detailed to allow predictions in the manner of Aristotle's model."* [63]

Another concept derived from the ancient Greeks was the concept of the 'eternal Universe'. Possibly through observing the stars, which never appeared to change position, (as opposed to the planets that did), it was inferred that the Universe was eternal. This was the view of secular science in the Western World until the early 20th-Century. In 1915, when Einstein discovered that his General Theory of Relativity predicted that the Universe should be expanding or contracting, rather than being in a steady state, he initially introduced a 'fudge factor', which he called the 'Cosmological Constant' to reinstate a steady state Universe. However, evidence built up through the 20th-Century and in 1965, with the accidental discovery of the Cosmic Background Radiation (as predicted by George Gamow in the 1940s), the Big Bang Theory became accepted as fact and scientific consensus finally agreed that the Universe had a beginning and hence, was finite rather than eternal.

In addition, Aristotle believed in the immutability of species. He was a keen observer and student of anatomy. He concluded that species rarely changed. In contrast, Anaximander (610–546 BC) another Greek philosopher, believed that species could change. He believed that life originated in the sea and descendants of fish later moved onto the land. He did not describe a mechanism but he certainly described the principle by which simple organisms changed into more complex forms. In the 1st-Century BC, the Roman philosopher and poet, Lucretius, extended these ideas when he described the principle of survival of the fittest in his poetic work: *De rerum natura (On the Nature of Things)*, as a mechanism that could change living organisms. Fowler and Kuebler observed: *"Lucretius's argument contained all the Darwinian essentials: a random set of variations that produced various life-forms coupled with the differential selection of these forms, such that those most fit to survive in their respective environments flourished."* [65]

Darwinian Theory

As already noted, the Aristotelian worldview persisted until the European Renaissance, when new technology in the form of the telescope provided new information and hence, new evidence to change the paradigm of ancient Greece. All over Europe, enquiring minds were processing the new information and extending our understanding of the Universe. Much of this information and its sequelae became items of great controversy. One in particular, was the Copernican concept of a heliocentric universe, which ran contrary to academic understanding at that time. After Copernicus and the astronomers came Newton, the philosophers such as Sartre and then Charles Darwin. Rarely has one person created such controversy. Few figures in history can lay claim to radically changing the way that our society thinks but Charles Darwin is certainly one. Although the ideas were not new, as Anaximander had expressed them in the 6th-Century BC, Darwin was able to unify and popularise the Theory of Evolution by proposing a credible mechanism through random variation with natural selection. Darwin's understanding of genetics and information science was non-existent

64. Hawking, S. & Mlodinow, L.: *The Grand Design*. Bantam Press, London (2010) p.21.

65. Fowler, T.B. & Kuebler, D.: *The Evolution Controversy*. Baker Academic, Grand Rapids (2007) pp.43–5.

(they hadn't been discovered yet) but his ability to take the principle described by Anaximander, whereby life forms progress from the simple to the complex by small steps and give it a mechanism, created a unifying theory which now appeared plausible and rational.

In their recent book, *The Evolution Controversy*, which surveys the competing theories of Evolution, Intelligent Design and Creationism, Fowler and Kuebler wrote: *"In about 1830 natural selection, as a conservative principle, was recognised by Edward Blyth, Charles Lyell, and others. This principle — in essence, the survival of the fittest — was used as an example of divine providence, ensuring the conservation (survival) of species. Charles Darwin took that idea, and by conjoining it with ever-present random variations exhibited by organisms, endowed it with a new function, the promotion of innovation. In doing so he created a theory intended to explain the growing body of facts about the history of life on Earth."* [66]

Darwin was not the only one to espouse a belief in naturalistic evolution, however, it was his ability to tease out the essence of the ideas of others and then combine them into one theory that made him famous. His grandfather, Erasmus Darwin (1731–1802) and his father, Robert (1766–1848), opened his mind to other ways of viewing the reigning biblical paradigm of Creation but it was the ideas of uniformitarian gradualism developed by Charles Lyell (1797–1875), a geologist, and a study of population behaviours published by Thomas Malthus (1766–1834) in 1798 entitled: *An Essay on the Principle of Population* that helped him construct his centrepiece theory. Despite this background, it is clear that he wrestled with the concepts for two decades to bring it to the point where he could publish *The Origin of Species* in 1859. It was Charles Darwin's original observations of the Galapagos finches and many other species in their respective habitats around the World which provided the details to flesh-out his theory and make it credible. Darwin observed the differing conditions on neighbouring islands in the Galapagos and how the finches varied from one island habitat to another. In an isolated area such as the Galapagos, he concluded that the finches must have been related at some time in the past, but had since varied over time. He argued that their physical environments altered their physical forms. He was also impressed at how well each bird was adapted to survive in its own environment. He inferred that the differing environments caused the finch species to change to survive. Although he did not have a detailed mechanism, as biochemistry had not yet been discovered, he could see that the variations were there and the coincidence seemed more than just chance. He hypothesised that there was a gradual accumulation of small changes which were either beneficial or neutral to the survival of the individual. This created a pool of variation amongst individuals within a species and the differences gave some individuals within each species an advantage over their neighbours. Changing environmental conditions over time, made some physical attributes more beneficial than others in the race to survive. The most successful produced more offspring and thus, perpetuated their advantageous (even if minute) variations. Even though each change was small, if they had taken place over millions of years, he surmised that the accumulation of many small changes would account for all the species that ever lived on Earth. There must, therefore, he deduced, have been countless intermediate species that had lived and died as one species morphed into the next over the aeons of time that geology suggested had elapsed since life first appeared on the Earth.

Darwin was an avid pigeon fancier and he observed that changes occurred *within* species over time. When he linked this clearly observable fact to his

66. *Ibid.* p.23.

knowledge of fossils and the similarities *between* species, it led him to extrapolate that the small changes which he noted *within* species over a relatively short period of time might happen *between* species over a much longer period of time.[67]

Darwin's Theory of Evolution, as the unifying theory of biology, is broad in its scope and has a number of elements which need to be separated to understand them clearly. They encompass:

1 Change over time
2 Universal Common Descent (common origin)
3 Natural Selection (*"survival of the fittest"*) [68]
   a. Random mutation (allowing for increased variability)
   b. Heritability (preservation of advantageous changes)
   c. Differential reproduction (multiplication of the fittest)

Fowler and Kuebler wrote: *"...Darwin thrust a new function onto natural selection. He made it the promoter of innovation, so that working in conjunction with random changes in hereditary material, natural selection would choose those changes that were beneficial to the organism. Darwin of course knew nothing of genes and molecular genetics, but he surmised that new, slightly better organisms constantly arose, and that these would have an advantage over their counterparts...For all the impact that it has had on science, it is remarkable that* The Origin of Species *contained no experiments. Rather, it comprised a simple and elegant thesis, supported by a mountain of circumstantial evidence and observations."* [69]

Darwin had observed that not all individuals within a given species are identical and that this individuality was inheritable. He surmised that when resources became scarce (which has commonly occurred throughout history as a result of droughts or floods), some individuals within a species possessed characteristics which gave them an advantage, allowing them to survive adverse events when others did not. In *The Origin of Species* Darwin wrote: *"As many more individuals of each species are born than can possibly survive, and as consequently there is a frequently recurring struggle for existence, it follows that any being, if it vary however slightly in any manner profitable to itself, under the complex and sometimes varying conditions of life, will have a better chance of survival and thus be naturally selected. From the strong principle of inheritance, any selected variety will tend to propagate its new and modified form."* [70]

Darwin called this his *"special theory"* which is now referred to as *'Microevolution'* but as previously noted, this concept was neither new nor controversial. Darwin became controversial when he extended his theory to explain the derivation of *all* species from a single common ancestor which is now referred to as *'Macroevolution'*. Fowler and Kuebler continue: *"Although Darwin admitted that only limited change had been observed in animal husbandry, he posited that this process could be extrapolated over vast amounts of time, so that monumental change could occur."* [71] In *The Descent of Man*, Darwin wrote: *"The main conclusion here arrived at...is that man is descended from some less highly organised form."* [72]

The ideas within Darwinism speak powerfully to both the scientific as well as to the non-scientific community (such as, economists and social modellers). We see 'evolution' or progressive change in social structures and cultural organisations all the time. Every manager on Wall Street or military commander knows what *"survival of the fittest"* means. In biology, however, there are many misconceptions due to the constraints imposed by the immense complexity of living organisms,

67. Strobel, L. & Vogel, J.: *The Case for a Creator.* (Student Ed.) Zondervan, Grand Rapids (2004) p.30.

68. This phrase was first used by Herbert Spencer (1820-1903) in his *Principles of Psychology* which he published in 1855, four years prior to the publication of *Origin of Species.*

69. Fowler, T.B. & Kuebler, D.: (2007) p.55.

70. Darwin, C.: *The Origin of Species.* (1859) in Fowler, T.B. & Kuebler, D: (2007) p.56.

71. Fowler, T.B. & Kuebler, D.: (2007) p.56.

72. Darwin, C.: *The Descent of Man.* (1871) *in* Blanchard, J.: *Evolution Fact or Fiction.* Evangelical Press, Darlington, England (2002) p.5.

making it difficult to comprehend the whole picture simultaneously. As an example, Darwin's hypothesised fossil record which needed to demonstrate many *"tiny steps"* has not materialised in the way Darwin hypothesised that it would. This will be discussed in more detail in Chapter 5.

When *The Origin of Species* was published in 1859, the gene theory of inheritance had not yet been published. Gregor Mendel did not print his paper until 1866 and then it was another 35 years before it came to the attention of the wider scientific community. Based on what was generally accepted knowledge at the time, Darwin surmised that *any* change or mutation was inheritable. This was a popular theory previously proposed by the French naturalist, Jean-Baptiste Lamarck (1744–1829).[73] Later, when the gene theory was fully understood, Darwin's theory was modified during the first half of the 20th-Century and the Neo-Darwinian Theory of Evolution was born.

Assumptions

When I was in medical school, we were taught about many methods of investigation but the most powerful instrument in the armamentarium was the 'retrospectoscope'. Looking back, it is clear that Darwin made some assumptions, not all of which have been proven true with the passage of time. He had to do this, however, to be able to formulate a cohesive theory and subsequent researchers have been able to amend Darwin's thesis of 1859 in the light of new discoveries without undue problems until quite recently. But it does bring up the issue of assumptions and why they occur at all. Wouldn't it be better if we never assumed anything; waiting until we are sure before publishing new ideas? The answer is that life would become unmanageable as, even to make the most minor of decisions, requires assumptions. To some degree we all live lives based on assumptions. For example, we assume that the Sun will rise again each time it sets to create a new morning, that water in the kettle will continue to boil at 100°C (if we live at sea level) or that, under the effect of gravity, water will always run downhill. These assumptions are based on laws of the Universe which we presume are constant and inviolate. We need to make these assumptions to overcome the immense complexity of the World so that we can function from day to day. If we had to recalculate every action every minute of every day, life would become impossible. Assumptions are also short cuts we take when we do not have enough evidence to know everything about a situation. They may be right and they may be wrong, so we need to be ready to discard them and move on if subsequent evidence shows that our assumption was invalid. It needs to be appreciated that assumptions are rarely devoid of preconceived, ideological bias. When considering the bigger issues, there should be no 'a priori' agenda or motive other than a comprehension of the truth.[15] It is science's job to sift the assumptions and glean the evidence to expose and exclude any errors and hence, help us to understand our world. Every discipline depends on multiple assumptions every day, from engineers calculating stresses in new buildings to neurosurgeons working out the best way to remove a brain tumour. Most of these assumptions are founded on substantial bodies of evidence so the margin for error is small, yet, assumptions are common and necessary. Further, as already noted by Fowler and Kuebler, the formulation of a cohesive theory, even if partially based on assumptions, can allow predictions to be made to aid in the search for new evidence and expedite new discoveries. If we had to wait until everything was known about a new concept before it could be

73. Fowler, T.B. & Kuebler, D.: *The Evolution Controversy.* Baker Academic, Grand Rapids (2007) p.50.

published or trialled, much of the predictive power of theories would be lost and new discoveries would be delayed.

As Darwin also had the problem of insufficient information/evidence to comprehensively prove his Theory of Evolution, he had to make assumptions too. That being the case, what evidence did Darwin have for his Theory of Evolution? The conclusions he had observational evidence for were:

1. Natural selection and survival of the fittest happens (Galapagos finches).
2. Random mutations occur (lambs can be born with two heads).
3. There is great similarity between many different species (forelimbs of mammals).
4. The age of the Earth appears vast, allowing time for evolution to occur (Geologic column).

Therefore, when Darwin formulated his Theory of Evolution, what were the assumptions that he made? Although he had much less scientific information 150 years ago, he could still make educated guesses regarding the information that he was not privy to in 1859. Some of the assumptions he made include:[74]

1. Evolution would preserve even the slightest beneficial mutation until it took over the entire population (inferred from selective pigeon breeding).
2. Inheritance would preserve any changes to an individual throughout life (Lamarck).
3. Cells were simple bags of undifferentiated protoplasm—chemicals and water (light microscopic observation).
4. Organisms on Earth have originated through common descent (similarity of species).
5. The fossil record would reveal the transitional forms confirming common descent (fossils were being discovered all the time).
6. There is a logical (biochemical) pathway from the primeval soup to man. (This is a requirement if Darwinian evolution has occurred.) [75]
7. There has been time enough in the history of the Earth for evolution of species to occur.
8. Naturally occurring random processes can provide all the new information for speciation on Earth. (Points 7 and 8 are requirements for Darwinian Theory to have occurred.)
9. The physical conditions in the past have not changed significantly (in a statistical sense) over the course of Earth's history — often referred to as uniformitarianism.[76] (This assumption is necessary to allow predictions about the geologic column and is disputed by some creationists who maintain that cataclysmic events in the past have distorted the evidence.)

Although Darwin's evolutionary theory is still the reigning paradigm in mainstream scientific literature, as will be discussed over the next four chapters, the first three assumptions have been shown to be incorrect and the last six are extrapolations that are yet to be proven; as Fowler and Kuebler noted, much of the evidence is circumstantial. While scientific consensus considers that Darwin's

74. Lennox, J.C.: *God's Undertaker. Has Science Buried God?* Lion Books, Oxford (2007) p.162.

75. Josephson, B.: Letter to the Editor. *The Independent*. London. Jan., 12, 1997 *in* Lennox, J.C.: *God's Undertaker. Has Science Buried God?* Lion Hudson, Oxford (2007) p.156.

76. Fowler, T.B. & Kuebler, D.: *The Evolution Controversy.* Baker Academic, Grand Rapids (2007) pp.139–141.

theory has not been *proven* incorrect, to date, it has not been confirmed beyond all reasonable doubt either which is why it is still referred to as a *'theory'* even after 150 years. The reason it continues to be the reigning paradigm, despite the deficiencies listed above, is that competing theories have not garnered enough scientific evidence to convince a majority within the scientific community that there is a better explanation.

The Neo-Darwinian Synthesis

To overcome these holes in Darwinian Theory and to include subsequent discoveries, in the 1940s, Darwin's ideas were refined by the American palaeontologist, George Gaylord Simpson (among others). As already stated, Darwin was hampered by a lack of detailed mechanisms for inheritance. Mendel's work was not widely known until after 1900, and the term 'gene' was not coined until 1909, when the Danish botanist and geneticist, Willhelm Johanssen (1857–1927) proposed the term. Gregor Mendel (1822–1884) an Augustinian monk crossed 30,000 plants in his garden in Moravia to investigate the characteristics of inheritance. He studied the common garden pea (which no doubt he also ate for his dinner!) He discovered that the plant characteristics he was studying were always inherited in full rather than being blended, so that when he crossed short and tall stemmed plants, the next generation was either short or tall but not intermediate in size (if the occasional mutation was disregarded). He published his findings in a paper in 1866 but his results went largely unnoticed and certainly unacknowledged until German botanist, Carl Correns (1864–1933), Austrian agronomist Erich von Tschermak (1871–1962) and Dutch botanist, Hugo Devries (1848–1935) recognised their significance and Mendel's work came to wide, public recognition shortly after 1900.[77]

This awkward form of inheritance that produced yellow or green peas but not a blend of the two was initially confounding to Darwinian theory and even Thomas Huxley (1825–1895), a strong and loyal supporter of Darwin had questioned Darwin's proposal that: *"...natural selection acts solely by accumulating slight, successive, favourable variations, it can produce no great or sudden modification;"*[78] which is sometimes expressed as *"evolution proceeds in many small steps"*.[79] However, the greater understanding that came with gene theory vindicated Darwin's initial assertion. Darwin believed in *"directed evolution"* or 'Orthogenesis' but it *"lacked a mechanism [and] it also seemed to become less plausible when new evidence led to a revised picture of horse evolution."*[80] So by the 1930s, Darwin's theory needed updating and from 1936 to 1947, Darwinian Theory evolved (itself) into the 'Neo-Darwinian Synthesis'. With the understanding of genes and Mendelian inheritance, in 1949, Simpson wrote: *"Adaptation has a known mechanism: natural selection, acting on the genetics of populations...It is not quite completely understood, as yet, but its reality is established and its adequacy is highly probable."*[81] Simpson considered that directed evolution was not required but that a purely mechanical model provided a better explanation. Biologist, Jonathan Wells writes: *"As Simpson put it, he favoured the view that evolution 'is dependent only on the physical possibilities of the situation and on the interplay of organism and environment'...[further]...'Man...is the result of a purposeless and natural process that did not have him in mind.'"*[82] Simpson's work, in formulating the Neo-Darwinian Synthesis, to take into account all the new discoveries which had occurred since Darwin died in 1882, was so successful that since the 1940s

77. Fowler, T.B. & Kuebler, D.: *The Evolution Controversy.* Baker Academic, Grand Rapids (2007) pp.60–1.

78. Darwin, C.: *The Origin of Species.* (1859) reprinted by Penguin Classics (1985) p.444.

79. Fowler, T.B. & Kuebler, D.: *The Evolution Controversy.* Baker Academic, Grand Rapids (2007) pp.52–9.

80. Wells, J.: *Icons of Evolution.* Regnery Publishing, Washington, DC (2002) p.198.

81. Simpson, G.G.: *The Meaning of Evolution.* Yale Univ. Press, New Haven (1967) p.132 *in* Wells, J.: *Icons of Evolution.* Regnery Publishing, Washington, DC (2002) p.198.

82. *Ibid.* p.203.

no other theory has ever looked like threatening its supremacy, despite a growing disquiet regarding significant facets of the theory. However, the objections are increasing.

One of Simpson's notable critics, Philip Johnson writes: *"...the neo-Darwinian picture of evolution [is] governed at all levels by some combination of physical law, chance and natural selection"* [83] which necessarily require undirected processes. As time has progressed, however, evolution has sometimes been personified in statements such as the assertion that: *"evolution is seen as a grand tinkerer"* [84] which implies that there is some conscious process involved. 'Grand wanderer' or better 'a grand wandering process' would give a more accurate impression. This is important because the words we use, frame the thoughts which will lead us to the next discovery. Good science is enhanced by precise thinking and expression.

With our current knowledge of biology, we now know that there were four critical pieces of information that Darwin lacked:

1. Genes as a mechanism of inheritance;
2. The immense complexity of living cells;
3. Biochemical knowledge;
4. An understanding of the entity of Information.

Without an understanding of these four critical areas, it was next to impossible for Darwin to properly understand the mechanisms and processes within biology and their implications. The true complexity of the cell was not realised until after 1953, more than 70 years after Darwin's death, when initially, x-ray crystallography and later, electron microscopes allowed the internal architecture of the cell to be discovered. As these new investigative tools came along, they provided new lenses through which to view the living microcosm. As a result, new discoveries have continued to be incorporated into the Neo-Darwinian synthesis. Two other very significant scientists who were influential in the reshaping and further development of Darwin's theory were geneticist Theodosius Dobzhansky (1900–75) and naturalist Ernst Mayr (1904–2005).[85] Dobzhansky supported the new discipline of population genetics which relied upon mathematical models developed in the 1930s by Sir Ronald A. Fisher (1890–1962)[86] and J.B.S. Haldane (1892–1964) one of the three leading architects of the Neo-Darwinian Synthesis.[87] Dobzhansky summarised our understanding of species development within Neo-Darwinian Theory (Macroevolution) when he wrote in 1951: *"There is no way towards an understanding of the mechanisms of Macroevolutionary changes, which require time on a geological scale, other than through a full comprehension of the Microevolutionary process observable within the span of a human lifetime, often controlled by man's will and sometimes reproducible in laboratory experiments…A geneticist can approach Macroevolutionary phenomena only by inference from the known Microevolutionary ones."* [88] Thus, Dobzhansky confirms that Macroevolution is dependent on extrapolation from Microevolutionary observations. To overcome the time periods required to demonstrate this supposition, mathematical modelling was subsequently incorporated.

The problem with mathematical modelling, however, is that the outcomes depend upon the assumptions made in devising the mathematical algorithms and the long timeframes involved do not allow direct human observation to discriminate between the competing mathematical models (and hence, the

83. Johnson, P.: *Wedge of Truth.* InterVarsity Press, Downers Grove. Illinois (2000) p.89.

84. Fowler, T.B. & Kuebler, D.: *The Evolution Controversy.* Baker Academic, Grand Rapids (2007) p.162.

85. *Ibid.* p.70.

86. *Ibid.* p.65.

87. Dawkins, R.: *The Greatest Show on Earth.* Bantam Press, London (2009) p.211.

88. Dobzhansky, T.: *Genetics and the Origin of Species.* (3rd Ed.) Columbia University Press, New York (1951) pp.16–17 *in* Fowler, T.B. & Kuebler, D.: (2007) p.70.

assumptions made in developing them.) Ernst Mayer observed that the rate at which new traits were incorporated into bird populations was dependent upon the size and isolation of the colonies and extrapolated that given long enough, this process could explain all of evolution. He confirmed this view when he wrote in 1970: *"The proponents of the synthetic theory maintain that all of evolution is due to the accumulation of small genetic changes, guided by natural selection, and that transpecific evolution [the origin of new species] is nothing but an extrapolation and magnification of the events that take place within populations and species."* [89]

The problem is that observable Microevolutionary change has information limits that were not appreciated when the mathematical models were designed and hence, they do not reflect what is seen in biological systems. Macroevolution is usually 'proven' by extrapolating the findings of Microevolutionary studies which can only be 'validated' if the information limits are not appreciated or ignored. Figure 1.5 summarises the implied flow of information within the Neo-Darwinian Theory.

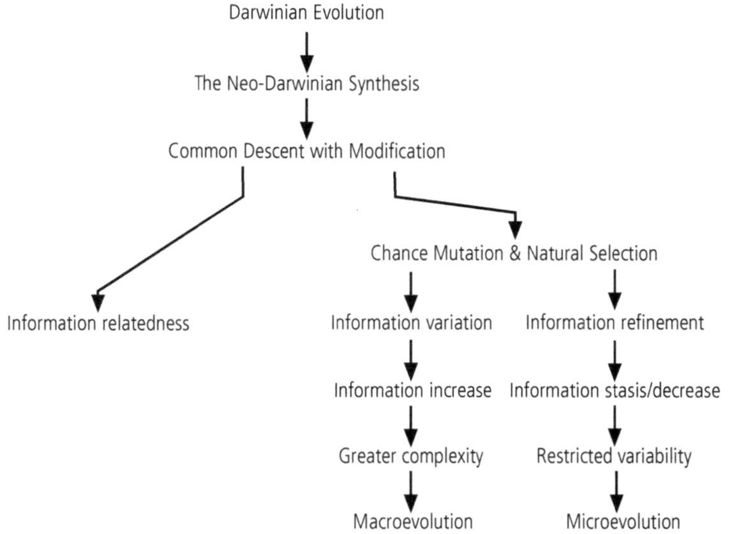

Figure 1.5: Summary of the Neo-Darwinian Synthesis and its implied Information flows. These inferred relationships will be examined and tested over the next four chapters.

Evidence for the Neo-Darwinian Synthesis

A significant amount of evidence has been cited to support the Neo-Darwinian Synthesis. This includes:

1. Observations of living species: for example Darwin's finches, peppered moths and bacterial antibiotic resistance;
2. Fossils such as the horse and bird evolutionary trees: which includes Archaeopteryx;
3. Embryology such as the similarity of embryos before birth;
4. Molecular biology;
5. DNA.

Richard Dawkins cites similar body plans as proof of common descent, however, in information terms, similarity simply demonstrates shared information;

89. Mayr, E.: *Populations, Species, and Evolution: An abridgement of Animal Species and Evolution.* Harvard Univ. Press, Cambridge, MA (1970) p.351 *in* Fowler, T.B. & Kuebler, D.: (2007) p.71

it does not explain how the information got there. Common descent is one explanation favoured by Richard Dawkins but the introduction of related information by another means would be another possible explanation. This will be investigated further in Chapter 3. Bilaterality is another feature which confirms that information is present within an organism, as, if there was no information, the result must be randomness and the two halves of the creature would be different or asymmetric. Thus, it is easy to demonstrate the presence of information but much harder to discern the source.

Fowler and Kuebler write: *"Note that all this evidence for common descent is circumstantial. In fact, it has to be, given the historical nature of evolutionary studies. Obviously, no one can observe humans and chimps descending from a common ancestor, but one can put together pieces of molecular and structural evidence left by this process to make a good case for common descent...*[similarly] *we cannot empirically demonstrate that eyes can evolve via natural selection in the same manner that we can demonstrate the evolution of antibiotic resistance. The former would occur on a scale that requires millions of years while the latter can occur in days or hours. This does not mean that Neo-Darwinian theory is incorrect, but it does mean that direct empirical verification is impossible...*[With] [a]*bsent empirical verification, one must hope to compile enough circumstantial evidence supporting the notion that small changes can accumulate to produce structures such as eyes and limbs. That is what Darwin attempted to do, and it is what most Neo-Darwinists continue to do today, taking advantage of DNA sequencing and fossil data that was unavailable to Darwin. Because this evidence is circumstantial, though, it is the subject of much debate both inside and outside Darwinian circles."* [90]

Due to the long time periods involved, a number of researchers have attempted to extend the mathematical models with computer modelling to simulate the extrapolations that result from Neo-Darwinian Theory but as mentioned earlier, this inevitably requires assumptions when writing the computer programs. Fowler and Kuebler note: *"The simulations are only as good as the assumptions they make, and it can often be difficult to ascertain if the assumptions made by the programmers actually reflect the relevant biology."* [91]

Neo-Darwinian Assumptions

To Darwin's nine assumptions discussed previously, three more have been added by the Neo-Darwinian theorists: [92]

1. Science is completely autonomous (to minimise bias as much as possible);
2. Naturalistic explanations are adequate to account for all that is observed in the Universe. (To keep science as rigorous as possible it has been limited to what can be physically observed or tested.);
3. Only naturalistic explanations are acceptable in science. (Again, this is in an attempt to keep science as rigorous as possible. The problem, as we shall see, is that it stifles some research.)

These extrapolations or assumptions will be discussed in the course of our journey into information. Fowler and Kuebler (who conclude that evolution is true) summarised the Neo-Darwinian case by saying: *"While there is solid evidence that these mechanisms can lead to small-scale change (Microevolution), a fact accepted*

90. Fowler, T.B. & Kuebler, D.: *The Evolution Controversy.* Baker Academic, Grand Rapids (2007) pp.147–8.

91. *Ibid.* p.159.

92. *Ibid.* p.139.

by all schools [of thought], *there is at present little direct evidence that they can explain* all *evolutionary change (Macroevolution); what evidence we have is primarily circumstantial. Phrased in other language, the question is whether random mutation and natural selection can work together to create novel complex biological information on a large scale. Neo-Darwinians argue that the available evidence supports their theory, and no other theory has a better explanation of this evidence, so they infer the truth of their theory by* inference to the best explanation." [93] The evidence and counter arguments are summarised in Table 1.4 below:

	Evidence[94,95]	Implication	Counter argument
1	Geologic Column	Suggests the long history of life on Earth is long enough for Macroevolution to have occurred.	Organisms appear suddenly without enough time to acquire complex information.
2	Fossil record	Fossils will be found showing the transitions of one life form into another.	Finely graded transitions are not found yet; it is circumstantial evidence; needs written pedigree.
3	Embryology	Embryos show the similarity of our evolutionary past.	Haeckel's drawings were grossly inaccurate.[96]
4	Random variation with Natural selection	Darwin's finches / peppered moths demonstrate adaptation of organisms to their environment.	There is agreement here. Most scientists agree that Microevolution is true.
5	Bacterial resistance	Proves Microevolution occurs and by extrapolation, Macroevolution.	This is information loss not gain which is Microevolutionary.
6	Selective breeding	If it works in captivity it demonstrates 'evolution in action'.	Selective breeding is not random[97] (again Microevolutionary).
7	Population genetics	Mathematical models can demonstrate Macroevolution over long timeframes.	Extrapolation of Microevolution into Macroevolution is invalid; the models are non-biological.
8	Anatomical similarities	Demonstrates body plan relatedness which confirms descent from a common ancestor.	Demonstrates similar information but not necessarily universal common descent.
9	Molecular biology	Biochemical pathways will confirm the path taken by evolution.	Unproven to date; more data is needed to demonstrate linkages.
10	DNA (genetics)	Common genetic code will demonstrate the information pathway from simple to complex organisms.	There is a confused DNA tree to date which does not support universal common descent.
11	Geographic distributions of flora and fauna	Variations in species and subspecies around the world demonstrate gradual global evolution.	Demonstrates migration and Microevolutionary survival of the fittest.[98]

Table 1.4: Summary of the Neo-Darwinian Synthesis

93. Fowler, T.B. & Kuebler, D.: *The Evolution Controversy.* Baker Academic, Grand Rapids (2007) p.344. (Emphasis in the original)

94. *Ibid.* p.345.

95. Lovett, R.A: Darwin's Evidence for Evolution. *Cosmos* 24 (2008) p.24.

96. Fowler, T.B. & Kuebler, D.: *The Evolution Controversy.* Baker Academic, Grand Rapids (2007) p.58.

97. *Ibid.* p.59.

98. Marsupials and monotremes (the platypus and echidna of Australia) became extinct where they had to compete with placental mammals but where they were protected by isolation, as in Australia, they were able to survive.

It is very important to avoid confusing observational data with mechanisms of causation. The existence of fossils which appear similar cannot prove their relatedness unless there is a 'written pedigree' to prove the relationship, as is required in pedigree breeding associations. Neo-Darwinists are hoping that the molecular biology and DNA sequences will provide this necessary evidence but the data to date suggests the lineage is 'mixed up'. This is dealt with in greater detail in Chapter 5.

Clarifying Ideas

The summary of the Neo-Darwinian Synthesis in Table 1.4 illustrates the extent of the dispute within scientific circles. How is it that thousands of eminent, university-trained people, many of them professors, can argue for 150 years about evolution and still be no closer to agreeing how to resolve the problem? To give a comparison, 150 years after Galileo or Newton there was no controversy over their theories, so why is it so for evolution? The answer is that many of the ideas have become muddled and to understand them properly the details need to be clearly identified with clarification of the concepts so that they can be analysed precisely. To aid this process, Table 1.5 identifies different levels of explanation which are divided into: Causes, Processes, Mechanisms and Results and all of them will be considered in turn.

Tier 4	Causation (= philosophy)	Neo-Darwinian Synthesis	Theistic Evolution	Intelligent Design	Biblical Theism
Tier 3	Processes (= hypotheses)	Random variation	Natural selection		Design
Tier 2	Mechanisms (= science)	DNA	Chemical Laws	Physical Laws	Information
Tier 1	Results (= observations)	Life		The Universe	

Table 1.5: Table of Explanations
This table categorises the differing levels of explanation within structured arguments.

Tier 4 in the Table of Explanations is the 'why' question. Causation is the ultimate premise of science, but if no one was there to observe the process, then we are reliant upon secondary sources of evidence, that are open to differing interpretations, depending upon one's personal philosophy or worldview; they may be able to be *disproven* but they cannot be *proven* by science. Note that the whole process of explanations depends upon intellectual enquiry but there are differing degrees of evidence behind each of the tiers. The problem with Tier 4 is that unless you fully understand the process that formed the entity, you cannot be sure of the causation. Secondary or circumstantial evidence can be corroborative but cannot attest beyond all doubt the causation and this is why there is so much continuing controversy. As a result, both sides of the debate are forced to resort to claiming: *"inference to the best explanation"*, but *"best"* is a relative term and its interpretation is completely in the mind of the beholder, as it is contingent upon which information is judged to be 'superior' and which facts will be considered inadmissible as 'evidence' all together.

The next level of explanation is 'Processes' which is the 'how' question. These are hypotheses which are again built-up from evidence but the interpretation is still dependent upon belief. Darwin postulated random variation with natural selection for the origin of species, while others have suggested design (more recently, 'Intelligent Design'). These explanations rarely have black and white outcomes and attract a lot of discussion. In contrast, Tier 2 explanations address mechanisms, which is the realm of science. These deal with laws that can be studied, tested and quantified. Tier 2 explanations cannot directly explain why life and the Universe exist. They are merely the tools that help us to uncover the evidence for which our enquiring minds crave an explanation (such as the why questions); this is why Tiers 3 and 4 are important. Tier 1 explanations are direct observations and relate to directly measured data that is often self-evident, such as the variation within life on Earth and the existence of the Universe. This is where science usually begins, with observation. It is very easy to muddle up the various tiers in Table 1.5 and much argument derives from confusion relating to the level of explanation and the degree to which the evidence corresponding to that level may be tested and verified. It will be noted that Information is in Tier 2 of Table 1.5. It is a mechanism or tool of science and is fundamental to everything we do.

Pure science starts at Tier 1, that is, from the bottom up but philosophy may start from anywhere (including Tier 4) providing the process follows logic, rigour and rationality. In practice, most people tend to use both a bottom-up and a top-down approach simultaneously. To give an example, everyone possesses a 'worldview' (Tier 4) which means that they approach a topic with a pre-existing system of ideas which is usually embedded in their hypothesis but which could equally be referred to as a 'bias' towards one idea or another. They then take direct observations from Tier 1 and work inwards from both Tier 1 to Tier 4 and from Tier 4 to Tier 1. This is how so many eminent scientists and philosophers can sincerely take the same data (Tier 1) and come up with such stunningly different conclusions (Tier 4). I believe that 99% of researchers and writers are 100% genuine in their beliefs and statements. The problem comes in the weighting of the evidence and hence, the interpretation of the facts that have been observed. To overcome these problems associated with the interpretation of the observations, a more powerful tool is required to discern reality from wishful thinking and information can do that by sharpening our understanding of the data that we observe.

Information is unavoidable

It has already been noted that this period in history has been dubbed: 'The Information Age' and we have observed that Information is everywhere. Once one's eyes are opened and our minds are attuned to looking for it, it becomes obvious that information underlies every facet of our existence. Figure 1.6 shows some of the multiple layers of information that have been required since the formation of the Universe at the 'Big Bang' (or some other equivalent process) and any study of information needs to account for these.

Despite this enormous volume of information the analysis need not be that complicated, as information can vary in only one of four ways. It can:

A. Increase;
B. Decrease;
C. Remain the same (stasis), or;
D. Change but be similar (equivalence).

By studying the information in living things and the Universe at large, much can be clarified and understood better. Advancing technology such as improved microscopes in biology and more powerful telescopes in cosmology are expanding the information available for study. The bottom line is: information is ubiquitous. We cannot properly understand any part of the Universe without it.

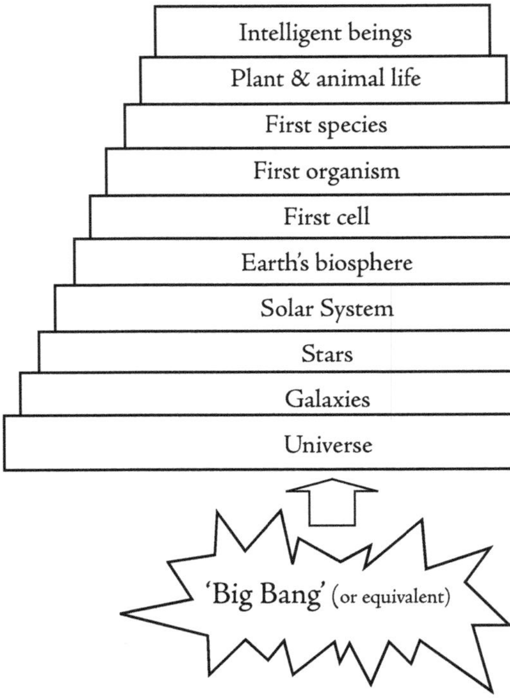

Figure 1.6: Layers of Information in the Universe Currently, the majority of scientists support the concept of the 'Big Bang' as a plausible explanation for the formation of the Universe. This will be examined in later chapters.

Purpose of this Book

The purpose of this book is to investigate information, to follow where the evidence and logic lead; to explore the relationship between information and the Universe by focusing firstly on biology and chemistry, then physics and cosmology, to better understand the origins of information. This especially applies to life and the implications for the proposition that life evolved on Earth. Information analysis will also be used to unravel the contradictions we find. We will be drilling down to basic units, to trace the step-by-step acquisition of information from simple to complex organisms, then from Earth's biosphere to our Solar System, the Milky Way and eventually to the beginning of the Universe.

Evolution implies change and Darwin used it to explain gradual, growing complexity. Today we realise that this increasing complexity required new information. Darwin saw this process of acquiring new complexity (information) in terms of random mutation with natural selection and many scientists view this as 'Evolution' but the information is lacking. There is no evidence that random mutation with natural selection can increase information to build complex organisms which will be demonstrated in the succeeding chapters. Natural selection *reduces* the information within a system, limiting future options and it, therefore, cannot *increase* complexity (see Figure 5.1). This process is sometimes referred to as Microevolution. It does not create new or increased complexity but adapts the organism's existing information by selecting from traits that are already stored in its DNA data bank. Random mutation is not the answer either. This will be dealt with in much greater depth in Chapter 5.

Some philosophical systems may be capable of explaining any and all facts within themselves - that is, they cannot be refuted and are, therefore, not 'scientific' within the formal definition of science. The danger in this, of course, is that such philosophical systems will become so enmeshed within the worldview with which they are associated that they can impede or prevent even legitimate criticism and thus, impede the progress of science itself. That this can happen with respect to Creationism is obvious. However, it can also happen in the case of the Neo-Darwinian school of thought. Consider the following statement made by George Gaylord Simpson in 1967: *"Although many details remain to be worked out, it is already evident that all the objective phenomena of the history of life can be explained by purely naturalistic or, in a proper sense of the sometimes abused word, materialistic factors. They are readily explicable on the basis of differential reproduction in populations [natural selection], and the mainly random interplay of the known processes of heredity [random mutations]. Therefore, man is the result of a purposeless and natural process that did not have him in mind."* [99] The breadth of this statement is likely to reap the ire of many theists; and in like manner the 'debate' continues.

Looking at life and the Universe through Informationist glasses, seeking any information we can glean, will give a new clarity to the evolution /theism/ intelligent design /creation debate. The other purpose of this book is to define some basic Laws of Biology and Information that may be agreed upon and from which, further research can proceed.

Conclusion

Therefore, in conclusion, information is not only common in the Universe, it is ubiquitous. Wherever one looks, there is information. In addition, the deeper one looks the more information there is to be found due to the multi-layering effect. However, the information is often inconspicuous, as it is intangible and so it often goes unnoticed. This unexpected finding regarding information, despite its profound significance, is the origin of the title of this book. Several paradoxes have already been described and more will be discussed as we uncover them; the unobtrusive nature of information is just one more. By refocusing on the *information content* itself, in contrast to the vehicles that transmit the information, we can gain a greater understanding of ourselves and our Universe. This distinction has been well expressed by the neurophysiologist and 1981 Nobel laureate in Physiology, Roger Sperry who stated: *"The meaning of the message will not be found in the physics and chemistry of the paper and ink"*.[100]

We know that the ancients studied the sky and tried to understand the stars and planets but their study of cosmology was severely hampered by their lack of ability to collect accurate information until the development of an efficient telescope in 1608. Aristotle and others used the scientific principle of '*inference to the best explanation*', to try to explain the observations that they had made but as history has shown, there was a long period when science was following a flawed paradigm which impeded further progress. As science has gained better tools, such as knowledge of the Law of Gravity, a much deeper understanding of the Universe has become possible, as knowledge and understanding build upon themselves. This phenomenon is also true of information. Over the last four hundred years, physics and chemistry have made giant strides but there are still glaring holes and questions in biology which standard Neo-Darwinian Theory has not been able to answer to date, such as: the origin of the first cell, the existence of life itself, the mechanisms of thought and the entity of consciousness. Most people who unquestioningly accept Neo-Darwinian Theory, do so with an incomplete understanding of the immense complexity of living organisms. The truth is found

99. Simpson, G.G.: *The Meaning of Evolution.* (Rev. Ed.) Yale University Press, New Haven (1967) p.345 *in* Fowler, T.B. & Kuebler, D.: (2007) p.40.

100. Sperry, R.: *from* Dawkins, R. & Lennox, J.C.: *The God Delusion Debate.* (DVD) Fixed Point Foundation, USA (2007).

in the fine detail and the way that detail fits together. The aim of this book is to use information, as dispassionately as possible, to examine all sides of the debate and let the evidence speak for itself.

The other point to note is that no theory in history has generated the degree of controversy that evolution has. The academic elite were certainly perturbed enough by Galileo's views to force him to recant before the Inquisition in 1633, although this may have had more to do with Galileo's other activities and beliefs (including mocking the Pope on the title page of a previous book published a year earlier) than his cosmology, but even so, there has not been 150 years of ongoing argument in the highest academic circles on any other scientific topic to match the vehemence of the evolution debate.[101] As more details of the inner workings of the cell emerge, including the results of the genome and ENCODE projects, many had hoped that the new information would settle some of these matters but much of the new information has only resulted in stoking the fires and if anything, the intensity of the controversy is growing.

One of the principal purposes of this book is to clarify ideas. This has been done through the integration of concepts from many different sources. The methodology has been: to look for information within each area of study, to discern principles or laws that can be generally agreed upon, before extending our study to biology and the physical Universe. The path followed has been to focus on mechanisms (how) rather than causes (why), although where appropriate, both areas are addressed. By using the lens of the information paradigm, I believe that we can gain a much greater understanding of the Universe and of ourselves.

101. Gonzalez and Richards attribute the burning at the stake of the Dominican monk, Guilermo Bruno, in 1600 to his heretical views on the Trinity, the incarnation and other doctrines rather than his heliocentric views. See Gonzalez, G. & Richards, J.W.: *The Privileged Planet.* Regnery Publishing, Washington DC (2004) p.242.

2

Exploring the Information Paradox

In the beginning was the bit.[1]

Hans Christian von Baeyer

Even though the Information Age began 30 years ago, information is still an enigma. Anthropologist, Terrence W. Deacon of the University of California, Berkeley writes: *"...although we use the concept of information almost daily without confusion, and we build machinery (computers) and network systems to move, analyse, and store it, I believe that we still do not really know what it is."* [2] He goes on to say: *"Information is by definition something in relation to something else, but in colloquial use the term can refer either to what is conveyed or what provides the conveyance."* [3] In Chapter 1, it was seen that information is a discrete third entity in the Universe after matter and energy. However, information differs from matter and energy in that, although it is ubiquitous, it is intangible and not directly detectable with the five human senses, making it harder to understand the entity itself. It is also important, as Deacon points out, to distinguish the information from its carrier. However, often the only avenue available to discovering this elusive entity is by investigating the properties and behaviour of its carriers or vehicles. In this chapter it will become apparent that, despite its complexity and intangibility, it is still eminently amenable to study and classification.

In the past, information has often been viewed as a concept or a useful tool to help us think about a particular situation but information is more than this. It is commonly said that *"information is power"* but why is it powerful? The reason is, it allows us to make sense of the world around us and hence, make predictions that help us to anticipate the future and so make the most use of the opportunities available to us. Usually, we do this intuitively without consciously realising what we are doing. A typical example would be: to know when and where to cast your fishing line to catch a fish for dinner. When the First Fleet arrived in Sydney Harbour in 1788, their total lack of knowledge about how, when and where to plant crops in an arid climate in the Southern Hemisphere, where all the seasons are 'back to front' compared with the Northern Hemisphere, meant that the new colony nearly starved to death in its first four years. The indigenous aboriginal population had copious information about nutritious food sources within the Australian bush but the British settlers did not initially understand how to access this information. The Australian aboriginal people have never had a formal written language, however, they retain most of the information they require for day to

1. von Baeyer, H.C.: In the Beginning was the Bit. *New Scientist* (169)2278 (17 Feb 2001) p.26.

2. Davies, P. & Gregersen, N.H. (Eds.): *Information and the Nature of Reality. From Physics to Metaphysics.* Cambridge Univ. Press, Cambridge (2010) p.147.

3. *Ibid.* p.159.

day living in their memories.[4] Thus, the human mind can also be a rich store of information.

As would be suspected in the Age of Information, much human endeavour in the Western World is now involved in activities dealing with information. Most of the financial and service sectors of our economies, including most white-collar jobs, are selling knowledge (information) and skills ('know-how'). For example a lawyer, your doctor, the teacher or a computer programmer all spend most of their time working with information. Further, Western societies appear to emphasise the value of that information by paying higher wages to people with an occupation that processes or delivers information. For example, a politician or an accountant is usually paid more than a manual labourer.

Measuring Information

British physicist, William Thomson, (1824-1907) better known as Lord Kelvin, after whom 'absolute zero' and the Kelvin temperature scale are named, had a famous dictum which stated:

"In physical science a first essential step in the direction of learning any subject is to find principles of numerical reckoning and practicable methods for measuring some quality connected with it. I often say that when you can measure what you are speaking about, and express it in numbers, you know something about it; but when you cannot express it in numbers, your knowledge is of a meagre and unsatisfactory kind; it may be the beginning of knowledge, but you have scarcely, in your thoughts, advanced to the stage of Science, whatever the matter may be." [5]

Understanding depends upon the frame of reference or paradigm from within which it is viewed. Lord Kelvin, being a physicist comes from a mathematical discipline, so this assessment may be considered a little harsh by anyone who is not a mathematician or a physicist but it contains a kernel of truth, in that it demands rigour in the way we deal with any topic. So, can information satisfy Lord Kelvin's requirement to *"express it in numbers"*, that is, digitally? We saw in Chapter 1 that Shannon or syntactic information is very amenable to mathematical manipulation but semantic or meaningful information poses a problem, so to some extent, it depends on what the information is and how it is organised.

The Organisation of Information

Since information is ubiquitous, any entity in the Universe can be assessed to determine its information content; take sound, for example. Sound data can be haphazard and disorganised if it is played in random frequencies which we interpret as 'noise' or it can be structured and harmonious which we may interpret as 'music'. Musical information is generated and arranged in many different ways too. It can be produced by a group of musicians during an improvised 'jam' session when there is no conductor or composer but the sound still possesses an easily recognisable 'melody'. The most beautiful music, however, is highly structured, written by a composer, led by a conductor, rehearsed and played by professional musicians, such as in a staged performance. A symphony is built up of many parts (instruments/systems), which all have their own musical script (information). Each individual musician's script is combined in one body of information, called

4. Some information is recorded in pictorial form in paintings and on implements such as boomerangs.

5. Thomson, W.: *Popular Lectures and addresses, Vol. 1.* MacMillan & Co., London (1889) p.73.

the conductor's score. Instruments must begin and end playing at precise times to achieve the composer's vision for the musical piece. The instruments themselves are constructed of many parts which have been carefully designed, manufactured and assembled, then tuned to create precisely the right pitch: (for example, woodwinds have reeds and keys to make the notes.) Yet opening or closing each stop going up or down the tube that forms a saxophone, to divert the air one way or another does not appear to have much relationship to music or a melody but when co-ordinated with all the other parts of the instrument and played by a competent musician, it makes music. Note that there are a myriad of ways to create noise, yet only a more limited number of ways to make musical harmony, like islands of beauty in an ocean of sound.

Each component of each instrument is itself made up of levers, springs, pads and screws that are precisely manufactured. The screws and spindles, in turn, are themselves made of raw materials such: as metal, plastic, cork or reed but again precisely specified by the instrument's designer. Thus, it can be seen that there is a vast amount of information at every level of musical systems too.

In the case of an orchestra, the information is *aperiodic*, unlike a crystal lattice, which is formed from a regular array of molecules; although a crystal can grow to be very large, it is repetitive and stores little information. A crystal lattice, as seen in table salt (sodium chloride), is equivalent to an orchestra composed entirely of violins playing the same two notes over and over and over again, with no variation. The crystal lattice is easier to represent in mathematical terms due to its uniformity and the small amounts of information it contains, whereas to describe the orchestra in mathematical terms would be extremely difficult and from a superficial analysis, it would seem impossible. This difficulty in writing a mathematical formula to explain the complex relationships between vast amounts of information in the orchestra also mirrors the problems we have in trying to understand the way we think. Computers think in syntactic information (count the bits) but humans think in semantics (which involves meaning).[6] To test this, try to remember a 10 digit string of numbers where there is no meaning or pattern such as 8160259349 and compare that with the first 20 words of the much more complex soliloquy by Hamlet which we met in Chapter 1: "*To be or not to be that is the question. Whether 'tis nobler to suffer the slings and arrows of outrageous fortune or...*" (and you may be able to go on a lot further). When there is meaning, despite the increased complexity, our minds can remember much larger quantities of information. One of the problems is that scientists do not know exactly how our brains think. It appears that the brain has a 'machine code' under the semantic thought processes. But I am skipping ahead and there is a lot we have to cover before we can consider the way our brains handle information. We will return to thinking and consciousness in Chapter 10.

Now as already noted, energy flows and matter are not enough to describe life, which requires detailed information,[7] yet strangely, to gather or construct information requires work and information is costly to acquire, so energy must be expended to do it.[8] Similarly, to create order requires energy input too, which is a restatement of the Second Law of Thermodynamics.[9] This is the 7th paradox: *information is neither matter nor energy but, in our physical world, it requires effort (which involves energy) to collect or construct it.* To test this, consider how much work is required to collect or generate information for a research project or write a computer programme. It does not grow, like a sand dune or a snow drift on a

6. Lennox, J.C.: *God's Undertaker. Has Science Buried God?* Lion Books, Oxford (2007) p.143.

7. Woodward, T.: *Darwin Strikes Back*. Baker Books, Grand Rapids (2006) p.117.

8. Davidson, J.D. & Rees-Mogg, W.: *The Great Reckoning: How the World will change before the Year 2000*. Pan Books, London (1994) p.298.

9. Missler, C.: *Cosmic Codes*. Koinonia House, Coeur d'Alene, Idaho (2004) p.318.

windy day but requires conscious activity to compile. This is why books do not write themselves and your homework never mysteriously materialised overnight either.

Further, as information requires storage involving matter or energy, the storage mechanism is subject to the Law of Entropy. And yet, if the storage mechanism is disrupted, the information being massless does not change. The information may be lost to us so that we can no longer retrieve it, due to the loss of its physical repository but the information itself being massless is unchanged. This is Paradox 8. Thus, *the vehicle carrying the information is subject to entropy but not the information itself.*

Hardware, Software and Information

After considering the organisation of information, the next task is to examine the manipulation or management of information. Our principal model or tool for doing this is the computer. In Chapter 1, we observed that when a computer is first manufactured it can be plugged in and will 'run' but it will not function as a processor until it is loaded with the appropriate 'software' information. The 'hardware' is indispensable as the physical platform on which the software runs; it also defines the environment in which the software is active (delineates its boundaries). Software is the set of instructions (information commands) which tell the hardware what to do. Hardware is the physical apparatus that allows the non-physical information carried by the software to operate. As the hardware and software have to work very closely together, they must match each other. An example would be the difference between an Apple computer and a Microsoft Windows computer. Trying to run the software specifically written for one family of computer hardware on the other type will not work. Although the DVD (the vehicle) carrying the software instructions may be identical for both types of software and the resultant program that runs on the relevant computer appears almost identical on the screen, the instructions must be specific to each type of hardware. And yet, although the software written for each type of computer platform is unique to that platform, the underlying message that is being conveyed by the respective software packages is identical! Therefore, we must distinguish not only between the hardware and the software but also between the software and the message or information that is being carried by the software. In a different analogy, in the car industry, most of the vehicles on the road have components that perform the same function and they all have engines and wheels. However, they are usually slightly different and if a part breaks, such as an indicator light, then usually a specifically branded part, such as, a Mercedes-Benz lens cover needs to be used to replace it, although sometimes an Audi part might also perform the same function. This is the case in biological systems too. Each species has its own unique genetic code which specifies development, replication and general activity, even though a lot of the behaviour can be almost alike within similar species. As the software is unique, it must have developed simultaneously or in parallel with the physical bodies of living organisms that it runs on. But which came first, the software, the hardware or the information?

To help us to understand this difficult question, let us look at a more familiar example first. As noted in the conclusion to Chapter 1, the information printed on this page is extrinsic to the paper and the ink that was used to print it.[10] But which came first, the technology to manufacture the paper and ink or the thought which generated the ideas (the information), so that there was something meaningful to record? In this case, it is the paper and writing technology but if we were to look at a Beethoven symphony on a compact disc (CD) then the ideas (the musical score) came first and the CD was only invented in the last 30 years. From these examples, it can be seen that it does not matter which came first; the information and the medium on which it is recorded are completely independent of each other. They only need to come together at the point where the information is stored in its respective tangible medium. However, it is important to remember that the intangible information requires a tangible platform on which to be recorded or through which it may be communicated (such as a radio broadcast). Therefore, before the Beethoven symphony was recorded on the compact disc, it had to be written down by the composer in a tangible format. This is also a requirement for living systems, where the intangible information stored within the genetic code needs to be recorded in a tangible form, in the DNA, so that it is available to the next generation.

Another way to demonstrate this intriguing property of information is through its expression in different languages (types of software). For example, when Julius Caesar invaded Britain in 45 BC he would have used the word: *"bellum"* but Napoleon in France in 1800 would have been more familiar with the word: *"guerre"* while in English today, we use the word: *"war"*. Although they are all different, they mean the same thing. In the same way that identical information can be expressed differently, the physical medium upon which the respective words are written can also be changed. Caesar may have written on papyrus (reeds) or on vellum (sheepskin) whereas Napoleon wrote on paper with a quill pen and today I am typing on a computer using an electronic word processing programme. It can be seen that the medium (vellum, paper or computer), which is the hardware, can change but the information carrier which is the language (for example Latin, French, English or even Braille), which is the software, can change independently as well and no meaning is lost. Even the time can change (over 2000 years has elapsed since Julius Caesar invaded Britain) but the meaning remains the same. When imparting information through language, however, it is important to realise that the words that are 'carrying' the information only have meaning because specific words have been assigned unique meanings. For example, *"let"* in tennis, retains its mediaeval meaning: "to impede" or "hinder" but in general usage today the English word *"let"* has the opposite meaning, which is: "to allow". If I make up a new word such as: *"infobits"* in Chapter 1, as the author I can define its meaning as I choose. From the above examples, it can be seen that the information itself is independent of both the medium (platform) and the language (carrier). To detect the information, the platform must have material properties so that we can perceive it with our physical senses. In contrast, the vehicle and the information are non-material. Further, as the information does not have mass, it cannot be measured in length, width or height and nor does it depend on time. It has no physical characteristics and yet it still exists and it does have the interesting characteristic that it is transmissible through space and time, such as via television, radio or through the written word.

From these observations, it is clear that information is more than just the medium (hardware and software) which carries the message. A classic

10. Sperry, R. *from* Dawkins, R. & Lennox, J.C.: *The God Delusion Debate.* (DVD) Fixed Point Foundation, USA (2007).

demonstration of this can be seen in movies, where the bearer of bad tidings tries to distinguish themselves from the information they are carrying by saying: *"Don't shoot me, I'm only the messenger!"* In this case, the person is the 'hardware' and they have an unpleasant task to perform. The 'software' is the language that they speak. Some common examples are illustrated in Table 2.1.

	Computer	Novel	School	Musical symphony
Hardware	Processor box and screen	Paper	Blackboard and chalk or whiteboard and marker	Paper Score
Software	Programme	Language	School syllabus	Staves system
Information	Data	Story	Knowledge	Music

Table 2.1: Illustrations of Hardware, Software and Information

An interesting problem to contemplate is posed by the following scenario: if a great composer, such as Bach or Beethoven, had composed one more symphony at the end of their life but before the piece was played, the score was lost in a house fire and the composer died, so that the symphony could not be written down again. There was obviously information in the musical score which is now lost. What happened to the information recorded on the paper (the hardware)? Is it gone forever or can someone else rediscover it somehow? The intuitive answer is that it is lost forever, burnt in the fire, when its physical carrier was destroyed but there may be another possibility which will be discussed further when we investigate memory in Chapter 10.

Quality versus Quantity

So is there a best way to measure information? Is it the degree of complexity (such as the number of links or items as Shannon described), the significance or meaning (semantic information) or the amount of effort to generate it? As seen in Chapter 1, Claude Shannon described how discrete packets or parcels of information such as electronic signals in a wireless system might be defined. He developed a method to quantitatively measure (that is, to count) pieces of information, which formed the Shannon Theory of Information. This is also called *syntactic information.*[11] Shannon's theory is good for measuring the volume or quantity of information but is incapable of measuring the meaning or value of information. Information which contains meaning is called *semantic information.* It is possible to have a large amount of data which is meaningless and hence, useless, so any complete measure of information must include both the quality of the information as well as the quantity. The problem is that, at present, there is no objective way to mathematically quantify the meaning or quality of a piece of information. Table 2.2 summarises these two fundamentally different functions of information.

11. Lennox, J.C.: *God's Undertaker. Has Science Buried God?* Lion Books, Oxford (2007) pp.141–2.

Type	Descriptor	Function	Notes
Syntactic	Quantitative	Counts the number of pieces of information.	Shannon Theory
Semantic	Qualitative	Comprises the meaning within information.	Difficult to describe mathematically

Table 2.2: The Functions of Information

Transmission of Information

It was previously noted that information can be transferred from one person to another, such as: the musical score proceeding from Beethoven's mind onto paper and then to his orchestra. Information is also transferable over time; if it was not, then we would not know of the exploits of Julius Caesar. This can only happen, however, if there is effective transmission of information via its carriers. During the Dark Ages in Europe, a lot of the ancient Greek and Roman writings were lost to European scholars until they were rediscovered during the Renaissance in the Middle East and Spain, where they had been preserved by Muslim scholars. Without this remote repository of the information, we would not have known about much of ancient Greece and Rome, such as Euclid, who will become important when we look at Relativity Theory in Chapter 8.

However, as already stated, the possession of information does not automatically mean that it is readily recognisable, even if it has great importance or value. A classic example is where one painting is painted on top of another older, more valuable painting, such as: *The Taking of Christ* by the 16th-Century Italian painter, Caravaggio, which disappeared for two centuries until it was found in Dublin in 1990. The lack of recognition of the true value of the information may have occurred through ignorance or deliberate deception but the misattribution will continue until an investigator reveals the real situation. Alternatively, we may identify a piece of writing as being likely to contain meaning, yet be unable to decipher that meaning. This was very evident in the study of the Mayan civilisation. After the European invasion of Central America in the 16th-Century, the ability to read the Mayan hieroglyphs was lost until very recently, when they were re-deciphered once more. Most scholars agreed that there was significant meaning (information) in the hieroglyphs and a significant body of writing had been preserved until the present day (effective transmission had occurred) but we had lost the ability to decipher the meaning of the information.

So in summary, the quality of semantic information depends on a minimum number of steps which include:

1. It has encoded meaning;
2. Effective transmission occurred;
3. Ability to identify the information exists;
4. Ability to interpret the information exists.

Another everyday example of the above requirements can be illustrated while sending a text message on my mobile phone to a friend. If I have sent a random string of letters and numbers, there is no meaning in the transmission. My friend will receive a string of letters and numbers and even though they can read and understand my language, if I have not written a meaningful statement in the first place, there is no semantic information. There is syntactic information but apart from the fact that I sent a communication, there is no other information in this communication except perhaps to identify me as the sender and that I was there to send the information at that time. Alternatively, if I type the message: "Meet you for lunch at 12 o'clock." but forget to push the send button on my phone, there was no transmission and although there was meaning in the message that I typed on my phone, my friend could not receive it. When I do press the send button, assuming that I have the correct number in my mobile phone, my friend will receive the message. If, however, my grandfather is the receiver and does not appreciate that the mobile phone has received a text message, then the communication fails at stage three (he failed to identify the information). Alternatively, if my friend is Portuguese, he may be able to read the message but be unable to interpret it if he does not understand English. From the above examples, it can be seen that to accurately measure how much meaning is contained in an information transmission, we must first have the ability to appreciate and catalogue the information it contains.

Information in Biological Systems

It is now time to look at biology and the information stored in the DNA in a living cell. Richard Dawkins says: "*Biology is the study of complicated things…Physics books may be complicated, but…The objects and phenomena that a physics book describes are simpler than a single cell in the body of its author. And the author consists of trillions of those cells, many of them different from each other, organised with intricate architecture and precision-engineering into a working machine capable of writing a book…Each nucleus…contains a digitally coded database, larger, in information content, than all 30 volumes of the Encyclopaedia Britannica put together. And this figure is for each cell, not all the cells of the body put together.*"[12] Thus, our biological information system is ultra-compact and *extremely* large.

But is this anything unusual? Maybe this type of arrangement is common in the Universe. After all, crystals are ordered, can be very large and are self-organising as well, so why not proteins and life, too? We have already seen, however, that there is a difference in the information content between a crystal lattice and a protein. The highly ordered repeating pattern of the crystal lattice is self-organising but incorporates very little information, whereas the protein is very irregular, highly structured but cannot *self-organise* (see Chapter 3). In contrast, the protein contains a vast amount of information as described by Richard Dawkins. What is more, the 'functional' information, which 'runs' on intrinsic software, has multiple roles as demonstrated in Table 2.3.

Although it can be seen from Table 2.3 that systems in the Universe are intricate, their complexity derives from layered information that is built-up from simple entities through the incremental addition of information. To visualise this, compare a novel with a symphony and then life. Although all are very different, the

12. Dawkins, R.: *The Blind Watchmaker* W.W. Norton, New York (1996) pp.1–3 *in* Johnson, P.: *Wedge of Truth*. InterVarsity Press, Downers Grove, Illinois (2000) p.126.

Table 2.3: A Comparison of Information in Computing and Biological Systems
Computers are the basic tool used for information management today. They can manipulate, store and reproduce data (information). This is achieved through software that operates via a machine code allowing the computer to function as a processor. In addition, there is the information that allowed the manufacturer to build the fabric of the computer not to mention the knowledge of electronics required by the inventor to devise the design of the microchip. These functions are also observed in biological systems.

Level	Role of Information	Computer Analogy	Biological Equivalent
1	Transient / Dynamic information	Information entered on the computer	Current information (such as your present thoughts)
2	Operational programmes	Resident programmes that are necessary to use the computer (such as word processing)	Neural routines within the brain (such as how to walk or talk)
3	Messaging systems	Machine code to run the computer (also the operating system)	The neuro-endocrine system (as in blood pressure regulation)
4	Structural elements (platform fabric)	Information to specify and make the computer hardware	Proteins /organic molecules to build the body
5	Architecture / plan	Knowledge to devise the microchip /computer and write the software	Genome and epigenome / DNA and RNA
6	Purpose	Concept of why it has been made.	Chance or design

information within each entity is layered in a similar way. However, also note that there needs to be an extra layer of information to manage the other information. Even though the musicians in an orchestra, collectively, have the whole score (specifying the notes and style in which the music is to be played) a conductor is still required to bring the music to life. There is a similar layer of co-ordinating information which has recently been found in living cells, which is performed by the *epigenome*. (See Table 2.4) The main difference between the three systems in this table is that in the case of the narrator and the conductor, the controller is external to the story and symphony respectively, but the epigenome is intrinsic to life and is contained within the cell's DNA. The epigenome will be discussed in greater detail in Chapter 3.

		Chronicle	Music	Life
		Narrator	Conductor	Epigenome
+ Information	→	Storyline	Symphony	Life
+ Information	→	Book	Score	Cell
+ Information	→	Chapter	Movement	Chromosome
+ Information	→	Page	Sheet	Gene
+ Information	→	Sentence	Phrase	DNA
+ Information	→	Word	Chord	Triplet (codon)
Information	→	Letter	Note	Nucleotide

Table 2.4: Layered Information illustrating the Structure of Systems
Note that to progress from the bottom to the top of the table, each successive layer requires the addition of *new* information.

Architecture of Life

Information is not only multilayered in the three dimensions of physical space but note that information is sequential in the fourth dimension of time as well. The design specifications for a car not only need to say which components are required and how they are to be made but also how they are to be assembled and the order in which it must be done. Placing the steering wheel where a rubber tyre was supposed to be located will not benefit the car or the motorist for very long and putting an indicator lens cover on the car before the light bulb is inserted will not work. This principle applies not only to inanimate objects but to life also; life pre-specifies all these processes and more. It not only stipulates the parts, how they are manufactured and their arrangement but as we have seen, their 'operating systems' and protocols, too. In a car, this would be equivalent to equipping it with the ability (information) not only to build itself but to plan and conduct all its own servicing. In living things this manual is encoded within the DNA or RNA. In simpler life forms, such as insects that only live for a few weeks, the information appears to be mainly 'hard-wired', meaning that the information is encoded as there is no time for learning. In longer lived species, however, such as mammals, there is more time to develop 'soft-wiring' which we call learning. Hard-wiring is immediately available, efficient and reliable but it is costly in versatility; soft-wiring is flexible but is costly in the time required to learn/acquire it.

Locating Maximal Complexity

Figure 2.1 shows how information in our World and Universe ranges from the infinitesimally small to the extremely large. It is interesting to note that the complexity of information peaks at intelligent life, which is almost exactly in the middle of the size scales within the Cosmos. It is common to associate greater complexity with larger objects but in our Universe this is not always so.[13] Blue whales are more complex than bacteria, however, inanimate objects such as an iceberg, even though it may be much larger than a blue whale, can be fully described in terms of the basic physical laws of nature (water and gravity), whereas biology has additional characteristics, such as thought processes and the phenomenon of life itself, which are not completely understood simply from the biochemistry and physics of their constituent elements. There appears to be underlying information which contributes to the characteristics of living entities that is in addition to their tangible, physical features which will be discussed in later chapters.[14]

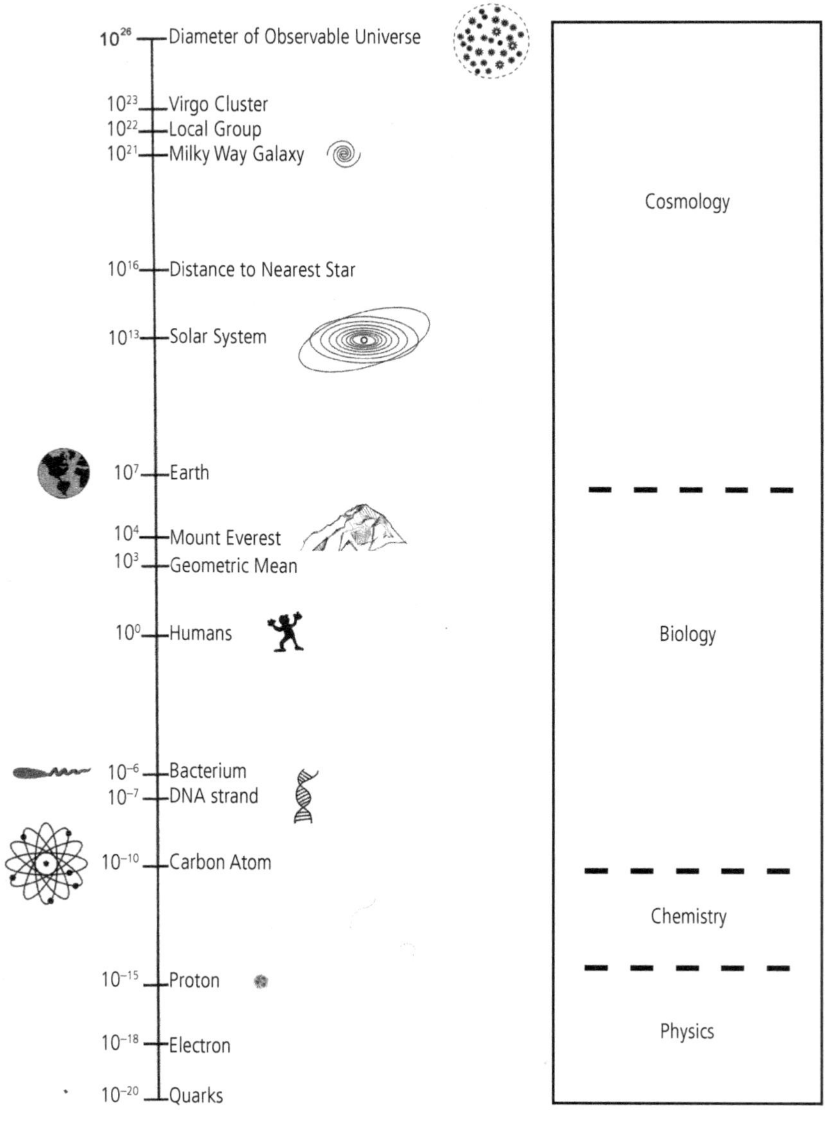

Figure 2.1: Size Scales in the Universe
Gonzalez and Richards write: *"Although Earth and human beings are often considered 'insignificant' compared to the vastness of the cosmos, the Earth-human size range actually indicates the geometric mean of extremes of size in the universe. We are near the middle of the size scale ranging from quarks to the unobservable universe, when figured on a logarithmic scale (in powers of 10)...Approximate sizes and distances are indicated in metres.)"* [13] *[Adapted from Gonzalez, G. & Richards, J.W.: The Privileged Planet. Regnery Publishing (2004).*

13. Gonzalez, G. & Richards, J.W.: *The Privileged Planet*. Regnery Publishing, Washington DC (2004) p.216.

14. *Ibid.* p.212.

Lennox writes: "*The galaxies are unimaginably large compared with you. However, you know that they exist, but they don't know that you exist. You are more significant, therefore, than a galaxy. Size is not necessarily a reliable measure of value, as any woman can tell you as she looks at the diamonds on her finger, and compares them with lumps of coal.*" [15] This introduces the concept of material versus non-material characteristics which will be discussed further in Chapter 10.

Systems

We have already observed that many entities in the World have complex structures that are composed of interrelated information. When groups of entities or processes act together synergistically, they are called systems. The study of systems is a fundamental part of science and engineering (as well as many other disciplines) and it will be helpful, at this stage, to briefly look at how engineers classify systems. This will help us to better understand what we are observing in more complex entities, such as living communities. Systems can be classified according to their degree of complexity, the simplest of which is an open loop.[16]

1. **Open-Loop System** (such as: a thermometer, watch, glucometer, oximeter or calculator)
   This is a simple assessment device that measures a variable and comprises a means to interpret the result (for example: a watch or thermometer). It requires a person to interpret, correct and maintain it. (See Figure 2.2)

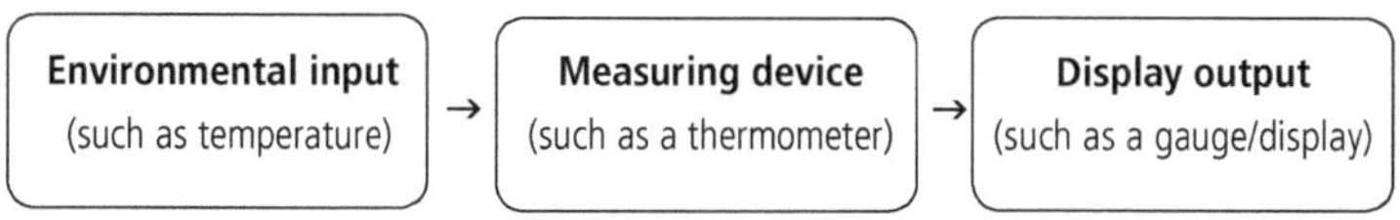

Figure 2.2: Open Loop System

2. **Closed-Loop System** (such as: a thermostat in an oven or refrigerator)
   The second simplest system is a closed-loop system. In a closed-loop system, the output is fed back to automatically adjust the input. In Figure 2.3, the thermometer measurements in the previous open-loop system have been connected to a device that automatically reads them and adjusts the energy supply to the oven to keep the temperature constant.

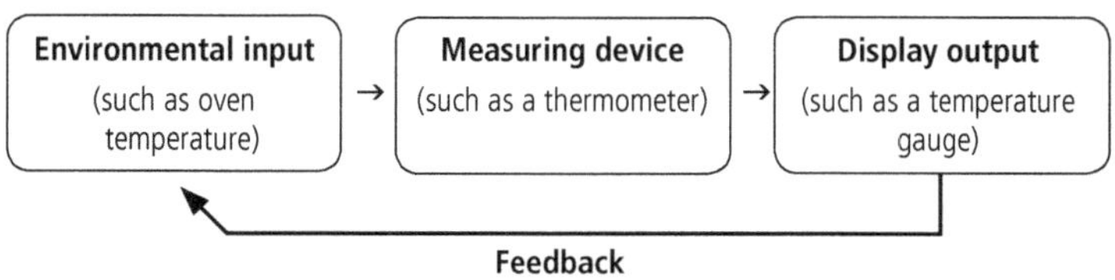

Figure 2.3: Closed Loop System

3. **Self-Adaptive System** (such as: a plane's autopilot)
   The third type of system is a self-adaptive system. Self-adaptive systems monitor their own functioning. They require pre-programmed variables but they have the capacity to self-adjust. These systems are composed of a number of sub-systems which are interlinked and require each sub-system to be functioning satisfactorily for the total system to operate in the optimum way.

15. Lennox, J.C.: *Seven Days that divide the world: the beginning according to Genesis and Science*. Zondervan, Grand Rapids (2011) p.99.

16. Missler, C.: *Cosmic Codes*. Koinonia House, Coeur d'Alene, Idaho (2004) pp.316–8.

4. **Self-Modifying Systems** (such as: a computer)
   The fourth level of complexity is a self-modifying system. Self-modifying systems not only measure the variable they were designed to regulate (such as the temperature) but they also monitor themselves and their own output and then adjust their own performance. They are able to detect the accuracy of their own performance and the task that they were designed to perform and then automatically adjust their own functioning. Many computers can do this. If they do detect a problem they can warn the operator to check for faults.

5. **Self-Diagnostic Systems** (such as: computer systems on commercial aircraft)
   Fifth-level systems include the ability to self-diagnose problems that may arise. A typical example is an engine management system on an aeroplane where the system can detect the presence of a fault in one of the engines and advise the pilot of the location and nature of the problem (such as a fire), so that the pilot can take appropriate action.

6. **Self-Protecting Systems** (such as: electricity grids)
   Sixth-level systems are self-protecting. After detecting that there is a problem they can automatically take evasive action to protect themselves. This capacity is built into electricity grids to protect them from damaging fluctuations in the power supply, such as a lightning strike or a major power failure. This was demonstrated a few years ago when the North-Eastern sector of the United States of America was automatically shut down by protective mechanisms when a significant number of generators malfunctioned. Most people will also have experienced a power failure after a lightning strike on their power lines. This can be due to damage from the lightning strike itself but it is more often due to isolating switches shutting down that part of the network and earthing the excess electricity.

7. **Self-Repairing Systems** (such as: computer operating systems such as *Windows*)
   Seventh-level systems can self-repair. After detecting that there is a problem, the system can interrogate its memory and correct the defect without external intervention. To a certain extent the Windows operating system on personal computers can correct software malfunctions but only science-fiction robots, such as R2-D2 in *Star Wars* can achieve hardware self-repair at this stage.

8. **Self-Programming Machines** (as seen in robots such as C-3P0 in *Star Wars*)
   The eighth level of design is now at the cutting edge of technology where the machines can program themselves. The best-known examples are from science fiction, such as C-3P0 in *Star Wars*.

9. **Self-Replicating Systems** (such as: a bacterium or an amoeba)
   At the ninth level, systems can self-replicate. At present this is only seen in living systems, such as bacteria or amoebae.

10. **Self-Aware Systems** (such as: the human brain)
Level-ten systems are self-aware and can appreciate their own existence. They are only seen in science-fiction and the human brain. The degree to which other animals are self-aware is discussed further in Chapter 10.

Level	Features	Examples
1	Open-Loop System	Thermometer, watch, glucometer or calculator
2	Closed-Loop System	Thermostat in an oven or an air-conditioner
3	Self-Adaptive System	Aeroplane autopilot
4	Self-Modifying Systems	Computer
5	Self-Diagnostic Systems	Computer systems on commercial aircraft
6	Self-Protecting Systems	Electricity grids
7	Self-Repairing Systems	Some computer operating systems or science fiction robots, such as R2-D2 in *Star Wars*
8	Self-Programming Systems	Robots such as C-3P0 in *Star Wars*
9	Self-Replicating Systems	Bacteria or an amoeba
10	Self-Aware Systems	The human brain

Table 2.5: Systems Design Summary[16]

Although this process is intricate and (again) multilayered, it is not mysterious. As systems analyst, Dr Chuck Missler says: *"A computer can be viewed as simply a calculator that pushes its own buttons."* [17] However, each stage requires a quantum increase in information from the stage below it. Thus, there is a progressively increasing information requirement from Level 1 to Level 10. Systems Level 9 is the minimum entry point for living organisms as, if they cannot reproduce, any beneficial traits cannot be transferred to the next generation. This phenomenal degree of sophistication (Systems levels 1 – 9) had to be possessed by the first organism. In comparison, the best systems mankind can build, even after millennia of technological discoveries, only partially satisfy Systems Level 8. It is interesting to note that William Paley's watch was Systems level 1 but the hand that picked it up was Systems level 10.[18] Part of our journey will involve the investigation of this escalation in information from Level 1 to Level 10.

Order in Complexity

Thus, the vast quantities of information in both the biological and non-biological spheres occur in hierarchies, patterns and other organised ways. If the Universe was formed by chance, it would be logical to expect that the Universe would be without order, without logic and incomprehensible. The fact that the Universe contained order surprised Albert Einstein and prompted him to write: *"The most incomprehensible thing about the universe is that it is comprehensible."* [19]
As already discussed, complexity is a measure of the number of the pieces of information that are required to precisely define an entity combined with their arrangement. The information may or may not have been pre-defined. A library is an example of complexity with pre-defined order and hence, is an illustration of purposeful information. A rubbish tip is a model of complexity without pre-defined or purposeful information. Mathematician, William Dembski calls the former: *"specified complexity"*.[20] Another way to view these situations is that extra

17. Missler, C.: *Cosmic Codes*. Koinonia House, Coeur d'Alene, Idaho (2004) p.23.

18. William Paley (1743–1805) was an English philosopher who published his *Natural Theology* in 1802, in which he introduced an argument for the existence of God through an analogy to a watch. Richard Dawkins took up this theme in *The Blind Watchmaker* (1986).

19. Einstein, A. *in* Lennox, J.C.: *God's Undertaker. Has Science Buried God?* Lion Books, Oxford (2007) p.58.

20. Dembski, W.A.: *The Design Inference*. Cambridge Univ. Press, Cambridge (1998) *in* Gonzalez, G. & Richards, J.W.: *The Privileged Planet*. Regnery Publishing, Washington DC (2004) p.298.

information was required to be input to create the order in the library, whereas no extra information was required to create a rubbish dump (the arrangement happened by chance). Complexity with information may represent design, such as in the Dewey Decimal indexing system found in many libraries, or it may represent chance occurrence, such as a stack of dumped books, like an old encyclopaedia at the tip. There is historical information in the encyclopaedia, which represents previous information storage, an artefact of a previous usage, but it was chance that the individual volumes of the encyclopaedia ended up in a stack together at the tip. No information was incorporated in the way that they were dumped. So, for the complexity to contain meaningful information, the information must relate to and value-add to the whole. The information in the books at the tip does not add more meaning to the tip itself (although as artefacts they may be a valuable record of the civilisation that generated them).

So, is there a way to discern whether a stack of books was placed intentionally or not? This question is important because it allows us to distinguish between related and coincidental facts when we observe a complex system (which is most of the time). Table 2.6 illustrates how the number of related facts affects the degree of certainty that may be deduced regarding the relationship between the constituent parts of any system. In the same way that the intersection of two lines defines a point, the intersection of two facts also identifies an intersecting relationship. The intersection of three lines *confirms* a point and in clinical medicine we use the intersection of three facts to confirm the identity of a patient (for example name, date of birth and address). The chance of having the wrong identity with these three authenticated pieces of information is remote. However, if we were to have four pieces of intersecting information, the chances of identifying the wrong patient are now extremely remote. So, what if we had 10, 20 or 100 pieces of intersecting information? This process can also be used to predict the likelihood of information turning up by chance in any randomly observed situation. If we were to observe two intersecting facts then there is a significant chance that the situation may just be coincidence, which is to say that it occurred randomly or 'by chance'. But if we have three intersecting facts then the likelihood that this happened 'by accident', which is to say that it was unplanned, is remote. If, however, we had 10, 20 or 100 contingent facts then the chance that a relationship is unrelated becomes unbelievably small and we would say that there was 'no chance' that the relationship between the facts could have happened by accident. Something must have caused it to happen, which is another way of saying that information was required to bring that situation about. This is summarised in Table 2.6.

Number of Related Facts	Description	Certainty of the Information *or* Relationship
1	Isolated	No relationship / no complexity
2	Linked	Unspecified (possibly chance)
3	Complicated linkage	Probable
4	Simple complexity	Very likely
10	Marked complexity	Certain
20	Intricate complexity	Confirmed
100	Vast Complexity	Beyond doubt

Table 2.6: Measuring Complexity

In the case of the encyclopaedia at the dump, each volume has many related facts (alphabetic arrangement, where all the 'C' topics are together in one book, a consistent layout, many relating to each topic so that there are easily 10 and possibly more than 100 related facts per volume making it certain that the information within the volume is related and hence, they did not appear in the encyclopaedia by accident. Whereas, the fact that the books are in a stack and at the dump is only two pieces of information and leaves the relationship unspecified. It may or may not have occurred by chance. So, there may or may not be information intrinsic to this situation. However, if I were to disclose that the volumes of the encyclopaedia are *not* actually discarded under a pile of rubbish but in alphabetic order on an old bookshelf at the entrance to the dump labelled 'Pre-loved Items', there are now four simultaneously related facts which create simple complexity and there is most likely to be meaningful information involved in this situation. That is, they have probably been pulled out of the general rubbish and arranged in order on the bookshelf in the recycling area of the dump.

The Generation of Information

We have observed that information is a real entity that can change outcomes when it is introduced to a system (such as software causing a computer to function) or prevent outcomes from occurring when it is removed (such as the inability of a cell to replicate without its DNA) but where does the information come from? The model of the Universe that is accepted by most scientists today states that everything began at the 'Big Bang', before which there was nothing. If this is correct, at some stage after the formation of the Universe, the information that we now observe in the Cosmos and on Earth must have been generated. How did this occur: for the Universe generally and for living cells specifically? From an informational frame of reference there are only three possibilities: either the coding of the information occurred by chance, which required no additional input; or, a law of the Universe required that the coding of the information *had* to happen, which represents intrinsic information within the law that caused it to happen (where did that come from?) or, external coding of the information occurred; extrinsic information was introduced into the system. These three possibilities are illustrated in Figure 2.4.

There are some problems, however. Firstly, chance only produces random disorder. Order which produces functionality is not created by selecting a portion of chaotic randomness. Next, Philip Johnson observes: "*Law produces the same simple pattern over and over again – highly ordered, repetitive sequences like crystals… The heart of the problem is that physical laws are simple and general, and by their nature they produce the same thing over and over again. The law-governed processes can produce simple repetitive patterns, as in crystals, but they can't produce the complex, specified sequences by which the nucleotides of DNA code for proteins…*"[21]

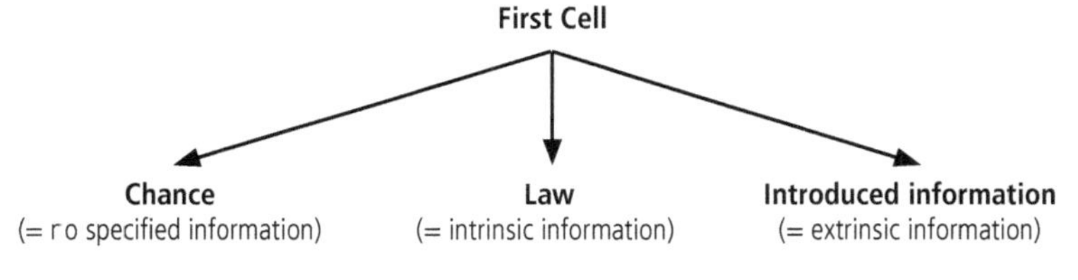

Figure 2.4: The Generation of Information
There are only three choices. This illustration uses the example of the first cell but it could equally be any other information containing entity in the Universe including the Universe itself.

Indeed, information in living things is novel, which is the requirement for a working cell.[21]

So, can information be generated by random means as T.H. Huxley asserted in his Oxford debate with Bishop Wilberforce in 1860; the equivalence to monkeys bashing on typewriters? [22, 23] The answer is a qualified yes: *meaningless* (Shannon or syntactic) information can be generated by random means but *meaningful* (semantic) information cannot. That is, there is some useless information in the random sequence of the letters typed by the monkeys but they lack any functionality. Thus, the meagre amount of syntactic information generated by chance cannot explain the immensely intricate, semantic information that is observed in a living cell or the Universe at large. That leaves intrinsic and extrinsic information as possible explanations. However, as our environment has such a profound effect on everything on Earth, the information associated with our physical environment will be considered separately to other intrinsic information. Table 2.7 shows these three variations of information in natural systems: 1. Intrinsic, 2. Environmental and 3. Extrinsic.

21. Johnson, P.: *Wedge of Truth*. InterVarsity Press, Downers Grove, Illinois (2000) pp.53-4.

22. Lennox, J.C.: *God's Undertaker. Has Science Buried God?* (2nd Ed.) Lion Books, Oxford (2009) p.63.

23. Gitt, W.: *In the Beginning was Information*. (English 3rd Ed) CLV, Bielefeld, Germany (2003) p.205.

Type of Information	Example	Notes
Intrinsic	A crystal lattice (such as table salt - NaCl)	The crystal lattice needs no new information. It is specified by the information that defines the bonding properties of the atoms.
Environmental	A Rayleigh-Benard convection cell (may sometimes be observed when cold cream is added to a bowl of hot soup.)	The pattern is created by the physical conditions specified by the shape of the bowl, temperature of the liquids and the order of their combination.)
Extrinsic	Proteins	The order of the components within the protein is neither specified by the intrinsic properties of the constituent amino acids, nor by the environment of the cytoplasm but by an extrinsic source of information (stored in the DNA or RNA).

Table 2.7: Types of Information in Natural Systems

From the example of the Rayleigh-Benard convection cell,[24] it can be observed that this phenomenon is dependent on information that is intrinsic to its physical environment and hence, environmental information is a subset of intrinsic information. Although the effect appears spontaneous, the spontaneity is an illusion caused by information contained within (intrinsic to) the physical surroundings.

Science is very good at dealing with intrinsic and environmental information where laws can be described, patterns identified and experiments conducted and repeated. Science struggles with the origin of extrinsic information such as that found in DNA and for the information responsible for the formation of the Universe. The Theory of Darwinian Evolution is one explanation to account for the appearance of information in DNA but it has been controversial since it was first proposed. To complicate matters further, often in natural systems, a mixture of information is involved in determining environmental outcomes. Fowler and Kuebler quote one example where: *"...in the 1950s,...Russian chemists Boris Belousov and A.M. Zhabotinsky discovered a chemical reaction, the Belousov-Zhabotinsky (BZ) reaction, which generated spontaneously ordered patterns that change over time. As the major reaction proceeds, one of the reactants, bromous acid, is generated as one of the products. This positive feedback is coupled with a separate inhibitory reaction and the oscillation between these two generates an ordered pattern in which concentric waves move out from different regions of the dish, annihilating one another on contact. The oscillation keeps the reaction from reaching equilibrium, and*

24. John Lennox explains that Rayleigh-Benard convection may be seen where: *"...heat flowing smoothly through a liquid suddenly changes into a convection current that reorganises the liquid so that a honeycomb pattern made up of hexagonal cells appears, looking just like the famous rock formations at the Giant's Causeway in Northern Ireland..."* in Lennox, J.C.: *God's Undertaker. Has Science Buried God?* Lion Books, Oxford (2007) p.124.

as a result the reaction is able to generate a dynamically ordered pattern spontaneously. This pattern appears to arise for 'free' since it is not specified by the molecules within the reaction." [25] However, although the system is free of an external energy source, it is not 'free' of an information source. There is both intrinsic and environmental information in the reaction.[26] Intrinsic information is encoded in the chemical shells and composition of the molecules involved but there is environmental input in the dish, ambient temperature et cetera. To test this, try repeating the reaction after substituting a different element for bromine in the bromous acid or make a different environment (would it work in a frozen or gaseous state?) Thus, both intrinsic and environmentally specified information are significant in this reaction too.

Interpretation of Information

From the above discussion, it is observed that the information that is resident within natural systems is not always obvious; the information often lies unrecognised. To become useful to the informationist, the information must be detected so that it can be interpreted and then analysed. Although this process sounds obvious and simple, it is not always intuitive or easy. Georgetown University theologian, John F. Haught writes: *"In order for a message to come across as informative it must include novelty...The amount of information in a message varies in direct proportion to the improbability of its content. If I anticipate that all the messages from a given source will be the same, then when one of them is delivered it carries little or no information as I already know what it is going to say. There is an apparent paradox involved here: the more informative a message is, the less immediately comprehensible it may be, at least as far as the communication of meaning is concerned. If so, then a maximally informative message, if we could conceive of the finite universe as its carrier, would be accompanied by the least redundancy. It would be something like that which cosmologists call a singularity, and as such inaccessible to conventional science, as scientific method in its concern for predictability is most at home in the realm of what appears probable or predictable rather than improbable or unpredictable. So a message might be unreadable by conventional science to the degree that it is improbable – that is to say informative."* [27, 28] This is our 9th paradox. The implication of this inverse relationship is that the most significant information may be the least obvious or intuitive. Collecting the most useful information, thus, requires diligent investigation and often necessitates searching 'outside the square'. Information research may, therefore, conflict with the reigning establishment, whether it is scientific or religious but failure to engage with this information may leave much undiscovered since, as Haught argues, this is where the most fruitful discoveries are most likely to lie hidden; and there are many unexpected discoveries to be made in the chapters that follow, so we will move on. But before we do so, we will summarise the enigmatic findings to date.

Summarising the Information Paradox

Thus, information is not as simple or obvious as it first appeared. The paradoxes noted to date are:

1. Relatively little is known about the entity of information itself, even though most people acknowledge that we are living in the 'Information Age'.

25. Fowler, T.B. & Kuebler, D.: *The Evolution Controversy.* Baker Academic, Grand Rapids (2007) p.313

26. The main reaction is: $HBrO_2 + BrO_{3-} + 3H^+ + 2Ce^{3+} \rightarrow 2\,HBrO_2 + 2Ce^{4+}\,H_2O$ - Fowler, T.B. & Kuebler, D.: (2007) p.313

27. Davies, P. & Gregersen, N.H. (Eds.): *Information and the Nature of Reality. From Physics to Metaphysics.* Cambridge Univ. Press, Cambridge (2010) pp.304–5.

28. Singularities occur where one variable becomes infinite (see Chapter 8).

2. Information is neither matter nor energy; yet matter and energy can convey or store information as long as the physical arrangement within the matter or energy has meaning.
3. Matter is not meaning; yet meaning can be conveyed or stored by matter (or energy) as long as the physical arrangement within the matter (or energy) contains information.
4. Information has no mass, but it is real as it changes function and hence, the outcomes of matter and energy. Information is a discrete (yet non-material) entity in its own right and must be considered separately from matter and energy.
5. According to the majority of scientists today, the information generation process to make a mousetrap is different from the information generation process to make a man. The simpler one is intentionally manufactured with information, matter and energy from a plan and the other evolved by inducting information, matter and energy by chance.
6. Despite information being ubiquitous, it is often inconspicuous, as it is intangible and so it often goes unnoticed.
7. Matter and energy flows are not enough to describe life, which requires detailed information, yet strangely, to gather or construct information requires work. Thus, information is costly to acquire and energy must be expended to collect and maintain it.
8. Information requires storage in matter or energy so it is subject to the Law of Entropy. And yet, if the storage mechanism is disrupted, the information being mass-less, does not change. Thus, the vehicles that carry information are subject to entropy but not the information itself.
9. *"The more informative a message is, the less immediately comprehensible it may be, at least as far as the communication of meaning is concerned...Thus a message might be unreadable by conventional science to the degree that it is improbable – that is to say informative."* [27]

A number of these concepts run counter to our familiar, material paradigm, and sound startling, especially when they are listed together but they will become more comfortable as we use the tool of information to investigate life, the Earth and our Universe over the next eight chapters.

Conclusion

John Lennox observed that the premise that there is information in our Universe is supported by: *"...the ordered nature of the universe, its mathematical intelligibility and fine-tuning..."* [29] Further, this information is ubiquitous, highly organised, multilayered and sequentially arranged. We began by trying to measure information and have discovered that there are a number of scales by which information can be quantified; such as the number of pieces incorporated, the number of layers involved, the number of functions, whether it has meaning, the effort required to generate it and so on. Further, useful information can either be intrinsic or extrinsic to an entity. The most surprising observation is that information has the ability to generate great complexity which can be studied by using systems analysis, suggesting that it is not haphazard but arranged by some process. Also, as complexity grows, the accuracy of the information and

29. Lennox, J.C.: *God's Undertaker. Has Science Buried God?* Lion Books, Oxford (2007) p.169.

the precision with which it is compiled becomes increasingly critical as a single incorrect keystroke in a complex computer program may make the system unreliable at best and fail completely at worst. So, the most sophisticated systems, such as the computers on the space shuttles, had error checking routines to overcome these risks. In Chapter 3 we will discover that living cells have error checking capability connected to error correcting routines too.

These findings then beg the question: how was the information in the first living organism generated? This is where we will begin in Chapter 3, with the formation in the first cell.

3

Information and the First Cell

An honest man, armed with all the knowledge available to us now, could only state that in some sense, the origin of life appears at the moment to be almost a miracle, so many are the conditions which would have to have been satisfied to get it going.[1]

FRANCIS CRICK

In Chapter 1, T.H. Huxley described how inanimate molecules might be imagined to join together to form the first cell and when we study life, this would seem to be a good place to begin our investigation into the information within living things, as the cell is the basic unit of life. Bernd-Olaf Küppers, Professor of Natural Philosophy at the Friedrich Schiller University at Jena wrote: *"Ever since the elucidation of the molecular basis of living systems, we have known that all elementary processes of life are governed by information. Thus, information turns out to be a key concept in understanding living matter."* [2] The logical question therefore is: what information was required for the formation of life? Stating it in another way, what information is required to convert inanimate atoms and inorganic molecules into functional biochemical structures and then to animate the first living cell? This is the process of going from chemistry to biochemistry to biology, which is the process described by molecular or biochemical evolution.

Information Flows within Cells

As already established, Darwin devised his ideas of *"variation and natural selection"* [3] in an age when the fundamentals of the Universe were defined in terms of matter and energy but we now recognise the importance of the extra entity of *information*.[4] The multiple levels of embedded information that were previously noted in many areas of the Universe are especially important in living systems, which require vast amounts of data to describe life, such as the approximately 25,000 genes in the human genome.[5] They, in turn, specify the thousands of different proteins required for a typical cell to function.[6] As we discovered in Chapter 1, in Darwin's day, the cell was assumed to be an uncomplicated drop of organic materials in a membrane. In 1905, Ernst Haeckel (1834–1919) wrote that the cell was composed of: *"...homogeneous and structure-less globules of protoplasm."* [7] It was not until Watson and Crick described the double-helix structure of DNA in 1953[8] that the incredible richness of this information and the efficiency with which it is utilised, could begin to be fully understood.

1. Crick, F.H.C.: *Life Itself: Its Origin and Nature*. Simon & Schuster, New York (1981) p.88.

2. Küppers, B.-O.: *Information and the Origin of Life*. Cambridge, MA: the MIT Press (1990) *in Information and Communication in Living Matter, in* Davies, P. and Gregersen, N.H. (eds.): *Information and the Nature of Reality: from Physics to Metaphysics*. Cambridge University Press (2010) p.170.

3. Darwin, C.: *The Descent of Man*. John Murray, London (1871) p.613

4. Meyer, S.C.: The origin of biological information and the higher taxonomic categories. *Proceedings of the Biological Society of Washington* 117:213–239 (2004).

5. Sarfati, J: *By Design*. Creation Book Publishers, Brisbane, Australia (2008) p.169.

6. Andrews, E.: *Who Made God? Searching for a Theory of Everything*. EP Books, Darlington, England (2009) p.181.

7. Haeckel, E.: *The Wonders of Life*. Translated by McCabe, J.: Harper, London (1905) p.111 *in* Meyer, S.: *Signature in the Cell*. Harper One, New York (2009) p.47.

8. Watson, J.D. & Crick, F.H.C.: Molecular structure of nucleic acids: A structure for DNA. Nature 171: 737–738 (1953).

However, there is not only the question of what information was involved but also, how did its carrier, the first DNA, come into existence? In 1958, Crick described the progressive transfer of information from the DNA, protected in the nuclei of eukaryotic cells, to messenger RNA, which allows the transfer of information out of the nucleus.[9] This arrangement enables the subsequent production of proteins in the cytoplasm of the cell. This process became known as the *'Central Dogma of Molecular Biology.'* In a subsequent paper in 1970, Crick restated the principle as: *"information cannot be transferred from protein to either protein or nucleic acid."*[10] However, he added the proviso: *"Although the details of the classification proposed here are plausible, our knowledge of molecular biology, even in one cell — let alone for all the organisms in nature — is still far too incomplete to allow us to assert dogmatically that it is correct."*[11]

Many objections have been raised, such as retroviruses[12], transposons[13] and prions[14]; but half a century later, the Central Dogma has not disappeared from the literature or debate.[15, 16, 17, 18] Retroviruses can take RNA into the nucleus and change the DNA code but the *cytoplasmic proteins cannot affect the sequence of nucleic acids in the nucleus which codes the source information.* Transposons can transfer DNA between different chromosomes[19] but again the proteins from the cytoplasm have not altered the nucleic acids. Prions are different. They can appear to 'replicate' without DNA or RNA by refolding (altering the shape) of the prion protein molecules from their host. However, each host manufactures their own

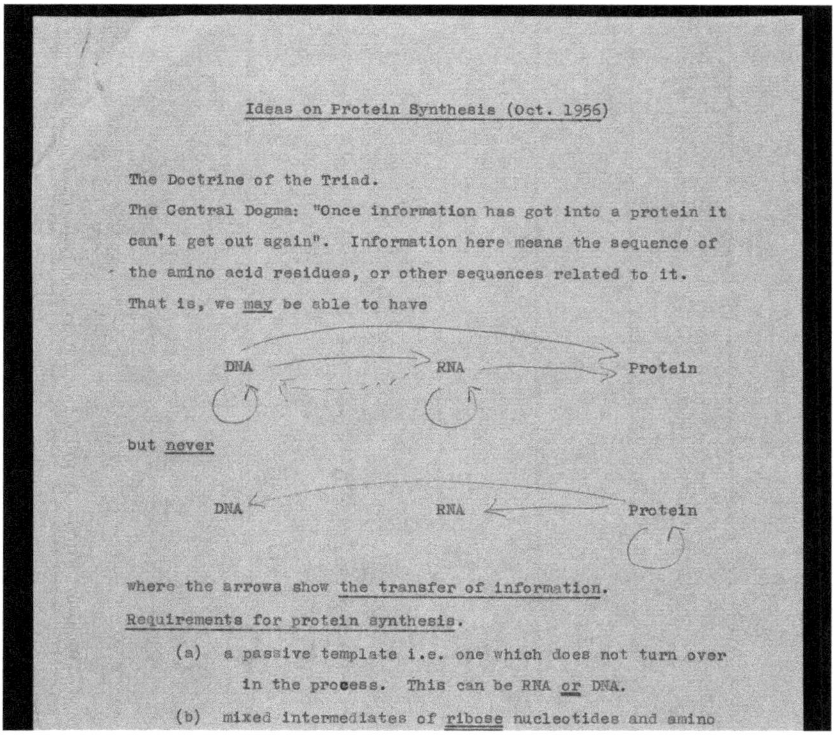

Ideas on Protein Synthesis (Oct. 1956)

The Doctrine of the Triad.
The Central Dogma: "Once information has got into a protein it can't get out again". Information here means the sequence of the amino acid residues, or other sequences related to it.
That is, we may be able to have

DNA → RNA → Protein

but never

DNA ← RNA ← Protein

where the arrows show the transfer of information.
Requirements for protein synthesis.
(a) a passive template i.e. one which does not turn over in the process. This can be RNA or DNA.
(b) mixed intermediates of ribose nucleotides and amino

Figure 3.1: Crick's diagram from "Ideas on protein synthesis" (October 1956). In the accompanying notes, Crick wrote: *"The Doctrine of the Triad. The Central dogma: 'Once information has got into a protein it can't get out again'. Information here means the sequence of the amino acid residues, or other sequences related to it." (From Crick, F.H.C.: Ideas on Protein Synthesis. (1956) in the Wellcome Library, London. Reproduced with permission.)*

9. Crick, F.H.C.: Ideas on Protein Synthesis. Symposium of the Society for Experimental Biology XII, 139-163 (1958).

10. The Central Dogma of Molecular Biology has sometimes been understood to refer to the "one gene = one protein" principle but this is not the way Crick defined it in his 1970 paper. See Note 11.

11. Crick, F.H.C.: Central Dogma of Molecular Biology. *Nature* 227:561–563 (1970).

12. Hayes, B.: Experimental Lamarckism. *American Scientist* 87:494–8 (1999).

13. Mockler, T.C. & Ecker, J.R.: Applications of DNA tiling arrays for whole-genome analysis. *Genomics* 85:1–15 (2005).

14. Ng, S.B.L. & Doig, A.J.: Molecular and Chemical basis of prion-related diseases. *Chemical Society Reviews* 26:425–432 (1997).

15. Birney, E.: *Inaugural Francis Crick Lecture of the Royal Society.* Institute of Science in Society. (2003).

16. Arkin, A.: Setting the standard in synthetic biology. *Nature Biotechnology* 26(7):771–4 (2008).

17. Maynard Smith, J.: *The Concept of Information in Biology. in* Davies, P. and Gregersen, N.H. (eds.): *Information and the Nature of Reality: from Physics to Metaphysics.* Cambridge University Press (2010) p.129.

18. Hoffmeyer, J.: *Semiotic freedom: an emerging force, in* Davies, P. and Gregersen, N.H. (eds.): *Information and the Nature of Reality: from Physics to Metaphysics.* Cambridge University Press (2010) p.186.

19. Smit, A.F.A. & Riggs, A.D.: Tiggers and other DNA transposon fossils in the human genome. *Proceedings of the National Academy of Sciences of the USA* 93:1443–8 (1996).

prion proteins from prion protein genes and the prion proteins do not change the sequence code in the nucleus.[20]

This DNA that resides in the nucleus of each eukaryotic cell (or the cytoplasm of prokaryotic cells) is the universal coding material for all life.[21] When the cell needs to create a new protein, it unzips the correct piece of DNA and copies it to messenger RNA (mRNA). The copy is transferred to the cytoplasm where special machines, the ribosomes, use the RNA sequence to manufacture the protein required. In terms of the information carried by the DNA and RNA, *this is a one way process*. Although retroviruses, using reverse transcriptase, can splice pieces of RNA back into the nuclear DNA, *proteins* cannot be unfolded, sequenced and translated back into code to be reinserted into the nucleus to make new DNA. New DNA production (such as in the manufacture of a new cell) is made by splitting the entire strand of DNA in the nucleus and copying it in a process called replication. Richard Dawkins likens this one-way process to produce proteins from DNA to baking a cake. Dawkins writes: "*A recipe captures something of the truth, and it is an analogy that I sometimes use, to explain why 'blueprint' is not appropriate. Unlike a blueprint, a recipe is irreversible. If you follow a cake recipe step by step, you'll end up with a cake. But you can't take a cake and reconstruct the recipe - certainly not the exact words of the recipe - whereas...you could take a house and reconstruct something close to the original blueprint. This is because of a one-to-one mapping between bits of house and bits of blueprint. With conspicuous exceptions such as the cherry on top, there is no one-to-one mapping between bits of cake and the words, say, or sentences of its recipe.*" [22] This one way process which was first described by Crick is reproduced using current terminology in Figure 3.2:

20. Dalsgaard, N.J.: Prion diseases. An overview. *APMIS* 110:3–13 (2002).

21. Joyce, G.F.: The antiquity of RNA-based evolution. *Nature* 418:214–221 (2002).

22. Dawkins, R.: *The Greatest Show on Earth*. Bantam Press, London (2009) p.221.

The Law of Coding

From the preceding review of the Central Dogma, it can be seen that 50 years after Crick first described this principle, there are still no exceptions to the dictum that, although information can be transferred from RNA to DNA, there are *no exceptions* to the rule that: *proteins cannot change the information in the DNA code that produced them* to cause a change in the proteins produced. DNA methylation can affect the expression of DNA by adding a methyl group to the Cytosine – Guanine base pairs of a new DNA strand after it has been replicated but this does not change the information subsequently transmitted to the proteins by the base

Figure 3.2: Information Flow in Cells.
This diagram illustrates our current understanding of Crick's original drawing from 1956 and shows how accurate Crick's original concept was. The main differences are: that we now know that RNA performs a regulatory function in cells and that DNA cannot produce proteins without RNA transmission of the information first. The other major difference is that the protein product can give a negative or inhibitory feedback to the RNA production process. However, this does not change the original information in the DNA, meaning that the genomic information flow is still one way.

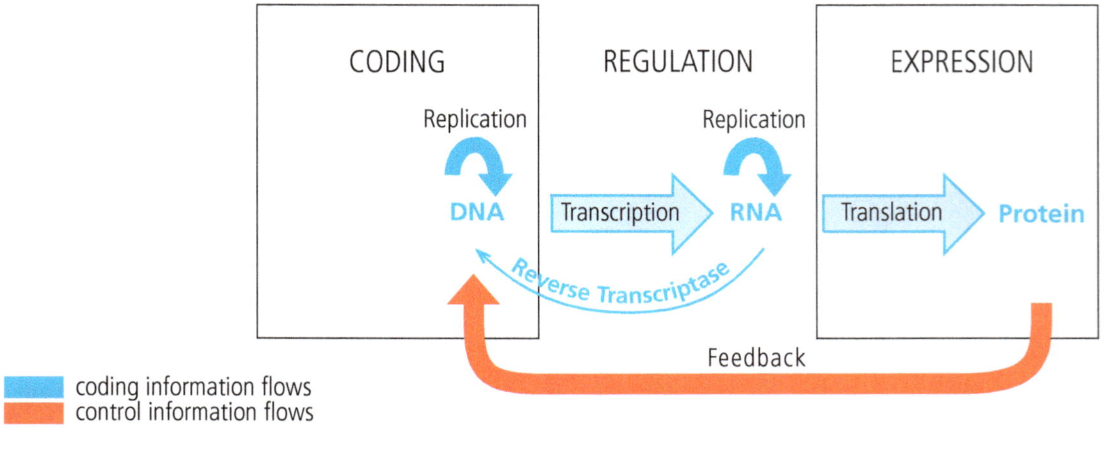

code order and neither does it change the original DNA.[23] DNA methylation is a modification that normal DNA undergoes after synthesis; it is a control mechanism associated with gene function and expression.[24] Proteins can also vary the expression of the DNA via prions or through feedback via control proteins (see Figure 3.2) but they cannot change their own source code. Errors in copying can change the DNA code during replication and this is how variation can enter the system (see Figure 3.3). This, however, is not caused by the proteins produced by the code. Theoretically, a faulty ribosome (a mixture of RNA and protein) may cause faulty replication but these proteins are not coded in the nucleus, so the protein products do not change the code that made them as ribosomes have their own DNA within them and not within the nucleus. Therefore, the proteins that are described by and derive from the DNA must be created *after* the DNA has been specified or coded (that is, come into existence).

In the same way, the splicosome, a specialised group of 150 proteins together with five molecules of RNA, form a molecular machine which cuts the messenger RNA into segments that are then recombined to make different sequences in a 'cut and paste' fashion. In Figure 3.4, this can be seen in the intermediate stage of information transfer and is part of the production process within the regulation phase of protein synthesis. Once again, it can be seen that the products of this process do not affect the nuclear DNA in the initial or coding phase of protein synthesis. So, altering the products by splicing cannot alter the DNA which generated it. The complexity of the splicing process suggests it is

23. Li, E., Beard, C. & Jaenisch, R.: Role for DNA methylation in genomic imprinting. *Nature* 366:362–5 (1993).

24. Razin, A. & Riggs, A.D.: DNA methylation and gene function. *Science* 210:604–10 (1980).

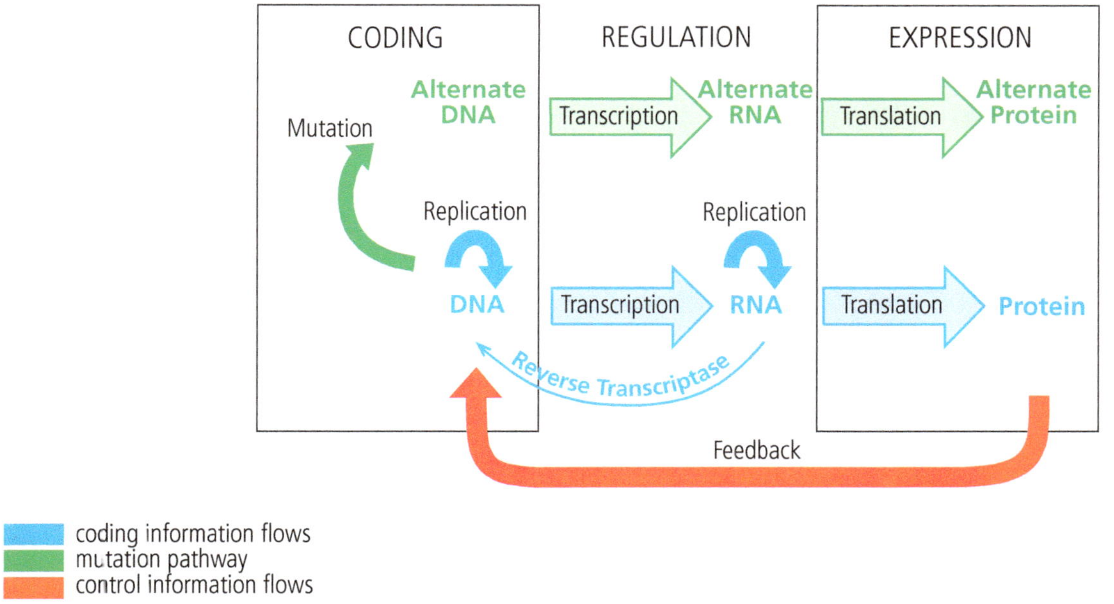

Figure 3.3: Information Flow in Cells showing the Mutation Pathway.
Alternate protein comes from alternate DNA not from existing proteins. Mutation causes alternate coding that is also at the initial level of information transfer. Transcription from DNA to RNA is an intermediate level of information transfer which is important for regulation within the system, while translation from RNA to protein synthesis is the final stage. Expression in the final stage is a one way process. Feedback control systems are a post-production level of information movement which is another form of cell regulation. If RNA was the initial information store, as is the case in some viruses, then RNA would become the initial level of information transfer/ storage for that system.

under direct control of some instruction source, which may be from the genome but is more likely to originate in the epigenome.[25] The epigenome will be discussed in greater detail later in the chapter but it is enough at this stage to note that the epigenome has a major role in regulation and expression of the genomic information during the production of proteins.[26]

Superimposed on top of the basic processes described above, is a complex regulatory function controlled by microRNA (miRNA). Micro-RNAs are single-stranded RNA sequences that control gene expression through the regulation of messenger RNA in the cytoplasm. They are about 21 to 23 nucleotides in length and are manufactured in a multistage process. They derive from DNA in the nucleus of the cell but as the RNA segments are not translated into protein, they are called 'non-coding RNA', which can be a confusing term as it still codes information. A less confusing term would be 'non-*protein*-coding RNA'. Micro-RNA has multiple functions in the cell. It inhibits protein translation causing down-regulation of the messenger RNA transcripts but without causing them to become degraded. Micro-RNA can also facilitate cleavage and/or methylation of the messenger RNA as illustrated in Figure 3.4.[27]

If a functional protein was able to form by chance, as has been extrapolated from the simulation of the 'pre-biotic soup' by University of Chicago chemists, Stanley Miller and Harold Urey in 1953,[28] this nascent protein still does not have the RNA or DNA information store which is the hallmark of all living cells today.

25. Wang, Y. et al.: A complex network of factors with overlapping affinities represses splicing through intronic elements. Nature Structural & Molecular Biology 20:36-45(2013) Published Online 16 Dec. 2012.

26. Lennox, J.C.: *God's Undertaker. Has Science Buried God?* Lion Books, Oxford (2007) p.133.

27. Lin, H.: piRNA (26 Oct 2007) *ScienceDaily in* Andrews, E.H.: *Who Made God? Searching for a Theory of Everything*. EP Books, Darlington, England (2009) p.236.

28. Miller, S.L.: A production of amino acids under possible primitive earth conditions. *Science* 117:528–9 (1953).

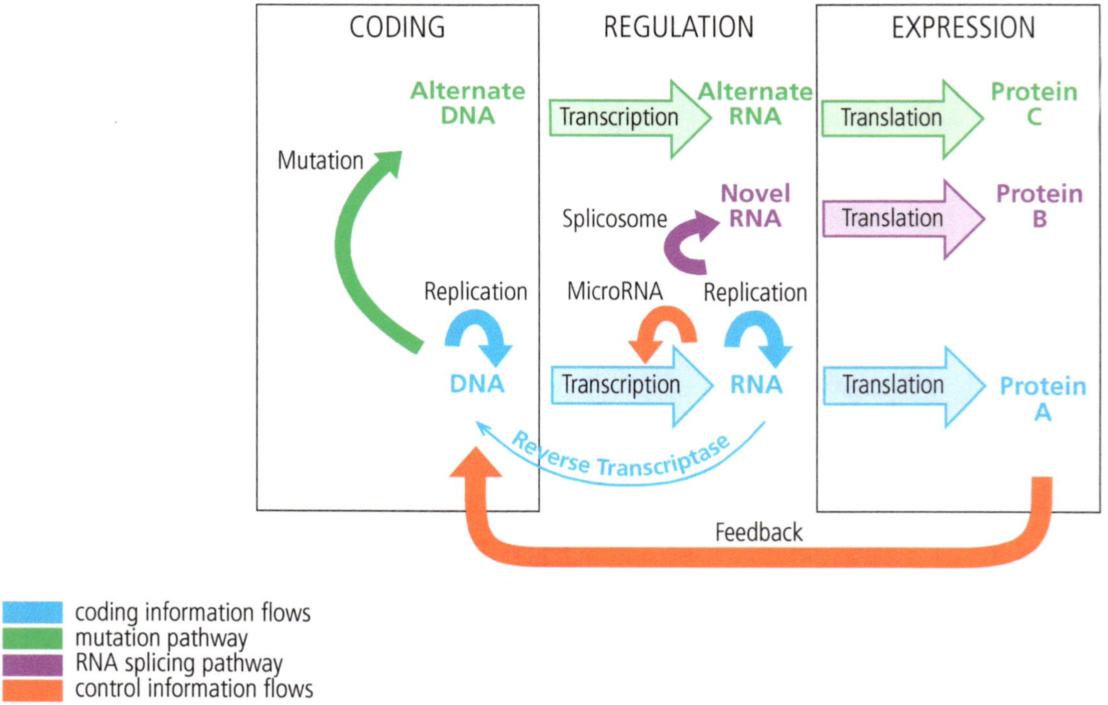

Figure 3.4: Information Flows in Cells showing other Intrinsic Pathways.
The splicosome splits segments of RNA allowing them to recombine in novel arrangements while micro-RNA is part of the cell's internal regulatory system stopping overproduction of unnecessary products. Neither of these processes alters the information in the DNA in the initial (coding) phase of protein production.

Therefore, this new protein must be mass produced by a biochemical template or alternatively, learn to cleave itself by some (yet undescribed) pre-biotic process, so that it can avoid extinction. It is possible to theorise that the templates that made such early proteins have now disappeared but the number of proteins required for the most basic functioning of the most rudimentary one-celled organism would have needed many templates for even the simplest cell to function.[29] Yet this phenomenon has not been described in natural biological systems to date. There is neither historical nor contemporary evidence that functional proteins boot-strapped themselves up from pre-biotic conditions, to encode the requisite information, to be in a position for subsequent reverse engineering, to code their DNA or RNA. But, even if they had achieved this feat by some process that is yet to be discovered, reverse engineering of proteins has not been observed in natural biological systems either. It is also significant to note that Miller was only able to produce isolated amino acids[30] without the integrated information content required for protein production and hence, functional ability.

Another problem with the Miller-Urey experiment was that it could not simulate the vast periods of time in Earth's pre-history. In order to overcome this difficulty Richard Dawkins[31] in 1986 and Bernd-Olaf Küppers[32] in 1987, used computer simulations of this process to build proteins but pre-specified the outcomes in their 'purifying' process. In doing so, they introduced external information in specifying the desired endpoint. In a natural system, where the information is not pre-specified, each protein contributing to the formation of the first functioning cell would have to spontaneously evolve the required complementary functions for a proto-organism independently, through chance variation, prior to natural selection. These complementary functions must include:

(a) Coding of information (so that the gains made are not subsequently lost);
(b) Energy management (to allow the proto-organism to function) and
(c) Some form of homeostasis mechanism/machinery (to stop it disintegrating again).

In summarising the process of evolution from prebiotic soup, Hulbert Yockey concluded: *"Although at the beginning, the paradigm was worth consideration, now the entire effort in the primeval soup paradigm is self-deception on the ideology of its champions."*[33]

Yet, even if evolution could have created the primordial proteins for a primitive proto-cell, there is still no mechanism in a natural cell whereby the organism can take a protein and reverse engineer it to create new DNA code from the protein molecules that carry out the functions of the cell. To create new DNA, another piece of information code, that is, DNA (or RNA) is required *first*. I have summarised this observation as the Biological Law of Coding which states: ***"In natural biological systems, function cannot directly alter DNA coding."*** Through natural selection, the form of the protein has an *indirect* effect on the DNA as, if the protein leads to the survival of the organism, then the DNA will survive to reproduce but the DNA code will either survive or die without being directly altered by the protein molecule. The obverse of this law is: ***"In natural biological systems, coding of information always precedes the synthesis of functional proteins."*** This is logical as natural selection can only act on the form and function of existing proteins and can only indirectly influence the DNA source code as

29. Mushegian, A.R. & Koonin, E.V.: A minimal gene set for cellular life derived by comparison of complete bacterial genomes. *Proceedings of the National Academy of Science of the USA* 93:10268-73 (1996).

30. Miller, S.L.: Production of Some Organic Compounds under Possible Primitive Earth Conditions. *Journal of the American Chemical Society* 77:2351 (1955).

31. Dawkins, R.: *The Blind Watchmaker.* Longman, London (1986) pp.47-9

32. Küppers, B.: On the Prior Probability of the Existence of Life, *in* Kruger et al.: *The Probabilistic Revolution.* MIT Press, Cambridge, MA (1987) pp.355-369.

33. Yockey, H.P.: *Information Theory and Molecular Biology.* Cambridge University Press (1992) p.336.

described above. Even in the case of mutation, where an incorrect transcription or translation occurs as in Sickle Cell Disease, there was source code with specified information prior to the mutation occurring. Schematically this can be represented as an extension of Crick's diagram of 1956 as shown in Figure 3.5:

Living systems are information rich, which is a requirement for them to function and reproduce but the information must have entered the system at some point and information is a prerequisite to initial information transfer (see Figure 3.5). In modern Neo-Darwinian evolutionary theory, organisms generate information *"in feedback loops via replication and selection"*[34] as the species randomly mutates, before individuals in that species compete over subsequent generations. This happens whenever a mutated protein ('Protein C' in Figure 3.5) has an advantage over the initial protein ('Protein A'). In the Darwinian model, the species gradually incorporates these randomly acquired changes into the genome over a very long period of time, which equates to new information and hopefully offers a survival advantage. The system can be considered to be 'learning' through environmental experience. This is not the learning that we know from school, which is an active process. It is not even passive learning such as the disinterested student who only goes to school because he has to. It is the *accidental* learning from bumbling around in the dark and discovering things by tripping over them. It could be termed 'blind learning'. As indicated above, this information *must* be acquired randomly as, if it was intentional then there would have had to have been a pre-set endpoint that was already specified which would have required predetermined information. So, can random 'learning' in biological systems, as described above, account for the information that is found in living creatures?

34. Eigen, M.: The Origin of Genetic Information: Viruses as Models. *Gene* 135:37–47 (1993).

Figure 3.5: The Basic Biological Information Pathway.
Information enters the system as a prerequisite via coding and is stored (in this example) in DNA. Coding of information is the initial level of information transfer/storage.

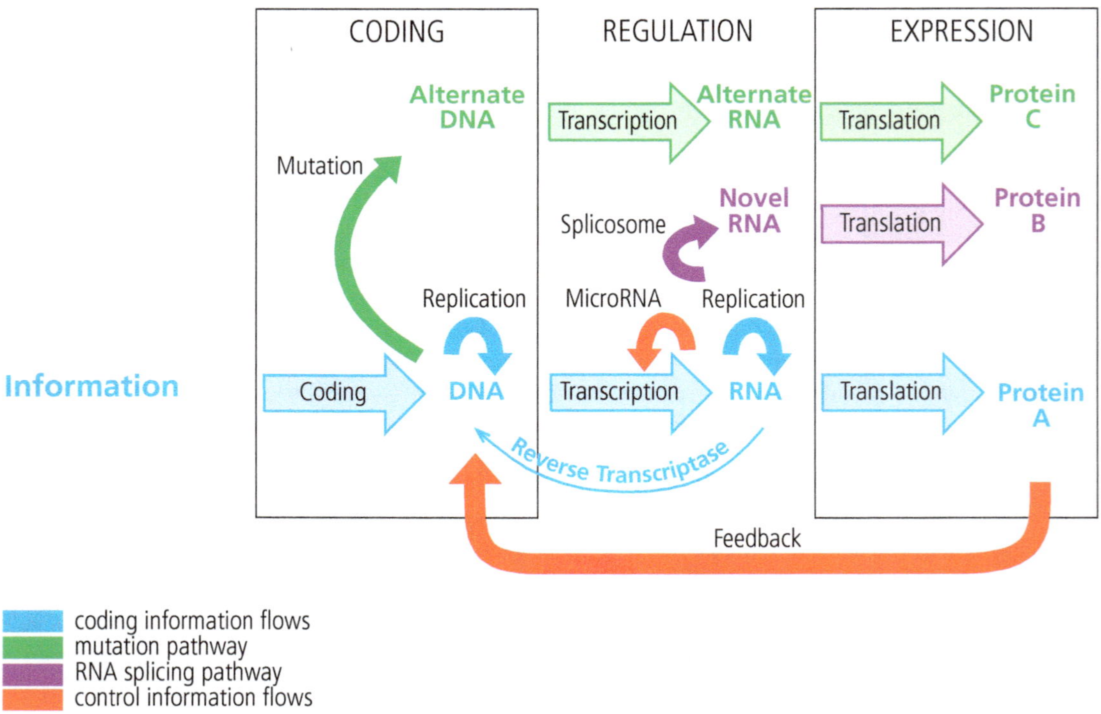

If a protein was created by a natural process, such as in the pre-biotic soup,[35] without that protein first being specified by a DNA (or RNA) sequence, then there is no mechanism possible in natural systems to reverse engineer a matching DNA/RNA sequence to reproduce a similar protein (or organism) for the next generation and all living things use either DNA or RNA to store their biological information to reproduce.[21] This is equivalent to trying to develop the next model of your motor car by driving it around, accumulating scratches and dents from occasional accidents and expecting the results not only to be an improvement on your current car but also imagining that the information specifying the alterations in your current model to magically appear in the engineer's blueprints, from which he built your existing car. It just does not happen. Your car would need a 'scratch-and-dent logger' that recorded all the random alterations to the car, that were then radioed back to your manufacturer's blueprint to automatically incorporate the changes for the next generation of vehicles to be produced. No such device exists for your car and neither does it exist in living cells. Without an information storage system, the information will be lost and there is no evolutionary advantage. The first requirement for 'biological learning', as described above, is a pre-existing information store (a DNA sequence in this case).

Implications for the First Cell

Thus, information flows demonstrate that for the first cell to evolve, it had to develop its nucleic acids (DNA or RNA information code) before its constituent proteins. This contravenes the process of Neo-Darwinian evolution, as natural selection acts only through the survivability of the organism (or its constituent proteins if it is a proto-organism). If you doubt this, consider how one string of DNA code could be selected over another to confer increased survivability for the resultant entity before it was translated into its protein derivatives? Benefit is conferred by the derivative protein, the active entity, not the unexpressed source code that is its memory. The other problem with this necessity to have information in the form of DNA or RNA first, is that outside a cell and especially in the presence of water, the nucleic acid chains are very unstable and rapidly denature. The only way to maintain DNA or RNA in nature is within a living cell.[36] Therefore, the protective cell membrane had to predate the DNA but the DNA must predate the membrane it codes for! It might be imagined that little membrane-like bubbles (called 'micelles') could form by chance and that the DNA or RNA was able to develop in the protective coat of the bubble by some process yet to be described. Then at a later date it developed the DNA code to produce its own cell membrane or wall; but apart from the problem of generating the information encoded in the DNA, we do not observe this micelle process happening spontaneously in nature today. If this process could occur, we should see cells in the process of doing this in nature and it is just not there. So, the only alternative left is that the DNA and membrane appeared contemporaneously.

The necessity for simultaneous generation of both *proteins*, which require a storage mechanism, with a *source code*, from which to manufacture the very same proteins, is the classic 'chicken and egg' problem.[37] This is a closed loop system as described in Chapter 2. Other mechanisms, such as transposons, can splice pieces of DNA in and out of genomes but they do not explain where the first nucleic acid sequence came from (including the transposon's nucleic acids, which also require transposase and/or reverse transcriptase to function).[38] Thus, some process had

35. This is the presumption for the creation of the first cell in contemporary evolutionary theory. See: Cavalier-Smith, T.: Obcells as Proto-Organisms: Membrane Heredity, Lithophosphorylation, and the Origin of the Genetic Code, the First Cells, and Photosynthesis. *Journal of Molecular Evolution* 53:555–595 (2001).

36. Lindahl,T.: Instability and decay of the primary structure of DNA, *Nature* 362(6422):709–715 (1993) *in* Sarfati, J: *By Design*. Creation Book Publishers, Brisbane, Australia (2008) p.178.

37. Vaneechoutte, M.: The Scientific Origin of Life: Considerations on the Evolution of Information, Leading to an Alternative Proposal for Explaining the Origin of the Cell, a Semantically Closed System. *Annals of the New York Academy of Sciences* 901:139–147 (2000).

38. Mizuuchi, M. & Mizuuchi, K.: Efficient Mu transposition requires interaction of transposase with a DNA sequence at the Mu operator: implications for regulation. *Cell* 58:399–408 (1989).

to have coded the appropriate information into the first strand of DNA or RNA for the first cell to have come into existence. If this was not so, there would have had to have been simultaneous evolution of nucleic acid coding in parallel with their biologically active derivative proteins, which is extremely improbable, even if a time frame of 4 to 5 billion years is considered.[4] As a separate but parallel problem of equal importance to the above, is the origin of the cytoplasmic machinery which is essential to convert the information that is stored in the DNA (or RNA) into functional proteins. Evolutionary biologists John Maynard and Eors Szathmary write: *"The existing translational machinery is at the same time so complex, so universal and so essential, that it is hard to see how it could have come into existence, or how life could have existed without it."* [39] Thus, there is a need to explain the information in both the source code and the translational machinery in the cytoplasm.

An alternative suggestion is that the first DNA may have come to Earth through interstellar space as in the ice of comets[40] and hence, there may have been 8 to 15 billion years to evolve the DNA.[41] The excess of L-isomers of amino acids found in the Murchison Comet, has been proffered as support for this theory but the excess is small and would not account for the 100% homochirality found in *all* living things on Earth.[42, 43] If the formation of life was random, it would seem reasonable to expect that 50% of organic compounds would be left-handed and the other 50% would be right-handed. The fact that 100% are left-handed *without exception*, suggests that left-handedness was specified or selected by some process at the beginning of life on Earth.[42]

The next problem is that even the simplest cell has many requirements. Minimal gene sets for the simplest organism have been estimated to be at least 80 genes (though probably closer to 250 genes) coding for a minimum of 100 proteins.[44] This equates to between 30,000–90,000 base pairs as each protein probably required 100–300 amino acids, a total of between 10,000 and 30,000 amino acids in specific sequence.[44] From this data, it can be seen that the amount of information to be coded, even for the simplest one-celled organism, is still large. Even though there is disagreement, at currently estimated mutation rates,[45,46] some researchers suggest that it might have been possible to amass large strings of DNA in this timeframe *providing there was a mechanism to achieve it.*[47,48] Even if it was not on Earth, however, there are still all the same problems of assembling complex DNA sequences and the same chicken and egg dilemma with encoding information in the DNA, RNA or some other information carrier somewhere else in the Universe. This theory removes the 'pre-biotic soup' from Earth to another location in the Universe but does not solve any of the problems previously discussed, such as the chicken and egg dilemma, homochirality and in particular, how did sufficient *meaningful information* become encoded in the first nucleic acid sequences to build the first cell? To overcome the problem of the

39. Lennox, J.C.: *God's Undertaker. Has Science Buried God?* Lion Books, Oxford (2007) p.131.

40. Wallis, M.K. & Wickramasinghe, N.C.: Interstellar transfer of planetary microbiota. *Monthly Notices of the Royal Astronomical Society* 348:52 (2004).

41. Bolte, M. & Hogan, C.J.: Conflict over the age of the Universe. *Nature* 376:399–402 (2002).

42. Sarfati, J.: *By Design.* Creation Book Publishers, Brisbane, Australia (2008) p.175.

43. Organic compounds, (those based on carbon), can exist in left or right handed configurations. This is referred to as chirality. All living organisms on the earth are left-handed and *cannot react with right-handed molecules.* This property is used in some medical treatments.

44. Koonin, E.V.: How Many Genes Can Make a Cell: The Minimal-Gene-Set Concept. *Annual Review of Genomics and Human Genetics* 1:99–116 (2000).

45. Brown, W.M., George, M. & Wilson, A.C.: Rapid Evolution of Animal Mitochondrial DNA. *Proceedings of the National Academy of Sciences of the USA* 76:1967–1971 (1979).

46. Ho, S.Y.W., Phillips, M.J., Cooper, A. & Drummond, A.J.: Time Dependency of Molecular Rate Estimates and Systematic Overestimation of Recent Divergence Times. *Molecular Biology and Evolution* 22:1561–8 (2005).

47. Oberbeck, V.R. & Fogleman, G.: Estimates of the maximum time to originate life. *Origins of Life* 19:549–560 (1989).

48. Orgel, L.E.: The Origin of Life- How long did it take? *Origins of Life* 28:91–6 (1998).

generation of meaningful information, British chemist, Leslie Orgel (1927–2007) proposed 'Directed Panspermia' in 1973 but this is another way of saying that the information for the coding of life had to come from a source external to the Earth.[49] Imagining that information generation, which could not be conceived to have occurred on Earth, had mysteriously happened somewhere else in the vastness of the Universe is woolly thinking. Why should the laws that apply to information generation on Earth be any different elsewhere in the Universe? For this reason, multiple universes ('the Multiverse') with Kronin's eternal inflation[50] do not solve the problem of information generation either, because what is impossible in one universe is impossible in the next unless one speculates that other universes may have special laws that can generate the requisite information. However, there is no more evidence for this than there is for the Multiverse itself. The last problem with an alternative universe with special properties, such as the ability to generate information-containing DNA, is that there is no mechanism for the transfer of the DNA into our universe (see Chapter 8). As Norwegian biologist, Albert Voie states: *"the [evolution of the] structure of life has a probability of zero."*[51]

To overcome the chicken and egg impasse, it has been suggested that the problem might be solved if the initial functional molecules encoded for their own replication as RNA is able to do. As a result, an 'RNA World' has been proposed where RNA molecules both coded for the information and acted as the catalytic molecules simultaneously.[52] This model is very attractive as it appears to satisfy both the function and storage criteria simultaneously but there are still some unanswered questions.[36,53,54,55] These include, firstly, synthesising the RNA and then, preventing its subsequent degradation through cross reactivity with other molecules soon after their formation.[56,57,58] Many of the molecules essential for life, such as =C=O, interact to destroy each other. In particular, phosphate containing molecules, such as ATP, DNA and RNA, precipitate with calcium which is prevalent in the sea. In living cells, there are mechanisms to prevent these processes destroying the cell or to repair the damage if it should happen but outside a living cell, in an uncontrolled environment, there is nothing to conserve these essential requirements for life. So, for the first cell to come into existence, there needed to be a very precisely controlled microenvironment to allow it to happen.[59] Secondly, the specific 3',5'–phospodiester linkages required for all known RNA in living organisms today, would be diluted by other compounds such as 5',5'–pyrophosphate and 2',5'–phospodiester linkages resulting in RNA that did not resemble life as we know it.[60] Thirdly, the newly formed RNA needs transcription machinery in the form of ribosomes or similar apparatus to replicate, which would have to simultaneously evolve.[61,62] Fourthly, the first ribo-organism would have required a diverse range of catalytic ability to provide a significant number of the functions that proteins provide in cells today, which must have

49. Crick, F. & Orgel, L.E.: *Directed Panspermia*, Icarus 19:341-6 (1973) *in* Sarfati, J.: *By Design*. Creation Book Publishers (2008) p.150.

50. Koonin, E.V.: The cosmological model of eternal inflation and the transition from chance to biological evolution in the history of life. *Biology Direct* 2:15 (2007).

51. Voie, A.: Biological function and the genetic code are interdependent. *Chaos, Solitons and Fractals* 28(4):1000–4 (2006).

52. Gilbert, W.: Origin of life: The RNA world. *Nature* 319:618 (1986).

53. Mills, G.C. & Kenyon, D.: The RNA World: A Critique. *Origins & Design* 17:1 (1996).

54. De Duve, C.: *Vital Dust: Life as a Cosmic Imperative*. Basic Books, New York (1995) p.23.

55. Meyer, S.C.: DNA and the Origin of Life: Information, Specification, and Explanation *in* Campbell, J.A. & Meyer, S.C.: *Darwinianism, Design and Public Education* (eds.) Michigan State University Press, Lansing, Michigan (2003) pp.223–285

56. Shapiro, R.: Prebiotic cytosine synthesis: A critical analysis and implications for the origin of life. *Proceedings of the National Academy of Sciences of the USA* 96:4396–4401 (1999).

57. Waldrupp, M.M.: Did Life Really Start out in an RNA World? *Science* 246:1248–9 (1989).

58. Thaxton, C., Bradley, W. & Olsen, R.: *The Mystery of Life's Origin*. Philosophical Library, New York (1984).

59. Sarfati, J.: *By Design*. Creation Book Publishers (2008) pp.175–180.

60. Joyce, G.F. & Orgell, L.E.: Prospects for understanding the RNA World. *in* Gesteland, R.F. & Atkins, J.F. (eds.): *The RNA World*. Laboratory Press, Cold Spring Harbour (1993) pp.1–25.

61. Szostak, J.W., Bartel, D.P. & Luisi, P.L.: Synthesizing Life. *Nature* 409:387–390 (2001).

62. Pohorille, A. & Deamer, D.: Artificial Cells: prospects for biotechnology. *Trends in Biotech.* 20: 123–8 (2002).

been subsequently lost, as the RNA molecules with catalytic activity today have only limited activity.[63] Yet, there is no evidence that this diverse range of catalysing RNA ever existed. James Gibson of the Geoscience Research Institute in California concluded: *"Even if RNA were produced, it could not survive nor could it form the basis for a naturalistic origin of life. Some other mechanism must be sought to explain the origin of life."* [64]

To overcome the complexity required for an RNA World, a simpler peptide nucleic acid (PNA) precursor model has been proposed. Nucleic acid strands may have been incorporated into a micelle or lipid proto-container to form a primitive 'cell'. Since the probability of formation of the desirable PNA is low, the model requires *"many PNA variants"* [65] but these are not found in living systems today, so there must have been a process to convert PNA to RNA or DNA though this process has never been observed. To have RNA templates instead of PNA ones, as described previously, requires transcription machinery[61,62] which would have had to have been coded prior to their formation and hence, cannot explain the first cell. (This constitutes another closed loop.) To have a DNA, RNA or PNA information store in isolation, without ribosomes, is analogous to having an encyclopaedia on DVD but no computer or DVD player to access the information. The complex double helical structure of DNA makes an initial DNA template even less likely than an RNA or PNA one and once again, requires the transcription machinery.

If this list of problems seems insurmountable, Scottish chemist, Graham Cairns-Smith from the University of Glasgow lists 19 biological problems associated with the formation of the first cell![66] New York University Professor of Chemistry, Robert Shapiro summarised the situation well when he wrote: *"A profound difficulty exists, however, with the idea of RNA, or any other replicator, at the start of life. Existing replicators can serve as templates for the synthesis of additional copies of themselves, but this device cannot be used for the preparation of the very first such molecule, which must arise spontaneously from an unorganised mixture. The formation of an information-bearing [RNA chain or equivalent] through undirected chemical synthesis appears very improbable."* [67] However, from the preceding analysis of information flows, we have seen that the Law of Coding shows that it is not just improbable but *impossible.*

This is not a definitive list of all the problems. Another significant issue in living systems is: their dependency on in-built, error-correction mechanisms. John Lennox writes: *"One very important implication of the existence of alternative splicing and error repair mechanisms is that DNA would appear to depend on life for its existence rather than life on DNA thus calling in question the common notion that life originated in an RNA to DNA to life sequence (the RNA world scenario)."*[68] Alternate splicing and repair mechanisms are discussed in greater detail in Chapter 4.

The corollary of the Law of Coding is that only after the information is specified in the form of RNA or DNA and the first functional cells derived (that is, life had already come into existence), could mutation create natural variation on which natural selection could operate as described by Charles Darwin. Therefore, evolution by natural selection could not have created the first cell. That is not the end of the difficulties however, as there are problems with surmising that Darwinian evolution became a significant influence later too (see Chapter 5).

Paul Davies amplifies the problem Darwinian Evolution has in this respect by comparing the organisation seen in life with undirected chemistry when he

63. Hager, A.J., Pollard, J.D. Jr & Szostak, J.W.: Ribozymes: aiming at RNA replication and protein synthesis. *Chemistry & Biology* 3:717–725 (1996).

64. Gibson, L.J.: Did Life Begin in an "RNA World"? *Origins* 20:45–52 (1993).

65. Rasmussen, S., Chen, L., Stadler, B.M.R. & Stadler, P.F.: Proto-Organism Kinetics: Evolutionary Dynamics of Lipid Aggregates with Genes and Metabolism. *Origins of Life and Evolution of Biospheres* 34:171–180 (2004).

66. Cairns-Smith, A.G.: Genetic Takeover and the Mineral Origins of Life. Cambridge University Press, NY (1982) *in* Sarfati, J.: *By Design*. Creation Book Publishers (2008) pp.182–4.

67. Shapiro, R.: *A replicator was not involved in the origin of life*. IUBMB Life (A Journal of the International Union of Biochemistry and Molecular Biology) 49:173–5 (2000) *in* Sarfati, J.: *By Design*. Creation Book Publishers (2008) p.187.

68. Lennox, J.C.: *God's Undertaker. Has Science Buried God?* Lion Books, Oxford (2007) p.134.

writes: "*If we accept that the genome [the totality of genes] is random and information-rich, then appealing to non-random chemistry to make life is a clear contradiction... The whole point of the genetic code...is to free life from the shackles of non-random chemical bonding. A genome can choose whichever amino-acid sequence it wants, oblivious of the chemical preferences of molecules.*" [69]

Self-organisation

Fowler and Kuebler also acknowledged some of the problems associated with the formation of the first cell when they wrote: "*Cells are dynamically organised structures that contain much complex information. They are composed of many integrated parts organised in a specific fashion to allow for survival in a dynamic world. The organisational work that goes into making a cell function is staggering: tens of thousands of genes, proteins and metabolites interact extensively with one another in order to create the dynamics of a living cell. The field of proteomics, which has as its task the categorisation of all the protein interactions within the cell, is merely beginning its daunting quest to gain insight into how cells are organised. To make matters more complex, many of the cellular components such as mitochondria, ribosomes, and flagella are tightly organised and regulated structures that could be studied by proteomics in their own right.*" [70] As random chance would seem incapable of creating this bewildering degree of complexity, even in the simplest form of life, a number of scientists have proposed that some process intrinsic to the Universe has caused life to appear, such as the 'laws of nature'.

Two such scientists are Russian-born chemist and Nobel laureate, Ilya Prigogine (1917–2003) and Belgian Professor of Philosophy, Isabelle Stengers, who suggested that life might be self-organising; that due to the non-linear effects of Chaos Theory (see Chapter 9), organic molecules will intrinsically attract and form life.[71] The beauty and symmetry seen in crystal formation and snowflakes are some of the best-known examples. However, as we have seen, although crystals can be extremely large, they are repeating patterns that do not contain a lot of information. In contrast, life requires an immense amount of information to specify it and nearly all the components of life are highly irregular and *asymmetric*. The problem is that self-organising systems do not create novel new information. Stephen Meyer observes that if there were laws or rules for generating information, there would be little variation in information at all. A universal law to produce information would cause regular patterns, such as crystal lattices, rather than result in the generation of the exquisitely varied information required for life. Therefore, there *cannot* be laws to cause self-organisation as they are a contradiction in terms. He summarises this point by saying: "*Self-organisational theorists explain well what does not need to be explained. What needs explaining is not the origin of order...but the origin of information.*" [72] It is important here to recognise that there is a difference between the presence of the raw building materials and the completed entity. The house is just a stack of bricks and timber until information is introduced, firstly by the architect and subsequently by the builder, who arrange them in specific configurations to design and then construct the house. The information that is incorporated to define the relationships between the individual pieces of building material by the architect and builders (think infobits) is the entity that makes the difference.

To get around this problem of an architect and builder, philosopher of biology, Michael Ruse suggested that the first cell may have used the catalytic

69. Davies, P.: *The Origin of Life*. Penguin (2003) p.237 *in* Andrews, E.: (2009) p.202.

70. Fowler, T.B. & Kuebler, D.: *The Evolution Controversy*. Baker Academic, Grand Rapids (2007) p.312.

71. Prigogine I and Stengers I: *Order out of Chaos*. University of Michigan. Bantam Books (1984).

72. Meyer, S.: *The Return of the God Hypothesis*. Seattle, Discovery Institute Centre for the Renewal of Science and Culture (1998) p.37 *in* Lennox, J.C.: (2007) p.125.

properties of certain crystals to expedite the first cellular reactions[73] but Edgar Andrews, Emeritus-Dean of Engineering at London University writes: "*It is known that certain crystalline minerals such as zeolites do display catalytic activity and can facilitate polymer formation (chain-building), but they only work under carefully controlled conditions. Some technical publications have explored this route to the formation of polymers of amino-acids and bases. For example, computer modelling has been used to predict how amino-acids might behave within the microscopic pores of zeolites, while studies of the absorption of bases on graphite have shown that some are absorbed more strongly than others. But the relevance of such results to the origin of life is both remote and speculative, and to date nothing resembling a protein or nucleic acid polymer has ever been produced by such methods.*" [74] This also ignores the whole problem of where did the *information* come from!

Richard Dawkins cites the process of virus protein coat construction as an example of self-organisation or "*self-assembly*". He writes: "*The adenovirus... has exactly 252 capsomeres,... arranged in an icosahedron. The icosahedron is that platonic perfect solid that has 20 triangular faces. The capsomeres are arranged into an icosahedron not by any kind of master plan or blueprint but simply by each one of them obeying the laws of chemical attraction locally when it bumps into others like itself. This is how crystals are formed, and, indeed, the adenovirus could be described as a very small hollow crystal. The 'crystallisation' of viruses is an especially beautiful example of the 'self-assembly' that I am touting as a major principle by which living creatures are put together. The T4 'lunar lander' phage also has an icosahedron for its main DNA receptacle, but its self-assembled quaternary structure is more complex, incorporating additional protein units, assembled according to different local rules, in the injection apparatus and the 'legs' that are attached to the icosahedron.*" [75] Among a number of problems, are the questions of: what specified the material that forms the coat of the virus and if they can self-assemble, why do we not find little icosahedrons mushrooming everywhere of their own accord? The biggest problem, however, is not the icosahedron receptacle but the information in the RNA that it contains, which cannot be explained by the above process. There is also the small matter of what specified the "*laws of chemical attraction*" that caused the capsomeres to form into the icosahedron? (See Chapter 6.)

The final nail in the self-organisation coffin is the problem of thermodynamics. Edgar Andrews writes: "*Crystallisation is driven exclusively by thermodynamics and is fully understood in terms of the tendency of all physical systems to minimise their 'free energy'. Living cells must also obey thermodynamics, of course, but what happens when molecules get together to create molecular machines is quite different from crystallisation. To start with, such machines involve special combinations of different molecules. Far from lowering the free energy of the system, mixing different polymers together actually raises it. This means that assembling molecular machines cannot happen spontaneously (if it did, it could be done in a cake mixer). Building a molecular machine is like making water run uphill - something that can't happen of its own accord but can be done by using a pump...That calls for a higher layer of organisation which must itself be organised.*" [76]

Sources of Information in the First Cell

We have previously ascertained that cells contain and store information in their DNA/RNA, which is the ubiquitous information store for all organisms on Earth. In animal (eukaryotic) cells this information is stored in the nucleus and

73. Ruse, M.: Interview *on Expelled*. (DVD) Stein, B.: Condios Entertainment (2009).

74. Andrews, E.: (2009) p.198.

75. Dawkins, R.: *The Greatest Show on Earth*. Bantam Press, London (2009) pp.223–4.

76. Andrews, E.: (2009) pp.206–7.

provides structure, management and catalysts to accelerate biochemical reactions (see Chapter 4); but that is not the only source of cellular information. A second source of information comes from the cytoplasm and specifies the cellular machinery. Especially important are the nano-machines, which allow the cell to function, along with the mitochondria, which house the power plants to provide energy through ATP, so that life can oppose the Law of Entropy. Other proteins are required for the replication of DNA and RNA. Both sources of cellular information need to be identified and tracked. There are five distinct categories of information-bearing molecules within the cell which are summarised in Figure 3.6.

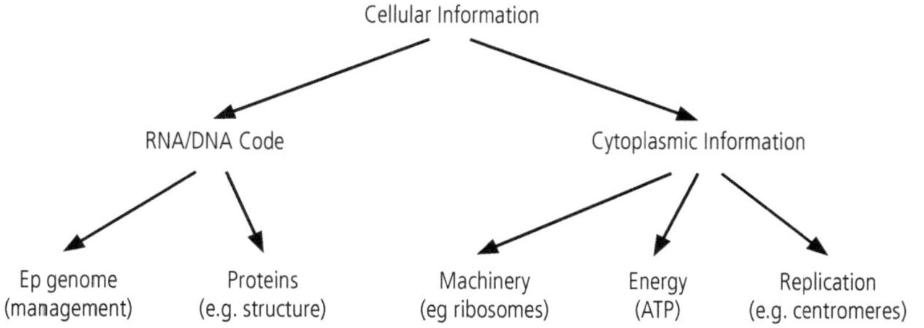

Figure 3.6: Categories of Intracellular Information. Each of these categories will be dealt with in turn, however replication will be covered in Chapter 4, as it could be argued that it was not required for the formation of the *first* cell.

Information Generation

So, if cells contain so much information, where does it all come from? In Figure 3.5 it can be seen that information is required prior to the generation of DNA, but that still does not identify where this information came from. Since information is a discrete entity in its own right and needs to be generated by some mechanism, it would be easier if the source had a name even though its identity is not yet known. The name I have given it is the '*Biological Information Generator*' (see Figure 3.7). As we shall see later, there are other information generators too. Science has provided little information to explain the generation process to date. Different scientists explain it in different ways. Darwin's hypothesis requires the process of random variation with natural selection to generate the information and this mechanism is supported by contemporary scientists, such as Richard Dawkins, while other scientists, such as Michael Behe, have invoked 'Intelligent Design'. Some scientists, such as Sir Fred Hoyle (1915–2001), have stated that the process is unknown.[77] Theists would name God as the Biological Information Generator. This will be investigated in Chapter 12.

Variation Selection and Regulation

In Figure 3.7, it can be seen that information generation must occur at Stage 1, prior to the entry of the information into the system. Darwinian evolution purports to explain information generation through random mutation at Stage 2 and natural selection at Stage 4. This process could be imagined to work, if there was already a vast amount of information in the system that is simply rearranged without the need to input *new* information. This process can occur and is termed 'Microevolution'. As already seen, however, the Law of Coding clearly shows that the first DNA in the first cell cannot be generated in this manner as information flow is a one way process. Thus the Neo-Darwinian Theory is left without a

77. Lennox, J.C.: *God's Undertaker. Has Science Buried God?* Lion Books, Oxford (2007) p.110.

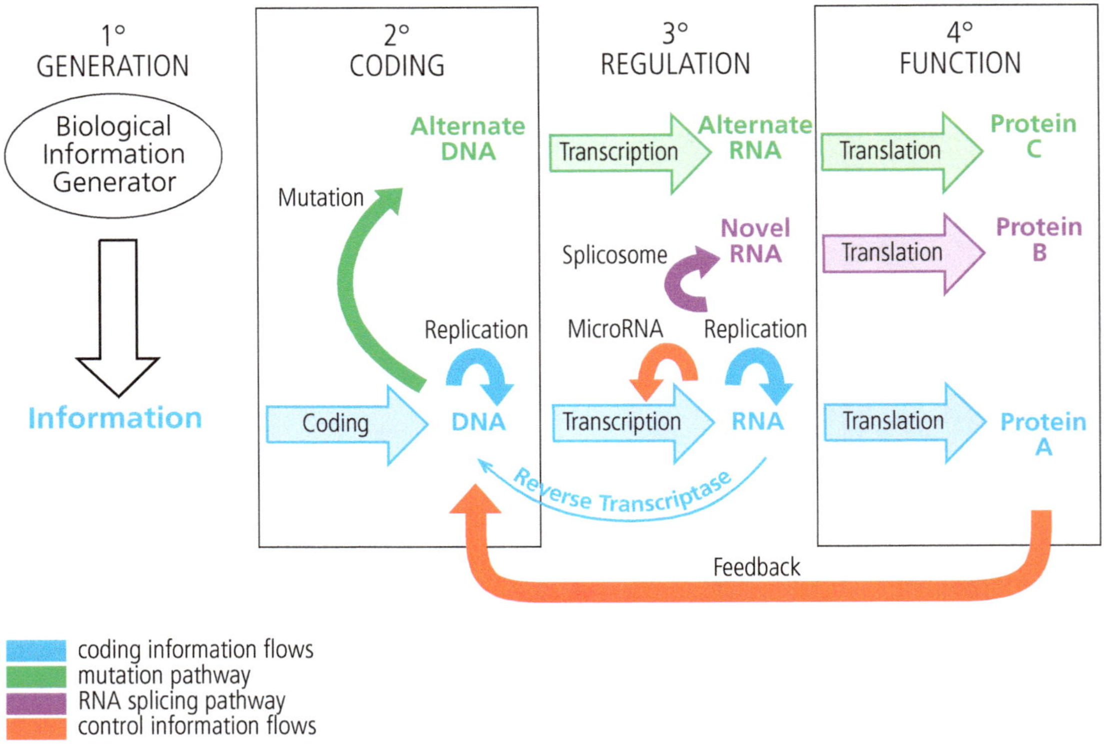

Figure 3.7: Information Generation in Biological Systems
Information must be generated by some process before it can enter the system to be stored in DNA. As this information generation process is not explained well by science at this stage, the mechanism has been illustrated as the 'Biological Information Generator'.

mechanism to explain information generation and hence Macroevolution is impossible (See Chapter 5).

Therefore, if random mutation with natural selection does not generate ever more complex information for the system as Darwin's supporters surmise, what does this process do, as it was noted in Chapter 1 that it plays a significant role in biological systems? In fact, random mutation with natural selection has two significant functions. They act as:

1. A variation selector;
2. A variance regulator.

1. Variation Selection: The first function of natural selection is to allow species to adapt to their individual microenvironments. This occurs through selection from amongst the characteristics already contained in the collective genome of each individual species; those characteristics that are most suited for that microenvironment lead to the survival of the organism that expresses them. It relies upon the information having been previously programmed into the organism's nucleic acids. In contrast, random mutations, at least initially, lead to the damage and/or deletion of genetic material with the loss of information rather than its generation. This is particularly noticeable in small, geographically isolated populations, such as remote islands and will be discussed further when we examine fitness landscapes later in this chapter.

2. Variance Regulation: As the majority of new mutations are not beneficial to the cell and a significant number of them can be lethal, random mutations are a significant risk to any organism. Some mutations such as sickle haemoglobin may give an individual an advantage in a specific microenvironment but at the expense of lost overall efficiency. Bacterial resistance to antibiotics is a perfect example of the adaptation of an organism to a microenvironment: one that contains antibiotics. Resistance can be achieved in a number of ways, such as: by turning off the pumps which import nutrients (including the antibiotic), accelerating the excretion of waste (as well as the antibiotic) or altering replicating machinery to avoid interference from the antibiotics (see Chapter 7 for more details). These result in the sacrifice of cell efficiency and the loss of information to gain a short-term advantage, which is to evade the antibiotic and survive. However, bacteria do not have in-built switches to turn this vital machinery off temporarily, so the only way to shut down these cell functions is to break the machines permanently. If that bacterial population subsequently leaves that antibiotic-containing microenvironment, the broken pumps or other adaptations that it has developed to survive the antibiotics, do not return to their former efficiency, so the antibiotic-resistant colony is usually outcompeted for resources and in the long-term will not survive. To illustrate this point, imagine an automatic gearbox in a car where one cog 'mutates' (for example break in two), conferring on the gearbox the short-term gain of decreasing friction and higher maximum speed when going downhill. Unfortunately, with only one change, the overall efficiency of the gearbox will decrease and losing the second gear will be a hindrance to general performance of the vehicle, such as going uphill. An improvement would depend upon a number of *complementary* point mutations. This is discussed further in the next chapter.

Over many generations, the build-up of random mutations are increasingly likely to damage the cell's operating machinery, so there needs to be a mechanism to remove the damaging mutations before they compromise the species gene pool. Natural selection performs this function of *variance regulation* to minimise mutations which may be harmful to that cell line and in so doing, natural selection causes variation within species to return towards the mean, rather than away from it. James Shapiro writes: "*It has been a surprise to learn how thoroughly cells protect themselves against precisely the kinds of accidental genetic change that, according to conventional theory, are the sources of evolutionary variability. By virtue of their proofreading and repair systems, living cells are not the passive victims of the random forces of chemistry and physics. They devote large resources to suppressing random genetic variation and have the capacity to set the level of background localised mutability by adjusting the activity of their repair systems.*" [78]

John Lennox adds: "*There is evidence that the repair mechanisms may be even more sophisticated than this. In Nature (434, 2005 p.505) Robert Pruitt reports the astonishing fact that certain genetic mutants of weedy cress (Arabidopsis thaliana) produce normal offspring which somehow acquired their genetic information from normal ancestors other than their parents. This should be impossible as it contradicts the received wisdom of Mendelian genetics. Pruitt makes the suggestion that RNA templates inherited from earlier generations may be engaged in repairing the DNA of the mutant genes and bringing it back to its ancestral norm.*" [79] This must involve some form of controlling information, which may be associated with the epigenome (our genetic manager), switching previously encoded control mechanisms on or off to permit the repairs described. These weedy cress mutants, that have the capability to self-repair, must have been pre-programmed with error correcting software which requires an extra layer of information. The presence of

78. Shapiro, J.: *A Third Way.* (alternative to Darwinism & Creationism). *Boston Review* 22(1):32–33 (1997) p.13 *in* Lennox, J.C.: *God's Undertaker. Has Science Buried God?* Lion Books (2007) p.134.

79. Lolle, S. J., Victor, J. L., Young, J. M. & Pruitt, R. E.: Nature. 434:505–9 (2005) *in* Lennox, J.C.: *God's Undertaker. Has Science Buried God?* Lion Books, Oxford (2007) p.188, Ref. 12.

the epigenome adds yet another layer of information on top of the error correction programming and the original source code. The sophistication of this variance regulation is exquisite and will be explored further under the topic of Redundancy in Chapter 4.

Cytoplasmic Machinery

Having genetic information safely and successfully packaged in a suitable code (RNA/DNA or equivalent) is only one part of developing the first cell. When looking at life, as in the computer analogy in Chapter 2, simply having the code for the word-processing program and the machine language on compact discs (the 'software') is insufficient to produce a functional machine. There is still a requirement to have a fully operational computer platform (the physical box or 'hardware' which we recognise as 'the computer') on which to run the software. In a living cell, this is equivalent to the cell membrane, cytoplasm and organelles (including the intricate nano-scale machines that digitally process the information stored in DNA) or in the case of membranes, to operate ports which control what comes in and what goes out. In addition, in living cells it is also necessary to have the instructions to auto-construct and self-manage the system (which is the supervising 'epigenome'). This is equivalent in our computer analogy, to having an inbuilt repair, maintenance and management robot that is pre-programmed to firstly: build and secondly: to diagnose and manage software and hardware repairs automatically, which is way beyond the ability of our current computers. How did all this information become programmed to specify, build and then maintain these four-dimensional protein machines? (The fourth dimension is time as the cell is pre-programmed to grow and change over time.)

Another unsolved problem is the development of the machines which fold the proteins, called 'chaperones', as proteins only become active when precisely folded. If the code came first then there is a translation problem as described in 1971 by French biologist and Nobel laureate, Jacques Monod (1910–1976): "*The code is meaningless unless translated. The modern cell's translating machinery consists of at least 50 macromolecular components* which are themselves coded in DNA: the code cannot be translated otherwise than by products of translation." Stephen Meyer noted that "*Scientists now* [2009] *know that translation actually requires more than 100 proteins.*" [80] So did the proteins come first or the chaperones and nano-machines that produce them?

The other problem is that living cells are approximately 70% water and many organic molecules are hydrophobic, that is, they do not like water. Most of the chemical reactions that sustain our cells, with their mixture of hydrophobic and hydrophilic molecules, are very slow or non-existent except in the presence of catalysts or accelerator proteins, which assist the reactions without being consumed in the process. Without the catalysts, the organic processes would be too slow for cells to respire and survive. So which came first, the chemical reactions or the catalysts? If the catalysts came first, it must have been a very lucky accident, as the catalysts are specific to each individual reaction. If the catalysts came second, then the components of the cell were waiting around for a very long time for sufficient catalysts to be present in the one place at the one time for cell respiration to commence. (See the Krebs cycle in Chapter 4) The odds appear to be so small, it is hard to imagine. Richard Wolfenden, Professor of Biochemistry at the University of North Carolina and a member of the U.S. National Academy

80. Monod, J.: *Chance and Necessity: An Essay on the Natural Philosophy of Modern Biology*. Vintage, NY (1972) p.143 *in* Meyer, S.C.: *Signature in the Cell*. HarperOne, NY (2009) pp.133–4.

of Sciences wrote in 1995: "*Without catalysts, there would be no life at all, from microbes to humans,...It makes you wonder how natural selection operated in such a way as to produce a protein that got off the ground as a primitive catalyst for such an extraordinarily slow reaction.*" [81] The British philosopher of science, Sir Karl Popper (1902–1994), summarised the situation concisely when he wrote: "*What makes the origin of life and of the genetic code a disturbing riddle is this: the genetic code is without any biological function unless it is translated; that is, unless it leads to the synthesis of the proteins whose structure is laid down by the code. But...the machinery by which the cell...translates the code consists of at least 50 macromolecular components which are themselves coded in the DNA. Plus the code cannot be translated except by using certain products of its translation. This constitutes a baffling circle; a really vicious circle, it seems, for any attempt to form a model or theory of the Genesis of the genetic code.*" [82] Popper's conundrum is illustrated in Figure 3.8 below.

The information required to specify the complexity observed in cytoplasmic machinery is vast and this topic will be discussed further in the next chapter.

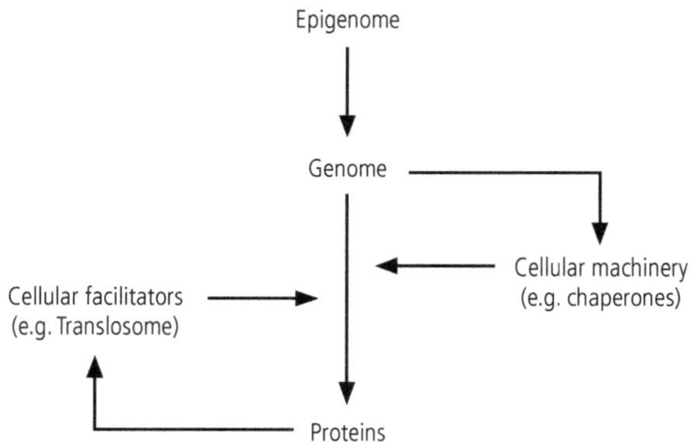

Figure 3.8: Popper's Conundrum: This information flow diagram illustrates Popper's riddle above. Popper did not know about the epigenome but it only adds to the dilemma. Note that this is another (double) closed loop system, as discussed in Chapter 2.

Energy Management

For the first organism to exist, there had to be energy for it to live on. Photosynthesis is the main mechanism for energy generation in biological systems, including plants, algae and cyanobacteria. The process of photosynthesis is intricate and depends upon an extremely complex 'catalytic core'[83] of Mn-Ca-O_2-Mn within the chloroplasts of green plants. The catalytic core assists (catalyses) the cleavage of water molecules with the aid of light photons, before donating back electrons to stabilise the process. The energy requirement for this process can be gauged by the extreme heat released in the Hindenburg disaster of 1937, when the reverse process occurred. The live footage showed vast amounts of water being formed as the hydrogen in the airship burned and it appeared to be 'raining'. The information required for the chloroplast complex, including the catalytic core and its support proteins, is enormous and requires precision construction. As the core complex is universal to photosynthetic organisms, it does not appear to have evolved over time and, in fact, there is no evidence it evolved at all. Even the cyanobacteria, which are considered by many biologists to be very ancient and hence presumed to be 'primitive', contain this mind-bogglingly complex nano-machine. This means that the information has been present throughout the entire history of green plants. The corollary of this observation is that after the initial

81. Wolfenden, R.: *Without enzymes, biological reaction essential to life takes 2.3 billion years.* Univ. of North Carolina website *at* http://www.med.unc.edu/biochem/news/without-enzyme-biological-reaction-essential-to-life-takes-2-3-billion-years-unc-study/

82. Popper, K.: *Scientific Reduction and the Essential Incompleteness of All Science, in* Ayala, F.J. & Dobzhansky, T. (Ed.): *Studies in the Philosophy of Biology: Reduction and Related Problems* University of California Press, Berkeley (1974) p.270 *in* Sarfati, J: *By Design*. Creation Book Publishers (2008) p.165.

83. *Synechococcus* elongatus at 3.8Å resolution *Nature* 409(6821):739–43 (2001) *in* Sarfati, J.: *By Design*. Creation Book Publishers (2008) pp.126–7.

generation of the information which built the first chloroplast, there has been information stasis.

Photosynthesis converts carbon dioxide (CO_2) and water (H_2O) into carbon to carbon (C–C) bond energy through a series of steps, then by glycolysis to **a**denosine **di**phosphate (ADP) and finally, **a**denosine **tri**phosphate (ATP) via a further number of steps. As plants are predated by most other organisms and very few plants predate anything, it is generally accepted that plants are at the bottom of the food chain but which came first, the colossally complex chlorophyll-producing plants or scavenging one celled organisms? The sheer complexity of chlorophyll suggests that, from an informational point of view, if evolution produced life then the scavengers should have preceded the chlorophyll-producing plants, which already contained their most complex molecule at the beginning of their life on Earth and which has remained unchanged ever since. Plants therefore, appear to have greater complexity (information content) than an amoeba, even though plants are now below the scavengers in the food chain. In evolutionary terms, this is absurd unless the hunters originally had another food source which is now extinct, leaving no trace. There is also the trifling question of where did the information in the chlorophyll molecule come from since the evidence suggests that it has not evolved?

The next problem is that ATP is the universal energy source in *all* living things. It is manufactured by a highly efficient nano-machine called '*ATP-synthase*' which is found in mitochondria and chloroplasts, so it is in all plants, bacteria and animals. Since all life needs ATP-synthase, then it must have formed with the first cell or very early after the formation of life but ATP-synthase is another extremely complex machine and is required for the cell's self-sustainability (its energy supply). So how could it have evolved so early in the history of life on Earth? It is significant to note that the first plants contained both the photosynthesising catalytic core and ATP synthase!

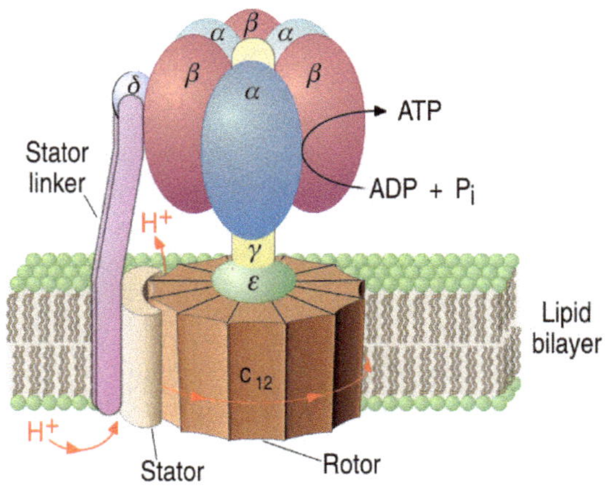

Figure 3.9: The ATP- synthase Generator[84]
ATP is the universal energy storage for all life. *[Copyright Apologetics Press (2005). Reproduced with permission.]*

The human body generates and consumes about half its own weight in ATP every day which must be continually produced by trillions of these nano-machines. Michael Mayer, Assistant Professor of Chemical and Biomedical Engineering at the University of Michigan has studied the ATP-synthase molecule and discovered that it operates like a micro hydro-electric plant.[85] As the protons

84. Charles McCown at Apologetics Press (2005) http://www. apologeticspress. org/apcontentaspx?category=12 &article=1281

85. Mayer, M.: Univ. of North Carolina website: nano tsunami. com at www.voyle.net

(hydrogen atoms) flow across the mitochondrial membranes they turn the ATP synthase turbines, which operate a series of ratchet gear mechanisms that then convert the lower energy molecule ADP to the higher energy molecule ATP. In the same way that hydro-electric turbines generate energy (electricity) to power our towns and cities, ATP-synthase 'generates' energy in the form of ATP to power our cells. The ATP-synthase motor rotates at over 9000 rpm and is considered to operate at nearly 100% efficiency.[86] It is more accurate to say that the ATP-synthase does not actually 'generate' energy but rather converts the higher concentration of hydrogen ions on one side of the membrane into the higher energy molecule ATP, in which the energy is stored. This is a conversion process rather than a generation process but that is actually what a hydroelectric plant does; it converts potential energy stored at a higher altitude into electricity using gravity. It will also be observed that the ATP-synthase turbines are not run by water but by hydrogen (acid) ions and they do not produce electricity but generate high-energy organic molecules so they could be called: Organic acid-turbines!

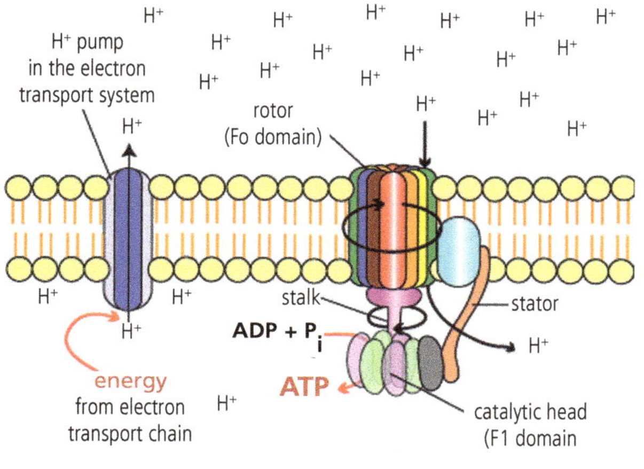

Figure 3.10: The Energy Cycle within Mitochondria.[87] The hydrogen ions (H+) cycle back and forth across the mitochondrial membrane to generate high energy ATP but note that it involves another closed loop. *[Reproduced with permission courtesy of Professor G. E. Kaiser.]*

The biochemical pathways described above are particularly noteworthy but there are scores more that are just as essential for the functioning of even the simplest cell. To develop the information to specify all these pathways via Neo-Darwinian principles would have required precise code acquisition ability, with each successive step needing to convey some additional survival advantage for the organism, otherwise the change would not have been incorporated into the gene pool. Natural Selection would have removed any disadvantageous mutations, as was seen in Chapter 1 but random mutation is not an efficient mechanism to generate useful variants for Natural Selection to choose from. Even if endosymbiosis (the fusion of two organisms to create a single more complex one) was true, there still needs to be an explanation as to where each of the merging organisms acquired their own smaller amount of information (see Chapter 5). Endosymbiosis is the Darwinian explanation for the induction of mitochondria into cells. However, after explaining the origin of the information, there also needs to be an explanation as to how the fusion process was able to occur because nearly all observable fusion processes are lethal or deleterious to the hybrid organism such as Trisomy 21 (Down's syndrome), Trisomy 5 (Cri du Chat) and Klinefelter's (XXY). None of the human Trisomies is an improvement upon the original

86. Stryer, L.: The world's smallest molecular motor. *Biochemistry* 18.4.3, *in Creation* 31(4):23 (2009) Ref. 11.

87. Kaiser, G.E.: Chemiosmosis and ATP Synthesis at *http://student.ccbcmd.edu/~gkaiser/biotutorials/energy/atpsynthase_il.html*

genomic DNA code. Although *physical* duplication or fusion may be possible via transposons, the problem is the acquisition of new epigenomic control sequences (think information management software again) to run the new hybrid organism that has resulted from the duplication or fusion process.

If mitochondria were once discrete organisms that evolved the ADP – ATP mechanism and were subsequently incorporated into more complex cells as has been hypothesised, how did the first mitochondrion acquire its complex information programme? At the moment there are more questions than answers. Fowler and Kuebler note: "*Endosymbiosis...involves the loss of informational content. A mitochondrion contains less information than a free-living prokaryote; in fact, mitochondria could be considered degenerative prokaryotes. For example, mitochondrial ribosomes include proteins that are encoded within the nucleus of the host cell, while prokaryotic ribosomes are self-sufficient. Because of this, if one wants an explanation of how the biochemical process of the TCA cycle...and the electron transport cycle emerged—both integral parts of the mitochondria - endosymbiosis does not offer any new information.*" [88] (See Figure 4.1) It is important to state at this point that just because DNA structure and ATP synthase are identical in all living organisms, this does not prove that all organisms are descended from bacteria. However, it does demonstrate common information and probably a common information generator and suggests that both DNA and ATP synthase were present when life first appeared on Earth and that they have not evolved since. The Information generators are discussed in Chapter 9.

The paramount importance of the energy management system to living organisms can be best appreciated by the survival time of organisms when the energy supply system becomes incapacitated. Many spy films have highlighted how lethal cyanide pills are to human beings. Cyanide blocks the ATP-synthase turbines in mitochondria and the individual dies within minutes! Without a continuous supply of ATP all life dies very rapidly.

Genetic Information Management (The Epigenome)

The general function of DNA has been known for a long time but the actual size and precise activity of the genome has been a mystery until very recently. Fowler and Kuebler write: "*The human genome consists of approximately 25,000…genes. If each gene encoded for 10,000 base pairs, more than enough information to build a protein, it would require only 300 million bases to make up the human genome. In contrast, the actual human genome consists of 3 billion bases, at least 10 times more DNA than is apparently needed.*" [89] It is also known that protein-coding DNA accounts for only 3% to 5% of the genetic material in human cell nuclei. The apparently surplus DNA was assumed to be the equivalent of genetic fossils or errors which had never been eliminated and has previously been referred to as '*junk DNA*'. However, a recent undertaking called the ENCyclopaedia Of DNA Elements (ENCODE) Project, which is a collaboration of 80 research institutions across North America, Asia, Europe and Australia, now suggests that 93% of the genome is transcribed to RNA.[90] The term 'junk DNA' was independently coined by both David Comings and Susumu Ohno in 1972. One of the findings of the Encode project is that the previously ignored RNA which had been thought to be a simple messaging system is equally as important as the protein-coding DNA. John Matic, Professor of Molecular Biology at the University of Queensland, Australia writes: "*The majority of the human genome [is operated] by a hidden RNA*

88. Fowler, T.B. & Kuebler, D.: (2007) p.323.

89. Fowler, T. & Kuebler, D.: pp.106–7 & Footnote 42: "*Much of this extra DNA consists of highly repetitive sequences, which vary in size and are repeated thousands of times throughout the genome. Humans have a repetitive sequence called the Alu sequence, which is about 300 bases in length and is found at roughly a million sites in the human genome. This particular repetitive sequence, which does not encode any gene although it is 90% similar to the 7SL RNA gene, rather surprisingly accounts for one-tenth of the human genome [however]...The sequences in the Alu segments are not identical. This RNA Molecule is involved in forming a complex that recognises and binds to specific sequences on proteins that are destined to enter the endoplasmic reticulum.*" Therefore maybe it has a function in transport or waste management which has not been understood yet. If it is completely redundant, why is it so extensively conserved?

90. Sarfati, J.: *By Design*. Creation Book Publishers (2008) pp.216–7.

regulatory system that directs differentiation and development." [91] This makes sense in developmental terms. There is a medical aphorism which says: '*use it or lose it*'. If 95% or more of the genome was useless DNA then the organisms that could eliminate the bulk of their 'Junk DNA' would have had a survival advantage. The fact that what has been considered 'superfluous DNA' in the past, has been preserved for so long, suggests that it has a significant role; this includes the 'Junk DNA' which is now referred to as '*untranslated regions*' or UTR's.[90] The other astonishing finding of Encode is, as Matic explains: "*Genes appear to overlap each other and share stretches of molecular code. Moreover, genes and non-genetic DNA appear to work in close, if mysterious, conjunction and also seem to communicate across relatively vast genomic distances in ways not understood. Scientists have long understood that small numbers of RNAs act as 'dimmer knobs' that adjust the intensity of genes, thus regulating biological processes.*" [91] For the correct assembly of the cell components, information is required to specify the instructions for the assembly machines, the timing of cellular activities and placement of parts. Information is also required to control the production, so that there is enough product when it is needed and not too much when it is not needed, so that the cells are not clogged with useless raw materials or intermediates. This undertaking is equivalent to writing out every instruction for the manufacture of a computer but not just so that a human being can follow the manual and successfully build one but so that the computer will construct itself from a supply of raw materials with no outside intervention (and a computer is *much* simpler than a cell!) In the case of plants, the only allowable raw materials are carbon dioxide, water and sunlight with a few trace elements in a plot of dirt. Thus the construction manual is unavoidably enormous! If the vast majority of the genome that is not transcribed into proteins but only into RNA was part of the control processes of the epigenome, then the presence of so much 'excess' DNA begins to make sense.

Australian biologist, Alex Williams explains that the epigenome represents: "*Meta-information* [which] *is information about information...*[It] *is, by definition, entirely dependent upon the information that it relates to* [and] *without it, the basic information is of no use...Information and meta-information are mutually interdependent – each is useless without the other.*" [92] Thus, the epigenome as well as the protein-producing DNA are both essential when assembling organisms. The genes store the information required to manufacture the building blocks but the epigenome stores the information which explains how to arrange them. If we were considering houses instead of organisms, the genome would be specifying how to manufacture the bricks, floors, walls and roof materials. The epigenome is like the builder who specifies the way the building materials will be interconnected. So, in the same way that one builder using identical materials may construct many different houses, a slightly different epigenome can build a markedly different body plan using the same materials, the same laws, identical tools and similar construction methods. This explains why there can be 96%–98%, congruity in the genes between human beings and chimpanzees and yet the body plans are so radically different. It also explains why the human genome with 25,000 genes has only 28% more genes than significantly less complicated organisms such as sea anemones, yet there is vastly different complexity.[93] It is interesting to note, that in human beings only 25,000 genes produce around 100,000 proteins, so there is considerable overlap of the genetic material to produce a fourfold output from the genetic code. This extraordinary feat is achieved by overlapping the gene sequences which can be read in one direction on the first strand of DNA but *in the*

91. Matic, J.: DNA Unravelled, *in The Boston Globe* (24 Sep 2007).

92. Williams, A.: How Life Works. *Journal of Creation* 22(2):89–90 (2008).

93. Putnam NH, et al.: Sea anemone genome reveals ancestral eumetazoan gene repertoire and genomic organization. *Science* 317(5834):86–94 (2007).

reverse direction on the mirror image strand opposite! This is the equivalent to taking Richard Dawkins' cake recipe, looking at it in a mirror and being able to read the perfect instructions to make lasagne; then cutting and pasting the wording a couple of times and coming up with chicken curry, followed by a soufflé! Human beings have not achieved this, even with the aid of powerful computers. So, the idea that such sophisticated genomic architecture can occur by random chance, even if it is aided and abetted by blind Natural Selection, completely beggars belief. This epigenomic information pathway is shown in Figure 3.11.

As previously discussed, the epigenome controls the DNA through a number of mechanisms including methylation and it results in the differentiation of tissues so that some form muscle and others form bone.[94] This cell specialisation is not relevant in an amoeba but it becomes increasingly important as life forms become more complex. Epigenetics is the reason why identical twins have different finger prints. Further, epigenetics is the reason why the differences between identical twins increase as they age.[95,96] The intricacy of the epigenome is still being discovered but its effects can cross generations. Mouse studies have shown that mice fed low protein diets give birth to jumpy offspring, while lambs born to mothers with Vitamin B_{12} and folic acid deficiency were fatter, insulin-resistant and had higher blood pressure in middle age.[97]

The epigenome also explains how characteristics are able to skip generations and why variable penetrance[98] occurs, which are occurrences that cannot be understood by classical Mendelian genetics alone. Another significant problem that derives from an overlapping genetic code is that a change in one gene is likely to affect more than one protein or epigenetic control sequence. This effect is called pliotropy. It explains why certain inherited abnormalities are often seen together, which in medicine is called a 'syndrome'. If a mutation occurs in an overlapping region which controls several genes, although the mutation may be beneficial in one protein it is likely to be detrimental in the others and the end result will be a crippled organism, which is likely to be eliminated, so there is an inbuilt resistance to change.[99] If a mutation that causes collateral damage to the organism does occur, then a second compensatory mutation may possibly minimise the harm that was caused by the first mutation but this is rare (see Figure 4.11). Note also from Figure 3.11, that there is no mechanism for either the DNA or the derivative proteins to alter their epigenome, so complementary (random) mutations must occur synchronously in the epigenetic DNA and its matching genomic DNA. Mutations in the protein-producing genome without epigenetic control are likely to remain in the DNA doing nothing because there are no controls to activate them. If a DNA segment already under epigenetic control mutates, there is a requirement that the matching epigenetic DNA mutate in synergy with the mutated genomic DNA, to avoid aberrant results when the gene is expressed. Otherwise there will be a loss of synchrony between the epigenetic and genomic DNA. This means that successful mutation requires synchronous parallel random changes in both the genomic and epigenetic DNA. From the Neo-Darwinian frame of reference, the chance of this is vanishingly small.

Richard Dawkins points out in his book: *The Greatest Show on Earth*, that: although bacteria frequently transfer genetic material (via transposons), multicellular organisms rarely do, except during sexual reproduction and then, only within their own species.[100] If multi-celled organisms did swap DNA in the way that bacteria do, a different epigenome (new instructional information) would

94. Farmer, M. & Wieland, C.: *Clash over Origins: Creation vs. Evolution* Debate (DVD) Creation Ministries International, Brisbane, Australia.

95. Sarfati, J.: *By Design*. Creation Book Publishers (2008) pp.210–7.

96. Junk DNA now looks like Powerful regulator *in Science Daily* (24 April 2007).

97. Society for the study of reproduction-US conference, (July 2009) *in Australian Doctor* (31st July 2009).

98. In variable penetrance, an inherited characteristic is expressed to different degrees in succeeding generations.

99. Fowler, T.B. & Kuebler, D.: (2007) pp.180–2.

100. Dawkins, R.: *The Greatest Show on Earth*. Bantam Press, London (2009) p.33.

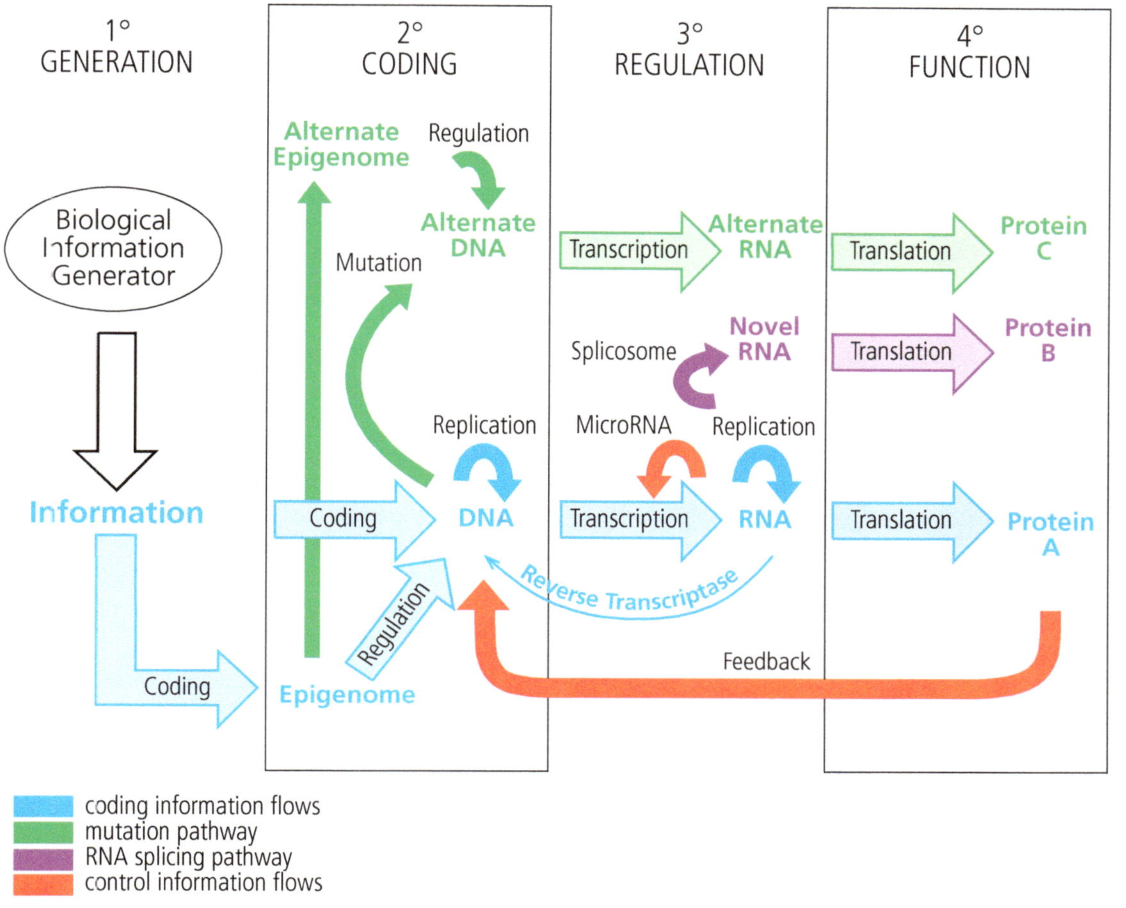

Figure 3.11: Information Flows in Cells illustrating the Epigenome.
It can be seen that the epigenome results in a double coding process: firstly for the DNA and secondly for the epigenome. Both are essential. To elucidate this, if you want to deliver a message in a letter, you first need the information inside the envelope but the letter cannot be delivered until you tell the post office how and where to deliver it (for example surface mail, air mail or sea mail along with the address). Thus, having the basic amino acid sequence information is not enough; the instruction manual is also required *simultaneously* if the information in the DNA is to be useful. It could be argued that the letter could have been written but that the delivery instructions were not produced for some period of time. However, there is a problem with this suggestion; as the information in the letter is useless until it has the delivery instructions, random variation over the generations will denature and corrupt the information stored in the letter unless natural selection regularly opts for that phenotype and hence, keeps the information 'pure'. So, under Neo-Darwinian Theory, the DNA and its epigenome must develop synchronously and *this 'co-evolution' of the genome and epigenome must occur at every step* as an organism mutates. Minute change, where there is no net increase in information (as measured in infobits), such as the change in a base pair, is unlikely to require an epigenomic change. Therefore, *Microevolution* is not a problem. However, to build the information required for macroevolutionary change inevitably requires complementary changes in the epigenome as well.

be required to drive the new DNA mix. Without this alternative epigenetic-genomic DNA combination, the new mutated DNA would be unrecognised by the existing epigenome which controls it. This is similar to having a new printer for your computer but trying to use the software from your old printer to make the new printer work. Unless you have the specific driver for the new printer, even though you have a fully functioning computer and a fully functioning printer that are correctly connected physically, the lack of the appropriate software to tell them how to tango, means that they both sit there and do nothing. In fact, this may be the ultimate definition of a species—the inability of the epigenome to coordinate the male and female DNA contributions to the next generation. It explains why, despite 96–98% of chimp and human DNA being identical, the two species cannot interbreed; the critical control software in the epigenome is species-specific. As already stated, it also explains why random variation with natural selection permits Microevolution but does not allow Macroevolution, because Macroevolution would require not only new information to specify the new genetic trait but also, the specifically written software 'driver' (complementary epigenetic DNA), *at the same time*. Michael Behe, in his book *The Edge of Evolution*, showed how rare an event this double simultaneous mutation is when he discussed the development of chloroquine resistance in the malarial parasite.[101] So it is possible that simultaneous complementary mutations may occur if only one base pair change is required in the genome and only one mutation is required in the epigenome. However, the sophistication of multi-celled organisms is much more likely to require strings of new epigenetic DNA information code, rather than single letter changes and this is another reason why macroevolutionary change is not observed. This will be discussed in greater depth in Chapter 5.

Thus, cells are self-organising, self-regulating, automated machines which auto-construct as well as auto-destruct if the processes go wrong. All these functions must be specified, not the least of which are the embryonic instructions, defining how the organism is to develop from a single cell. Embryogenesis is the most exquisitely beautiful and complex progression, with a symphony of hormones and molecules orchestrating the development of all the parts of a new human being. Yet, apart from the macroscopic anatomy that can be seen, such as eyes and fingers, each cell requires information specifying its internal transport mechanisms, communication systems, repair procedures, refuelling apparatus and waste removal. This microscopic world is discussed in Chapter 4.

Now, if you think that the answers to the previous questions were challenging, the best is yet to come! The most perplexing question of all is: How did the epigenome and the synchrony between the epigenome and the genomic DNA develop, because synchronous, double-coding is required? A relatively new branch of the Neo-Darwinian Synthesis has begun to address this difficult question. **Evo**lutionary **Devel**opmental Biology (Evo Devo) looks at small changes in the developmental genes (the epigenome), such as, the Hox genes and studies how tiny alterations in the control genes can cause large changes in body plans in only one or a few generations.[102] The Hox and other developmental genes are acting as genetic switches, which turn gene expression on and off. Sean Carroll, Professor of Molecular Biology and Genetics and an investigator with the Howard Hughes Medical Institute at the University of Wisconsin wrote in 2008: *"A large portion of the genes in the toolkit code for transcription* [control] *factors. These bind to regulatory DNA sequences (often called enhancers) to 'switch on' other genes. That's how genetic switches work."* [103] However, the Hox genes are only the control genes that pull the levers. There needs to be a master mechanism to tell the Hox gene when to become active, as well as a product producing gene attached at the other end of the lever. Whether there is an interplay of Hox-type genes in a feedback network that acts as the 'manager' or a master control gene, it can be seen that information is needed to code for the multiple layers of managing 'software'. Evo Devo involves the expression of existing body plan information, not the generation

101. Behe, M.: *The Edge of Evolution*. Free Press, NY (2007) p.63.

102. Fowler, T.B. & Kuebler, D.: (2007) pp.302–5.

103. Carroll, S.B., Prud'homme, B., Gompel, N.: Regulating evolution. *Scientific American* 296(5):38–45 (2008) *in Creation* 31(4):45 (2009).

of information for radically new body plans de novo. Mutations in the *epigenome* can have large effects but it is not simple. Fowler and Kuebler write: *"The main problem with such large changes, in terms of their usefulness for evolution, is that they are usually detrimental to the organism. In fact, none of the known Hox Drosophila mutants is considered to be a successful improvement over the wild-type form. This may be because of the limits of laboratory research, or it may be because large-scale changes that cause organisms to move away from their local optimal region of 'design space' normally are not advantageous."* [104] Jerry Coyne, Professor of Biology at the University of Chicago commented: *"these examples represent the loss of traits, rather than the origin of evolutionary novelties."* [105] US cognitive scientist, Daniel Dennett made this clear when he wrote: *"Large leaps sideways in a fitness landscape will almost never be to your benefit; wherever you currently find yourself, you are where you are because this has been a good region of Design Space for your ancestors — you are near the top of some peak or other in the space—so, the bigger the step you take (jumping randomly, of course), the more likely you are to jump off a cliff."* [106] (See Figure 3.12)

104. Fowler, T.B. & Kuebler, D.: (2007) p.306

105. Coyne, J.A.: Switching on Evolution – how does evo-devo explain the huge diversity of life on Earth? Nature 435(7045):1029–1030 (2005) *in Creation* 31(4):44 (2009).

106 Dennett, D.C.: *Darwin's Dangerous Idea: Evolution and the Meanings of Life.* Simon & Shuster, NY (1995) p.288 *in* Fowler, T.B. & Kuebler, D.: (2007) p.306.

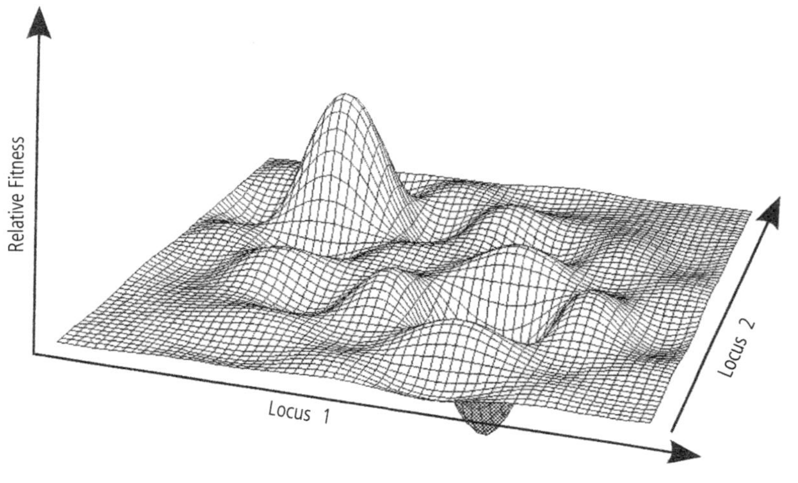

Figure 3.12: A Fitness Landscape
Fitness peaks and valleys are clearly visible; but as the environment is changing continually, the fitness landscape for each individual organism is constantly changing too, albeit very slowly and the changes are usually small. Beneficial gene combinations today may be less so tomorrow, which is why some species have become extinct.[107] *[Reproduced from Fowler, T.B. & Kuebler, D.: The Evolution Controversy. Baker Academic, a division of Baker Publishing Group (2007). Used by permission.]*

If the epigenome alone is not complicated enough, there are some who see even more sophisticated complexity. John Matic has proposed a *'genetic network',* which implies a further extension of the integration of information.[108]

The most significant implication of Evo Devo though, is that, because the developmental genes are almost identical from snails through to humans, the principle of common descent in evolutionary theory predicts that there must have been a shared common ancestor prior to the appearance of snails, which has been dubbed: 'Urbilateria' (derived from 'bilateral' or the universal ancestor of all organisms with twofold symmetry). This hypothesised organism must have contained all the Hox genes for all the higher organisms which were yet to evolve. Sean Carroll wrote: *"The genetic potential was in place for at least 50 million years, and probably a fair bit longer, before large, complex forms emerged."* [109] These genes include the programming for the development of eyes, heart and limbs which Urbilateria did not possess and whose descendants would not use for millions of years to come. Carroll writes: *"It is intriguing to ponder just what so many genes were doing in Urbilateria."* [110] The answer is that no mechanism exists in current Neo-

107. Fowler, T.B. & Kuebler, D.: (2007) p.68.

108. Matic, J.S.: *Non-coding RNAs: The architects of eukaryotic complexity.* EMBO Reports 2:986–991 (November, 2001) *in* Sarfati, J.: *By Design.* Creation Book Publishers (2008) p.214.

109. Carroll, S.B.: *Endless Forms Most Beautiful: The New Science of Evo Devo.* W.W. Norton & Company, New York (2005) p.139, Ref. 4 *in Creation* 31(4):44 (2009).

110. *Ibid.* p.144 *in Creation* 31(4):44 (2009).

Darwinian Theory for such sophisticated pre-emptory information development and under the existing paradigm, this deduction is impossible. The only possible conclusion from this information, is that Evo Devo cannot be the explanation for the formation of the epigenome, leaving the Neo-Darwinian Theory without an answer to this problem too.

The Minimum Information requirement for the First Cell

Therefore, in summary, it is very clear that the amount of intrically integrated information in an ordinary cell is immense and I am sure Darwin would have been amazed at what his 'protoplasm' contained. However, despite the mind-boggling amount of information, all the above detail can be simplified to one of two processes for the first cell. Either the *replicator* came first or the *metabolic pathways* arrived first. Richard Dawkins championed the Replicator-First model in 1976, in his book *The Selfish Gene*, when he wrote: *"At some point a particularly remarkable molecule was formed by accident. We will call it the Replicator. It may not have been the biggest or most complex molecule around, but it had the extraordinary property of being able to create copies of itself."* [111] John Lennox explains the alternative view on the formation of the first cell when he writes: *"Robert Shapiro and others suggest a 'metabolism' or 'small molecule' first scenario for the origin of life, that is, one that does not initially contain a mechanism for heredity and therefore involves small molecules rather than large information-bearing molecules like DNA or RNA."* [112] This dichotomy is very elegant but from an information standpoint, it is clear that *both must be first*! A replicator's function is to remember and reproduce the metabolic information stored in that initial cell. However, without the metabolic mechanisms, there is nothing for the replicator to do. In contrast, the metabolic mechanisms must have a replicator to reproduce their information from the first generation or there is no second generation. Thus, both replicators and mechanisms had to be present at the start and not in small measure but in vast array; in fact the first cell had to be completely competent in its first generation! The minimal requirements for the existence of the first cell are summarised in Figure 3.13.

From Figure 3.13 it can be seen, that the piecemeal 'learning' in biological systems dictated by random variation with natural selection, is completely inadequate as an explanation for the integrated information required to form the first cell, because all the functions listed, including the software (black arrows) and hardware (white arrows), needed to be present when that first cell 'sparked' into life. When constructing the first cell for the first time, the bit by bit approach does not work. This falsifies the Neo-Darwinian dictum of gradualism: *'one minute step at a time'*. Life is an all or nothing entity, as it is impossible to be 'partly' alive. You are either alive or you are not.[113]

This question of how did the first cell 'spark' into existence is the conundrum which lies at the heart of the origin of life. The ultimate test for this process was the Miller-Urey experiment of 1953, which was initially hailed as 'proof' that life began in the 'primordial soup' but subsequent research has shown that assumptions on atmospheric composition were wrong and even the modest experimental results were optimistic. Biochemist, Professor Trevor Palmer wrote: *"Robert Shapiro, Professor of Chemistry at New York University, pointed out in his 1988 book,* Origins, *that, contrary to what is sometimes stated, there was no synthesis of nucleotides and, indeed, only two of the component bases (adenine and guanine)*

111. Dawkins, R.: *The Selfish Gene* (1976) *in* Lennox, J.C.: *God's Undertaker. Has Science Buried God?* Second Ed. Lion Books, Oxford (2009) pp.125–6.

112. Shapiro, R.: A simpler origin for life. *Scientific American*, pp. 24–31 (25 June 2007) *in* Lennox, J.C.: *God's Undertaker. Has Science Buried God?* (2nd Ed.) Lion Books, Oxford (2009) p.131.

113. In this context, to be 'alive' can be thought of as the ability to reproduce. The difficulty surrounding the definition of: 'What is life?' is addressed in Chapter 10.

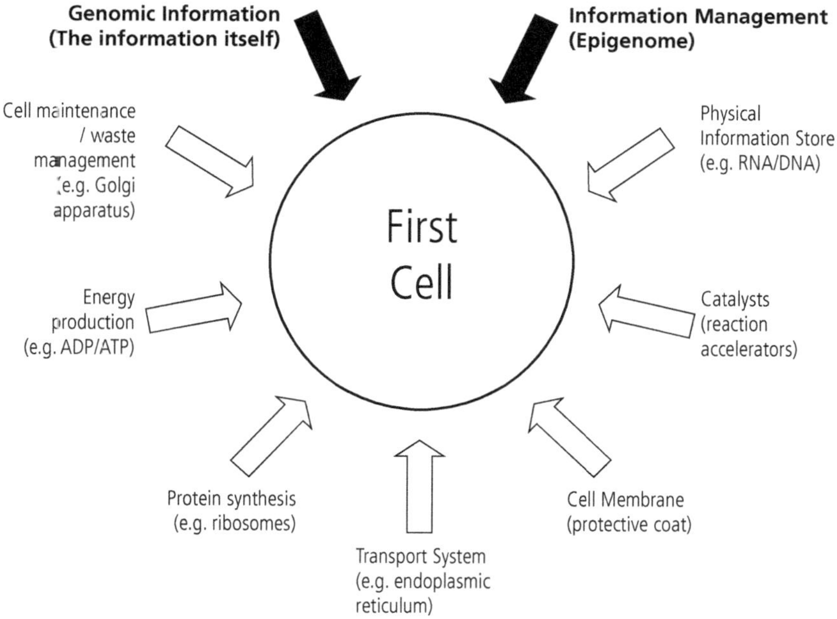

Figure 3.13: The Minimum Information requirement for the Formation of the First Cell Material components of the cell have white arrows and non-material components have black.

were produced to any significant extent. Also, only two of the required 20 amino acids were formed in appreciable amounts, which was reduced to one when the experiment was repeated later with a less-reducing atmosphere, reflecting changing views about the primeval Earth." [114] The evolutionary paradigm falls apart at every turn.

Conclusion

Each of us is limited by the information available at our point in history. Charles Darwin described and explained what he saw within the confines of the knowledge-base and technology available and the Theory of Evolution appeared logical to him at that time. The Dutchman, Antonie van Leeuwenhoek (1632–1723) was the first to describe the appearance of a single cell under the microscope in 1676. The technology of microscopes remained essentially unchanged for the next 250 years, so Darwin's view of the cell was essentially unchanged from Leeuwenhoek's. German scientist, Ernst Haeckel, was the pre-eminent microscopist of Darwin's time and Darwin was undoubtedly influenced by him. Haeckel defined the paradigm when he looked down his microscope at a single cell, which he described as *"a simple little lump of albuminous combination of carbon"* [115] With the technology of the time, it was considered the best hypothesis to explain the information available.

As a consequence, life was imagined as many simple blobs of protoplasm multiplied many millions of times to create living organisms[116] but with our current knowledge of biochemical pathways and our expanded understanding of the modes of transmission of information in living systems, there needs to be another explanation for the emergence of the first cell. From the above discussion, it becomes obvious that the original notion of a *'simple cell'* was a misnomer. The *'simple cell'* does not exist and it never could have. RNA models are not a viable alternative and reverse engineering of DNA from cytoplasmic proteins

114. Palmer, T.: *Controversy - Catastrophism and Evolution.* Kluwer Academic, New York p.266 *in* Bendewald, J.: *Evolution Shot Full of Holes.* Evidence Press, Madison, Wisconsin (2004) p.89.

115. Farley, J.: *The Spontaneous Generation Controversy from Descartes to Oparin.* John Hopkins University Press, Baltimore (1979) p.73 *in* Behe, M.J.: *Darwin's Black Box: The Biochemical Challenge to Evolution.* Free Press, New York (2006) p.24.

116. Meyer, S.C.: DNA and other Designs. *First Things* 102 (2000).

does not happen. Random variation with natural selection may explain the adaptation of an existing species to its environment (such as the development of bacterial resistance to antibiotics) but it cannot explain the origin of the first cell. Francis Crick commented: "*What is so frustrating for our present purpose is that it seems almost impossible to give any numerical value to the probability of what seems a rather unlikely sequence of events...an honest man armed with all the knowledge available to us now, could only state that in some sense, the origin of life appears at the moment to be almost a miracle, so many are the conditions which would have had to have been satisfied to get it going.*" [1] This paper was written in 1981 but the intervening quarter of a century has only magnified the problem as the previously unimaginable complexity of life has progressively been uncovered and, in particular, the discovery of the mechanical nano-world inside every living cell. Professor Chandra Wickramasinghe, Director of the Cardiff Centre for Astrobiology wrote in 2000: "*The emergence of life from a primordial soup on the Earth is merely an article of faith that scientists are finding difficult to shed. There is no experimental evidence to support this at the present time. Indeed all attempts to create life from non-life, starting from Pasteur, have been unsuccessful.*" [117] Edgar Andrews quotes the geneticist and evolutionary biologist John Haldane (1892–1964) who wrote: "*If the minimal organism involves not only the code for its one or more proteins, but also 20 types of soluble RNA, one for each amino-acid, and the equivalent of ribosomal RNA, our descendants may be able to make one, but we must give up the idea that such an organism could have been produced in the past, except by a similar pre-existing organism or by an agent, natural or supernatural, at least as intelligent as ourselves, and with a good deal more knowledge.*" [118] Meaningful information needed to be introduced into the system, either in the form of DNA or transferred to the DNA, for the first cell and current biochemical evolutionary theory has not even come close to explaining this.

117. Britt, R., Panspermia Q and A: leading proponent Chandra Wickramasinghe, (27 October 2000) *in* Matthews, M.: Life Discovered in Space. *at* http://creation.com/life-discovered-in-space

118. Haldane, J.B.S.: *Data needed for a blueprint of the first organism.* published posthumously *in* S.W. Fox (Ed.) *The origins of prebiological systems and of their molecular matrices.* Academic Press, New York (1965) p.12 *in* Andrews, E.: (2009) p.174.

4

Information and the First Organism

...what appears to me as supremely significant is that a combination capable of giving rise to life and mind should exist at all.[1]

Christian de Duve

If the premise that Darwinian Evolution is true is to be accepted, then once the first cell came into existence, no matter how simple it was, it must have developed a mechanism to reproduce before it died or all the information that it had acquired would have been lost. Once that was achieved, it then needed to incorporate increasing amounts of information so that it could grow and evolve into the first *complex* organism. After cells become grouped together, there is a need for specialisation with an escalation in the complexity of their arrangements, which requires increasing amounts of information, to manage the way that the individual cells co-ordinate with each other. The comparison between a single-celled organism, such as an amoeba and a multi-celled organism, such as a rabbit, is similar to the comparison between a blacksmith who works by himself and a massive steelworks with thousands of workers, which needs a sophisticated management structure. So, in this chapter we will be investigating the additional information that would have been required for the transition from a single-celled organism to the formation of the first multi-cellular creature.

Requirements for living organisms

To summarise our journey to date, in the previous chapter on the appearance of the first cell, we found that all living cells need:

a. A coding system to store information so that gains are not lost;
b. Homeostasis mechanisms to prevent the organism disintegrating again because the Law of Entropy decrees that it will fragment if it is not actively maintained; and
c. An energy management system, as opposing the Law of Entropy is only possible by injecting energy into the system.

The fourth requirement for the ongoing existence of a cell, as stated above, is *reproducibility* or the ability to make copies of itself. Initially, we will

1. de Duve, C.: *Life Evolving.* Oxford University Press, NY (2002) p.199 *in* Lennox, J.C.: *God's Undertaker. Has Science Buried God?* Lion Books, Oxford (2007) p.74.

limit discussion to the simpler process of asexual reproduction or mitosis. Sexual reproduction or meiosis is investigated in Chapter 5.

In living things today, reproduction does not happen by chance but is a rigorously controlled and very detailed process. In prokaryotes, such as bacteria, where there is asexual reproduction, special apparatus (such as centromeres to separate the contents of the daughter cells) and enzymes (such as DNA polymerase to unzip the DNA for replication) are necessary and they also would have had to evolve. The biggest problem here is that instead of millions of years to achieve this ability, the first cell must successfully accomplish this intricate and complex task within its own life time, which is totally outside the capability of the Neo-Darwinian model as it stands today. However, centromeres and DNA polymerase are only the tip of the iceberg. Michael Denton wrote: *"Although the tiniest bacterial cells are incredibly small, weighing less than 10 $^{-12}$ gms, each is in effect a veritable micro-miniaturised factory containing thousands of exquisitely designed pieces of intricate machinery..."* [2] To better understand Denton's statement, we will begin by looking at the machinery of reproduction.

Reproduction (Mitosis)

Although cell division by mitosis is the simpler form of replication, it still involves a myriad of intricate steps which must be followed precisely to achieve successful replication. It begins by unravelling, then splitting before duplicating the DNA. Centromere formation follows and in eukaryotes, such as plants and animals, dissolution of the nuclear membrane is necessary before cell division is completed. Finally, the nuclear membrane is reconstructed for both daughter cells.

DNA strands are very long and thin and to package them into the cell without tangling requires careful folding. The DNA is wound around special proteins called histones and nucleosomes. Human DNA is nearly two metres long but only twenty nanometres wide. This is so fine that a single strand of DNA can only be seen with the most powerful microscopes but if it was magnified, so that the DNA strand was as thick as fishing line, it would be almost 200 km (125 miles) long. Even so, it is packed so efficiently that the fishing line would fit inside a basketball. To achieve this feat, the cell has winding machines that wind and unwind the strands of DNA at 8000 rpm.[3,4]

As discussed above, reproduction must have been one of the first functions that the first cell achieved yet cell division and reproduction is one of the most complex operations discovered. In addition to numerous activities within the cell, which are required to be executed in precise order with exquisite precision, involving vast amounts of genetic material, there is also the complex cell machinery necessary for the process to operate successfully. We are still learning about the many machines required for successful cell division. Apart from the histones and nucleosomes, replication requires another small motor protein called helicase which splits the DNA strands to allow copying. The helicase requires ATP to run, so were the helicase and ATP present first or the DNA? Replication requires all three to be present at once.[5] Cell division also requires linear motors (kinesins) to tractor the cells apart to split them[6] but if such complex machinery is required so early in the development of cells, where and when did the information generation take place to make these machines?

And there is more! If you are not yet convinced there is a fundamental information problem in the formation of the first organism, bear in mind that the

2. Denton, M.: *Evolution: A Theory in Crisis.* Adler & Adler, Chevy Chase, Maryland (1986) p.250.

3. Missler, C.: *Cosmic Codes.* Koinonia House, Coeur d'Alene, Idaho (2004) pp.322–4.

4. Sarfati, J.: *By Design.* Creation Book Publishers (2008) p.169.

5. *Ibid.* p.166–7.

6. *Ibid.* p.140.

DNA needed to develop start codes and stop codes[7] to specify which bits to copy plus RNA polymerase to perform the copying function;[8] and this all happened at the *beginning* of the development of life on Earth. Further, it is not just the molecular machines that have great antiquity but also the physical DNA/ RNA code-carrier itself. As Fowler and Kuebler note: *"...the same codon corresponds to the same amino acid in all organisms from bacteria to maple trees to humans... The main exceptions to this are found in the coding of genes in the mitochondrial genome. However, while mitochondria employ a few codons that differ from the nuclear genome, the majority of the codons are identical. Similar exceptions occur in single-cell protozoans known as ciliates."* [9] From these observations, it would appear that the standard code, which is seen in almost all life on Earth today, was there at the very beginning, with only minor variations described.[10]

There are also significant problems in developing the epigenetic coding to initiate the replication process but even if we ignore that, to develop a functional centromere alone would have taken at least 100 steps and probably more. Each step needed to maintain a functional configuration of the molecules in the evolving nano-machine, so that during the time it took to develop these 100 stages, natural selection did not remove non-productive bits (assuming it could have been spread over a number of generations). However, finding 100 intermediate molecules, all of which are useful for other purposes, defies belief and the centromeres are only one of more than fifty components required *simultaneously* for cell division. Instead, all this information needs to be pre-programmed in the DNA or RNA as, once again, it simply cannot form by evolution from its own functional proteins, as this would break the Law of Coding, as discussed in Chapter 3. Thus, there must be some other process to build complexity (increase information) within the cell to this level.

Finally, one new function and possibly two might evolve *simultaneously* but three or more are impossible (see 'Co-dependant Information' at the end of this chapter) and fifty or more is beyond contemplation. The chance of this happening spontaneously was colourfully described by British physicist, Sir Fred Hoyle (1915–2001) when he wrote: *"Imagine 10^{50} blind persons each with a scrambled Rubik cube, and try to conceive of the chance of them all simultaneously arriving at the solved form. You then have the chance of arriving by random shuffling of just one of the many biopolymers on which life depends."* [11] If I was buying a ticket in a lottery, this would not be it!

Cell Structure

It is now time to move on to the detailed information within the cell. We will assume that the insurmountable problems observed above have been somehow overcome and that our cell survived its first cell division and is becoming more complex. John Lennox gives us some appreciation of the inner workings of a cell when he writes: *"It is hard for us to get any kind of picture of the seething, dizzying complex activity that occurs inside a living cell, which contains within its lipid membrane maybe 100 million proteins of 20,000 different types and yet the whole cell is so tiny that a couple of hundred could be placed on the dot in this letter 'i'."* [12] Within the cell, with so many proteins in such a confined space, the consistency of the intracellular fluid has been estimated to be similar to the density and viscosity of yoghurt.[13] Yet, the activity is frenetic, with transport links (the cytoplasmic reticulum) being absorbed and rebuilt constantly as materials are moved around

7. Missler, C.: (2004) pp.321–2.

8. Sarfati, J.: (2008) p.166.

9. Fowler, T.B. & Kuebler, D.: *The Evolution Controversy.* Baker Academic, Grand Rapids (2007) p.101 & footnote 33.

10. There are 18 different codes described to date but they affect only a tiny group of organisms. How they developed is yet to be adequately explained. (See Meyer, S.C. et al: *Explore Evolution.* Hill House Publishers, London (2009) p.58.)

11. Hoyle, F.: The Big Bang in Astronomy. *New Scientist* 92:521–7 (1981) *in* Sarfati, J.: (2008) p.159.

12. Lennox, J.C.: *God's Undertaker. Has Science Buried God?* Lion Books, Oxford (2007) p.117.

13. Williams, A.: Mutations: evolution's engine becomes evolution's end! *Journal of Creation* 22(2):61 (2008).

within the cell. In Figure 3.13 we listed the minimal requirements for the formation of the first cell. Table 4.1 below expands that list as we examine the essential elements in a *typical* cell.

In plants and bacteria, the structure is provided by the rigid cell wall but in animals the structure is provided by the cytoskeleton made of micro-tubules. Ribosomes are like the card readers on old mainframe computers which required punch cards to operate. The ribosome reads the RNA sequence and translates it into the protein requested by the cell. Mitochondria are micro-power plants that produce energy for the cell, as discussed in Chapter 3 and kinesin is a motor protein which provides transport in all living things. The Golgi apparatus is an organelle that is involved with waste removal.

All these machines are specified by information, just as the parts of a computer are. Prior to the production of a new computer, the information is assembled and the plans are then sent to the factory for manufacture. The design work is done first, rather than the factory foreman being told to 'work it out' as he goes along, which is the currently accepted Neo-Darwinian theory for the formation of the first cell and first organism. In response to this hypothesis, Sir Fred Hoyle retorted: *"The notion that not only the biopolymers but the operating program of a living cell could be arrived at by chance in a primordial organic soup here on the Earth is evidently nonsense of a high order."* [11]

Requirement	Cellular Elements
Cellular Memory (hardware)	DNA/RNA
Genomic information (software)	Information within the DNA/RNA that specifies the cell
Information management	Epigenome
Machinery to read the memory	Ribosomes
Energy	Mitochondria (producing ATP)
Protein folding machinery	Chaperones
Cell structure	Cell wall (plants/bacteria) /cytoskeleton (animals)
Transport	Kinesin and cytoplasmic reticulum
Protection	Cell wall (bacteria/plants) or membrane (animals)
Waste removal	Golgi apparatus
Import/export control	Port assembly
Reproduction	Centromeres (amongst many others)
Reaction accelerators	Catalysts

Table 4.1: Requirements for a Typical Eukaryotic Cell
Most cells require all of the functions listed. There are more functions in many cells, especially specialised cells, such as actin and myosin in muscle to generate contraction.

Protein Synthesis

As already described in Chapter 1, the cell functions like a busy factory or a bustling town where the workers are proteins. To produce proteins, the epigenome in the nucleus of the cell decides whether a particular protein is required. This process is not yet understood but it may be through a feedback loop. Information in the epigenome sends a message via RNA to the genome, which then copies a section of the master DNA blueprint to create messenger RNA (mRNA). The mRNA is transported to the cytoplasm (effectively from the manager's office where the blueprints are stored, onto the factory floor) where the protein manufacturing machines called ribosomes take the mRNA information and translate it into an amino acid sequence. Ribosomes are universal to all living organisms.[14] In 1999, Roger Garrett, Professor of Molecular Biology

14. Smith, C.: *Codes and Creation.* (CD) Creation Ministries International, Brisbane, Australia.

at the University of Copenhagen wrote: *"The ribosome...is probably the most sophisticated machine ever made. All of its components are active and moving, and it is environmentally friendly, producing only GDP and phosphate."* [15] In eukaryotes (nucleated cells), ribosomes contain 73 different proteins with four ribosomal RNAs which are integrated into one functioning unit. Even in simpler bacterial cells which have no nuclear membrane, (prokaryotes), the bacterial ribosome is constructed from 50 different proteins with three different ribosomal RNAs.[5]

From the above it can be seen that the ribosome is integral to the functioning of every cell and would appear to have been necessary for the survival of the first cell too, as functioning proteins cannot be built without them and every living cell contains them. The conundrum is that: as the ribosome is one of the most complex nano-machines ever discovered (although chlorophyll and ATP-synthase have similar degrees of complexity), how did the first cell acquire the information to build such complex machinery, which then appears to have remained unchanged for the next 500 million years? This does not sound like Neo-Darwinian Evolution!

Then there is the problem with the chaperones. The mRNA determines the sequence of the amino acids in the protein and the ribosome arranges them but the newly constructed protein is unusable until it undergoes further processing, because proteins will only work when folded into a specific three-dimensional shape. Thus, once the protein sequence is complete, the amino acid chain is transferred to a chaperone, a nano-machine,[16] where the protein is precisely folded to create a unique shape which gives the protein its function. How does the chaperone know which way to fold each protein? There must be more information, either built into the chaperone or intrinsic to the molecules being folded, which specifies what the chaperone needs to do. This creates a problem regarding which came first, as, if the DNA/protein sequence evolved prior to the chaperone, the protein is likely to fold up into the wrong configuration and be useless but if the chaperone came first, it would have had no function and hence no selection pressure to evolve under Darwinian rules.

Finally, we have traced protein manufacture from the blueprint to the factory floor but that is not the end of the manufacturing process. A lot of the molecules in cells are very complex and the raw proteins need further post-production processing, by linking them with sugar or lipid molecules before they can be put to their final use. In the cell, this is performed by the endoplasmic reticulum (sometimes called the cytoplasmic reticulum), which was first described in 1945 by Keith R. Porter, Albert Claude, and Ernest F. Fullam.[17,18] Endoplasmic reticulum is composed of interconnected tubules and enclosed spaces called cisternae, where different molecules are manufactured before being transported to where they are needed.

Internal Cellular Transport

As previously stated, the intracellular environment is considered to have the viscosity of yoghurt, so it would take a long time for chemicals to move around the cell if they had to get there by random diffusion. Edgar Meyhöfer, Associate-Professor of Biomedical Engineering at the University of Michigan describes how every living plant and animal cell has an internal highway system of microtubules on which a transport molecule or motor protein called kinesin travels. The microtubules create a network which collectively forms the

15. Garrett, R.A.: Mechanics of the Ribosome. *Nature* 400(6747):811–2 (1999) *in* Sarfati, J.: (2008) p.167.

16. Sarfati, J.: *Refuting Evolution 2*. Answers in Genesis, Acacia Ridge, Qld, Australia. (2002) p.108.

17. Aubisson, P. et al: *Biology in Context—The Spectrum of Life*. (2nd Ed.) Oxford Univ. Press, Melbourne (2004).

18. Porter, K.R., Claude, A. & Fullam, E.F.: A Study of Tissue Culture Cells by Electron Microscopy. Methods and Preliminary Observations. *J. Exp. Med.* 81:233–246 (1945) *in Proceedings of the American Philosophical Society* 150(4): 687–696 (2006).

cytoskeleton. It is similar to the road network in your town and the kinesin is like a robotic truck which literally walks along the 'roads'. Unlike the roads that you are used to however, the microtubules are made as they are required and then reabsorbed and recycled. And instead of burning petrol, kinesin uses ATP for fuel (one ATP molecule per eight nanometre step).[19] Although kinesin and the microtubules are essential cell components, they can work against the cell if hijacked by invading pathogens, such as cholera, which depends upon the kinesin to infect cells.

Cell Maintenance

Just as towns need factories to manufacture goods, they also need a post office and waste management systems. We have seen that the ribosomes are the 'factories' in our cells and Professor Emeritus Holmes Rolston III of Colorado State University introduces the need for cell maintenance systems when he writes: *"In wild nature, in view of entropy, only minimal levels of complexity can arise automatically. More complex biomolecules will not be reproduced often and reliably enough by spontaneous assembly. Advanced levels will require maintenance of the formed and functional structures and processes...Maintenance will require directed assembly and repair because otherwise the spontaneous breakdown rate will overwhelm the spontaneous construction rate. Spontaneous formation is a starter, but if you have to start from scratch each time, you do not get far."* [20] Ribosomes produce new proteins to replace damaged ones, while waste removal is performed by the Golgi apparatus (sometimes referred to as the Golgi body or Golgi complex). It is an intracellular organelle that acts as a packaging plant and transhipment point, where the packages of finished macromolecules are dispatched to their final destination. They are packaged in vesicles (small fluid-filled chambers like liquid bubbles) which can be used to export waste products as well. Vesicles that are carrying molecules marked for export, are transported to the cell membrane and fuse with it, emptying their contents outside the cell. The Golgi apparatus represents yet another layer of complexity in our already fantastically complex, 'simple' cell and is found in most eukaryotic (animal and plant) cells. On electron micrographs they appear as multi-layered stacks of membrane and were first described by the Italian physician, Camillo Golgi in 1898. The Golgi apparatus typically has four to eight folds in each stack which is often located peripherally in the cell.[21]

In addition, there are proteasomes, which are large protein molecules in all eukaryotes and archaia, which break down and remove excess or damaged proteins. They are like the garbage trucks of the town, which keep the neighbourhood clean and tidy.

Complex Biochemistry and Catalysts

Organisms also depend on interlocking biochemical information systems, which have similarities to irreducibly complex organelles, such as the ribosome. To delve into the intricate biochemistry of living cells would require a number of large textbooks on biology and organic chemistry but for the purposes of this investigation, it suffices to look at just two examples. The first is catalysts and the second is the Citric Acid Cycle. There are many other possible choices but these

19. Cai, D., Verhey, K.J. & Meyhöfer, E. Tracking single Kinesin molecules in the cytoplasm of mammalian cells. *Biophysics Journal* 92(12):4137–44 (2007).

20. Rolston, H.III: Care on Earth: generating informed concern, *in* Davies, P. &Gregersen, N.H.: *Information and the Nature of Reality: from Physics to Metaphysics.* Cambridge University press (2010) p.209.

21. Fabene, P.F. & Bentivoglio, M.: 1898–1998: Camillo Golgi and 'the Golgi': one hundred years of terminological clones. *Brain Research Bulletin* 47(3):195–8 (1998) *at* http://en.wikipedia.org/wiki/Golgi_apparatus.

are two instances which affect every living cell and are typical of the problems encountered in contemplating the formation of the first complex cellular organism.

Each living cell is filled with many, many chemicals. When I was at school, I could never understand why all the chemicals in the cell did not inappropriately interact with each other, causing catastrophic agglutination and the death of the cell. Richard Dawkins succinctly explains: *"Many reactions in cell chemistry are so slow that, without the right enzyme, for practical purposes they don't occur at all. But with the right enzyme that happens very fast, and can churn out products in bulk."* [22] These enzymes or catalysts are special facilitator substances, which may be a chemical or a protein that can speed up the chemical reaction between two other substances *"by as much as a billion or a trillion times"* [22] and yet they are not consumed in the process. Biochemist Jonathan Sarfati elaborates: *"Catalysts do not affect the equilibrium, but only the* rate *at which the equilibrium is reached. They work by lowering the activation energy, which means decreasing the energy of a transitional state or reaction intermediate."* [23] As a result, the catalyst can repeat the procedure many times and thus a tiny amount goes a very long way. Each reaction requires its own *unique* catalyst so large numbers of catalysts are needed but each in only tiny quantity. In a normal cell, the supply of catalysts is tightly controlled, so that in the absence of the catalysts most cellular chemicals do not interact at all but if they do, the amount of reaction is negligible and hence the cell is safe from runaway chemical catastrophe. Thus, enzymes (catalysts) are vitally important to coerce biological processes within the cell to occur within realistic timeframes. Therefore, rather than suffering from runaway chemical reactions, the first cell must have sat around for literally aeons, waiting for its chemical reactions to occur. In practice, the cells without catalysts would have died before the necessary minimal set of catalysts could have evolved, unless they somehow inherited a genetic (information) legacy of a working cellular environment, which is what happens today. (This is the basis of Pasteur's Law as discussed in Chapter 1, where we discovered that: 'only life begets life'). Once again, an information injection would have been required before the system could have sparked into life.

Dawkins' suggestion that low efficiency catalysts would have sufficed to begin the first cellular reactions, before natural selection made the processes more efficient over time, ignores the necessity for dozens, if not hundreds of catalysts to be present right from the start, even in the most minimal cell and the lower the efficiency of the initial catalysts, the more important it would have been to have *all the catalysts present* because reactions in general would have been slower. Once again, there is a requirement for vast amounts of information in the very first cell. In Dawkins own words: *"details matter"*. [24] It was therefore necessary for the simultaneous development of multiple chemical accelerators, even in the first cell, which guarantees that all efforts to find new life in the 'primordial soup' will be unsuccessful. Without prior injection of information, as required by the Law of Coding, *nothing* can happen. This applies even without taking into account any feedback or epigenetic control of the catalysts that cells have today.

The second illustration of complex biochemistry is the Citric Acid or Krebs Cycle which was described by the German-born physician and biochemist Hans Krebs (1900–1981), in 1937, for which he shared the Nobel Prize in Medicine in 1953. It forms part of the biochemical chain that utilises sugar within the cell and forms a closed loop system.[25,26] Figure 4.1 summarises some of the intricate biochemistry involved in harnessing energy to run the cell and although

22. Dawkins, R.: *The Greatest Show on Earth*. Bantam Press, London (2009) p.238.

23. Sarfati, J.: *By Design*. Creation Book Publishers (2008) p.157 footnote 34.

24. Dawkins, R.: (2009) p.216.

25. Fowler, T.B. & Kuebler, D.: (2007) pp.61–5.

26. Lennox, J.C.: *God's Undertaker. Has Science Buried God?* (2nd Ed.) Lion Books, Oxford (2009) pp.131–2.

the detail is vitally important for the efficient functioning of living cells, at this point in our journey, the concept is more important than the detail.

Under Darwinian evolutionary rules, each of these biochemical links or relationships had to evolve unplanned and there are eight organic molecules in the Citric Acid Cycle, which folds back on itself to form a closed loop, thus allowing it to 'cycle' like a rotary engine. Each of the steps requires its own catalyst to facilitate the movement of the molecules around the ring. US mathematician William Dembski has calculated the probability of just one protein such as these happening at random and the probabilities are unbelievably small. He states: *"Even the formation of a single functional protein of 100 to 200 amino acids has a probability of 10^{-100}"* [27] and the Citric Acid Cycle represents just one small biochemical sequence within the vast biochemical array found within every living cell. Even after establishing a functioning Citric Acid Cycle, there are still dozens of other biochemical pathways that are essential for the survival of even the simplest cell, leaving aside the structure of the cell itself, which also had to be constructed.

Fusion

To overcome the difficulties encountered in the formation of the first organism, a number of solutions have been suggested, such as the endosymbiosis mentioned in Chapter 3. It has been noted that mitochondria have slight coding differences in their DNA and it has been suggested that they were originally autonomous bacteria which came into a symbiotic relationship with another larger, more complex cell. It is possible to theorise that two simpler entities or modules blended together to give an instantly functioning and reproducing cell but the sheer complexity of the process of reproduction makes this seem extremely unlikely, bordering on the impossible. This is analogous to taking a normally functioning car and a normally functioning boat and by putting them in a workshop and fusing them together you can create an amphibious vehicle. While engineering this process, one must remember that the factory technicians who are carrying out the 'army-duck' conversion are blinded and have not got the slightest idea what they are doing. Even if the cytoplasmic structures and their coding came ready-made from another cell or entity, the information imbedded in the coding still had to have been specified at some earlier stage and the coding in the epigenome to produce an army-duck rather than a car and a boat will need to be adjusted for it to work in this new situation. Expecting the blind engineers to fuse the cytoplasmic structures and simultaneously retro-write the DNA/RNA as well as any epigenetics for the management systems to control the whole process, once again, defies belief; especially as we know that it breaks the Law of Coding twice (as neither retro-writing of the DNA nor the epigenome can happen under Darwinian rules). Stephen Meyer calls this the 'displacement problem', where the source of the information is attributed to some other entity (such as the fusion of the two cells) but no genuine process is described to generate the information in the first place.[28]

Duplication

As fusion does not appear to be an effective method for building new cellular information, duplication followed by progressive mutation has been suggested.

27. *Unlocking the Mystery of Life*. (DVD) Illustra Media (2002).

28. Meyer, S.: *Signature in the Cell*. Harper One, New York (2009) pp.267–270.

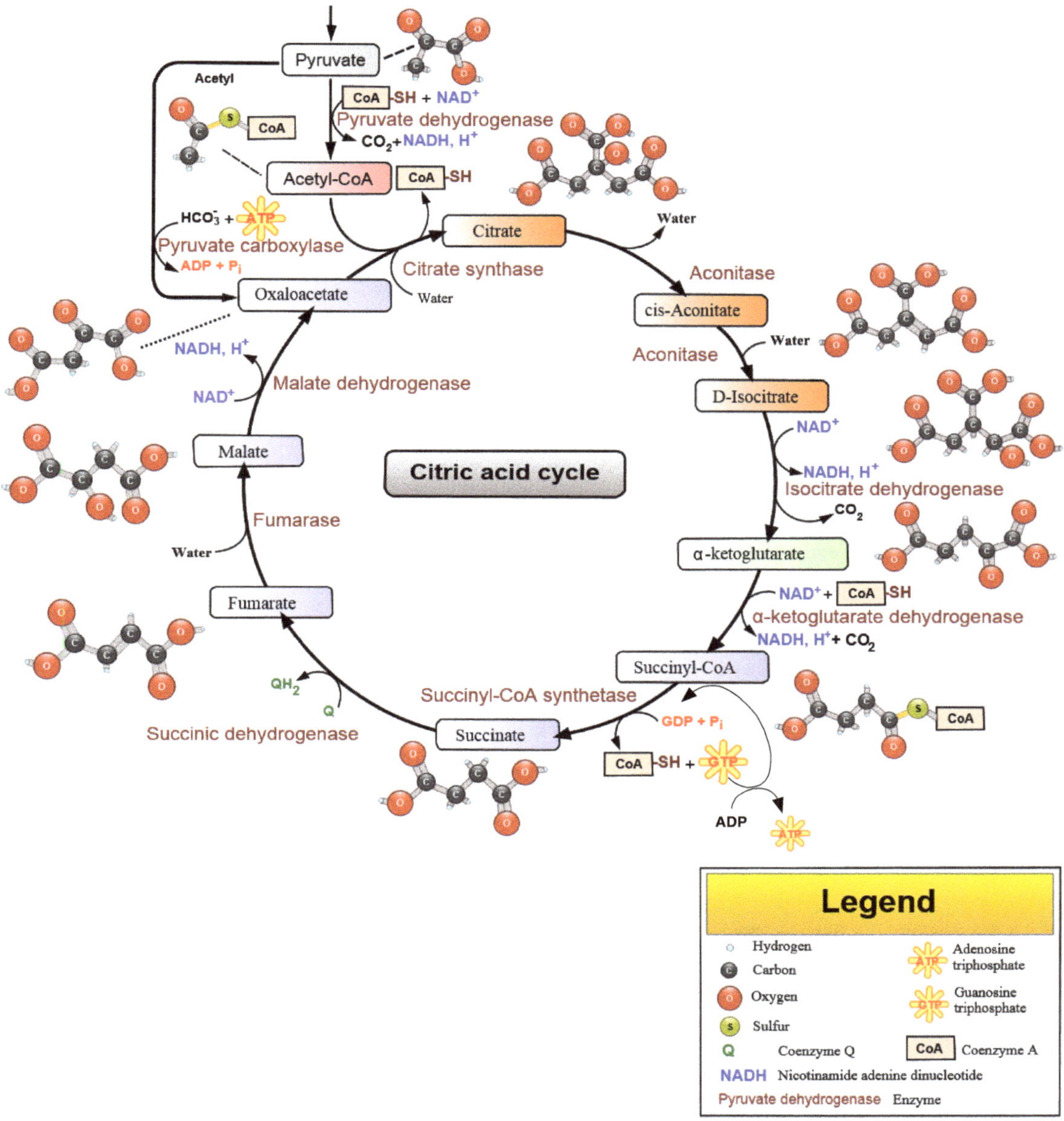

Figure 4.1: The Citric Acid Cycle[29]
This elegant biochemical closed-loop system is integral to the energy production of ATP in every cell and yet it is just one of the myriad of essential biochemical relays in all living cells. *[Reproduced from Wikimedia Commons under the GNU Free Documentation Licence.]*

Darwinian Evolution can increase the gross volume of information in a system (the total number of base pairs) by duplicating an organism's genome through mutation to create redundancy and then allow the redundant copy to mutate. There is no *net* increase in *information* when there are two copies of the same piece of genome but once the copy mutates then there is an original, useful DNA

29. Wikimedia Commons - http://en.wikipedia.org/wiki/File:Citric_acid_cycle_with_aconitate_2.svg

sequence, which can maintain the function of the cell and the redundant mutated copy, which is almost certainly non-functional (at this stage). Taking any working machine, such as an internal combustion engine and changing one piece at random such as pulling off a distributor lead will probably impair and possibly incapacitate the engine. However, if there is a copy of the original genetic material, then the cell can keep the non-functional piece of DNA, which can mutate further without destroying the cell's ability to function on a day to day basis. It is like buying a new engine for your car and using it to run the vehicle, while keeping the old one in the back of the vehicle for a rainy day, to tinker with in your spare time. In this case, duplication and mutation has increased the total information in the system but it is not all functional. The increase is in *non-functional* information at this stage. Neo-Darwinian Theory maintains that the extra engine in the back of your vehicle will continue to mutate over a long period of time, until it gains some new function, when it becomes *functional* information which has now increased the functional complexity of the whole.

So far so good; the main problem with this scenario is that, as the gene undergoing development is not expressed but rather the normal gene is expressed to keep the cell functioning normally, there is no selection pressure for the redundant engine to turn into anything in particular and we all know that an old piece of machinery left in the back shed does not turn into a super-powerful tractor by neglect (not even a three-wheeled, dodgy version that half goes!) The Second Law of Thermodynamics decrees that old machinery will fall to pieces by degrees, unless some active process builds it up or maintains it. However, let's suppose there is some unknown process in the Neo-Darwinian mechanism which has so far gone undetected that could satisfy the Second Law of Thermodynamics when there is no direct selection pressure from natural selection to refine the redundant motor and let's assume it did morph into a new functional structure. Could this be the solution to increasing information to build more complex organisms? Well not quite, as that is still not the whole story. What I haven't told you yet is that the vehicle you are in is in a literal do or die race to the finish line. Only vehicles that win the race will get through to the next round of races to pass on their information to the next generation (and gain another opportunity for their redundant motor in the back of their vehicle to mutate into something useful). The question then is: will your vehicle carrying a disabled engine in its boot be able to win the race to make the next round, as the other vehicles, which are not encumbered with a redundant engine, will be lighter and faster? To mix metaphors, natural selection acts like the umpire in a Wimbledon tennis tournament; you must win each round or you are eliminated and the umpire's decision is final! If you do not win every time then you cannot carry on competing. Thus, the question is: can the process of duplication and redundancy with progressive mutation work fast enough and efficiently enough to evade the winnowing effect of natural selection, which acts to exclude non-functioning components? Total *functional* complexity is an asset but total *non-functional* complexity is definitely a liability and the balance between the two will determine if our mutated car will win the evolution race or not. Is there a way of working out which process will prevail? In determining the answer to this question, there is one more factor that needs to be taken into account.

The biggest snag comes when the epigenome is encountered. Let's assume that against all odds and ignoring all the reasons listed previously for why it cannot happen, that we do reach the summit of Mount Improbable and our mutant motor has slowly evolved into a light-weight winch mechanism in the back of our car. Even in this unlikely event, it still needs to develop a control system so that it can be operated appropriately. Activating the winch to drag us backwards when we are racing for the finish line will not help us win the race but having a winch to haul us out of a bog after we inadvertently took a wrong turn, might just save us from elimination from the race prematurely. So how could the appropriate control mechanism evolve? There would have to be an unused switch on the dashboard of the car to which the winch could be linked to activate our new machinery when it was required and to keep it switched off when it was not advantageous to deploy it. The new dashboard switch is equivalent to redundancy in the epigenome which just happens to have the right type of switch to control the winch. This has to happen in a timely manner so that it is all connected and functioning prior to us landing in the bog. Landing in the bog with a winch that we cannot control will not save us and will probably make us sink faster due to the extra weight. However, a switch alone is not enough. The most complicated piece of the process is the development of the automated decision-making process so that the winch switches on when we land in the bog before we sink irretrievably out of sight but then switches itself off again once we have escaped the danger. To achieve the necessary co-ordinated final result, the parallel but completely independent redundancies in the genome and the epigenome must link up to create a functional new unit or sub-system. The question is: will it all happen in time before natural selection eliminates the prototype winch as too much excess baggage to win the race and make the next round? Theoretically, if there was endless time, then any combination would be possible and 4.6 billion years, which is the estimated age of the Earth by the majority of scientists, seems a long time but developing just the Citric Acid Cycle alone shows the number of precise molecular steps that must be achieved and there has not been enough seconds in the history of the Universe to try the number of combinations required to achieve a workable result, even allowing for the winnowing process of natural selection to narrow down the selection process. I know that this claim will be disputed by Neo-Darwinists so let me point out that the age of the Earth is not the most significant point regarding time. It is that we are all involuntarily entered in the 'Race for Survival' and competency, efficiency and time are all of the essence; the problems are that:

1. Neo-Darwinian Evolution cannot refine a characteristic that is not expressed as it cannot be selected for (*because* it is not expressed);
2. Carrying the dead weight of half evolved accessories will not make us '*the fittest*' so that we can survive (especially if there are more than one or two partly evolved traits);
3. Retro-fitting an epigenetic control to a new survival feature defies the Law of Coding even if,
4. Macroevolution could evolve the new feature (the winch) in the first place — which it cannot (as it also defies the Law of Coding).

Points 1, 3 and 4 are *impossible* and point 2 is simply *unlikely*. When you add three impossibles to an unlikely it does not mean it could happen if we wait long enough or have enough universes in which to experiment. It means that even if we wait forever it will *never* happen, even in an infinite number of universes!

Duplication is *not* the solution to macroevolutionary information gain, even though it appeared to be the best hope for success.

Correction and Repair

Scientists have always been amazed at the precision of the reproductive process in living organisms. Recent discoveries have shown that apart from the sophisticated systems which allow duplication in the first place, there is also an error correction mechanism. Lennox writes: *'The incredibly precise duplication of DNA is not accomplished by the DNA alone: it depends on the presence of the living cell. In its normal surroundings in the cell, the DNA replicates with roughly one error in 3 billion nucleotides (remember the human genome is about 3 billion nucleotides long). However on its own in a test tube the error rate rises dramatically to about 1 in 100. When, still in a test tube, appropriate protein enzymes are added, the error rate sinks to about 1 in 10 million. The final low error rate depends on the addition of yet more proteins in the form of 'repair' enzymes that detect and correct errors."* [30] So these successive layers of replication and error correction show that life puts an exceedingly high value on maintaining the integrity of its information systems. This suggests that there is a need for extreme accuracy in the copying process, because if it was not required, then it would not be preserved as it is obviously a complicated and hence 'expensive' process. But also, from a common-sense perspective, if multiple mutational errors regularly corrupted the code, such a supremely sophisticated information system would become debased and it would not function correctly. And yet, even then, some errors do creep in, which we call 'mutations'.

Marcel-Paul Schützenberger (1920–1996), a French mathematician and physician, commenting on the simple substitution of bases due to mutations concluded: *"...evolution could not be the accumulation of such typographical errors."*[31] A greater understanding of his comments can be gained by extending his typographical analogy. The creation of a new species would be like a mediaeval manuscript turning into a completely different document after many generations of copying, despite the best efforts of the proof readers to remove all mistakes. It could be envisaged that some copying errors would occur in later copies of the mediaeval transcripts but it is unlikely that they would improve the original text or that there would be enough of them, in the light of the scrutiny of the proof readers, to cause significant change in the original document, even after many generations of copying. (Remember that there was an average of one-base pair change per generation in the human genome). Thus living organisms set a high value on maintaining their functioning genome, as the lives of their descendants *literally* depend upon the integrity of the process, yet Darwin saw these errors as central to his theory, so we need to study them in more detail. Despite all the problems already encountered, could mutations still be the answer to increasing complexity and information gain? We now need to examine this central plank of Neo-Darwinian Theory.

Mutations

There is some variation between different researchers but generally, as stated above, scientists agree that new mutations are rare. Estimates of mutation rates vary from 1 nucleotide in every 100 million to one in 3 billion, which is a 30 fold difference. If *thirty* different changes happened every generation it would take 100

30. Lennox, J.C.:. (2009) p.143.

31. Schützenberger, M.P.: The Miracle of Darwinism. *Origins and Design*, 17(2):10–15 (1996) *in* Lennox, J.C.: (2007) p.163.

million generations or at least 2 billion human years to systematically replace every nucleotide in the human genome, if different bases changed each time (which in the real world does not happen by random chance). However, it is very clear that change by mutation is necessarily slow. If only *one* change occurred each generation then it would take 60 billion human years or four times the estimated age of the Universe! [32,33] Darwin thought that these mutations would cause the species to continually change, yet most mechanisms in living organisms cause natural selection to move back towards the norm rather than evolve new forms by drifting further from the norm as Darwin supposed. Factors in human beings and mammals that affect this include:

a. The extreme accuracy of replication and changes are only inheritable (continue in subsequent generations) if the changes occur in the gametes or reproductive cells which are the ova or sperm;
b. Cell mechanisms repair most mistakes in transcription;
c. The immune system eliminates most damaged (altered) DNA, whether it may have a survival advantage or not;
d. The new mutation must be selected as the gamete used for reproduction; (most sperm and ova do not fertilise to produce new off-spring (so there is great wastage); the average ejaculate in the human male contains 20 million sperm, of which usually only one (although rarely two) will fertilise an egg. In a woman, of the thousands of eggs or ova stored in her ovaries, only 500 (one per month for 40 years) will typically be released with the potential to be fertilised, of which usually less than 10 will be fertilised to produce offspring. As a result, even the rare potentially beneficial mutation is unlikely to pass to the next generation;
e. The new mutation must give the cell/organism an advantage *at every step* of its mutation pathway or it will be eliminated by natural selection.

Every new mutation must run the gauntlet of these five steps to become established in the genetic pool otherwise it will be eliminated again by natural selection! The five step process has strong survival advantage as 99 percent of mutations are deleterious and without it, all organisms would become extinct within a few generations due to the accumulated mutation-load. When the operating systems in our computers become corrupted, we reboot them and start again with the same machine but with living organisms, the result is death, so that model is discarded and the process needs to start again with another individual from the same species (with fewer defects).

Microbiologist, Dr Lee Spetner wrote: *"All point mutations that have been studied on the molecular level turn out to reduce the genetic information and not to increase it."* [34] They do not add any *meaningful information*. Note that this is different from survivability of the organism which is a completely different issue. In 1998, when Richard Dawkins was asked to give an example of a mutation which had added meaningful information, he was unable to do so.[35] In addition, most changes in the genome are deleterious, making the system less functional (although sometimes more variable). The reason for this is that given by Daniel Dennett in Chapter 3. If an organism already occupies the optimum position in its design space, then every direction is downhill until another pinnacle in

32. Behe, M.J.: *The Edge of Evolution*. Simon & Schuster, New York (2007) p.11.

33. Lennox, J.C.: (2007) p.134.

34. Spetner, L.: *Not by Chance*. The Judaica Press, Brooklyn, NY (1997) p.138.

35. Keziah: *From a Frog to a Prince*. Answers in Genesis, Acacia Ridge, Qld (1998). Also see: Johnson, P.: *Wedge of Truth*. InterVarsity Press, Downers Grove, Illinois (2000) p.40.

the design space is found. This may be 10, 100 or 10 million base-pair changes away; so one, two or even five base-pair changes are inevitably deleterious to the organism, which as a result, limits change. Fitness landscapes are analogous to the geography of the Galapagos Islands, which have different varieties of tortoise on neighbouring islands. Even though the islands are relatively close together, the islands are far enough apart to prevent the tortoises crossing the ocean separating the islands, which prevents interbreeding and change. This same principle applies to mutations within living organisms; each major type of organism is stuck on its own optimum design-space 'island' and is unable to cross to adjacent islands of viability in its fitness landscape.[36] On their own individual island, each tortoise can vary to some degree which demonstrates Microevolution but crossing the 'ocean' to reach the next 'island' would be macroevolutionary and is not observed. No matter how the tortoises vary, they still remain tortoises. Thus, single base pair changes may increase the Shannon or syntactic information but not the meaningful or semantic information. If the mutation results in major damage to the genome, the genome is either repaired or the organism is selected against (it dies). Most babies with serious mutations die at or before birth but if the change is only slight, the mutation may remain in the gene pool to build up in the genome, as seen in sickle cell trait.[37] This fixed boundary between Macro and Microevolution is illustrated in Figure 4.2.

The progressive accumulation of slight changes in the genome was the mechanism envisaged by Darwin to generate the complex information seen within the biological kingdoms. As has been shown, however, most change incapacitates an already functioning cell or system. The exception, as discussed earlier in the chapter, is: if the mutation occurs in a duplicated gene that is not initially expressed but lies dormant, where it can undergo further mutation ('information enhancement') at a later date. Last time I used the analogy of a car to illustrate the race for survival, into which every individual in every species is involuntarily entered but sometimes non-biological analogies are not reflective of living systems, so let's consider a biological example to ensure that we are appreciating life as it really is. As an example, let us assume that a gene for a cell pump or injector mutates by duplication, as is suggested for the development of the bacterial flagellum. At this stage we shall assume (as Richard Dawkins does in *Climbing Mount Improbable*,[38] that by some lucky chance, the unexpressed gene mutation is not selected against by the usual processes discussed above and it undergoes 50, 100 or even 1000 mutations. So that it is now radically different from the original function it had when the original gene duplication occurred and that it then gets expressed (by chance) as a bacterial flagellum. Remember that this structure had to emerge randomly as the gene was not expressed previously so it could not be assisted by natural selection. It is now able to function in its new form, except that it now requires epigenetic control to work, as was the case for our redundant motor which progressively mutated into a winch in the previous example. Unless it has complementary epigenetic information with suitable decision making software to switch the flagellum on and off appropriately, it is just a liability which trails limply along behind the bacterium and required valuable resources to build but which does not enhance survival and the bacterium will then be selected against.

There is just no way that the information in the genome specifying the new bacterial flagellum can build the epigenetic control or a decision-making mechanism except by pure chance and it has already been shown that the time estimated for the history of the entire Universe is insufficient for this to occur

36. Michael Behe in *The Edge of Evolution* (p.218) showed that Microevolution can change organisms up to the level of species, with possible changes between Genera, Families and Orders of organisms but not beyond that (see Figure 4.2).

37. Sanford, J.C.: *Genetic Entropy and the Mystery of the Genome*. Ivan Press, Lima, New York (2005).

38. Dawkins, R.: *Climbing Mount Improbable*. Norton, New York (1996).

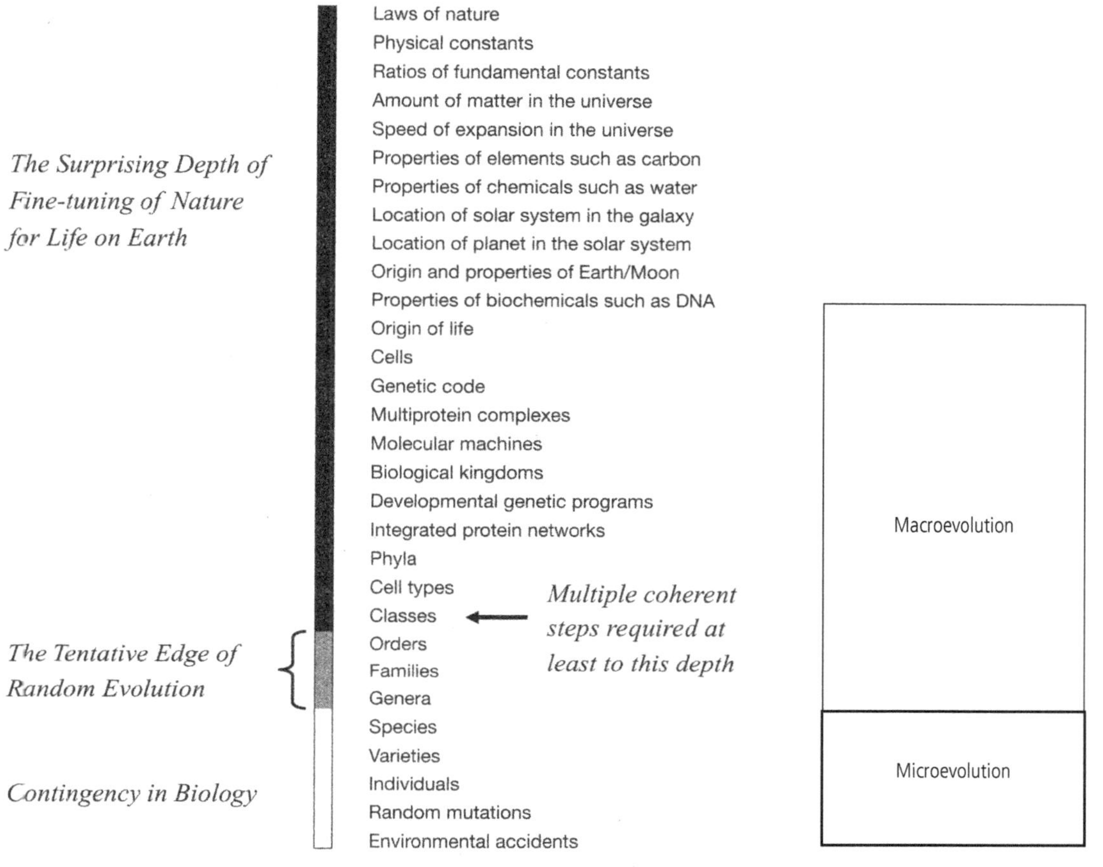

under Neo-Darwinian principles. It is analogous to the development of a switching device on a train line to connect a new siding to the main line. When the workmen finish installing a new switching mechanism in the rails to allow the trains to be diverted onto the sideline, it will not move or be useful until more workmen install a control system in the switching box so that the stationmaster can pull the lever to divert a train onto the sideline. Without the lever and pulleys to connect to the switchbox interchange mechanism to complete the system and make it fully operational, it lies dormant and useless and the stationmaster can gaze at it longingly and wistfully but it will not help him until the complementary switching gear (which requires information) has also been installed. In the case of the bacterial flagellum, we also need to evolve the 'stationmaster' to decide when it is appropriate to activate the switch and when it is not. To imagine that the epigenetic control mechanism will independently evolve in the same timeframe as the immensely complex flagellum and that they will match each other so that the system works before the whole project is selected against by natural selection defies credulity. Remember that a non-functioning flagellum would be more of a hindrance than none at all and would lead to elimination, as its owner will be slower to evade predators with a non-functioning tail dragging behind, even if it was initially very short. It will not be able to gather food as efficiently either. The only other alternative is that the epigenome just happens to fortuitously have a

Figure 4.2: The Boundary between Microevolution and Macroevolution.[39]
This boundary, shown beside Michael Behe's diagram above, is dictated by the constraints of the information in the epigenome. Species (and below) can vary to some extent within defined limits through random mutation with natural selection but higher categories cannot change in the same way. *[Reproduced from Behe, M.J.: The Edge of Evolution. Simon & Schuster (2007). Used with permission.]*

39. Behe, M.J.: *The Edge of Evolution.* Simon & Schuster, New York (2007) p.218.

multitude of redundant control mechanism ready for just such an event as the spontaneous evolution of the flagellum, which is irrational thinking, unless there was pre-programmed information to prepare the control mechanisms in advance. Even if there is enough 'junk' DNA lying around to supply the complementary epigenetic information, as we will see in the section on co-dependent information at the end of this chapter, the precise intersection of four required conditions does not occur naturally in the Universe. It can be achieved but it requires information input. Even if none of the preceding problems with evolution existed, in this case:

a) The flagellar motor had to evolve randomly against the odds (itself a multistage event);
b) The epigenetic control mechanism also had to evolve to complement the motor;
c) The gene for the flagellum must be expressed at the right time, after it is finished evolving but before random mutation breaks it up again;
d) The bacterium in which it appears must be in an environment (for example a watery medium) that will give a bacterium with a flagellar motor an advantage, so that it can preferentially compete and hence take over the gene (information) pool. The flagellum is not likely to be helpful if the bacterium is in a dry sandy soil.

These four events must occur simultaneously if the structures are to eventuate. The cell injector/flagellar motor is a particularly complex example but illustrates the problem of:

1) The Genome: functional genomic information to create the change;
2) The Epigenome: complementary epigenetic information for functional control;
3) Timing: appropriate timing so that the expression occurs at precisely the right time in the development of the organism to give a benefit before elimination by natural selection;
4) The Environment: the process must occur in an environment that is conducive to the change, giving an advantage that allows it to compete and take over the gene pool.

In contrast to duplication mutations, point mutations may give some survival advantage (such as sickle cell trait in malarial areas) but they come at the cost of disrupting the finely-tuned, cellular machinery that already exists, which was not the case with the duplication mutations discussed previously. However, the point mutation has the advantage, which the duplication mutation lacked, that it is expressed and can be selected for, which allows natural selection to help the refining process but is this any better at building sophisticated nanotechnology or any other enhancement for that matter? Because point mutations damage cells, such as those that confer antibiotic resistance in bacteria, Michael Behe likens these point mutations to a retreating army that blows up its own bridges and burns its own crops rather than surrendering them to the advancing enemy. The retreating army loses some assets but limits the enemy advance in the hope that it will live to fight another day.[40]

The sickle cell gene is an effective ploy against the malarial parasite but is still just another example of depleting information and breaking valuable cellular

40. Behe, M.J.: (2007) p.42.

assets. If the host subsequently escapes their malarial environment (for example, by migrating from central Africa to Europe, where there is no malaria), they will be at a disadvantage, with damaged cellular machinery and if one of their offspring gets a double dose of the gene, they suffer sickle cell disease/anaemia. Even in malarial areas, where the sickle cell variation is advantageous for people infected with the malarial parasite, this affected *population* has not lost the 'normal' gene within the population. Indeed, it is dependent on the normal gene within the wider population to create heterozygous children (with one normal and one abnormal gene) rather than homozygous children (with two abnormal genes), who rapidly die without medical intervention. If this variation ever did take over the entire gene pool, then all of the offspring would be homozygous and suffer sickle cell anaemia and die, eliminating the entire human population.[41] If the variation was truly superior, it would have replaced the 'normal' gene pool as Darwin predicted, which it has not done and cannot do, as there is only survival value in the heterozygous offspring; these heterozygous offspring will inevitably have some normal descendants through Mendelian inheritance who lack protection from malaria.

Point mutations change the information in the DNA and hence the relationship between the data points; thus point mutations alter the Shannon information, which creates a variation in the gene pool. However, the important question is: does this new relationship between the data points constitute an increase or a decrease in the total *meaningful* information in the DNA? How can we measure this? The way Darwin suggested that we should measure it is: does the change add to the survival of the organism and hence to the species? In Darwinian Theory the changes also need to increase complexity (if the amoeba is to eventually produce a man). However, most mutations do not satisfy both criteria, that is, increased survival with increased complexity. Stated in another way, a library does not build up its book stock by losing or destroying books on a regular basis unless it has an even larger acquisition programme that outstrips the book depletion process.[42] Just rearranging the books on the shelves or adding extra copies of the existing books does not increase the total meaningful information in the library either.

Further, to build variation, mutations need to be isolated, to prevent the newly altered gene being swamped by the original genes which, at least initially, vastly outnumber them in the population. This is another constraint hampering a mutation from taking over the gene pool; this time the environment imposes the restraint.[43] In the unique environment of a water-filled cave, the loss of sight in cave fish may develop once there is segregation from the light and other fish populations. The loss of sight is not adding information to the gene pool, although it can increase survivability while the fish remain in the darkness of the cave (by saving the 'cost' of maintaining functional eyes).[44]

So, in summary, neither point mutations of the genome, nor gene duplication with multiple (unexpressed) mutations will increase the meaningful information in an organism to the extent that a new type of organism is formed. Bacteria have been studied for well over one hundred years since the pioneering work of Pasteur and Koch in the 1860s and 1870s. Bacteria can replicate within hours and so the past century corresponds to the equivalence in human generations of millions of years of possible evolutionary development; and yet, in this period of observation, no radically new structures and no newly evolved species have been described. Richard Lenski's team from Michigan State

41. *Ibid.* p.26.

42. Spetner, L.M.: (1997) p.143.

43. Jones, S.: The End of Evolution? *Cosmos* 24:35–9 (2008).

44. Sarfati, J.: *By Design.* Creation Book Publishers (2008) p.19.

University has cultured six to seven generations of E. coli bacteria every day for over twenty years which equates to billions upon billions of bacteria and the equivalence of nearly a million human years of evolution,[45] yet no new species emerged. Michael Behe observes: *"The lab bacteria performed much like the wild pathogens: A host of incoherent changes have slightly altered pre-existing systems. Nothing fundamentally new has been produced. No new protein-protein interactions, no new molecular machines. As with thalassaemia in humans, some large evolutionary advantages have been conferred by breaking things. Several populations of bacteria lost their ability to repair DNA. One of the most beneficial mutations, seen repeatedly in separate cultures, was the bacterium's loss of the ability to make a sugar called ribose, which is a component of RNA. Another was a change in a regulatory gene called spoT, which affected en masse how fifty-nine other genes work, either increasing or decreasing their activity. One likely explanation for the net good effect of this very blunt mutation is that it turned off the energetically costly genes that make the bacterial flagellum, saving the cell some energy. Breaking some genes and turning others off, however, won't make much of anything. After a while, beneficial changes from the experiment petered out."* [46] Richard Dawkins highlighted the acquisition (after 33,100 generations) of the ability to metabolise citrate, which formed part of the culture medium of the experiment. This, he declared, was proof of evolution *"before our very eyes"*.[47] In reply, biochemist, Jonathan Sarfati wrote: *"But E. coli can already use citrate directly under anaerobic conditions,...such anaerobic respiration is less efficient than aerobic respiration, so there is good reason to have it switched off unless the oxygen required for aerobic respiration is lacking. But here, there is a loss of regulation allowing this anaerobic citrate respiration to work all the time, so it's informationally* downhill." [48] Even if we allow 125 years since the German paediatrician and bacteriologist Theodor Escherich first described Escherichia coli (E. coli) in 1885 and average only three generations per day, the 'evolutionary time' equates to: 125 years of observation x 365 days/year x 3 generations/day x 20 years per human generation = 2.7 million human-year-equivalents. If Neo-Darwinian evolution was really true, surely we should see a radically different species given this number of generations in which evolution could have operated. The problem is that mutation can only delete information or increase the random variability within the gene pool, which does not equate to novel information resulting in the generation of new structures within an organism. New information requires multiple complementary changes over many generations with *additional* epigenetic control, which has *never* been observed.

Law and Chance

Philip Johnson summarises Evolutionary Theory's problem with information when he writes: *"Genuinely creative evolution thus requires a mechanism capable of creating immense amounts of complex specified aperiodic genetic information. Random mutation is not such a mechanism, nor is natural selection, nor is any physical or chemical law. Laws produce simple repetitive order, and chance produces meaningless disorder. When combined, law and chance work against each other to prevent the emergence of a meaningful sequence."* [49] So, although the combination of law and chance is supposed to increase information, it does not entrain new meaningful information into the genome.[50] It adds random variations but even billions of years cannot convert the random variations into finely honed nano-machines if no additional information is added to organise them.

45. Dawkins, R.: *The Greatest Show on Earth*. Bantam Press, London (2009) pp.117–130.

46. Behe, M.J.: *The Edge of Evolution*. Simon & Schuster, New York (2007) p.142.

47. Dawkins, R.: *The Greatest Show on Earth*. Bantam Press, London (2009) p.126.

48. Sarfati, J.: *The Greatest hoax on Earth? Refuting Dawkins on Evolution*. Creation Book Publishers. Atlanta , Georgia (2010) pp.66–7.

49. Johnson, P.: *Wedge of Truth*. InterVarsity Press, Downers Grove, Illinois (2000) p.127.

50. *Ibid*. p.131.

Modularity

It is now recognised that activity within cells is organised into modules or discrete units like Lego blocks, under the supervision of control genes; complexity might therefore be envisaged to increase by combining different modules in novel ways to achieve new functions previously unobtainable by single genes.[51] This is the pattern that is seen in the motor vehicles produced by a global car maker such as General Motors, where vehicles in Country A will have one sort of gear box but another suspension system and yet another braking system but the electronics system and the airbags are the same as cars in Country B. The cars' designers choose the components that best fit the product that they wish to make in each country. To 'mix and match' components, the complete modules must be swapped, one for another, not an incomplete part of one for an incomplete part of another. In physics, the term for modularity is a 'quantum' (from the Latin meaning an *"Amount; share* [or] *portion"*,[52]) while the process is called 'quantisation', which is the concept of dealing only in whole units. The process of quantisation is significant throughout the Universe and we will meet it again in later chapters regarding our physical Universe. However, to give a biological example, Mycoplasma genitalium is a parasitising bacterium that has one of the smallest genomes as it lacks information possessed by other fully independent organisms[53] but due to the modularity of biological systems, it can utilise the machinery stored in the host cells that it colonises. In the same way, viruses contain even less information than mycoplasma and travel ultra-light but they can thrive by tapping into the cellular machinery of the host cells they invade. If different organisms had completely different biochemistry then the principle of one organism eating or colonising another would be pointless, as the different biochemistries would make the 'food' incompatible with the carnivore/parasite's metabolism and the 'food' would, at best, have no nutritional value and at worse, be poisonous or even lethal. This would happen, for example, if some species were composed of molecules constructed from D-isomers instead of L-isomers. (See Chapter 1.)

So, if biological systems are built from discrete modules, in information terms, how small can the modules be? To adequately answer this question we first have to ask: what defines a module? In living organisms, a module is a functional unit that is built up from a number of smaller parts to perform a task which the smaller units cannot perform alone. So the smallest module is one that has the least number of parts (the least information) and yet still performs a function that is greater than the sum of its parts. If it cannot do this then there is no advantage in packaging the parts together and they may just as well operate as separate entities and enjoy the 'freedom' of being independent. Modularity is observed in many different species; the eye is a module that performs the function of vision but much smaller modules exist, such as ribosomes, mitochondria and the catalytic core in chloroplasts. It would have helped Darwin's cause if the basic unit of biology had been nucleotide-sized, as it would have been much easier to build up complex structures with smaller units but this is not the case. Individual modules appear to be controlled by the epigenome and so whole modules may be swapped one for another with retention of their function and associated epigenetic control (see Evo Devo in Chapter 3) but swapping individual nucleotides, as in the example of point mutations, only results in breaking up the smooth functioning of individual units rather than enhancing them.

51. Williams, A.: Astonishing DNA complexity demolishes neo-Darwinism. *Journal of Creation* 21(3):115 (2007).

52. *Concise Oxford Dictionary*. (5th Ed.) Oxford University Press, London (1973).

53. Wood, T.C.: Genome Decay in the Mycoplasmas. *Impact* 340 (Oct. 2001); www.icr.org/pubs/imp/imp-340.htm> *in* Sarfati, J.: *Refuting Evolution 2*. Answers in Genesis. Acacia Ridge, Qld (2002) p.155.

Co-ordinated Development

The development of a multicellular organism from a single, undifferentiated cell is intriguing. This area of medicine is called 'embryology', which describes all the processes by which the single fertilised egg cell changes to create a fully-competent adult. Edgar Andrews summarised it succinctly when he wrote: "*Just as information is the name of the molecular game, so organisation is the name of the cellular game.*" [54] For the first cell line to survive, natural selection required that it be the fittest and strongest and to overcome its neighbours, if that was what was required, however, to develop multicellular organisms, the individual cell must become subservient to the group (a form of 'cellular socialism'). Each cell instead of trying to out-compete his brother must now not only tolerate his brother but must work co-operatively for the good of the whole. These two processes of 'survival of the fittest' and 'cellular socialism' oppose one another and can be seen to be contradictory. Therefore, if evolutionary theory is true, then 'survival of the fittest' had to morph into the opposite notion of 'symbiosis' for the individual cells involved, where the cell 'learns' that it can achieve more if it works with its fellow cells (competitors) than if it just tries to overcome them or out-survive them. So, how did this fundamental change in competitive behaviour come about? Richard Dawkins in his book: *The Greatest Show on Earth* gives his explanation as: "*...how embryology works. It is all done by local rules, at various levels but especially the level of the single cell. No choreographer. No conductor of the orchestra. No central planning. No architect. In the field of development, or manufacture, the equivalent of this kind of programming is* self-assembly." [55]

Dawkins calls this 'bottom-up' rather than 'top-down' design. Top-down is like an architect designing a cathedral and bottom-up is like termites building a termite mound.[56] However, Dawkins is ignoring the pre-programmed information that is in the DNA of every living cell (including termites). I would suggest that the 'conductor of the orchestra' is the information in the epigenetic component of the DNA and that the 'architect' was the process that laid down the information in both the genome and the epigenome. *Superficially* there appears to be a self-assembly process, which is locally mediated at the cellular level but it is short sightedness to fail to recognise the embedded information within the cell as the cause for the 'local rules' which create this *perception* of 'self-assembly'. Dawkins tacitly confirms this principle of embedded information, when he describes the formation of the spinal cord through the simple growth of some cells more than others to cause local folding or buckling of the tissues. Dawkins writes: "*But, given that there are no hands to do the folding, by what mechanical process are these dynamic movements achieved? Partly, as I have already said, by simple expansion itself. Cells multiply all through a sheet of tissue. Its area therefore increases and, having nowhere else to go, it has little choice but to buckle or invaginate. But the process is more controlled than that...*" [57] If the process is 'controlled' though, where did the information come from to create that control? In the body, this co-operation between cells is mediated differently at different levels. At a cellular level it is mediated by various tissue factors that move from cell to cell turning different genes on or off and this appears to be the process by which tissue specialisation occurs to form different organs. At a higher, command level, the neuro-endocrine system uses hormones, which are protein catalysts (such as adrenaline and insulin), as well as nerve mediated messages, to control tissue development. All this activity requires information, information, information, which has to reside

54. Andrews, E.: *Who Made God? Searching for a Theory of Everything*. EP Books, Darlington, England (2009) p.205.

55. Dawkins, R.: *The Greatest Show on Earth*. Bantam Press, London (2009) p.220.

56. *Ibid*. p.217.

57. *Ibid*. p.229.

somewhere and the most promising location is in both the genome and the epigenome. There is just the problem of how it got into the DNA in the first place, as was discussed in Chapter 3. Yes, it is locally mediated but it is directed by a pre-specified information source.

Information Levels in Communication

Thus, the process of differentiation within living tissues is complex, with messaging from the DNA leading to local communication between cells as they segregate but how does this intricate process occur? The first thing to note is that the differentiation of tissues requires new information at a minimum of three levels. The first is required to define the different tissue types (such as liver cells or nerve cells). This information is located in the genes in the DNA and determines *what* forms – this is structural information. A second category of information is needed at the co-ordination level, supplied by the epigenome, to control the sequence in which the processes are to take place. This defines *when* it forms. A third layer of information is required for the initially undifferentiated cells to change and for the new organs to form in the right place at the right time, then stop; this governs *how* it forms. This phenomenally complicated process, which Richard Dawkins simply refers to as: *"local self-assembly"*,[58] is probably epigenetic in origin. Errors in development may be observed if any of these information systems malfunction. If normal tissue grows in the wrong place, it is called an hamartoma, which is an error of what forms where. One day while I was assisting at a hysterectomy in Reading, England, we were all astounded when the surgeon found a perfectly formed canine tooth in the wall of the patient's womb; it was a perfectly normal tooth but in completely the wrong place, which may explain why the woman was experiencing pain! Alternatively, if the normal development process goes wrong during the early stages of embryonic development a teratoma can form, which contains all three embryological germ-cell layers but without any recognisable form, as if the epigenome did not supply the information to organise normal development. The tissues appear to lack the information that specifies how to construct normal tissue architecture. In contrast to this growth abnormality, if uncontrolled growth starts after birth, in an established tissue and does not stop at the right time, it forms a cancer. In this case, there is a problem with either the information which stipulates when or how to form normal tissue.

When development goes irretrievably wrong in one cell within a multi-cellular organism (which inevitably happens in processes as multifaceted as complex organismal development) there needs to be a mechanism to recover the situation. A fourth type of information defines the feedback loops, which facilitate the detailed regulation that is found in biological systems. In cells, there is an auto-destruct process, such that if the organism's command mechanism senses that there has been an irretrievable breakdown, the 'self-destruct button' is pushed and the process of apoptosis destroys the aberrant cell. All these processes need to be defined by information, which had to be generated somehow. This is summarised in the Table 4.2 following.

These communication processes lie at the heart of embryogenesis. Richard Dawkins concludes his section on embryogenesis by saying: *"There is no overall plan of development, no blueprint, no architect's plan, no architect. The development of the embryo, and ultimately of the adult, is achieved by local rules implemented by cells, interacting with other cells on a local basis. What goes on inside cells, similarly,*

58. Dawkins, R.: *The Greatest Show on Earth*. Bantam Press, London (2009) p.243.

	Type	Information Source	Function
1	Cell formation	Genomic DNA	Stem cell formation
2	Co-ordination	Epigenetic DNA	Embryogenesis
3	Cell Differentiation	Epigenetic DNA	Organ development
4	Functional (e.g. feedback loops)	Possibly Epigenetic or intracellular	Efficient functioning of the cell/tissues.

Table 4.2: Types of Biological Information

is governed by local rules that apply to molecules, especially protein molecules, within the cells and in the cell membranes, interacting with other such molecules. Again, the rules are all local, local, local." [59] And yet Dawkins had previously stated only five pages earlier: *"So the whole course of embryonic development is controlled, via an intricate sequence of events, by genes...Whether or not a given gene is turned on in a given cell at a given time is determined, often via a cascade of other genes called switch genes or controller genes, by the chemical environment of the cell."* [60] Now firstly, we have already seen that genes (DNA) require information to specify what Dawkins terms the 'genetic recipe'. Secondly, the *"controller genes"* which Dawkins speaks about, also require information and finally, the *"local rules"*, which control the immediate cellular environment, are controlled by intrinsic information. So Dawkins has described *at least* three levels of information controlling embryogenesis. In Chapter 3, we discovered that the Law of Coding prevents random mutation or natural selection from providing the information for DNA so the information must have been generated in some other way but even a billion years of random mutation with natural selection cannot generate something that is impossible and as Dawkins says: *"...neither I nor anybody else can tell you the details of how it happened."* [61] It is clear that embryogenesis is not as simple as Dawkins' *"local rules"* statement might suggest! Once again, the bottom line depends upon multiple levels of information to direct these processes.

Internal or Micro-communication

Once organisms contain specialised cells, these cells need to be able to communicate with each other, to share information about the macro-environment without (for example which direction food may be found) and their micro-environment within, such as the need to eliminate waste products that have built up. Again, information is needed to specify the hardware components, the operations software and the language necessary for the communications system. Some information is chemically mediated, such as hormones, like insulin and some information is electrically mediated via nerve cells, as in round worms. In 'higher' organisms, both systems are integrated in the *neuro-endocrine system*, where they work in synergy. That is not the end of the story however. There are complex feedback loops in the human body between the individual glands, such as the female ovaries, which produce oestrogen and progesterone (or the male testes which produce testosterone), the pituitary gland at the base of the brain, which produces releasing hormones, such as luteinising hormone (LH) and follicle-stimulating hormone (FSH) and the hypothalamus in the base of the brain, which controls the pituitary through further hormones, like gonadotropin releasing hormones (GnRH). This interconnectivity of hormone systems provide very sophisticated control mechanisms which deliver exquisite regulation of the

59. Dawkins, R.: *The Greatest Show on Earth*. Bantam Press, London (2009) p.247.

60. *Ibid.* pp.242–3.

61. *Ibid.* p.249.

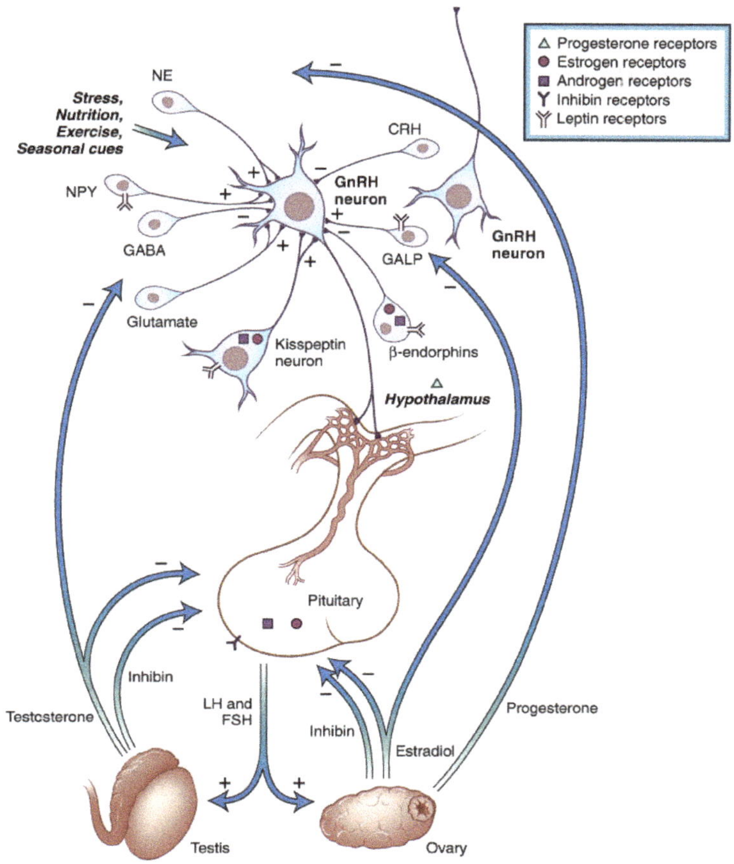

Figure 4.3: Regulation of the Hypothalamic-Pituitary-Gonadal Axis [62]
It is not necessary to understand the detailed names or specific relationships but to appreciate the significant quantity of information required to specify this system. However, for those who have an interest, the GnRH released from the hypothalamus has the role of controlling the axis. The LH and FSH released by the pituitary have a regulatory role and the oestrogen and progestogen/ testosterone from the ovaries/testes carry the active information. Note the negative feedback loops mediated by the oestrogen, progesterone, testosterone and inhibin that fine-tune the system. *[Reproduced under licence from Elsevier Images.]*

body's organ systems but it comes at the considerable price of vast quantities of specialised information. See Figure 4.3.

Thus, it can be seen that communication systems within an organism also require information at many levels. To achieve this, the organism needs the correct components to be specified, as well as the correct placement during the assembly of those components. This is in addition to the method of construction and on-going repair processes, which necessitate the specification of further information. However, communication not only occurs within single-celled organisms but between them as well. The following report referring to an article in *Nature* in 2008 reads: *"Cooperative behaviour occurs in unicellular organisms. For example, Salmonella typhimurium can arrange themselves in two ranks for invasion – the first rank launches a suicide attack and the second rank slips through the confusion in the defence caused by the first wave. Therefore, some communication between unicellular organisms occurs to allow for cooperation."* [63] Other examples, such as the inter-cellular communication between the individual amoeboid cells that make up slime mould and the inter-organismal communication exemplified in human speech will be discussed in detail in Chapter 5. The four types of communication are summarised in Table 4.3.

Lastly, communications information must be timely; the right information at the right time is important in the development of the embryo to achieve complete segmental development. It is also important for survival, such as the ability to detect a predator in time to allow escape.

62. Elsevier Images. http://www.elsevierimages.com/image/28267.htm

63. Ackermann, M. et al.: Self-destructive cooperation mediated by phenotypic noise. *Nature* 454:987–990 (2008) *in Journal of Creation* 23(1) 2009 pp.5–7.

	Type	Example	Required for
1	Intra-cellular	Protein synthesis	Cellular competency
2	Inter-cellular (single cell)	Salmonella typhimurium	Species co-ordination
3	Intra-organismal	Insulin or FSH release	Organism competency
4	Inter-organismal	Speech for hunting	Community competency

Table 4.3: Types of Biological Communication

Biochemical (Molecular) Evolution

All biological systems can ultimately be traced back to their biochemical building blocks and we know that Darwinian Evolution can only move one step at a time, so if Darwin's hypothesis is true, there must be a biochemical evolutionary pathway for every protein, every enzyme system and every organelle in every living creature, yet there is no detailed traceable trail of how evolution could have occurred at a molecular level for any of them. In the same way that Darwin predicted that there should be fossil 'links' between the species within the plant and animal kingdoms, when different species are compared, there should be identifiable biochemical 'intermediates' but evolutionary biologists have been unable to show any biochemical developmental pathways even though the Neo-Darwinian Theory requires that they existed. Michael Behe writes: *"Molecular evolution is not based on scientific authority. There is no publication in the scientific literature — in prestigious journals, specialty journals, or books — that describes how molecular evolution of any real, complex, biochemical system either did occur, or even might have occurred. There are assertions that such evolution occurred, but absolutely none are supported by pertinent experiments or calculations... Despite comparing sequences and mathematical modelling, molecular evolution has never addressed the question of how complex structures came to be."* [64]

Lennox comments: *"James Shapiro, a biochemist at the University of Chicago, also admits that there are no detailed Darwinian accounts for the evolution of any fundamental biochemical or cellular system: only a variety of wishful speculation."* [65] Microbiologist, Michael Denton amplifies Shapiro's statement when he writes: *"In terms of their biochemistry, none of the species deemed 'intermediate', 'ancestral' or 'primitive' by generations of evolutionary biologists, and alluded to as evidence of sequence in nature, shows any sign of their supposed intermediate status."* [66]

"The fact that lungfish, monotremes and all the other favourite links of evolutionary biology give no hint of their supposed transitional status at a molecular level is perfectly in keeping with the fact that there are many morphological features of their biology which have never been easy to reconcile with their supposedly transitional status and which have always suggested that they represent unique and isolated types." [67]

Fowler and Kuebler reviewed the Darwinian evolutionary explanation for the TCA (Citric Acid) Cycle and note that several organisms have incomplete 'cycles'. They write: *"While the presence of a variety of simpler systems in these organisms supports the notion that the TCA cycle could have evolved in stages from other distinct pathways, constructing a feasible pathway for building the TCA cycle from the ground up is still beyond our ability."* In conclusion they wrote: *"...there are some intriguing possibilities as to how natural selection may have been able to build such systems. Without denying the real potential of these possibilities, one must admit that we do not at present know whether or how natural selection actually could have built the systems in question."* [68]

64. Behe, M.J.: *Darwin's Black Box: The Biochemical Challenge to Evolution*. Free Press, NY (2006) pp.185–6.

65. Lennox, J.C.: *God's Undertaker. Has Science Buried God?* Lion Books, Oxford (2007) p.119.

66. Denton, M.: *Evolution: A Theory in Crisis*. Adler & Adler, Chevy Chase, Maryland (1986) p.293.

67. *Ibid*. p.294.

68. Fowler, T.B. & Kuebler, D.: (2007) p.165.

So there is no gradual evolutionary information progression demonstrated in molecular biology, which there must be if Darwin's theory is to be verified, and in information terms, the Neo-Darwinian Theory is found wanting again.

Integrating Functions in More Complex Organisms

Once organisms become larger than a small collection of cells, they inevitably contain specialised functions, since nutrient delivery and waste disposal are required for all living tissues and the central cells in a large clump will not have direct contact with the food source to allow the nutrients to defuse directly through their outer membranes. The patterns that are observed in increasingly complex organisms are similar to those seen when a sparsely populated farming community begins to grow. As the streets are constructed to access all the new house blocks, a town may be discerned. In a mammal, the lanes and highways are equivalent to the bloodstream which connects all the cells to the specialised organs, such as the lungs for oxygen, the intestines for food and the kidneys for waste disposal. After the roads, telephone wires are laid for communication (neurones or nerves), a police force is formed to protect the neighbourhood (the immune system) and the town council (the brain) begins to plan and co-ordinate activities. One of the activities of the town council will be to appoint an emergency response team (the clotting system) to effect repairs in the event of storm damage to the town. This is a basic scenario for any standard country town but some towns have specialised functions such as being near a coalmine or a natural harbour. These towns demonstrate specialisation to better meet the needs of their local economies and sometimes the wider community; they may have wharves, cranes, railway lines and storage facilities. Living organisms also possess specialised tissues and abilities, which are too numerous to list here, but some of the first ones we learn about as a child are the 'five senses'.

The Five Senses

In human beings, the traditional five senses are: vision, hearing, smell, taste and touch. They all scan our environment and import information for our brains to make the best use of the habitats in which we live. There are numerous other senses such as motion and balance sensors in the inner ear and temperature in the skin but due to the vast size of this topic, we will limit this analysis to the first three senses, namely: vision, hearing and smell as being representative of the rest.

The Eye

Our senses collect information from our surroundings and convert it into electrical or chemical signals which our bodies (usually our brain) can interpret. Perhaps the most complex and most studied of these is the eye, which first appears in the fossil record at the beginning or immediately before the Cambrian explosion. Trilobites are the first creatures in the fossil record with an identifiable eye and 95% of species living today have eyes.[69] The eye is especially notable because Darwin himself wrote about the significance of the formation of the eye in *The Origin of Species*. Thirty different types of eyes have been described and because of the differences between them, John Rennie, the editor of *Scientific American* wrote in 2002: "*It now appears that in various families of organisms, eyes*

69. Parker, A.: The Eyes have it. *Cosmos* 2:48–9 (2005).

have evolved independently." [70] However, there is little evidence to elucidate how the information to specify even one eye entered the genome, let alone for the process to be replicated multiple times.[70]

Neo-Darwinian explanations as to the mechanism of formation of the human eye are often similar to the scenario painted by Swedish zoologists Dan Nilsson and Susanne Pelger in their 1994 paper which began with a light sensitive layer that became covered by another layer of clear cells but where did the information come from for the initial two layers?[71] Michael Behe describes the detailed biochemistry required for light-sensitive layers and it involves vast amounts of information.[72] As discussed previously, there needs to be simultaneous development of the hardware (the anatomy and biochemistry) and several layers of software (the electro-chemical coding for the messaging, analysis of the signals and embryogenesis). These specifications must define the generation of a message in the retina, transmission of that message to the decoding centres (requiring nerves) and decoding equipment in the vestigial brain, as well as sufficient processing ability to conceptualise the significance of the signal transmitted. If an evolutionary advantage was to be gained, the organism must then have been able to initiate an evasive response once the information was interpreted and possibly danger was recognised; thus outgoing messages must be able to be generated, to allow appropriate action to be taken promptly, to escape from any would-be danger. All this information also needs to be reverse engineered into DNA or RNA! This is an eight step process which requires some parts to develop simultaneously. It is summarised in the Table 4.4 below.

Note that each of the information layers 1- 4 are multistage, complex processes. It could be hypothesised that, in their simplest forms, they might have been built 'one step at a time', such as the biochemistry required for layer 2 in Table 4.4, possibly involved just one photon recognising chemical initially, however the construction of that first chemical requires another intricate, multipart building process. It is also possible to hypothesise that information layers 3 to 6 might have existed before the development of the eye structure but layers seven and eight contradict the Law of Coding and illustrate that a Neo-Darwinian, step-wise process is inadequate to explain the structures we observe. Layers 1 and 2 are expanded in Table 4.5.

Each of these components is made possible by dozens of proteins, chemical pathways and/or electrical impulses (information instructions). One well studied example is the intricate multi-step process required for the recognition of photons in the retina of the eye involving the rhodopsin molecule.[73]

Thus, the amount of information required to build a human eye is truly immense. Although it is possible to speculate that the information accumulated sequentially, there is no evidence that it did. Fowler and Kuebler write: "*...data*

70. Rennie, J: 15 Answers to the Creationist Nonsense. *Scientific American* 287(1):78–85 (2002) p.83 *in* Sarfati, J.: *By Design*. Creation Book Publishers (2008) p.36.

71. Nilsson, D.E. & Pelger, S.: A pessimistic estimate of the time required for an eye to evolve. *Proc. R. Soc. Lond.* B 256: 53–8 (1994) *in* Sarfati, J.: *By Design*. Creation Book Publishers (2008) p.37.

72. Behe, M.J.: *Darwin's Black Box*. (2006) pp.18–21.

73. Aubisson, P. et al: *Biology in Context – The Spectrum of Life*. (2nd Ed.) Oxford Univ. Press, Melbourne (2004) p.478.

Hardware	1. Anatomy	Bi-layered eye development
	2. Biochemistry	Photon recognising chemicals (multi-stage feat)
	3. Transmission	Nerve connections
	4. Brain	Processing power
Software	5. Coding system	Messaging language to communicate with the brain
	6. Pattern analysis	Recognising the significance of the message
	7, DNA coding of 1- 6	To specify the processes within the eye above
	8. Epigenetic coding	To control the formation (embryogenesis) of the eye.

Table 4.4: The layers of information required for the formation of the first eye.

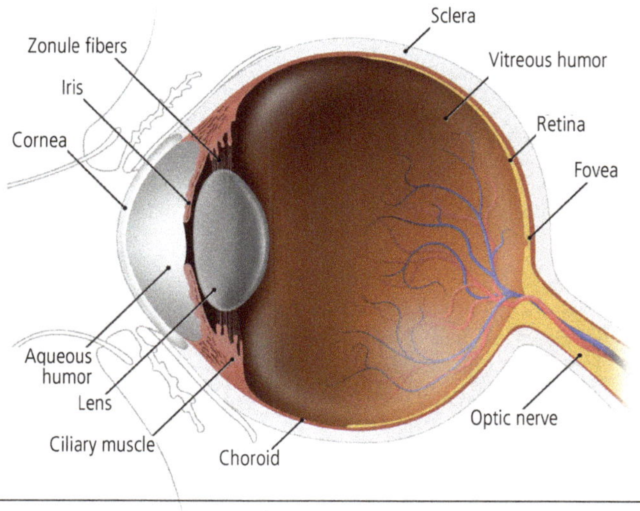

Figure 4.4: Anatomy of the Human Eye[75] *[Reproduced under licence from Thinkstock Photos.]*

Table 4.5: The Components of the Human Eye[76]
The detail within this table is provided for those who are interested in the extraordinary volume and intricacy of the information that is required to describe a human eye but it is not essential to understand all the terms or processes. It suffices to appreciate the complexity that is necessary for you to be able to read this page.

	Component	Function	Features
A	Lids	Retractable protection, cleaning and moistening	Blink reflex; smooth & striated muscle
B	Cornea	Transparency and protection	Fluid pumps
C	Iris	Light intensity regulation	Dynamic range 10,000,000 to 1; Ciliary reflex
D	Lens	Auto-focuses for a crisp image whether near or far.	One third refraction accommodation
E	Vitreous body	Light transmission	Reduces lens flare
F	Retina[77]	Detects 1 photon at 10^{-12} sec (Rhodopsin switch)	12 different types of ganglia; 137 million rods & cones; 10 billion calculations/second
G	Fovea	Permits detailed perception	Density of receptors
H	Cones	Improve daylight vision	Colour vision
I	Rods	Improve night vision	Low-light vision
J	Visual cortex	Creates real time colour vision	Integration of the 12 picture signals (see F above)
K	Dark adaptation	Enhances night vision	Arestin in retinal cells
L	Saccades	Allows perception of detail	Fine discrimination
M	Global vision	Efficiency & reduced distractions	Data selection
N	Ext. ocular muscles	Distance & speed perception	Binocular vision
O	Tear gland & ducts	Self-cleaning and nutrition	Bactericidal
P	Peripheral vision	Search function	Economical functioning

supporting the notion that natural selection could build the original photoreceptor is, at present, lacking." [74] And we have only scratched the surface of the detail within the human eye. Anyone with a mind to seeing the full complexity of a human being should consult an ophthalmology textbook to prove to themselves that the number of infobits in the human eye is astronomical. Despite these problems with developing the eye itself, it must be remembered that the eye is only one part of a larger system and it is necessary for the eye to be co-ordinated with its brain for image processing. Embryologically, the eye is a fusion of cells that will form the foetus's skin and brain. Cells from the layer forming the skin create the clear window at the front, while cells from the brain construct the photo-sensitive components at the back. This is a feat in itself, however, building the eye is only the start. As we studied electron-micrographs[78] of cells as medical students, our teachers were at pains to remind us that the black and white images were only snapshots in time of objects that were living and hence dynamic and forever changing. So once the eye has been built, it must begin to function and

74. Fowler, T.B. & Kuebler, D.: *The Evolution Controversy.* (2007) p.160.

75. Ali Kocakaya/iStock/ Thinkstock: http://www.thinkstockphotos.com.au/image/stock-illustration-human-eyeball/166214820

76. Sarfati, J.: *By Design.* Creation Book Publishers (2008) pp.24–31.

77. Dawkins, R.: *The Greatest Show on Earth.* Bantam Press, London (2009) p.236 (footnote).

78. Electron-micrographs are photographs taken with an electron microscope.

this is where we meet integrated systems because the eye is just an (extremely sophisticated) sensor of the brain. It collects raw data from the environment and begins processing it in the retina, which is a layer of cells at the back of the eye ball that is thinner than the piece of paper this text is printed on. The retina has 137 million of these photo-sensitive cells which perform 10 billion calculations per second as they process the images coming through the pupil.[79] To put this into perspective, John K. Stevens wrote in *Byte* magazine in 1985: "*To simulate 10 milliseconds of the complete processing of even a single nerve cell from the retina would require the solution of about 500 simultaneous non-linear differential equations 100 times and would take at least several minutes of processing time on a Cray super computer. Keeping in mind that there are 10 million or more* [now thought to be 137 million] *such cells interacting with each other in complex ways it would take a minimum of a hundred years of Cray time to simulate what takes place in your eye many times every second.*" [80]

The visual image is converted to an electrical signal that is transmitted along nerves, called the optic tracts, to the visual cortex in the brain for further processing and interpretation. The eye and the brain work as an integrated unit, to the extent that many neurologists classify the eye as a remotely located extension of the brain; without the brain, the eye is useless, which implies that the eye could not have preceded the brain. This principle of brain primacy must also apply for all the other senses, as well as the limbs, as they are also useless unless interpreted or driven by the brain. Thus, the nervous system, (if it is a simpler form like a jellyfish) or brain (for more complex creatures) must have preceded or at least formed at the same time as everything else. Then, the question which follows logically is: what is the purpose or function of an isolated brain or nervous system? It becomes patently obvious that systems must develop or be built as integrated units. The problem is that integrated systems come as large chunks of information or not at all. Even Charles Darwin had qualms about the veracity of his assumption that evolution resulted in the formation of the eye when he wrote: "*To suppose that the eye, with all its inimitable contrivances for adjusting the focus to different distances, for admitting different amounts of light, and for correction of spherical and chromatic aberration, could have been formed by natural selection, seems, I freely confess, absurd in the highest possible degree.*" [81] Having introduced just some of the problems of the visual system, let's try a simpler example and look at the ear.

Hearing

Vision detects that part of the electromagnetic spectrum corresponding to 'visible light'. In contrast, hearing is the interpretation of sound waves, which are vibrations that cause pressure variations in air to produce audible sound.[82] At this stage, we need to take a brief detour into sounds and waves and how sound is reproduced in our brains. In 1863, the German physiologist and physicist, Hermann von Helmholtz (1821–1894) showed that the ear was a frequency analyser. When you hear a sound, such as an orchestra playing a symphony, you are interpreting information that has been brought to your brain through the medium of sound waves. Your brain can decode the information and deduce not only that it is an orchestra performing but by separating out the sound frequencies that are being received, it can also identify the specific instruments that are playing. The musical sound waves pass along your ear canal to a special flexible membrane, called the eardrum, as seen in Figure 4.5.

79. Heenan, D.: The Skeptic's guide to God. Show and Tell. Sanctuary Cove, Qld (2008) p.18.

80. Stevens, J.K.: *Bite*. pp.287-299 (April 1985) *in* Menton, D.N.: *Can Evolution Produce an Eye?* Missouri Association for Creation. (1994) *at* http://reformednazarene.wordpress.com/nazarene-denomination-losing-its-way/can-evolution-produce-an-eye/

81. Darwin, C.: *The Origin of Species*. John Murray, London (1859) reprinted by Penguin Classics, London (1985) p.217.

82. *Andriessen, M. et al: Preliminary course physics one.* (2nd Ed.) Jacaranda, Milton, Qld (2004) p.16.

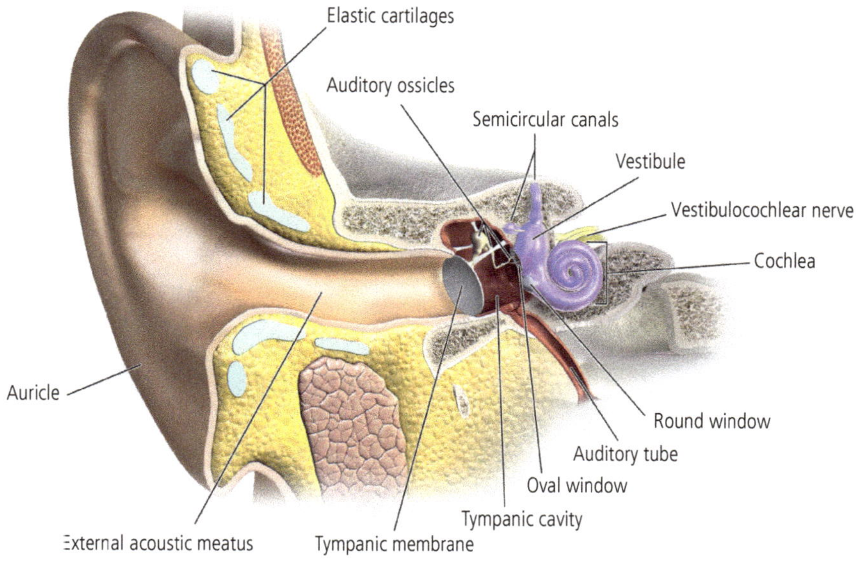

Figure 4.5: The Human Ear [83] *[Reproduced from Wikimedia Commons.]*

The eardrum is tensed just like the skin on a drum in the orchestra and can vibrate in the same manner that the drum did when it generated that sound. In the same way, multiple incoming frequencies cause corresponding vibrations in the eardrum that are transmitted by three tiny bones call the 'ossicles', which form the 'ossicular chain'. These tiny bones are hinged and lubricated in such a way that the vibrations are amplified as they are conveyed across the middle ear. The middle ear is an air-filled chamber that allows the drum to vibrate unimpeded. The ossicular chain transmits the sound vibrations to the cochlea, which is the sound interpreting centre in the inner ear. The cochlea looks like a turban shell. It is full of tiny hairs that are immersed in a fluid called 'endolymph' within the spiral of the cochlea. In adult human beings, the cochlea is about the size of a garden pea. As the sound travels through the endolymph, each individual frequency causes a tiny hair, which corresponds to that frequency, to vibrate or resonate between 20 and 20,000 times per second and there are 16,000 of these hairs. The longer cochlear hairs resonate at lower frequencies and the shorter hairs respond to the higher frequencies. Each microscopic hair is connected to its own nerve fibre, which travels through the auditory nerve to the brain. These components are summarised in Table 4.6.

If you could identify every hair in the cochlea, it would be theoretically possible to give each one a number and then record the series of numbers which your ear heard as the symphony was played by the orchestra. Your ear is therefore interpreting sound digitally. The fact that a digital representation of the information is conceivable means that it is theoretically possible to express the sound mathematically. The transmitted sound, which came to the eardrum as fluctuations in air pressure, has now been converted to digital frequency information for the brain.[82] The branch of mathematics associated with the manipulation of waves is called 'Fourier analysis'. So, to progress our understanding further, we need to take a minor detour into what engineers call 'Fourier transforms'.

83. Blaus, B.: Anatomy of the Ear. Wikimedia Commons. (2013) http://en.wikipedia.org/wiki/File:Blausen_0328_EarAnatomy.png.

	Component	Function	Features
A	Auricle (outer ear)	Location of sound direction	
B	External auditory canal	Protects ear drum	Amplifies sound
C	Ear wax	Cleans the ear	Smell deters insects
D	Hairs in the external canal	Protect against foreign bodies	Angled outwards
E	Epidermal cell migration	Automatic cleaning	Grows from the drum
F	Ear drum	Detects sound	Range 20 - 20,000 Hz
G	Middle ear	Air space for components to move	
H	Ossicles	Impedance matching from air to fluid in the inner ear	Amplify sounds
I	Ossicular muscles	Damp excessive noise (Stapedius & tensor tympani)	Range:1 million to 1 (jet engine to a whisper)
J	Eustacean tube	Equalises air pressures across drum	
K	Oval window	Admits sound to the inner ear	
L	Cochlea	Interprets frequencies	Double canal
M	Endolymph	Transmits sound	
N	Hair cells	Digitise the sound	16,000 graded lengths
O	Auditory nerve	Transmits signals to the brain	8th cranial nerve
P	Round window	Counter balances oval window	
Q	Semicircular canals	Interpret balance and movement	Sense Gravity

Table 4.6: The Components of the Human Ear [84]
Once again, understanding the medical terms is not as important as appreciating the scope and sophistication within the multi-layering of the information content.

Fourier Transforms

The Fourier transform is based on the discovery that it is possible to take any geometric shape which varies with time and break it down into an infinite number of sine waves and cosine waves which, when added together again, will recreate the original entity. Jean Baptiste Joseph Fourier (1768–1830) was a French mathematician and physicist after whom the Fourier series and Fourier's law of heat transfer are named.[85] In mathematical terms, a Fourier transform is a formula which can manipulate waveform information. In layman's terms, the Fourier transform is a tool which can convert waveforms into digital information and vice versa. The most familiar example is the way that a prism transforms or breaks up white light into the individual colours of the rainbow. Ordinary light, like sunlight, is called 'white light' by physicists and is a mixture of all the different frequencies (colours) of the visible spectrum. When the prism splits the frequencies apart, it is acting like a Fourier transform, separating a mixture of frequencies into their pure forms. (See Figure 4.6) This of course is the same phenomenon that produces rainbows.

Most people will also be aware that 'digital sound', as recorded on a modern compact disc (CD), is crisper and clearer than the older analogue form of sound recordings on vinyl records. This is because the digital sound, which is recorded as a series of numbers, is less easily distorted than the older analogue method of storage. The process of converting the numbers, which record the digital sound on a CD, into music once more, requires the application of a mathematical formula, which is called the 'Fourier transform' (see Figure 4.7). It is not necessary to understand the mathematics but to appreciate that the Fourier transform is simply a mathematical technique for turning one type of information (such as digital data that varies with time — referred to as the *'time-domain function'*), into another representation of the information, which may be a wave built up

84. Menton, D.: *The Hearing Ear and the Seeing Eye.* (DVD) Answers in Genesis, Hebron, KY (2003).

85. Rouse Ball, W.W.: `*A Short Account of the History of Mathematics'* 4th Ed., (1908) at http://www.maths.tcd.ie/pub/HistMath/People/Fourier/RouseBall/RB_Fourier.html.

Figure 4.6: A Prism as an example of the Fourier Transform *[Reproduced under licence from Vasyl Yakobchuk/ Hemera/Thinkstock Photos]*

from many different frequencies, (called a *'frequency spectrum'*). Sound engineers use this technique when they want to mathematically add sounds together. In a similar way, the cochlea in your ear transforms the air pressure frequency signal experienced by your ear drum, into digital information which is interpreted by your brain as a series of electrical impulses.[86]

One of the surprising discoveries in the last thirty years is that there is evidence to suggest that vision and hearing may be mediated by Fourier transforms. *"Berkeley neurophysiologists Russell and Karen de Valois discovered that the visual cortex processed patterns by Fourier transformations of the patterns."* [87,88] How the brain does this is not yet understood but this unexpected finding suggests that the brain has complex digital processing capacity. This subject will be explored further in Chapter 10, when we examine thought and memory.

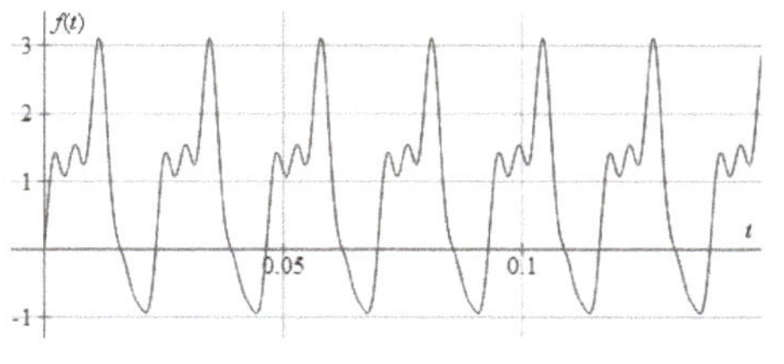

Figure 4.7: An example of a Fourier Transform[86] **Frequency** Spectrum Frequency information on the Y axis is plotted against time on the X axis. *[Reproduced with permission courtesy of Murray Bourne: Interactive Mathematics.]*

Smell

The sense of smell is another way that our brains can remotely sense the composition of our environment, in this case, chemically and our noses are capable of detecting thousands of different substances. If an animal has died in your vicinity, you can usually smell the rotting carcass after some days; you do not have to see or physically trip over it to know that it is there. Biochemist, Jonathan Sarfati, writes: *"...we can sense one component of rotten meat, ethyl mercaptan, at a concentration of 1/400,000,000th of a milligram per litre of air."* [89] In her book *Brain Sense*, Faith Hickman Brynie explains that: *"The sense of smell is hardwired into the human nose and brain. Molecules of odorant move up through the two nasal cavities until they come to the olfactory epithelium. This region in the nasal lining, about the size of a postage stamp, contains 10 to 20 million olfactory neurons. Spiny*

86. Bourne, M.: Interactive Mathematics – Fourier Series at http://www.intmath.com/fourier-series/7-fast-fourier-transform-fft.php

87. Missler, C.: *Cosmic Codes.* Koinonia House, Coeur d'Alene, Idaho (2004) p.330.

88. De Valois, R.L. & De Valois, K.K.: Spatial Vision. *Annual Review of Psychology* 31:309–341 (1980).

89. Sarfati, J.: *By Design.* Creation Book Publishers (2008) p.43–44.

projections (dendrites) of these neurons extend to the epithelial surface where they end in knobs. On the knobs are tiny hairs that float in a layer of mucus. The hairs contain odour receptors, one kind per cell, among the thousand or so that humans have. Odor molecules fit receptors the same way that keys fit locks. When an odoriferous molecule binds to its perfectly matched receptor, a nerve impulse begins." [90]

Sarfati continues: *"The sense of smell is actually responsible for most of the different 'tastes' of foods...The nose contains millions of receptors, of 500–1000 different types. They are in the yellow* olfactory epithelium, *that covers about 2.5* cm^2 *on each side of the inner nose. The different types of receptors are proteins folded so a particularly shaped odour molecule can dock. Each receptor is coupled to a* g-protein. *When the odour molecule docks with the receptor, the g-protein is released. This sets off a second messenger to stimulate a neuron to send a signal. This is transmitted by olfactory nerve fibres which enter either of two specialized structures* (olfactory bulbs), *stemlike projections under the front part of the brain. They sort the signals, and transmit them to the brain for processing.* [91] See Figure 4.8.

Figure 4.8: The Human Olfactory System [92]
The human olfactory system has between 10 and 40 million sensory receptor cells that connect to the brain via the olfactory bulb which allows us to identify between 3000 and 10,000 different aromas. *[Sagittal section illustrated by Patrick J. Lynch. Reproduced from Wikimedia Commons under Creative Commons licence; Olfactory bulb reproduced under licence from Photolibrary]*

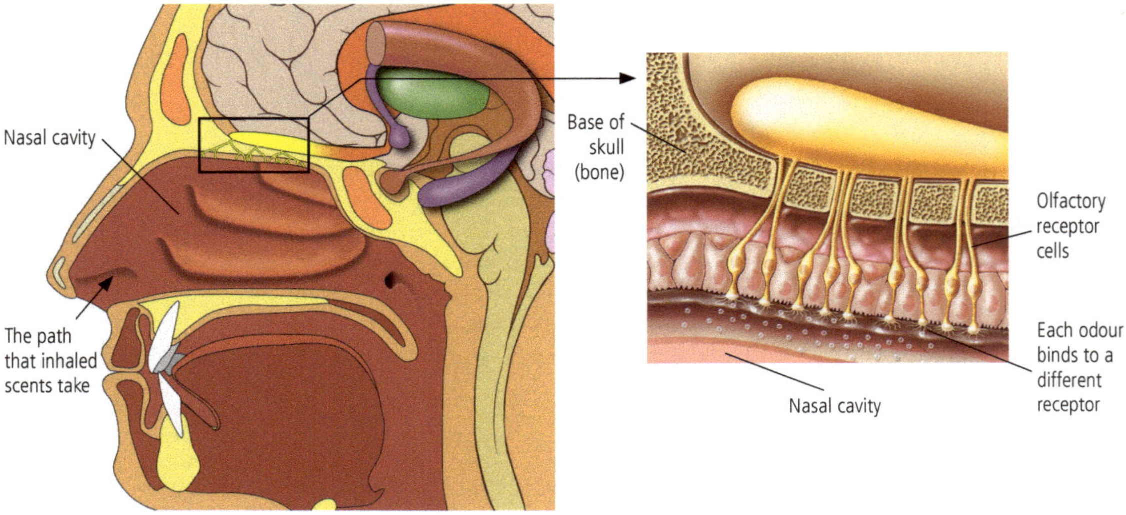

Brynie relates: *"In 1991 Columbia University scientists Linda Buck and Richard Axel discovered the receptor proteins that bind to odorant molecules. Buck and Axel received the Nobel Prize, and research on smell has boomed in the last twenty years, much of it directed towards the puzzling question of how humans can detect some ten thousand different odours using only one thousand different kinds of receptors."*[93] *"Conventional wisdom holds that molecules of odorants lock onto receptors in the nose. They fit there because of their size and shape. When they dock—like a boat in a narrow slip—they cause a subtle change in the shape of the receptor. That shape-change triggers a nerve impulse."* [94]

However, in 1996, Luca Turin, a biophysicist at University College, London, *"...proposed that the smell sensors actually detect the energy at which the different chemical molecules vibrate."* [95] *Turin...says shape is only one part of the explanation for how we smell. He says that smell receptors in the nose respond to the* vibration *of molecules after the docking. Perceived odors differ because the* frequency *of molecules' vibration differ—not just their size and shape. To support his assertion,*

90. Brynie, F.H.: *Brain Sense: The Science of the Senses and How We Process the World Around Us.* Amacom Books, NY (2009) p.39.

91. Sarfati, J.: *By Design.* Creation Book Publishers (2008) p.44.

92. Raloff, J.: *Dirty minds.* Cosmos 37:78 (2011).

93. Brynie, F.H.: *Brain Sense:* (2009) p.41.

94. *Ibid.* p.46

95. Sarfati, J.: *By Design.* Creation Book Publishers (2008) p.44.

Turin points out that some molecules that are almost identical in shape smell very different. Others that have different shapes, but vibrate at the same frequency, smell more or less alike." [94] Sarfati adds: *"Thus these receptors seem to work on the same quantum mechanical principles as vibrational spectroscopy."* [95,96] *"This means the smell receptors actually detect the energy of vibrational quantum transitions in the odour molecules, as first proposed by G.M. Dyson in 1937."* [97]

Brynie continues: *"In the 1920s and 1930s Malcolm Dyson, who had the misfortune to be gassed in World War I, experimented...with the class of compounds called isothiocyanates: mustard gas and phosgene. He started adding atoms to the compounds and changing their shape...Some of his molecules had similar shapes, but they smell different, while some with different shapes smelled the same. Dyson knew that molecules vibrate, but he had no way to measure their frequencies. Nonetheless, he proposed that the human nose must be responding to 'osmic frequencies.'...in the 1950s Canadian scientist Robert Wright breathed new life into the notion...Wright made some good arguments, and he continued espousing them through the 1970s. Unfortunately the range of 'smellable' frequencies he postulated was too narrow, and his ideas never garnered much serious attention."* [98]

"Then, in the 1990s, along came Turin. He suggested that the vibrational frequency of the molecule offered a better explanation for odour than did simple molecule shape. Turin rejected the idea that each type of smell receptor might respond to a different frequency. He asserted instead that each type responds to the same frequency range. *Some molecules have no odor because they are too large to fit into a receptor. Small molecules that vibrate but have no smell could, Turin said, lie below the threshold of frequencies that stimulate olfactory receptors."* [98] Sarfati adds: *"Turin proposes that below a threshold of 400 cm* $^{-1}$*, 'background noise' swamps the vibrational signal, so the nose cannot detect it."* [99]

One challenge to Turin's theory is the different smells of some enantiomers (optical isomers), as they have identical vibrational spectra. For example, R-carvone smells like spearmint, and S-carvone like caraway. The answer is: the spectra are identical only in an achiral medium, as in solution or gas phase. But the smell receptors are chiral and orient the two enantiomers differently. This means that different vibrating groups lie in the tunnelling direction in each enantiomer. Turin thinks that the caraway S-carvone is oriented so a carbonyl (C=O) group lies in that direction, so is detected; in the minty R-carvone, it lies at right angles, so is ignored. Turin supported this by manufacturing a caraway scent by mixing the minty carvone with the carbonyl-containing butanone ($C_2H_5COCH_3$)*.* [99]

"The precise chemistry of olfaction is still little understood. But Turin believes he has found a peptide (short protein) that could function as the electron donor together with NADPH. He has also found five amino acid residues coordinated to a zinc atom that could be the acceptor site. One warning sign of zinc deficiency is loss of the sense of smell, and zinc is often involved in biological electron-transfer reactions." [99]

In addition to smell, taste also converts chemical signatures in our environment into information that is intelligible to our brains but on a much more intimate level, as the object needs to be placed in your mouth first (which is not something that you want to do with a piece of meat to test whether it is rotten or not). The last of the five senses, touch also converts the physical composition of our environment into a digital signal that our brain can interpret but once again it requires intimate contact. Although all the senses are valuable and contribute different information, in general, their overall benefit is proportional to the amount of information that each sense collects. Thus vision, hearing and smell,

96. Sarfati writes: *"Turin proposed a mechanism where an electron tunnels from a donor site to an acceptor site on the receptor molecule, causing it to release the g-protein. Tunnelling requires both the starting and finishing points to have the same energy, but Turin believes that the donor site has a higher energy than the receptor. The energy difference is precisely that needed to excite the odour molecule into a higher vibrational quantum state. Therefore when the odour molecule lands, it can absorb the right amount of the electron's energy, enabling tunnelling through its orbitals."* [See Sarfati, J.: (2008) p45.]

97. Sell, C.: On the Right Scent, *Chemistry in Britain* 33(3):39–42 (1997) *in* Sarfati, J.: (2008) p.45.

98. Brynie, F.H.: *Brain Sense:* (2009) p.47.

99. Sarfati, J.: *By Design.* Creation Book Publishers (2008) p.46.

which can collect information remotely, often have a greater survival value than taste and touch. Some animals have other senses however which contribute large amounts of information about their environment and echolocation is the one we will examine next.

Echolocation

Echolocation is the ability to locate objects and estimate distances using very high-pitched sounds. Animals use it for locating food and navigating where vision is restricted, such as at night, in dark caves or underwater. Fowler and Kuebler write: *"Echolocation...is found in bats, two bird species* [oil birds and cave swifts], *dolphins,...*[toothed] *whales* [and shrews]. *Although the details of echolocation differ in each case, the basic principle is the same."* [100,101] Distance is estimated by the time delay of the echo that returns to the animal, while the size and shape of objects can be estimated by the scatter or distortion of the returning signal which bounced off the object of interest. Jonathan Sarfati adds: *"The pitch of the echo provides information on the objects speed via the Doppler Effect, named after the Austrian physicist Christian Doppler (1803–1853), who analysed sound from moving sources in 1842."* [102]

Bats and toothed whales produce ultrasounds which are sounds above a human being's ability to detect but still part of the electromagnetic spectrum. Other animals, such as the nocturnal, cave-dwelling oilbirds of South America, emit sounds that are in the audible range for human beings. The mechanisms are different but are similar to the way humans generate speech and sound. Horseshoe bats generate the sound pulses through specialised nasal structures whereas *"[t]he dolphin produces clicks by pressurizing the nasal system, then manipulating air through its sonic lips. The dolphin also allows an echo to return before emitting the next click... Echolocating cetaceans* [dolphin and whales] have *special oil-filled sinuses in the lower jaw which pass the echoes to the inner ear. They also have an amazing adaptation called the 'melon', a fatty protrusion on the forehead. This 'melon' is actually a sound lens – a sophisticated structure...to focus the emitted sound waves into a beam that the dolphin can direct where it likes. This sound lens depends on different lipids (fatty compounds) bending the ultrasonic sound waves travelling through them in different ways. The different lipids have to be arranged in the right shape and sequence in order to focus the returning sound echoes. Each separate lipid is unique and different from normal blubber lipids, and is made by a complicated chemical process that requires a number of different enzymes."* [103,104] It is clearly a complex system requiring a vast amount of information. Although there are a range of mechanisms, each one requires emission of a signal then the collection of the returning echo. For audible signals it would appear possible that echolocation is a simple extension of the range of hearing but for the bats and toothed whales that produce ultrasound, which organ evolved first and why? Why would a mechanism to detect ultrasound develop when there was no ultrasound generating device in the environment? Or why would a mechanism to generate ultrasound develop and be maintained when there was no way to detect the ultrasound and hence develop a benefit from it? Natural selection requires a function prior to development and there would appear to be none unless both the separate emission and detection devices evolved simultaneously but the chances of this happening at random seem ludicrously remote.

100. Fowler, T.B. & Kuebler, D.: *The Evolution Controversy* (2007) p.111.

101. Sarfati, J.: *By Design*. Creation Book Publishers (2008) p.51.

102. *Ibid.* p.47.

103. *Ibid.* p.48.

104. Echolocation-how stuff works- http://animals.howstuffworks.com/animal-facts/echolocation-Info.htm

The next fact to ponder is that dolphins and toothed whales have very different mechanisms for echolocation, despite appearing to be relatively closely related in Darwinian evolutionary terms while the only two bird species with echo locating ability are considered unrelated in evolutionary terms.[101] Thus, as in the case of the eye, vastly complex feats of biological engineering would have had to have evolved more than once which is hard to believe when the evidence for Macroevolution to build the first eye or develop an echolocation system, even once, is just not there. So the development of echolocation is just one more example where Neo-Darwinian evolution is incapable of giving a plausible scientific explanation. However, it is now time to look at other aspects of our first species.

Locomotion: The Flagellar Motor[105]

A number of bacteria including E. coli grow a long whip-like tail (called a flagellum) that can rotate like an electric motor. It can spin at up to 100,000 rpm and its intricate construction and complexity have sparked significant debate as to its origin.[106] The flagellar motor bears a close resemblance to the battery-powered engine in a toy submarine with the flagellum equivalent to the propeller. Michael Behe drew attention to the precise way in which the flagellar motor's forty proteins are fitted together to make a very efficient, discrete unit and he postulated that its multiple parts could not have been built up one protein at a time as Darwin's theory requires. Behe argues that as the function of the flagellum cannot be broken down into smaller components without losing its function, its ability to evolve by chance is nil and he has suggested the term: *"irreducibly complex"* [107] to describe this characteristic. In using this term Behe means that if the machine was simplified in any way then it would stop working, lose its function and hence lose its power to aid the survival of the bacterium and therefore fail to be conserved by natural selection. In reply to this argument, Behe's critics have suggested that different groups of proteins forming discrete parts of the flagellum may have been co-opted from other functional molecular machines within the cell. As an example, they cite a small pump/injector which is constructed from ten of the proteins that are required for the flagellar motor. However, there are problems with this suggestion. Firstly, the flagellar motor has forty proteins and the bacterial pump provides only one quarter of them but paramount is the derivation of the information to provide the epigenetic control for the flagellum, because apart from the instructions to build the motor, there needs to be operating instructions so that the bacterium can turn the flagellum on and off as necessary; even to drive it in reverse if that is appropriate. As was observed in Chapter 3, this is beyond the abilities of Neo-Darwinian mechanisms, due to the Law of Coding. Thus, there is a requirement for the issuing of new assembly and operating instructions.

A recent report provides an explanation for the similarity between the proteins in the flagellar motor and the bacterial pump/injector. Scott Minnich, Associate-Professor of Microbiology at the University of Idaho has researched the flagellar motor and reports that the bacterium needs cool temperatures to grow the flagellum, which does not form in temperatures over 37°C but forms the diminutive form of the structure, which acts as a pump or genome injector.[106] This suggests that there is an in-built switching mechanism, which changes the development of the bacterial structures, depending on the environment as the bacterium matures. Temperatures above 37°C are indicative of a food source

105. Meyer, S.C. et al: *Explore Evolution.* Hill House Publishers, London (2009) pp.115–124.

106. Macnab, R.M.: The Bacterial Flagellum: Reversible Rotary Propeller and Type III Export Apparatus," *Journal of Bacteriology* 181(23):7149–53 (1999) *in* Minnich, S.A. & Meyer, S.C.: Genetic Analysis of Coordinate Flagellar and Type III Regulatory Circuits in Pathogenic Bacteria. *in* Dembski, W.A.: (Ed.) *Darwin's Nemesis.* Inter-Varsity Press, Downers Grove, Illinois (2006) pp.214–223.

107. Behe, M.J.: *Darwin's Black Box.* (2006) p.65.

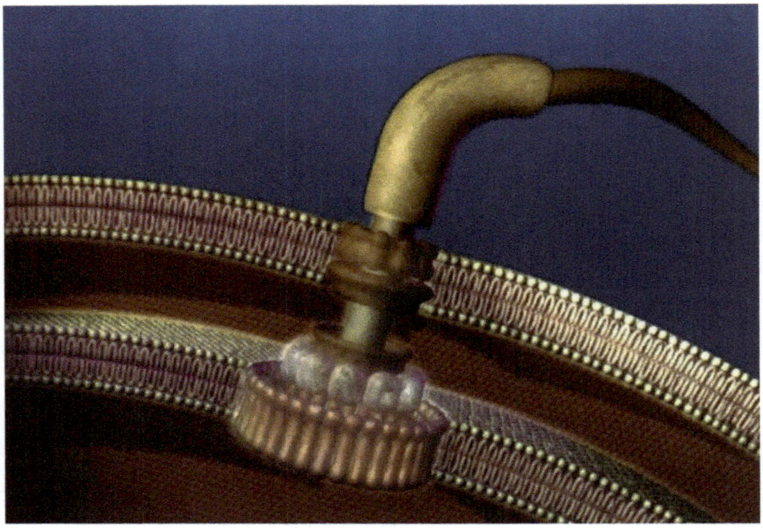

Figure 4.9: The Flagellar Motor[27]
[Illustration is from Unlocking the Mystery of life DVD (2002). Reprod with permission courtesy of Illustra Media.(27:30)]

and so the bacterium does not need the flagellum to travel any further but instead it needs the pump/injector to access the food which is already there. The information that controls the switch is likely to be resident in the epigenome.

Blood and the Clotting System[108]

The bacterial flagellum is surprisingly complex and specialised for a single-celled organism but human beings have a trillion cells and the more complex an organism is, the more specialised its subsystems need to be to maintain its functional integrity. The bloodstream is one such subsystem. The bloodstream is the nutrient supply line for the body and uninterrupted, effective functioning is imperative for mammalian and human life. Oxygen and nutrients, such as glucose and amino acids, are delivered and carbon dioxide, uric acid and other waste products are removed. Total circulation time is less than a minute and the total circulating volume in an adult human is only five litres of blood so the blood circulates at high speed and ceaselessly. If there is a defect in the system, such as a severed blood vessel, the blood spills out. Leaks need to be stemmed rapidly to prevent the individual from bleeding to death. To do this, the blood solidifies by a process called 'clotting' or 'coagulation'. It is a tightly regulated, highly complex process. The reason is that it must clot rapidly when needed (such as a cut finger) but remain viscous at all other times because if it were to clot in the blood vessels, it would stop the flow of life-giving nutrients and the individual may suffer injury or die.

Control of bleeding (haemostasis) is achieved by a progressive activation of special components in the bloodstream, which accumulate at the site of the injury. This is called the '*clotting cascade*' and is initiated by the damaged tissues. In the case of external trauma which results in, for example, a cut finger, the exposure of the damaged tissues to the air allows a phospholipid called 'tissue factor', which is present in the damaged tissue, to start a domino effect in the clotting factors which are permanently circulating in the bloodstream. The activation of the clotting cascade results in fibrinogen being converted to fibrin which allows a clot to form. This is called the tissue factor (or extrinsic) pathway. In addition, sticky

108. Behe, M.J.: *Darwin's Black Box*. (2006) pp.77–97.

cell fragments called platelets become activated and release granules which further promote the coagulation process. The platelets adhere to the damaged area to form a plug, with the long fibrin strands linking together to form a rough mesh. Red blood cells embed in the rough mesh to plug the leak. Alternatively, if there is no external trauma but there is damage to the lining of a blood vessel, a different pathway ensues, which is called the 'contact activation (or intrinsic) pathway'. Proteins which are usually isolated from the bloodstream, most notably von Willebrand factor, are then able to come into contact with the clotting factors in the bloodstream, forming a clot in a similar manner to that described in the tissue factor pathway above. The two activation pathways combine in a final common pathway to form the mature clot. This is a vast oversimplification, as the process requires a number of cofactors such as vitamin K, calcium and phospholipid and regulators such as protein C, antithrombin, plasmin, heparin and prostacyclin among numerous others. Anyone interested in the dozens of chemicals involved can consult a haematology textbook, however, the major clotting factors are summarised in Figure 4.10. All mammals have a similar coagulation system involving both cells (platelets) and proteins (the clotting factors).

That is not the end of the story however, as the clot is only an emergency and temporary repair. Once a definitive repair has been made by the body underneath the protection of the overlying mature clot, the reverse process of fibrinolysis begins to remove the scab through the effects of plasmin and other mediators, which is again, a complex and tightly regulated process. It becomes immediately apparent that there is a vast amount of information required to have this intricate system in a perpetual state of readiness, although most of the time it lies dormant and unused; yet the survival of an injured individual often depends upon it. It is equally apparent that there are many places where the process can

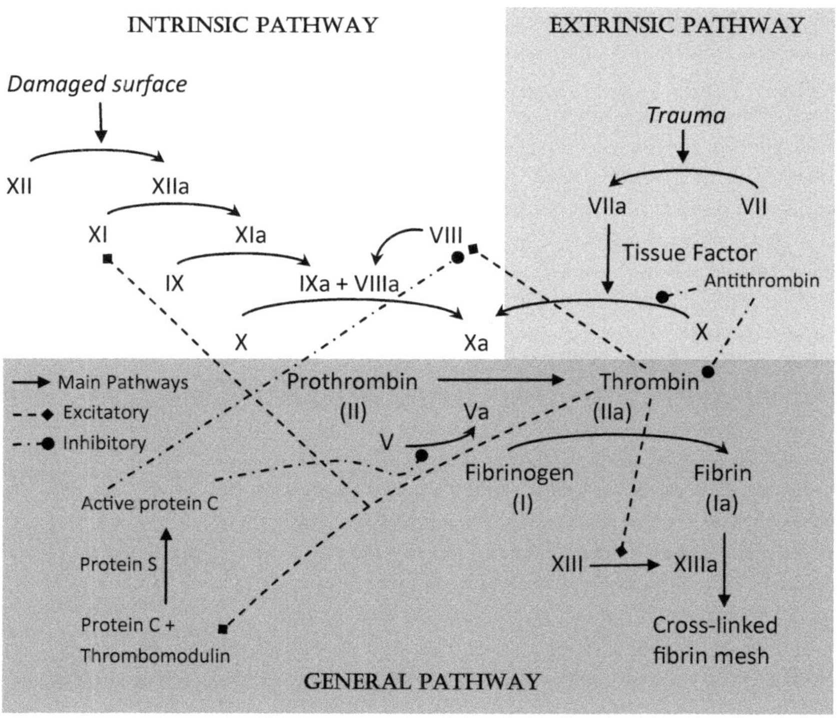

Figure 4.10: The Clotting Cascade [109]
The clotting cascade involves more than twenty different proteins, molecules and cell types to maintain the integrity of the cardiovascular system and lack of any one of them leads to life-threatening bleeding or clotting that may be fatal. It is inconceivable how this system could be either simplified or built up one step at a time.

109. Diagram redrawn from Braunwald, E. et al: *Harrison's Principles of Internal Medicine* (11th Ed.) McGraw-Hill Book Company, New York (1987) p.270 and http://www.manfred.maitz-online.de/Publications/Presentation. (31st March, 2013)

go wrong and lack of factors, such as Factor VIII, leads to haemophilia and bleeding problems, while platelet problems can result in von Willebrand disease. Alternatively, ineffective Protein C may lead to inappropriate coagulation and life-threatening deep vein thrombosis (DVT) in the leg or pelvic veins which, if left untreated, can travel to the lungs and block major blood vessels causing sudden death. The information coding that is required to specify this level of complexity rivals ultra-sophisticated computer programs, which need vast amounts of information just to exist in the first place, with further complex coding to maintain and operate them indefinitely. The system is permanently balanced on a knife's edge and to assert that the coagulation system occurred by chance appears absurd. As complex as the clotting system is, however, it is surpassed in human beings by even more complex systems; so now we will examine the human immune system.

The Immune System

The human immune system is a hugely complex and highly integrated network which defends our bodies from both internal and external attack. External attack can come from invading pathogens such as bacteria, viruses and fungi amongst others, while internal attack can come from mutations in the genome that set off a cancerous process that can consume the body, if not stopped. A cancer is simply a cell type which does not stop growing when it should and hence takes over the body, destroying the tissues around it like weeds taking over a field and smothering everything else that grows there. Biologist, Gerry Bergman, Adjunct Associate Professor at Medical University, Ohio and immunologist, Nancy O'Sullivan, Research Assistant Professor at the Department of Anatomy and Cell Biology, Wayne State University School of medicine, Detroit explain that: *"Two basic types of immune systems exist,* innate *and* adaptive. *Innate immunity is a germline* [DNA] *encoded system found in all multicellular plants and animals, including humans. This system's genes are expressed without major modification, and its receptors can recognise specific features of most pathogens, allowing it to respond to pathogens it has never encountered before. The adaptive immune system consists of two arms. Cell-mediated immunity against intracellular pathogens is carried out by white blood cells called T lymphocytes while humoral immunity against extracellular pathogens and toxins is mediated by antibody-producing B lymphocytes. The humoral adaptive immune system is able to produce enormous antibody diversity and 'adapt' to fight specific pathogens."* [110]

Even simple creatures such as worms have defence systems. Recently, researchers have discovered a communications system in worms that connects the worm's nervous and immune systems to help coordinate responses to invading organisms. Specific cells in the worm's nervous system coordinate the defence against toxic bacteria. *"Those first responders are part of the innate immune system, a kind of 'sixth sense' that is hard-wired and fends off invading microbes until the adaptive immune response is mobilised."* Alejandro Aballay at Duke University Medical Centre writes: *"It has been recognised for at least 20 years that there must be bi-directional communication between the nervous and the immune systems. But because of the complexity of the communication network it has been very difficult to prove this connection conclusively. The complexity of the nervous and immune systems of mammals, including humans, makes sorting out neural-immune communications a daunting task."* [111]

Bergman and O'Sullivan continue: *"All multicellular organisms have an elaborate innate immune system, and many animals have one of the two known arms*

110. Bergman, J. & O'Sullivan, N.: Did immune system antibody diversity evolve? *Journal of Creation* 22(2): 92–6 (2008).

111. Aballay, A.: Nervous System Rallies Immune System Forces against Invading Pathogens. *Howard Hughes Medical Institute News at* http://www.hhmi.org/news/20080919aballay.html (19 Sep. 2008).

of the adaptive system that produces enormous antibody diversity to destroy specific pathogens...The recent sequencing of the echinoderm sea urchin genome has found an 'unexpected sophistication' in this algae-eating life form, once labelled 'primitive'. Its immune system is actually more complex than the vertebrate system...As many as 5% of the sea urchin genes are involved in immune functions...The sea urchin even has interleukin and tumour necrosis factor genes previously known only in chordates." [110] Bergman and O'Sullivan review the immune systems of a range of organisms including worms, fish, sharks and vertebrates concluding: *"Various phyla use 'a remarkably extensive variety of solutions to meet fundamentally similar requirements for host protection.' The large discontinuity between the various means of generating immune system diversity in the animal kingdom makes it highly unlikely that one system could have evolved into another."* [110]

The complexity of the mammalian immune system is also vast. In an article in *Nature* in 2010, König et al identified 295 human cell factors that Influenza A strains had to overcome or evade to infect a host cell. The H_1N_1 'swine flu' belongs to this group of viruses.[112] We do not have the space or time to detail the full complexity of the human immune system although you can look it up in an immunology textbook if you are interested. To summarise it briefly however, as in other complex organisms, the human immune system has both cells and special proteins which fight infection. The cells tend to be non-specific in the way they fight invaders whereas the proteins, which are called 'immunoglobins', are like microscopic heat-seeking missiles which search out the enemy cells individually and destroy invading bacteria and viruses one by one. Each immunoglobin has a unique shape to lock onto one specific invading microbe. So the number of differently shaped immunoglobins is enormous. When the body is not under attack, the immune system retains memory cells of past infections, which are usually stored in lymph nodes which are like filters. The lymph nodes also act as libraries so that if an organism to which the person has previously been exposed by infection or immunisation should invade, the appropriately shaped protein can be located quickly and the 'library' in which the memory cell was stored is converted into an armaments factory which swells to mass produce the specific homing missile to neutralise the invading microbe. The extraordinary precision with which the immune system does this is specified by information in our genome and probably the epigenome as well. The origin of the information is still obscure, as was seen in the clotting system and the eye. However, before we address the source of the code, there is one entity which leaves even the sophistication of the clotting and immune systems in its wake and that is the human nervous system: the brain and the human mind.

The Nervous System

Improvements in microscopes and chemical staining techniques during the 19th-Century allowed researchers to discern the cells that make up the nervous system. Spanish anatomist, Raymon y Cajal, sliced tissue thinly and discovered individual nerve cells (called neurons). Cajal discovered that rather than being round like most other cells, nerve cells, were elongated with multiple bumps. These swellings formed the connections to neighbouring cells which are called *synapses*. Cajal began looking at other organisms and found that all vertebrates have brains made of neurons.[113] It has since been discovered that synapses are the same from human

112. König, R. et al: Human host factors required for influenza virus replication. *Nature* 463:813–817 (2010).

113. Grant, S: From little things, big things grow. *Univ. Sydney Alumni Mag.* Autumn 2009 pp.14–17.

beings to lobsters.[114] So, if Darwinian evolution is true, neurones must have been part of the living genome since the formation of crustaceans and not changed significantly since then. Although the networking of nerve pathways in mammals has greater complexity than crustaceans, such as lobsters, the basic neuronal structure remains the same. However, the pinnacle of neuronal architecture is the human brain and we need to spend some time wrapping our minds around its functioning now.

Brain Function and Thought

The human brain is thought to be the most complex entity in the Universe. It is estimated to have 100 billion (10^{11}) neurones or brain cells, each one of which has been estimated to have about 1000 connections or synapses. The total connectivity of the brain is therefore 10^{14} connections, which all fit very neatly in a space the size of a basketball! In this unimaginably dense and sophisticated network, how do we manage to retrieve anything and when we do, how do we do it so quickly? And to complicate matters further, for every neurone there are ten supporting cells. Science writer, Dr Karl Kruszelnicki explains: *"These trillion neurons are 'supported' by 10 trillion other cells. They include the astroglia (which make up the scaffolding that holds the neurons in place, and which also feed the neurons with nutrients). Another class of support cell is the oligodendrocyte, which makes the insulating myelin sheath that wraps around part of the neuron. There are also epididymal cells, which line the hollow chambers (called ventricles) in the brain. Other cells, the microglia, act as the brain's equivalent of the immune system. The neurons make up 10% of the cells in the brain, while the supporting cells make up 90%; and it gets more complicated yet. One strange finding from the autopsy examination of Einstein's brain was that he had more of these 'supporting' cells than average."* [115]

Dr Seth Grant, Professor of Molecular Neurobiology at the Welcome Trust Sanger Institute in Cambridge, England wrote in 2009: *"One human brain has a million times more connections then the entire World Wide Web…There are over 200 genes involved with learning and 1500 proteins associated with the normal functioning of a mammalian synapse…This makes the mammalian synapse the most complex of any cellular structure known."* [113] Grant's research suggests that our brain power is dependent upon the small gap between the nerve cells or neurones rather than the neurones themselves. These gaps are called synapses and hence the term: 'synapse proteins'. Grant discovered that invertebrates such as lobsters, have about half the number of vertebral synapse proteins but surprisingly unicellular organisms also contained 25% of mammalian synapse proteins too.[113] This may explain unicellular organisms' ability to process simple information, for example, relating to food sources and light. The bewildering complexity of the synapse may also explain some of the range of mental illness observed in human beings since the greater the complexity in a system, the more chance there is that something can go wrong.

In addition, each synapse can utilise between 100 and 150 specialised messenger molecules called neurotransmitters to communicate its message to the next neuron; this complexity is far greater than previously imagined.[116] Further still, synapses have the ability to change and adapt in response to stimuli, which is sometimes referred to as *plasticity*. This is the process we usually think of as 'learning' and is considered to be part of the mechanism involved in tolerance to drugs, such as morphine and benzodiazepines, for example, diazepam (Valium).

114 Wieland, C. & Farmer, M.: *Clash over Origins—Creation vs Evolution Debate.* (DVD) Creation Ministries International.

115. Kruszelnicki, K.S.: *Great Myth Conceptions.* HarperCollins Australia (2004) p.163.

116. Encyclopaedia of Mental Disorders (2009) http://www.minddisorders.com/Kau-Nu/Neurotransmitters.html

To put this in context, when a computer communicates between its microchips, it users only one type of messenger, an electron, to connect with the next entity in its internal network; imagine the power that might eventuate if your computer's microchips had 100 different electron-type messengers to choose from. This is on top of the 10^{14} connections described by Professor Grant. However, this is only a mechanistic model of brain functioning and gives no hint of how we are able to think. Philip Johnson writes: "[Materialist scientists] *struggle with what is termed the 'hard problem' of mind-science: how it is possible for material events in the brain to produce subjective conscious experience in the mind. If the brain is fundamentally a computer or group of computers, then a sensory input can produce an output (for example, lack of water can stimulate cells to invoke a mechanism to cause drinking). But computers do this without having any conscious awareness. Why aren't people like computers in this respect?*" [117]

So where do we start to figure out this problem of understanding thought as described by Johnson? We know that when we generate a thought, as the neurones converse with each other, they create tiny electrical-type impulses which can be graphed as waves on an EEG machine and we know that different regions of the brain are utilised, which can be imaged with a special type of scanner called a Positron Emission Tomography (PET) scanner. We also know that thoughts and emotions are intertwined; in fact a thought can generate an emotion.[118] Research in New Guinea in 1968 showed that all people experience the same emotions, which they expressed in a similar way through their facial expressions. Six emotions were recognised in all, including: happiness, disgust and fear and all were recognisable from facial expressions across cultures, even when there had not been any significant contact with that people group previously.[118] Thus the tangible physical structures of the brain and the intangible emotions and thoughts that are generated by the human brain appear to be similarly mediated in all races. This suggests that emotions, as well as brain structure, are derived from some source of stored information that expresses itself in a similar way in all cultures which is inherited rather than learned.

Stephen Pinker, Professor of Psychology at Harvard wrote: "*...brain science has shown that the mind is what the brain does.*" [119] The big question is, how does it do it? It is actually easier to say what mind/memory is *not* rather than what it *is*. Dr Karl Kruszelnicki wrote: "*...experiments [were] conducted by the neuropsychologist, Karl Lashley, in the 1920s. Lashley was trying to discover just where in the brain this strange thing called 'memory' was stored. He trained rats to run through mazes, and then measured how well they did as he removed more and more of the cortex of their brains. Lashley found that memory is not stored in one single place, but exists throughout the entire cortex, and probably a few other places as well.*"[120] Kruszelnicki continues: "*In the 1980s, Yorkshire Television in the UK showed a documentary called, Is Your Brain Really Necessary? It described the work of the late British neurologist, Professor John Lorber, who specialised in children's illnesses. He saw many cases of hydrocephalus, a condition involving the constant circulation of cerebro-spinal fluid around the brain and spinal cord. If too much fluid is produced, or if its outlet from the brain is blocked, then it can build up inside the skull. This excess fluid usually makes the skull grow bigger. However, sometimes it just makes the brain meat get thinner as the meat gets squashed up against the bony skull.*"

"*Professor Lorber discussed many cases where young people did not have much brain, but had normal intelligence. In one extraordinary case, a young man had only a one-millimetre thickness of grey matter in his brain, instead of the average 45*

117. Johnson, P.: Wedge of Truth. InterVarsity Press, Downers Grove, Illinois (2000) p.118.

118. Greenfield, S.: *Brain Story*. (Video) British Broadcasting Corporation, UK (2001).

119. Pinker, S.: "A Matter of the Soul." *The Weekly Standard*, Feb., 2nd 1998 *in* Johnson, P.: (2000) p.117.

120. Kruszelnicki, K.S.: Great Myth Conceptions. HarperCollins Australia (2004) pp.160–1.

millimetres. Even so, he had an IQ of 126 (the average is 100) and had gained an honours degree in mathematics!"

"This does not prove that most of the brain is useless. However, it does show that in some cases, the brain can recover from, or compensate for, quite major injuries."[121] So although the entire brain is important, memory storage can survive partial destruction of any part of the brain. If you think that this is confusing then you are not alone but there is some more information we can draw upon to clarify what has been observed.

In the same way that information is not the DNA within the cell but is an intangible entity that is stored in the order of the bases in the DNA strand, thought is not intrinsic to the neurones but seems to be associated with the order in which the neurones are activated within the brain. Life also appears to be a non-material, possibly extrinsic factor which is more than just the physical components of a cell, because to create life requires *dynamic* information which becomes active within a functioning cell. Recently, *Tyrannosaurus rex* DNA was extracted from a long dead bone. The DNA within the bone still contained some of its stored information but it was not alive. In a similar way, an electrical impulse in a peripheral nerve may contract a frog's leg when it is stimulated by an electric current (which many of us have done in science classes at school) but this does not constitute a 'thought' in that nerve cell. A thought requires a functional brain first (the hardware) but it also appears to require some extra intangible input in the same way that life requires an extra intangible input to create a living cell, not unlike the software in a computer.

Philip Johnson wrote: *"If you follow the materialist logic through to the end, as Pinker and Susan Blackmore have done, you will conclude that the personal self in which we put so much stock is no more than an illusion."* [117] Yet we are not an illusion so there must be something more than just the neuronal circuitry and biochemistry. But what else is there? Could we be looking at the wrong thing or in the wrong place because your thoughts as you read this page are not imaginary? What other evidence do we have? Well, we know that thoughts can be imaged as 'brainwaves' on an EEG machine and we know that the eye, the ear and the nose are frequency decoding devices, which operate digitally, so instead of thought being merely an electrical signal carried by neurones like an electric current running in an electrical circuit, could it be a frequency modulated process which is what we observe on an EEG? The work of Professors Russell and Karen de Valois of the University of California when viewed with what we have discovered already about information processing by our senses, suggests that the digital inputs from our senses and the digital throughputs of our neurones are converted through the mathematical process of Fourier transforms from intangible frequency information from our environment to digital, spatial information as it is interpreted by our (physical) senses and back again to frequency information for interpretation by our mind.[122] Thought might therefore be an intangible wave or frequency mediated process rather than just a physical, mechanistic one. The way this operates is obviously complex and additional background information is needed before we can pursue the subject of thought any further at present but we shall discuss it in greater detail in Chapter 10 once we have digested other essential background information. In our discussions above, I have not mentioned the embryogenesis or developmental mechanisms concerning the formation of the brain so there is a lot more information which has not been broached at this stage

121. *Ibid.* p.162.

122. Missler, C.: *Cosmic Codes.* Koinonia House, Coeur d'Alene, Idaho (2004) p.330.

but suffice it to say that apart from the baffling complexity of the anatomy and biochemistry of the brain, the order of construction is also important. Information must be applied in the correct order or it does not work. Again this will be dealt with in greater depth in Chapter 10.

Memes

Memes are a concept introduced by Richard Dawkins to convey the phenomenon of the collective memory of our culture, which is sometimes referred to as 'social capital'. It is a term that reflects the value to our society of the knowledge and understanding that has been gleaned over generations. Memes include information about behaviours learned by individuals and communities that are taught by parents to their children and handed on from generation to generation.[123] They include habits and values. Habits are learned shortcuts which are beneficial for our survival that allow us to function from day to day without having to rework everything from first principles again and again. In contrast, values, Mancur Olsen suggests, reflect what used to work, which implies that they may be out of date and may not be relevant today but as society tends to change slowly, values often have on-going beneficial behavioural information. Olsen's point however, is that there is no advantage in hanging onto information relating to circumstances that have changed and hence, the information contained within that societal value is no longer helpful. Values often relate to long-term solutions whereas habits are associated with short-term strategies.

The information in memes may be stored either in the brain or on other media, such as in books and is vital if human civilisations are to avoid repeating the mistakes of the past and have to rediscover that information from scratch.[124] These are culturally acquired rather than genetically (biologically) encoded. As such, memes are extrinsic information as they are learned, in contrast to intrinsic information which is internally coded. In computer terms, memes are like the information you entered into the word-processing package that you installed on your computer, which correlates with your experiences through your lifetime, as opposed to the machine language that is embedded in the central processing unit (CPU) by the manufacturer. Memes are part of the information from which we derive our world view and are fundamentally different from our ability to remember and our capacity for thought and reasoning.

Summary of the minimum requirements for the First Organism

In this chapter we have discussed some of the extra information that would have had to have been amassed if the first cell was to evolve into a multi-celled organism. Figure 4.11 summarises the information required to specify this first fully-competent, complex organism. However, these functions are present in nearly every living organism today also, so it would appear that they have been present from early in the history of life on Earth and throughout living history. The Burgess shale from the Cambrian Period includes many fossils even though they are completely soft-bodied animals. *"These remarkable fossils,"* according to Cambridge palaeontologist, Professor Simon Conway Morris, *"reveal not only their outlines but sometimes even internal organs such as the intestines or muscles."* [125] Thus, the fossil evidence confirms the intricate complexity of even the earliest organisms

123. Dawkins, R.: *The Selfish Gene*. Oxford University Press, Oxford (1976).

124. Johnson, P.: *Wedge of Truth*. InterVarsity Press, Downers Grove, Illinois (2000) p.109.

125. Conway Morris, S.: *The Crucible of Creation*. Oxford University Press, Oxford (1998) *in* Wells, J.: *Icons of Evolution*. Regnery Publishing, Washington, DC (2002) p.44.

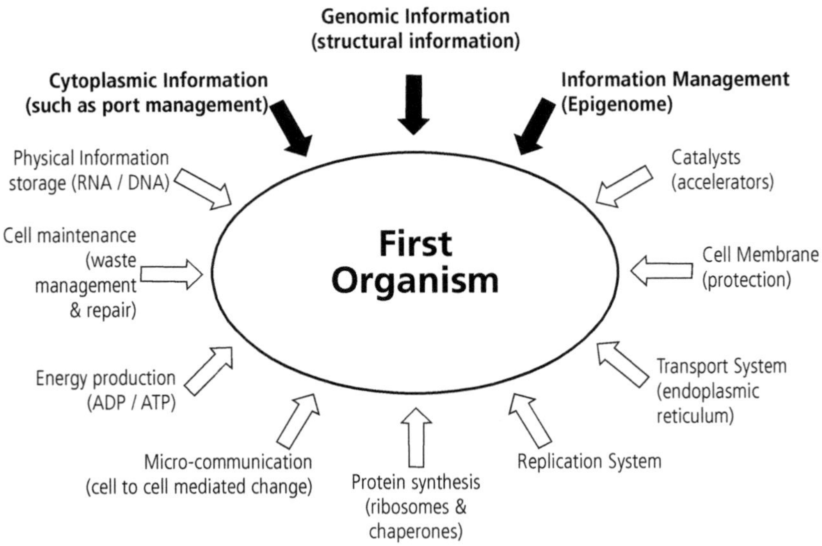

Figure 4.11: Minimum Information requirements for Formation of the First Organism
From the preceding discussion in this chapter it can be seen that this figure represents the minimum information requirements for a multi-celled organism, such as a hydra or a primitive jellyfish, yet the information is immense! More complex organisms require many more components, such as complex locomotor structures, information collecting devices (such as eyes), immune strategies, clotting routines, communications systems and information management infrastructure (such as a nervous system and brain).

and hence their information content. But where did this fantastic amount of information come from so early in the evolutionary timeframe? The Darwinian Model is at a loss to explain it. (See Figure 5.6)

The vast amount of operational infrastructure and hence information required for the first independently functioning organism is obviously enormous but it is impossible under Darwinian Theory for all this information to come together randomly at one place, at one time and just start working. It must therefore be presumed that some functions were acquired before others but it is difficult to work out which parts were initially 'unnecessary', as all the parts are indispensable for the organism to survive. It is therefore obvious that this immense number of functioning, information-laden systems did not just fall together one day in the Cambrian soup and start 'living'. Some process must have caused or assisted the compilation of the information, so what are our options?

Complexity Theory

To overcome this vast information requirement to specify the infrastructure for the first organism, some evolutionary theorists have developed Complexity Theory, where they hypothesise that self-organisation and auto-catalysis initiated life. Walter Bradley, Professor of Mechanical Engineering at Texas A&M University, explains: *"Stuart Kaufmann...of the Santa Fe Institute is the leading proponent of the most popular new paradigm. Accordingly complexity results naturally from the self-organising character of nature. In two recent books Kaufmann...argues that the origin of 'life' is essentially inevitable, not vastly improbable as Jacques Monod (1972), Francis Crick (1981), Robert Shapiro (1986) and...Thaxton, Bradley and Olsen (1984)...have previously argued. Kaufmann defines 'life' as a closed network of catalysed chemical reactions that reproduce each molecule in the network — a self-maintaining and self-reproducing metabolism that does not require self-replicating molecules. Kaufmann's ideas are based on computer simulations alone without any experimental support...If one has a system of 1,000,000 molecules then in theory it becomes highly probable that most molecules are catalysed in their production.*[126] Fowler and Kuebler write: *"Rather than positing that life originated as a simple self-replicating molecule, Kauffman believes*

126. Bradley, W.L.: 'Nature. Design or Designoid' *in* Dembski, W.A. (ed): *Mere Creation. Science, Faith and Intelligent Design.* InterVarsity Press (1998) p.44 .

that life originated wholesale, as an ordered dynamic network of molecules. According to Kaufmann: 'There are compelling reasons to believe that whenever a collection of chemicals contains enough different kinds of molecules, a metabolism will crystallize from the broth. If this argument is correct, metabolic networks need not be built one component at a time; they can spring full-grown from a primordial soup.' " [127]

If this was the case, then why didn't the Miller-Urey experiment in 1953 generate an amoeba or at least a lot more organic products than it did? This experiment and many like it have been repeated over the last fifty years and nothing remotely like life has ever looked like emerging. Even more poignant, why doesn't every stagnant pond generate new life forms today? Kaufmann continues: "*When the number of catalysed reactions is about equal to the number of [chemicals], a giant catalysed reaction web forms, and a collectively autocatalytic system snaps into existence. A living metabolism crystallizes. Life emerges as a phase transition.*" [128] This completely ignores the fundamental problem of where did the information come from to specify all this complexity? This is pure woolly, wishful thinking and has never been observed; it is a modern version of the 19th-Century protoplasm concept. In Miller-Urey type experiments, autocatalysis only results in a gelatinous mess, not thriving colonies of amoebae or bacteria or even viruses which are much simpler. Self-organisation and autocatalysis make an elegant theory but the evidence is completely absent. All 'self-organising' systems need information input *first* and the problems of information storage and replication have not been broached at all.[129]

Darwin looked at the mammalian eye 150 years ago, and even with the scientific knowledge of Victorian Britain, he wrote: "*If it could be demonstrated that any complex organ existed, which could not possibly have been formed by numerous, successive, slight modifications, my theory would absolutely break down.*"[130] The immense complexity of a 'simple' organism, let alone a functioning eye or immune system, leaves one wondering what Darwin would have thought today. As previously discussed, Michael Behe referred to this phenomenon of interlocking function as: "*irreducibly complex*" which in information terms is *co-dependent information*. Michael Behe argued that irreducible complexity applies to cellular organelles while Michael Denton had earlier argued that the same principle also applied to the building blocks of organelles, the proteins. What we have observed is that it applies at all levels and co-dependent information is a feature of all living things.[131]

Co-dependent Information

Simple arrangements of information, where one molecule links to another then another, might be imagined to develop along Darwinian principles, that is, gradual accretion over time but is there a level of information acquisition that could be defined as impossible to attain by this means? Behe in: *The Edge of Evolution*, describes the capability that malarial parasites have to develop antibiotic resistance through simultaneous mutations in their genome. He discovered that a single mutation is relatively easy, two simultaneous mutations (for example for chloroquine resistance) is rare and three simultaneous mutations has never been observed and is assumed to be impossible. It is interesting to note that in treating tuberculosis, it was found empirically that if three antibiotics with different mechanisms of action were used simultaneously,[132] bacterial resistance did not develop, even though the standard treatment course for tuberculosis was eighteen

127. Fowler, T.B. & Kuebler, D.: *The Evolution Controversy.* (2007) p.316.

128. Kauffman, S.: *At Home in the Universe: The Search for Laws of Self-Organisation and Complexity.* Oxford Univ. Press, Oxford (1995) p.62 *in* Fowler, T.B. & Kuebler, D.: (2007) p.317.

129. Fowler, T.B. & Kuebler, D.: (2007) p.320.

130. Darwin, C.: *The Origin of Species.* John Murray, London (1859) reprinted by Penguin Classics, London (1985) p.219.

131. Fowler, T.B. & Kuebler, D.: (2007) p.255.

132. Historically: rifampicin, isoniazid and ethambutol. A fourth antibiotic has now been added to the previous three, pyrazinamide due to developing resistances of the antibiotics when used individually.

months of continual antibiotics and hundreds of millions of people around the world have been having the treatment for the last sixty years. Tuberculosis resistance has now arisen through bacteria acquiring immunity to the antibiotics when they have been used individually, with the subsequent accumulation of these resistances through horizontal information transfer systems, such as transposons but resistance has not been shown to be a problem when all three antibiotics were used together to treat bacteria with no previous resistance to them. These findings suggest that three simultaneously intersecting pieces of co-dependent information can be considered to be unlikely to occur by chance. However, to be absolutely certain that an occurrence was not by chance, consider four simultaneously intersecting pieces of information. (See Figure 4.12) There could then be certainty that the occurrence was beyond random chance and hence we would have a suitable definition of exclusively co-dependent information; that is, it could not have occurred by chance. Yet Figure 4.11 illustrates that in the first organism, there are at least ten different co-dependent informational subsystems which must simultaneously connect with each other to form the first organism, not to mention the myriad of information that comprises the various biochemical routines which constitute each of the subsystems. If quaternary co-dependent information cannot occur by chance, then neither could the first organism and meaningful pre-specified information must have been required to build it. A definition for co-dependant information might therefore be: *functional outcomes that are dependent upon the combination of two or more pieces of information* where there are emergent properties that are only manifest by the combination of mutually discreet sub-systems.

Now Darwin argued that although the formation of structure ABCD in Figure 4.12 is impossible in one step, it is likely to occur if AB forms then ABC then ABCD. The problem here is, finding a viable functionality for each step along the way. This may sound simple in this example but real life biological examples involve a minimum of 256 proteins x 100 amino acids per protein which equals a minimum of 25,600 steps not counting any scaffolding proteins or RNA/DNA, ribosomes, chaperones et cetera needed for the construction process. Each of these tens of thousands of steps needs a viable function. Then, each unit in each protein system has to retro-write its DNA/RNA 'parent' abcd which (as was shown in Chapter 3), defies the Law of Coding, let alone the equally necessary task of retro-retro-writing the epigenome, which is required to control the production process and we haven't even started on embryogenesis yet! It can be seen that co-dependent information is required at many levels in biology, which is far beyond what Darwin and his colleagues envisioned 150 years ago. Even the notion of co-opting[133] functions already extant in other simple organisms (called

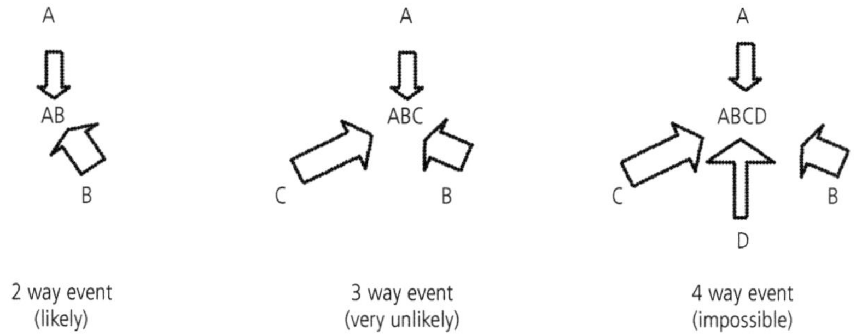

Figure 4.12: Chance Occurrences of simultaneous events in Nature
These diagrams illustrate the likelihood of a chance occurrence arising without the external input of information; that is, the incidence of an unaided or 'blind chance' event.

protists) for new capabilities, defines the Law of Coding as the epigenome then needs to be rewritten to reliably build and run the changes necessary for the next generation. So evolution of any co-dependent information system (such as: the bacterial flagellum, let alone the mammalian eye) by Darwinian rules, is impossible due to the Law of Coding, not to mention the more than 25,600 steps required to construct our most minimal organism.

Conclusion

So living organisms contain sequential layers of information which are highly integrated and they exhibit structured or organised complexity; and as yet, we have only just scratched the surface. This is not an anatomy textbook so I have only given a few examples but we could equally have chosen the liver, which we were informed in medical school would require an industrial complex as large as an oil refinery to completely replicate its functions outside the human body; or the human heart, which has the capacity and reliability to beat 70 times per minute for over 100 years. This is a phenomenal 3.6 trillion times in one human lifetime. Which machine or engine comes close to replicating this degree of reliability? The term 'simple cell' has often been used in biology in the past, however it can be seen from the immense amount of information required, the remarkable nano-machinery and exquisite integration, that this notion of a 'simple cell' is a myth.[2]

Michael Denton summarised the architecture of a cell by saying: "*Perhaps in no other area of modern biology is the challenge posed by the extreme complexity and ingenuity of biological adaptations more apparent than in the fascinating new molecular world of the cell...To grasp the reality of life as it has been revealed by molecular biology, we must magnify a cell a thousand million times until it is twenty kilometres in diameter and resembles a giant airship large enough to cover a great city like London or New York. What we would then see would be an object of unparalleled complexity and adaptive design. On the surface of the cell we would see millions of openings, like the port holes of a vast spaceship, opening and closing to allow a continual stream of materials to flow in and out. If we were to enter one of these openings we would find ourselves in a world of supreme technology and bewildering complexity... We would wonder at the level of control implicit in the movement of so many objects down so many seemingly endless conduits, all in perfect unison. We would see all around us, in every direction we looked, all sorts of robot-like machines. We would notice that the simplest of the functional components of the cell, the protein molecules, were astonishingly, complex pieces of molecular machinery, each one consisting of about 3000 atoms arranged in highly organised 3-D spatial conformation. We would wonder even more as we watched the strangely purposeful activities of these weird molecular machines,...*" [134]

Denton continues: "*Is it really credible that random processes could have constructed a reality, the smallest element of which - a functional protein or gene - is complex beyond our own creative capacities, a reality which is the very antithesis of chance, which excels in every sense anything produced by the intelligence of man? Alongside the level of ingenuity and complexity exhibited by the molecular machinery of life, even our most advanced artefacts appear clumsy...It would be an illusion to think that what we are aware of at present is any more than a fraction of the full extent of biological design. In practically every field of fundamental biological research ever increasing levels of design and complexity are being revealed at an ever-accelerating rate.*" [135]

133. Fowler, T.B. & Kuebler, D.: *The Evolution Controversy*. (2007) p.267.

134. Denton, M.: *Evolution: A Theory in Crisis*. Adler & Adler, Chevy Chase, Maryland (1986) pp.328–9.

135. *Ibid*. p.342.

Denton leaves us in no doubt about the intricacy of a functioning cell and hence the complexity of the organisms that are constructed from great numbers of these cells. Remember that the human body contains 70 trillion such cells! [136] Nor is there any doubt as to Denton's opinion as to the chance appearance of such complexity. However it is now time to temporarily put these problems regarding the information in the first organism aside as we consider the formation of the first species.

136. Lennox, J.C.: *God's Undertaker. Has Science Buried God?* 2nd Ed. (2009) p.136.

5

Information and the First Species

Rabbits come from slightly different rabbits, not from either [primeval] soup or potatoes. Where they came from in the first place is a problem yet to be solved, like much else of a cosmic scale.[1]

FRED HOYLE

Under the Darwinian paradigm, once the first multi-celled organism was fully formed, it needed to diversify into a species which is a group of organisms that are similar enough to interbreed and yet the individuals differ from each other in identifiable ways, such as, the many different breeds of dog. After that, Darwin said that the individuals within species drifted apart to the extent that they could no longer interbreed so that they became *separate* species. Darwin cited the vast number of similar yet differing species as evidence for his theory and the numbers are not small. The current estimate of the number of surviving species on the Earth is 8.7 million. Estimates generally vary between the 1.7 million species that have already been identified and 10 million, although numbers up to 100 million have been mooted.[2,3,4]

In Chapters 3 and 4, it was seen that information needed to be introduced to generate the first cell and the first organism and in this chapter we will see that further information needs to be present for the variation in species. Each step increases the complexity, which is visible externally as the physical structure of the organism (called the 'body plan') but in addition, inside the organism, there is extra information coded in the DNA for communication and control mechanisms which are largely invisible and have mostly eluded detection until the last half century. It can be seen that the information required to specify the development of an organism expands with each level of complexity also. Some of this DNA, which appears to be non-functioning in the adult, is required to navigate the organism through all the stages of embryogenesis, from a single fertilised egg to a mature adult, which was previously discussed in Chapter 4. In Chapter 3, we saw that this additional DNA has often been viewed as 'Junk DNA', purportedly the remains of millions of years of trial and error evolution and in human beings was thought to comprise about 97% of the total human genome, but it now appears that much of this little understood DNA is required for the orderly development of the growing organism. As the organism matures, some structures, such as the urachus, which connects internally to the umbilicus and nourishes the foetus in utero, are instructed to wither after birth, which is necessary to stop the new born baby from bleeding to death after delivery. This 'automated' process, which

1. Hoyle, F.: *The Mathematics of Evolution*, Weston Publications, Cardiff, University College Cardiff Press, (1987) p. 9 *in* Lennox, J.C.: *God's Undertaker. Has Science Buried God?* Lion Books, Oxford (2007) p.110.

2. All Species Foundation *in* Bendewald, J. & Sherwin, F.: Evolution Shot Full of Holes. Evidence Press, Madison, Wisconsin (2004) p.42.

3. Dawkins, R.: *The Greatest Show on Earth*. Bantam Press, London (2009) pp.329–330.

4. *Cosmos* 41:18 (2011).

orchestrates the involution of the urachus and many other developments, such as the re-absorption of the webbing between your embryonic fingers while you were still in your mother's womb, along with the involution of your thymus gland after the development of your immune system, are not yet identified but must be coded somewhere in the organism to allow each species to grow and reproduce reliably, generation after generation.

After considering the information requirements for the divergence of species, John Lennox comments: *"In addition there is a major problem with time since the time available for such 'random molecular shuffling' to occur is much shorter than many people think. According to current estimates there is relatively little time - less than a billion years after the formation of the earth roughly 4.5 billion years ago - for life to emerge (however it did so), since remains of single celled organisms have been found in the very oldest rocks."* [5] The problem is that no one has been able to describe how evolution might have achieved this incredible feat. Sir Fred Hoyle famously compared these odds against the spontaneous formation of life with the chance of a tornado sweeping through a junkyard and co-incidentally producing a Boeing 747 jet aircraft, which contains far fewer infobits than an amoeba.[6]

There are many differences between species. In Chapter 4 we examined the evidence for the evolution of the human eye and the immense complexity of its components and function but in 2002, John Rennie the Editor of *Scientific American* wrote: *"It now appears that in various families of organisms, eyes have evolved independently."* [7] In fact, there are more than thirty different eye designs in the animal kingdom, which are mutually exclusive and are therefore unable to be linked together in one chain of common descent. This is referred to as 'parallel' evolution. Dr Frank B. Salisbury, Professor Emeritus in Plant Science at Utah State University wrote: *"Even something as complex as the eye has appeared several times; for example, in the squid, the vertebrates, and the arthropods. It's bad enough accounting for the origin of such things once, but the thought of producing them several times according to the modern Synthetic Theory makes my head swim."* [8] And yet the multiple versions of eye architecture all exist, so how could this have happened? Geneticist, Professor Richard Goldschmidt (1878–1958) of the University of California, Berkeley introduced the terms: Microevolution and Macroevolution in his book in 1940, when he wrote that: *"the facts of microevolution do not suffice for an understanding of macroevolution...which...does not lead beyond the confines of the species, and the typical products of microevolution."* [9] This difference between Macro and Microevolution is fundamental to understanding the entire subject so it requires careful examination.

Microevolution

The discipline of genetics significantly changed the study of biology and in particular, inheritance and the Theory of Evolution. The American geneticist Thomas Hunt Morgan (1866–1945) studied fruit flies and made a substantial contribution to the understanding of genetic inheritance and mutation, which led to the development of the Neo-Darwinian Synthesis. Fowler and Kuebler write: *"Due to Morgan's influence, the accepted wisdom today is that most evolutionary change is the result of gradual genetic change."* [10] However, Fowler and Kuebler note: *"Morgan never actually witnessed the transformation of a fruit fly into a new species via the accumulation of minor mutations, only the emergence of new variants (some of which emerged quite abruptly) of a particular species of fruit fly. Even so, based upon*

5. Lennox, J.C.: *God's Undertaker. Has Science Buried God?* Lion Books, Oxford (2007) p.121.

6. Hoyle, F.: *The Intelligent universe.* Michael Joseph, London (1983) p.19 *in* Lennox, J.C.: (2007) p.123.

7. Rennie, J: 15 Answers to the Creationist Nonsense. *Scientific American* 287(1): 78–85 (2002) *in* Sarfati, J.: *By Design*. Creation Book Publishers (2008) p.36.

8. Salisbury, F.B.: Doubts About the Modern Synthetic Theory of Evolution. *American Biology Teacher* 33 (1971) pp.335–338 *in* Bendewald, J. & Sherwin, F.: (2004) p.36.

9. Goldschmidt, R.: *The Material Basis of Evolution*. Yale University Press (1940) pp.8 & 396 *in* d'Abrera, B.: *Butterflies of the Afrotropical Region.* Hill House Publishers, London (2009) p.539.

10. Fowler, T.B. & Kuebler, D.: *The Evolution Controversy.* Baker Academic, Grand Rapids (2007) p.64.

these observations, he extrapolated that if minor mutations were allowed to accumulate indefinitely, new species would arise. But just as Darwin never demonstrated speciation in the Origin of Species, *Morgan never demonstrated it in his work."* [10] Morgan was able to demonstrate that there is change in the genetic information but it was limited. As previously discussed, we now know that the changes Morgan observed occur within certain boundaries, which are defined by the information that describes that species. This is Microevolution. Thus, Microevolution is genetic or informational change over time *within a species*. Zoologist Pierre Grassé, former President of the French Academy of Sciences considered Microevolution to be fluctuations around a successful genotype, which vary due to local environmental conditions. His analogy was that of a butterfly within a greenhouse, which can fly back and forth within the external constraints beyond which it cannot move.[11] This is another way of visualising Jerry Coyne's fitness landscape described in Chapter 3. If an organism is already at or near the pinnacle of the optimal place in its fitness landscape, it will be deleterious to move too far in any direction. In addition, Microevolution lacks a mechanism to introduce new information. It can only cause deletion or duplication of information. In other words, it has no information generator.[12] These conclusions are supported by detailed studies of the Galapagos finches in the 1970's which showed cyclical changes in the finch beak sizes as a result of the changing food sources during the recurring droughts and wet seasons. Birds with larger beaks were more likely to survive in lean years when only the hardest seed were left to forage but in wet years when food was plentiful, birds with smaller beaks increased. However the birds did not change from being birds; the beak variations oscillated about a mean.[12]

So, if evolution does not generate new information for species change, what does it do? John Lennox wrote: *"Biologist and philosopher Paul Erbrich* [comments]: *'The mutation-selection mechanism is an optimisation mechanism.' That is, it enables an already existing living system to adapt selectively to changing environmental conditions much in the same way as gene[r]ic algorithms facilitate optimisation in engineering. It does not, however, create anything radically new."* [13]

To overcome this information generation problem, Simon Conway Morris suggested that once animals developed a significant degree of complexity, the body plans could be changed relatively simply by manipulating the control genes. You will remember that in the previous chapter, it was observed that life is modular and that the various functional units in living systems can be thought of as Lego bricks which can be added, removed or rearranged in different life forms. The control of this modularity is under the direction of the epigenome, which includes the Hox genes that are found in most organisms from flies to humans. In this way, small changes in the epigenome, such as turning a genetic switch on or off, can induce relatively large changes in the body plans or physical appearance of the organisms in subsequent generations.[14] This explanation accounts for the abrupt changes observed by Thomas Hunt Morgan. The only problem with this solution is: how was the original information in the genome and epigenome derived in the first place?

Darwin's solution to the information generation problem, as discussed in Chapter 1, was random variation. After acquiring the ability to replicate successfully, Darwin's original cell had to multiply but since copying processes are imperfect, errors would have crept in. Although we know that error rates in copying DNA and RNA are extremely low (about one in 10 billion),[15] they are not perfect. Darwin extrapolated that these errors would compound

11. Seiglie, M.: 10 Ways Darwin got it Wrong. *The Good News* 14(6):11 (2009).

12. Lennox, J.C.: (2007) p.99.

13. Lennox, J.C.: *God's Undertaker. Has Science Buried God?* (2nd Ed.) Lion Books, Oxford (2009) p.109.

14. Behe, M.J.: *The Edge of Evolution.* Simon & Schuster, New York (2007) pp.180–3.

15. Sarfati, J.: *By Design.* Creation Book Publishers, Brisbane, Australia (2008) p.163.

over successive generations, leading to significant change. Unfortunately, most mutations result in lethal variations, which leads to the extinction of that daughter line and hence the loss of that information. This is 'extinction of the unfittest'! Very occasionally a non-lethal variation occurs to broaden the gene pool and potentially increase the syntactic (Shannon) information within the system. Hence the term: 'random variation'. Remember this is different to the structured semantic information that imparts meaning.

In contrast to the haphazard process of random variation, natural selection is directed by the survivability of the organism within its particular environment and leads to greater order by culling less viable variants and hence there is a decrease in variability and a loss of information. The organisms within that species become specialised for their environment but in doing so they lose information rather than gain it. See Figure 5.1.

In Chapters 3 and 4, it was seen that random 'learning' by natural selection could not account for the information in the first cell or the first organism but could it account for the vast diversity that has occurred *since* the first organism?

There are two mechanisms that can affect Microevolution:

1. Splitting the existing gene pool (for example the wolf genome has divided to produce around 200 breeds of 'dog');[16]
2. Mutating the existing gene pool (for example achondroplasia to produce dachshunds).[17]

Microevolution through splitting the gene pool is obvious. A Chihuahua has all the 'small' genes and a Great Dane has all the 'big' genes. Mutation to produce microevolutionary change only requires change in the genomic DNA, such as cited by Richard Dawkins for achondroplasia in dogs and does not require a complementary change in the epigenome. Microevolution is relatively rapid and is seen in selective breeding programmes. This is the process at work in observable evolution or, as Richard Dawkins put it: "*before our very eyes*".[18] But there are limits to selective breeding.[19] Even where farmers are vigorously trying to enhance their herds with *active* selection (as opposed to *natural* selection), no matter how hard they try, sheep farmers never end up with woolly cows or elephants.

The reason for this is explained by Edgar Andrews who writes: "*...selective pressures in nature can only produce diversification by the loss of genetic information. If natural selection causes a species to evolve into two separate species, each of the new species must lack some of the genetic content of the parent genome. Otherwise nothing new has been produced. Natural selection, even when guided by geographical isolation, therefore, can never create new genetic information.*" [20] And so from an understanding of the mechanisms that drive Microevolution it can be seen why our sheep are still sheep, even if they are more productive and more valuable to the farmer. So what about Richard Goldschmidt's Macroevolution? What does it do?

Macroevolution

Macroevolution is *extra-species change*, whereas Microevolution was change *within* one species. Microevolution, as we have seen, is supported by observable evidence but Macroevolution or the transformation of one species into another has been controversial ever since *The Origin of Species* was published. John Lennox writes: "*...Darwin had been disturbed by a letter from the eminent botanist Joseph Hooker*

16. Dawkins, R.: (2009) pp.27–8.
17. *Ibid.* p.35.
18. *Ibid.* p.109.
19. Fowler, T.B. & Kuebler, D.: (2007) p.59.
20. Andrews, E.: *Who Made God? Searching for a Theory of Everything.* EP Books, Darlington, England (2009) p.222.

in 1852 in which Hooker argued that natural selection was in no sense a creative process."[21] Hooker and the equally eminent geologist Lyell considered Darwin had not disproved the reigning paradigm of the day that *"like does produce like"*. [22] Dawkins does not accept this however and says that although the chance mutations resulting in random variation are not enough on their own, when the refining power of natural selection is added, the power of Macroevolution is unleashed. Richard Dawkins writes: *"it is grindingly, creakingly, crashingly obvious that, if Darwinism were really a theory of chance, it couldn't work. You don't need to be a mathematician or a physicist to calculate that an eye or a haemoglobin molecule would take from here to infinity to self-assemble by sheer higgledy-piggledy luck."* [23] But can chance variation with natural selection produce a completely new species? Edgar Andrews suggests that the evidence is otherwise and points out that even the *"fruit fly Drosophila melanogaster, of which millions of generations have been bred in hundreds of laboratories and subjected to every conceivable indignity - from radiation to toxic chemicals - to induce and study mutations* [are not improved]. *The results have included a regiment of monsters (tiny but technically still monsters) with all manner of bizarre bodily features like extra legs where the antennae should be and missing hearts in embryos. But none of the flies was any better off for the experience, not even those which acquired an extra pair of useless wings. And, of course, the mutated Drosophila are still nothing but fruit flies."* [24]

Although evolution may appear to present a mechanism for information generation, it fails to deliver real evidence that it actually happens. Darwin observed Microevolution (small changes within a species) in animal husbandry and extrapolated it to large changes in the wild.[25] It may superficially appear logical but it is scientifically invalid. This becomes much easier to visualise when information change is graphed over time, as in Figure 5.1.

From Figure 5.1 below, it can be seen that Macroevolution on the left-hand side of the diagram requires an increase in information and diversity, which

21. Lennox, J.C.: p.104.

22. Hooker , J.: Letter 3831, CUL DAR 101: 61–2 *in* Lennox, J.C.: (2007) p.104.

23. Dawkins, R.: *Climbing Mount Improbable.* Norton, New York (1996) p.67.

24. Andrews, E.: (2009) p.229.

25. Fowler, T.B. & Kuebler, D.: (2007) p.59.

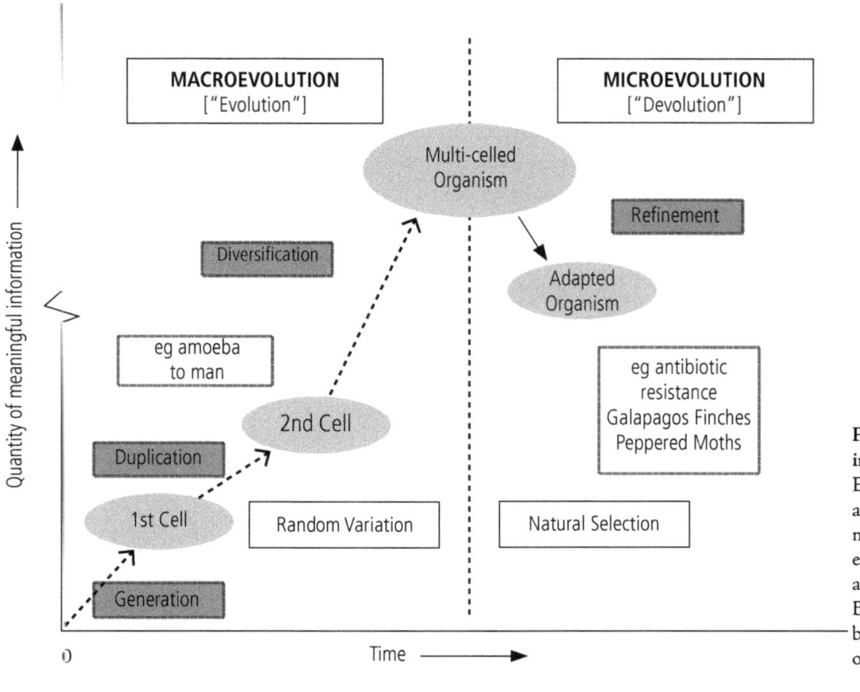

Figure 5.1: Information flows in the Evolutionary Process
Each upward arrow represents an input or gain of new, meaningful information and each downward arrow indicates a loss of useful information. But problems begin right at the beginning as random variation of nothing is still nothing.

was attributed by Charles Darwin to the mechanism of *"random variation"* and is traditionally thought of as the process of 'Evolution'. The second process on the right-hand side of Figure 5.1 is Microevolution, where organisms are refined and become adapted to their micro-environment. Darwin called this *"natural selection"* and it involves a loss of information. This can be thought of as degenerative Evolution or 'Devolution' as there is a net loss of information and in information terms it is the opposite process to Macroevolution. Edgar Andrews explains: *"Natural selection can only select what is present. It can never create what is absent."* [24] Natural selection of nothing *always* equals nothing no matter how long we wait or how much we may wish it were otherwise.

Further, from Figure 5.1 it can be seen that Macroevolution must pre-date Microevolution as there needs to be information in the system for Microevolution to occur. To increase the complexity within a system, the information introduced through random variation must outweigh the information loss through natural selection but does this actually happen? If this does occur, then greater variation will lead to increasing variety and species will drift apart as suggested by Darwin, creating the *"Tree of life"* but experimental evidence suggests that species tend to drift towards the norm or cycle back and forth about the mean rather than drift away from it. It has been noted that the beaks of the Galapagos finches cycle from larger to smaller as the seasons vary.[26] The same process has been observed in the British peppered moths, which were observed to become darker in the soot of the Industrial Revolution but subsequently lighter colours predominated again when the environment became cleaner. This is cycling or 'fine tuning', which is microevolutionary but is not evidence for an increase in meaningful information in the genome or Macroevolution. Philip Johnson writes: *"When evolution is defined as 'information creation'* [Darwin's supporters] *have nothing but speculation. Information-creating evolution is not empirical science at all because it has never been observed either in the wild or in the laboratory."* [27]

In addition, if a system is continuing to incorporate new information then new species should be relentlessly appearing but as life is subject to the Law of Entropy, like the rest of the Universe, the total amount of information in any system will gradually decrease and the total number of species and the variation within those species will decrease, which is the lamentation of ecologists the World over, with numerous extinctions reported every year. The problem with verifying this prediction that new species are continually in the process of being formed is that as long time scales are involved, the process may be too slow for human observation to measure accurately. On the flipside, new species are still being discovered but the question is, are they newly evolved or did they just escape our notice until now? New species tend to be discovered in inaccessible places (such as the Wollemi Pine in a wilderness area of New South Wales to the west of Sydney). These new discoveries have never evolved in established farm land where Macroevolution could be observed and hence validated in a controlled environment.

However, despite all these problems, if we assume that Macroevolution was possible and somehow did occur, even after gaining the machinery required to allow the cell to replicate successfully and then attain multi-cellular organism status, there would still have been problems for our fully competent organism as it diversified to create the first variation in species. From Figure 5.1, it can be seen that more information would have had to have been acquired for the single cell to graduate to a multi-celled creature as it climbed Darwin's tree. Under Darwinian

26. Johnson, P.: *Wedge of Truth*. InterVarsity Press, Downers Grove, Illinois (2000) pp.47–8.

27. *Ibid.* p.48.

rules, this could now be acquired by mutation with natural selection. If Darwin and Dawkins are correct with their assertions regarding the resulting Tree of life, then the slow genetic drift apart by individuals within the species should be verifiable, with similar information code sequences being found in related species but sequencing of the animal genomes has not produced a neat fit; to the contrary similar pieces of genome are shared in complex patterns. Laura Katz, US Professor of Biological Sciences, wrote: *"Accessing data from the genomes of organisms (individual genes) and analysing these data using sophisticated alignment and phylogenetic methods led to the expectation that we would be able to paint a clear picture of the evolution of eukaryotes...Instead of reconstructing a simple bifurcating tree of life. these gene genealogies have generated a complex picture of eukaryotic genomes whereby ancient lateral transfers (of individual genes or perhaps even entire genomes) has tangled the evolutionary history of eukaryotes. Resolution of these conflicting genealogies comes in recognising that eukaryotes are chimeric, containing genetic information from multiple ancestral lineages."* [28] Although Katz obviously intends chimeric to mean a *"thing of hybrid character"* others might choose an alternative definition offered by The Oxford Dictionary as a *"fanciful conception"*. [29]

Macroevolution requires co-dependent or contingent information, as discussed in Chapter 4, whereas Microevolution does not. Observable examples of Microevolution, such as Darwin described in the Galapagos finches and as observed in pigeon breeding, which lie on the right-hand side of Figure 5.1, do not justify, let alone prove that the extrapolation of Microevolution to Macroevolution, as seen on the left-hand side of Figure 5.1, is valid and for which no evolution-*"before our very eyes"* [18] -type evidence exists.[30] Philip Johnson observes: *"The famous textbook examples of variation in the moths in England or the beaks of finches on the Galapagos Islands involve only backward and forwards variation within a fundamentally stable species. No new features appear, and there is no directional change of any kind."* [26]

Thus we come back to the fundamental problem of how to introduce new information into the genome. Fowler and Kuebler identify three possible methods to increase information in a genome:[31]

1. Modification of an existing gene. This is modifying old information, which implies a loss of function of the old gene in favour of a new one, such as a bacterial mutation that disables an importation pump to gain antibiotic resistance. There is no net gain of information unless there is a compensating change in a further gene (a compensatory mutation). In practice, this results in a series of broken genes, not a net increase in information. In our design space map, if there are only one or two changes in our genome we lose our optimal position on a 'hill top' in our design space landscape so we need a *number* of mutually beneficial changes to find another peak, where we once again enjoy a relative advantage. Therefore, bacteria that have developed one or two mutations after exposure to antibiotics will survive for as long as the antibiotics remain in their environment, where they enjoy a competitive advantage but if the antibiotics are removed, the antibiotic resistant bacteria, which are now saddled with multiple pieces of broken machinery, will be out-competed by other wild strains around them as the mutations leave them in a less favourable position in their former design space.

2. Duplication and subsequent modification of an existing gene New information might be imagined to result if the modification conveys a new function.

28. Katz, L.A. The Tangled Web: Gene Genealogies and the Origin of Eukaryotes. *The American naturalist* 154: 137–145 (1999).

29. *Concise Oxford Dictionary.* (5th Ed.) Oxford University Press, London (1973).

30. Johnson, P.: (2000) p.70.

31. Fowler, T.B. & Kuebler, D.: (2007) p.173.

One could envisage a bacterium which could firstly:

a. duplicate an existing gene (mutation 1);
b. mutates the duplicate (for example the import pump gene) whilst maintaining the original gene intact (mutation 2);
c. develop a switching mechanism (epigenetic control gene) to switch between the new import pump gene in the presence of antibiotics and the original gene in the absence of antibiotics (mutation 3).

The problem lies in retro-coding the epigenome in step c, as the Law of Coding states that this cannot happen. This situation applies to any newly mutated gene as it will need epigenetic control to regulate its expression and function.

d. The epigenome needs to develop a feedback loop so that the epigenome knows when to turn the pumps on and off appropriately. (mutation 4) [See Figure 3.11]

This is now a four way random mutational event and in Chapter 4 we saw that this is impossible as a random occurrence. The complexity is now too great for it to happen without semantic information input, even if the Law of Coding did not exist. But wait you say: *'Cannot the four mutations happen sequentially and so it then becomes 'likely' or 'inevitable' given sufficient time?'* The answer is: 'no'. Even by ignoring the small problem of the Law of Coding, which never goes away, even allowing *unlimited* time, it will not happen. Behe showed that it took 70 years for the malarial parasite to develop a double mutation to gain resistance to chloroquine and we have up until now assumed that developing an epigenetic 'switch' requires only a single point mutation. Steps 2a and 2b are simple mutations but 2c and 2d are likely to require complex infrastructure development, so that a 'four stage' process is actually a *lot* more. The time required for this process would be trillions upon trillions of years, even without the Law of Coding and once that is taken into account as well, the answer is that it is impossible.

3. Modification of a non-coding section of DNA may involve altering a previous function or generating a new one, depending on whether the non-coding DNA had a previous function, such as in the epigenome discussed in Chapter 3. If the section of genome was previously non-functioning, meaning that it was truly superfluous DNA, then a new function has been added and an epigenetic switch must be developed for it before it is eliminated by chance or natural selection. Alternatively, if the non-coding sequence of DNA was part of the epigenome, then it can cause large jumps in phenotype as seen in the Drosophila Hox mutations, with the appearance of two extra wings in a single generation. However, Fowler and Kuebler note: "*The main problem with such large changes, in terms of their usefulness for evolution, is that they are usually detrimental to the organism. In fact, none of the known Hox* Drosophila *mutants is considered to be a successful improvement over the wild-type form.*" [32]

Thus, none of Fowler and Kuebler's three options advances Macroevolution, even though they can produce superficial changes in the phenotype through altering the DNA. The significant point is: unless the DNA

32. Fowler, T.B. & Kuebler, D.: (2007) p.306.

change can also include information to specify new epigenetic controls, there cannot be increased complexity and hence enhancement of the organism in the longer-term. Without epigenetic controls, the mutant organism will be out-competed by wild strains and eliminated by natural selection; however, if it could gain control of the newly acquired characteristic, the new mutation might become an asset instead of useless baggage that slows it down. Unfortunately, since the alteration required in the epigenome is unlikely to be just one change but a co-ordinated series of modifications, in practice, it never happens. This explains why none of the Hox Drosophila mutants have *ever* been an evolutionary improvement over the existing ones.

Thus, modifying existing information in organisms (as in bacterial resistance) does not help the case for Macroevolution but it is very useful for Microevolution (which does not require epigenetic change). The one apparent exception is beta-lactamase resistance in bacteria, which needs more detailed analysis and will be discussed further in Chapter 7.

The Geologic Column

Now that we have an appreciation of Micro and Macroevolution we are in a better position to assess the information that is available in our environment and we will begin with the geologic column, so that we can better understand the fossil record. When scientists dug into the Earth's crust, they observed that it was layered and contained fossils but the various layers contained different fossils. Darwin, and others before him, such as the British geologist, Charles Lyell (1797–1875), noted that the deeper layers had simpler life forms, many of which had become extinct; while the organisms with more complex body parts tended to be found in the more superficial layers. Because of the widespread occurrence of this layering pattern, it has been extrapolated (because human beings were not present when trilobites roamed the Earth and trilobites did not leave written records) that the layers correspond to sequential time periods. These successive layers have subsequently been used to date the fossils that have been found in them. This system of geological categorisation and its associated timescale is referred to as the 'Geologic Column' and is illustrated in Figure 5.2. As information has no mass and is therefore outside time, the information itself cannot be used to directly corroborate or disprove the timescale but there is no doubt that the fossils do exist in a layered pattern as shown in Figure 5.2.[33] To date information, we are dependent on the physical properties of the information carriers.

The Fossil Record

The distinguished French zoologist Pierre Grassé reminds us that: "*Naturalists must remember that the process of evolution is revealed only through fossil forms...only palaeontology can provide them with the evidence of evolution and reveal its course or mechanisms.*" [34] And the fossil record is large. Author and teacher John Blanchard continues: "*In 1981 a spokesman for the American Association for the Advancement of Science claimed that 100 million fossils, identified and dated, 'constitute 100 million facts that prove evolution beyond any doubt whatsoever'* " [35] However, if evolution is proven *'beyond any doubt whatsoever'* through the fossil record, then why is there so much dissent from academics and researchers in every discipline of science? The problem is, as all scientists acknowledge, that despite the large number of fossils

33. Meyer, S.C. et al: *Explore Evolution* Hill House Publishers, London (2009) pp.16–18.

34. Grassé, P.: *Evolution of Living Organisms.* Academic Press. p.4 (1973) [English translation (1977)] *in* Blanchard, J.: *Evolution Fact or Fiction* Evangelical Press, Darlington, England (2002) p.8. Also see: Johnson, P.E.: *Darwin on Trial.* (2nd Ed.) InterVarsity Press, Downers Grove. Illinois (1993) p.174.

35. Blanchard, J.: *Evolution Fact or Fiction* Evangelical Press. Darlington, England (2002) p.8.

already found, the fossil record is incomplete, which Darwin himself recognized was a problem for his theory. One of the key pieces of evidence cited by Darwin, as necessary for his theory, was the fossil record: "*...the number of intermediate and transitional links, between all living and extinct species, must have been inconceivably great. But assuredly, if this theory* [of Evolution] *be true, such have lived upon the earth.*" [36]

Despite his confidence, elsewhere in his book, Darwin ponders: "*...the distinctiveness of specific forms, and their not being blended together by innumerable transitional links, is a very obvious difficulty...the number of intermediate varieties, which have formerly existed on the earth,* [must] *be truly enormous. Why then is not every geological formation and every stratum full of such intermediate links? Geology assuredly does not reveal any such finely graduated organic chain; and this, perhaps, is the most obvious and gravest objection which can be urged against my theory. The explanation lies, as I believe, in the extreme imperfection of the geological record.*" [37] Darwin's claim, despite the fact that the fossil record did not demonstrate the 'intermediates' to prove the all-important links, seemed reasonable in 1859. Instead he predicted that they would one day be found. This expectation has been accepted for a long time but with the discovery of the pre-Cambrian sponge embryos at Cheng Jiang, China, that assertion is becoming harder to believe. If soft-bodied sponge embryos can be fossilised and survive for over 500 million years, according to the timescale expressed in the geologic column, then where are all the bony-skeleton fossils alluded to by evolutionary theory?[38]

Niles Eldridge's response to the paucity of the fossil record was: "*No wonder palaeontologists shied away from evolution for so long. It seems never to happen. Assiduous collecting up cliff faces yields zigzags, minor oscillations, and the very occasional slight accumulation of change over millions of years, at a rate too slow to really account for all the prodigious change that has occurred in evolutionary history. When we do see the introduction of evolutionary novelty, it usually shows up with a bang, and often with no firm evidence that the organisms did not evolve elsewhere! Evolution cannot forever be going on someplace else. Yet that's how the fossil record has struck many a forlorn palaeontologist looking to learn something about evolution.*" [39] Over the last 150 years, some intermediate forms have been unearthed but there are still significant gaps in the record and some suggested intermediates are out of order in the geologic column, which is another way of saying that they appear to be descendants of their own offspring! So the fossil record does not help clarify the situation but perhaps only muddies the waters and yet it must be accounted for in any explanation of the history of life on Earth.[40] Despite the great fanfare of Archaeopteryx (1861), Tiktaalik (2006) and Darwinius masillae (2009),[41] significant mistakes have been made such as Piltdown man (1913–53), Java man (1891–1956)[42] and Archaeoraptor (1999).[43] 'Intermediates' are still so sparse that Stephen Jay Gould wrote in 1977: "*The extreme rarity of transitional forms in the fossil record persists as the trade secret of palaeontology.*" [44] In 1999, Henry Gee, zoologist and chief science writer for *Nature* wrote: "*The intervals of time that separate fossils are so huge that we cannot say anything definite about their possible connection through ancestry and descent.*" Lee Strobel further quoting Gee writes: "*Consequently,...the conventional picture of human evolution is 'a completely human intervention created after the fact, shaped to accord with human prejudices...To take a line of fossils and claim that they represent a lineage is not a scientific hypothesis that can be tested, but an assertion that carries the same validity as a bedtime story — amusing, perhaps even instructive, but not scientific.'* " [46]

36. Darwin, C.: *The Origin of Species*. John Murray, London (1859) reprinted: Penguin Classics (1985) p.293.

37. *Ibid.* pp.291–2.

38. Chien, P. et al: *SEM Observation of Precambrian Sponge Embryos from Southern China, Revealing Ultrastructures including Yolk Granules, Secretion Granules, Cytoskeleton and Nuclei.* Paper presented to the North American Paleontological Convention, University of California, Berkeley. June 26-July , 2001 *in* Meyer, S.C. et al: *Explore Evolution* (2009) p.31.

39. Eldredge, N.: *Reinventing Darwin: The great Evolutionary Debate.* (1995) p.95 *in* Seiglie, M.: (2009) p.10.

40. Fowler, T.B. & Kuebler, D.: (2007) pp.86–88.

41. Dawkins, R.: (2009) pp.168 & 180.

42. Lubenow, M.L.: *Bones of Contention. A Creationist Assessment of Human Fossils.* Baker Books, Grand Rapids, Michigan (2007) p.99.

43. Strobel, L.: *The Case for a Creator.* Zondervan, Grand Rapids (2004) pp.58.

44. Gould, S.J.: Evolution's Erratic Pace. *Natural History* 86(5) (1977) *in* Lennox, J.C.: (2007) p.111.

Geologic column						Life
Eon	Era	Period		Epoch	Age MYBP*	Life forms /geologic events
Phanerozoic	Cenozoic	Quaternary		Holocene		
				Pleistocene	1.8	
		Tertiary	Neocene	Pliocene	5.2	Earliest *Homo*
				Miocene	23.8	
			Paleocene	Oligocene	33.5	First apes
				Eocene	55.5	First whales
				Paleocene	65	First horses
	Mesozoic	Cretaceous		Late	98.9	{ Asteroid impact? Extinction of dinosaurs
				Early	144	
		Jurassic		Late	160	First birds
				Middle	180	
				Early	206	First mammals
		Triassic		Late	228	
				Middle		
				Scythian	251	Permian mass extinction
	Paleozoic	Permian			290	First mammal-like reptiles
		Carboniferous		Pennsylvanian		
				Mississippian	354	First reptiles First amphibians First insects
		Devonian			409	First land plants
		Silurian			439	First fish with jaws
		Ordovician			495	
		Cambrian			543	First shelled organisms "Cambrrian explosion" First multi-celled organisms
Proterozoic	Precambrian				2500	
Archaean					3800	First bacteria
Hadean					4800	Origin of life? Oldest rocks Formation of earth

*MYBP Millions of years before present

Figure 5.2: The Geologic Column
The geologic timescale is depicted under 'Geologic Column' and the first appearance of specific organisms found in the fossil record is listed under 'Life'.[45] *[Reproduced from Fowler, T.B. & Kuebler, D. The Evolution Controversy. Baker Academic, a division of Baker Publishing Group. (2007). Used by permission.]*

Richard Dawkins assures us that more and more transitional forms are being discovered all the time and that it will only be a matter of time before most of the gaps in our information will be filled. The problem is that if I am trying to cross a chasm 1000 metres wide and 1000 metres deep and I decide to build a suspension bridge with ten centimetre wide slats, where one slat represents one piece of information, I will need 10,000 of the slats to complete my bridge but if I only have 100 slats and I am missing 9900 slats, I am never going to convince you to join me in walking across to the other side because we will never make it! This is the problem that Neo-Darwinian Theory has with the information provided by the fossil record at present and the situation is not getting better very quickly. This situation becomes particularly concerning when the amount of resources

45. Fowler, T.B. & Kuebler, D.: (2007) p.85.

46. Gee, H.: *In Search of Deep Time: Beyond the Fossil Record to a New History of life*. The Free Press, New York (1990) *in* Strobel, L.: (2004) pp.62–3.

that have been expended over the last 150 years to uncover the missing fossils is taken into account. Yes, Tiktaalik and Darwinius masillae added two extra slats (pieces of information) in the last couple of years but if we accept that every second slat is missing from the fossil record (was never fossilised) we still need at least another 5000 slats, not just two. It is difficult to estimate the number of steps or individual changes that would be required for reptiles to evolve into birds or a lemur-like primate to evolve into *Homo sapiens* but a *minimum* of 10,000 steps does not sound unreasonable. So, to cross these information bridges, Professor Dawkins has to walk on thin air for more than 95% of the distance as he traverses the interspecies chasms which divide the branches of the Neo-Darwinian fossil tree of life. To cross those bridges between the numerous fossil groups, based on sound informational evidence, the palaeontologists need to be uncovering one significant new fossil every year *for every major species transition* (of which there are at least ten so there needs to be at least ten new significant fossil finds per year) for at least the next 500 years (and even then, this will provide only one slat for every metre of our journey and represent, at most, 10 percent of a 'complete' Darwinian fossil record) so it will be a long time before we can prove the Neo-Darwinian Theory based on the fossil record using Darwin's own measure of a finely graded continuum.[47] However, there is some doubt as to how many fossils remain undiscovered. Michael Denton explains that 97.7% of living orders of terrestrial vertebrates have been fossilised and: *"Of the 329 living families of terrestrial vertebrates 261 or 79.1% have been found as fossils, and when birds (which are poorly fossilised) are excluded, the percentage rises to 87.8%* "[48] Thus the fossil record may not be as incomplete as Richard Dawkins suggests.

Biologist, Jonathan Wells continues: *"Thomas Henry Huxley...acknowledged, however, that 'we have no knowledge of the animals, which linked reptiles and birds together historically and genetically,' and that fossils 'only help us to form a reasonable conception of what those intermediate forms may have been.'* "[49] Wells comprehensively covers Archaeopteryx and related fossils and shows that there is still much controversy over the fossil record and the ancestry/descent of modern birds.[50] Archaeopteryx may not be directly on the proposed 'line of descent' from reptiles to birds but whether it is or not, it can be seen from the chasm and bridge analogy that many more related fossils are needed to mount a coherent case.

Professor Emeritus of Integrative Biology at Berkeley, James Valentine in his major study *On the Origin of Phyla* in 2004 wrote: *"Many of the branches [of the tree of life], large as well as small, are cryptogenic (cannot be traced into ancestors). Some of these gaps are surely caused by the incompleteness of the fossil record, but that cannot be the sole reason for the cryptogenic nature of some families, many invertebrate orders, all invertebrate classes, and all Metazoan phyla."* [51]

Even Stephen Jay Gould, Harvard Professor of Biology and Palaeontology wrote: *"The evolutionary trees that adorn our textbooks have data only at the tips and nodes of their branches; the rest is inference, however reasonable, not the evidence of fossils."* [52] Nils Heribert-Nilsson (1883–1955), Professor of Biology at Lund University, Sweden, after 40 years of study concluded: *"It is not even possible to make a caricature of evolution out of palaeobiological facts. The fossil material is now so complete that the lack of transitional species cannot be explained by the scarcity of the material. The deficiencies are real; they will never be filled."* [53,54] Sarfati comments: *"Contrary to cliché, absence of evidence is evidence of absence; it's just not* proof *of it. So it is legitimate to argue that the widespread absence of fossils documenting major*

47. There are at least 10 major transitions: bacteria to algae; algae to plants; bacteria to sponges; sponges to crustacea; crustacea to molluscs; molluscs to fish; fish to reptiles; reptiles to birds; reptiles and mammals and mammals to *Homo sapiens*.

48. Denton, M.: *Evolution: A Theory in Crisis.* Adler & Adler, Chevy Chase, Maryland (1986) pp.189–190.

49. Wells, J.: *Icons of Evolution*. Regnery Publishing, Washington, DC (2002) p.115.

50. *Ibid.* pp111–135.

51. Valentine, J.: *On the Origin of Phyla.* Univ. of Chicago Press, Chicago (2004) p.35 *in* Lennox, J.C.: (2007) p.113.

52. Gould, S.J.: Evolution's Erratic Pace. *Natural History* 86(5) (1977) *in* Blanchard, J.: (2002) pp.10–11.

53. Nilsson, H.: *Synthetische Artbildung,* Verlag CWK Gleerup, Lund, Sweden, (1953) p.1212 *in* Heenan, D.: (2008) p.9.

54. http://www.conservapedia.com/Nils_Heribert-Nilsson

evolutionary changes counts as evidence against evolution." [55] But before we conclude our discussion there are more observations that should be considered.

The Cambrian Explosion

Life on Earth has been divided into five Kingdoms, as depicted in Figure 5.3.[56] As discussed above, the remains of some of these organisms have been fossilised and are spread through many rock formations. In Figure 5.2 it can be seen that the Cambrian period is one of the oldest periods in the geologic column and lies near the lowest levels in most rock formations. Fossils have been found in these rock formations and those both above and below them. The importance of these strata is that the Cambrian rocks seem to mark a significant change in the fossilisation of living things. Fossil evidence suggests that only the bacteria and one-celled organisms, such as amoebae, existed for a significant period before the Cambrian period,[57] when more complex organisms appear in the rock layers in rapid succession in a comparatively short period. There was a massive increase in the number of species in what has been estimated to be a 5–10 million year period. Richard Dawkins comments: *"It is as though they were just planted there, without any evolutionary history."* [58] Despite Dawkins' protestations as to being quoted out of context, the observation that the Cambrian Explosion occurred is valid. The doyen of evolutionary biology, George Gaylord Simpson wrote: *"The earliest and most primitive members of every order already have the basic ordinal characteristics, and in no case is an approximately continuous series from one order to another known. In most cases the break is so sharp and the gap is so large that the origin of the order is speculative and much disputed."* [59] It is interesting to note that very little change in body plans has been observed in the geologic record before or since this time. This vast change in the number and complexity of the fossilised body plans lying in the rocks near the bottom of the geologic strata was a problem for Darwin who wrote: *"To the question why we do not find records of these vast primordial periods, I can give no satisfactory answer...the difficulty of understanding the absence of vast piles of fossiliferous strata, which on my theory no doubt were somewhere accumulated before the Silurian Epoch* [which includes the Cambrian, see Figure 5.2], *is very great...The case at present must remain inexplicable; and may be truly urged as a valid argument against the views here entertained."* [60] Stephen Jay Gould lamented: *"The fossil record had caused Darwin more grief than joy. Nothing distressed him more than the Cambrian explosion, the coincident appearance of almost all complex organic designs..."*[61] The reason that Darwin was concerned was that it defied the slow, gradual change which is demanded by Darwinian Evolution.

Looking at the situation in greater detail, there are three basic cell types or domains: Bacteria (Prokaryotes), Archaea (single-celled organisms including the photosynthesising cyanobacteria or blue-green algae) and Eucaryota (Eucaryotes which is all the rest).[62] From amongst all of the eucaryotes, there are only three animal phyla (major body plan types) that have been found in the Precambrian rocks: the Porifera (sponges), Cnidaria (jellyfish) and Ctenophora (comb-jellies) and 10 million years is not enough time for traditional Neo-Darwinian Theory to explain the approximately thirty animal phyla that first appeared during the Cambrian Explosion.[63,64] Thus at least 75 percent of the organism types that we find on Earth today first appeared in this relatively brief period, which is a big problem for Neo-Darwinian Evolution.

55. Sarfati, J.: *The Greatest hoax on Earth? Refuting Dawkins on Evolution*. Creation Book Publishers. Atlanta , Georgia (2010) p.106.

56. Bendewald, J. & Sherwin, F.: *Evolution Shot Full of Holes*. Evidence Press, Madison, Wisconsin (2004) p.116.

57. Ladiges, P., Evans, B., Saint, R. & Knox, B.: *Biology: an Australian Focus*. (4th Ed.) McGraw Hill, Australia (2010) pp.769–770.

58. Dawkins, R.: *The Blind Watchmaker*. Longman, London (1986) *in* Dawkins, R.: *The Greatest Show on Earth*. (2009) pp.147–8.

59. Simpson, G.G.: *Tempo and Mode in Evolution*. Columbia Univ. Press, New York (1944) pp.105–6 *in* Sarfati, J.: (2010) p.109.

60. Darwin, C.: *The Origin of Species*. (1859) reprinted by Penguin Classics, London (1985) pp.313–4.

61. Gould, S.J.: *The Panda's Thumb*. W.W. Norton & Co, New York (1980) pp.238–9 *in Journal of Creation* 23(3): 124 (2009).

62. Ladiges, P., Evans, B., Saint, R. & Knox, B.: (2010) pp.756–7.

63. Fowler, T.B. & Kuebler, D.: (2007) p.170.

64. The total number of animal phyla depends on the classification system used but most scientists consider there are about 40.

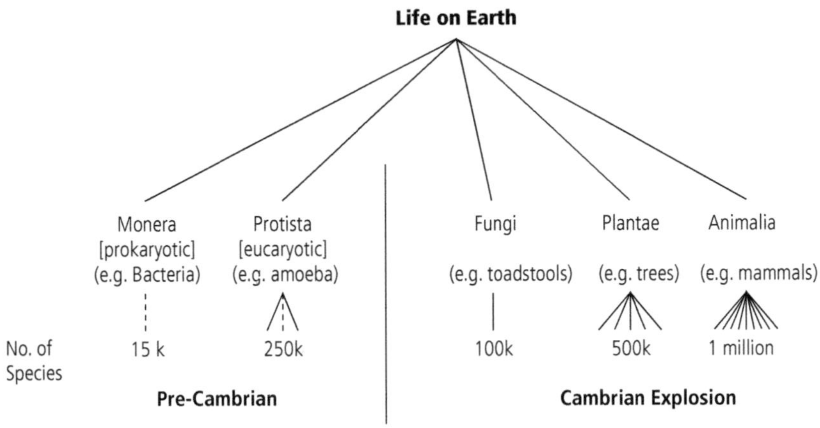

Figure 5.3: The Kingdoms of Life on Earth
Note that the Pre-Cambrian rocks account for 265,000 species while the Cambrian strata reveal another 1.6 million species, a 600 per cent increase in less than 10,000 years which is equivalent to the blink of an eye in geological time.[56,65]

So, in 1972, Niles Eldridge and Stephen Jay Gould formulated the Theory of Punctuated Equilibrium to try to account for this perplexing observation. This theory proposed that the more complex organisms might have been able to rapidly appear if organisms with simpler body plans were able to trade 'chunks' of their genomes, as bacteria do via plasmids when they gain resistance to multiple antibiotics. Remember that mitochondria have also been hypothesised to have once been discrete organisms which developed a symbiotic relationship with eucaryotic cells.[66] It has also been noted that control genes operate in a modular fashion (see Chapter 4) and apart from the possibility of plasmid-type exchange, variations in the control genes might have made profound changes to phenotypes in a single generation, as in the fruit fly experiments. Punctuated Equilibrium proposes that lateral information transfer can account for the great increase in the number of body plans in a relatively short period of time but is it sufficient to explain the vast increase in the amount of information in the Cambrian gene pool? Unfortunately mixing and matching Lego blocks or computer chips does not explain where the individual components originated. The information code in the DNA still had to have been written at some point and as was seen in the previous chapter, the more complex the organism, the greater is the degree of sophistication that is required within an organism's internal control mechanisms (information systems), such as hormonal messaging, which also need coding.

New living species are still being identified and it would seem likely that there are still numerous extinct species that have not yet been discovered so as noted previously, not all the information is available yet but even so, analysing the numbers of body plans in the Pre-Cambrian Era (see Figure 5.6), there was not enough genetic code to build the vast new array of body plans found in the Cambrian Period. It has been estimated that 265 thousand species increased to 1.6 million species and even a cursory survey of the body plans before and after the Cambrian Explosion will demonstrate the immense increase in complexity from one celled organisms, such as bacteria and fungi, to multi-celled molluscs, such as trilobites, let alone the fish, dinosaurs, birds and mammals that followed. On current evidence, there was a 600 per cent increase in species and an even greater increase in complexity. The information for this massive explosion in the DNA code, to specify all these new organisms, must have come from somewhere. Based on the currently accepted dating methods, there has been 500 million years elapse since the Cambrian Explosion, fifty to one hundred times longer than the

65. The period of the Cambrian Explosion is generally accepted to have occurred over 10,000 years. Some researchers suggest that it may have extended to 15,000 or even 20,000 years but this period is still far too short for Darwinian evolution to explain the vast and rapid increase in the fossilisation of species. (See Fowler, T.B. & Kuebler, D.: (2007) pp.170–1.)

66. Sagan, L.: On the Origin of Mitosing Cells. *Journal of Theoretical Biology* 14(3): 255–274 (1967).

period of the Cambrian Explosion itself and yet there are *no* new phyla since that time. Stephen Jay Gould commented: *"...so even the most cautious opinion holds that 500 million subsequent years of opportunity have not expanded the Cambrian range achieved in just 5 million years."* [67]

Palaeontologist, Niles Eldredge of the American Museum of Natural History and joint developer of the Theory of Punctuated Equilibrium amplifies Gould's previous statements when he says: *"...in the 60s as I tried in vain to document examples of the kind of slow directional change we all thought ought to be there ever since Darwin told us that natural selection should leave precisely such a tell-tale signal...I found instead that once species appear in the fossil record they tend not to change very much at all. Species remain imperturbably, implacably resistant to change as a matter of course — often for millions of years."* [68] A classic example of this is the fish, called the coelacanth, which is referred to as a 'living fossil'.

Stasis

So the fossil record reveals that once a species appears in the geologic column, it remains there unchanged for relatively long periods (millions of years in geologic time), with little or no change.[69] The 'living fossils' such as: the coelacanth, the nautilus shell, the ginkgo plant and comb jellies are the ultimate examples, as they have not changed from the fossil record.[70] Darwin predicted that gradual changes would be observed but instead, as we have just seen, there are sudden jumps from one body plan to the next, which had been noted by Darwin but dismissed as an illusion created by the paucity of the fossil record at that time.[71] It was this lack of evidence to demonstrate that smooth transitions had occurred in even a minority of lines of descent, which led Eldredge and Gould to propose their Theory of Punctuated Equilibrium.[72] Later Gould wrote: *"Stasis existed in overwhelming abundance, as every palaeontologist always knew. But this primary signal of the fossil record, defined as an absence of data for evolution, only highlighted our frustration – and certainly did not represent anything worth publishing."* [73] As a result, Eldredge and Gould saw populations within species not as fluid genes flowing relentlessly from one form of organism to the next but as stable entities. They therefore concluded: *"...if that stability is an inherent property both of individual development and the genetic structure of populations, then its power is immeasurably enhanced, for the basic property of homoeostatic systems, or steady states, is that they resist change by self-regulation. That local populations do not differentiate into new species, even though no external bar prevents it, stands as strong testimony to the inherent stability of species."* [74] In 2002, Gould summed it up when he wrote: *"...the tale itself illustrates the central fact of the fossil record so well — geologically abrupt origin and subsequent extended stasis of most species...Anatomy may fluctuate through time, but the last remnants of a species looked pretty much like the first representatives."* [75]

Evolution via punctuated equilibrium was avidly received by many palaeontologists, as it appeared to fit to the fossil evidence better, however, a new mechanism needed to be devised, as Neo-Darwinian Macroevolution is exceedingly slow and cannot be accelerated, so the concept of 'Species Selection' was proposed. Meyer et al explain: *"Traditional neo-Darwinism emphasises competition among* individuals *within the* same *species. Species selection stresses competition among rival species. This mechanism, say the theory's advocates, would produce rapid changes in body structure. Species selection acts on larger differences in biological form — differences between whole species, rather than differences between*

67. Gould, S.J.: The Evolution of life on Earth. *Scientific American* 89(1994) *in* Bendewald, J. & Sherwin, F.: (2004) pp.119–121.

68. Eldredge, N.: *Reinventing Darwin: The Great Debate at the high Table of Evolutionary Theory*. Wiley, NY (1995) p.3 *in* Lennox, J.C.: (2007) pp.112–3.

69. Freeman, S. & Herron, J.: *Evolutionary Analysis.* (3rd Ed.) Pearson-Prentice Hall, Upper Saddle River, NJ (2004) p.522 *in* Fowler, T.B. & Kuebler, D.: (2007) p.88.

70. Meyer, S.C. et al: (2009) p.25.

71. Darwin, C.: *The Origin of Species*. Penguin Classics, London. (1985) p.440.

72. Fowler, T.B. & Kuebler, D.: (2007) p.279.

73. Gould, S.J.: *The Structure of Evolutionary Theory*. Harvard Univ. Press, Cambridge, MA (2002) pp.759–780 *in* Fowler, T.B. & Kuebler, D.: (2007) p.282.

74. Eldredge, N. & Gould, S.J.: Punctuated Equilibria: An Alternative to Phyletic Gradualism *in* Models of Paleobiology, ed. Schopf, T.J.M.: Freeman, Cooper & Co., San Francisco (1972) pp.114–115 *in* Fowler, T.B. & Kuebler, D.: (2007) p.284.

75. Gould, S.J.: *The Structure of Evolutionary Theory*. (2002) p.749 *in Journal of Creation* 23(3): 124 (2009).

siblings. Evolutionary change would thus take place in bigger jumps. As a result, the branches of the Tree of Life would split off so abruptly that they would appear virtually horizontal.

Sudden species change means that fewer intermediate forms between species would have existed, and would only rarely be preserved. So, the fossil record would not preserve the 'punctuation', it would only preserve the 'equilibrium' — the new forms of life which remained stable over long periods of time." [76] Even as early as 1985 however, Valentine and Irwin noted significant problems when they wrote: *"The periods of stasis raise the possibility that the lineage would enter the fossil record, and we reiterate that we can identify none of the postulated intermediate forms. Finally, the large numbers of species that must be generated so as to form a pool from which the successful lineage is selected are nowhere to be found. We conclude that the probability that species selection is a general solution to the origin of higher taxa is not great, and that neither of the contending theories of evolutionary change at the species level, phyletic gradualism or punctuated equilibrium, seem applicable to the origin of new body plans."* [77]

But what is happening to the information within the gene pool in species selection? The proposed process is selecting between different body plan variations more aggressively but that does not create new information; it is a rapid form of Microevolution and as such cannot add information, so it cannot give a trilobite a compound eye or a reptile wings to fly. For the same reason, allopatric speciation, where a population becomes geographically separated and then diverges before being forced together again at a later date, can only select from the information in the existing gene pool and although it can explain punctuated Microevolution it cannot explain Macroevolution.[78] As a result, by 2002, even Gould stated: *"I recognise, that we know no mechanism for the origin of such organismal features other than conventional natural selection at the organismic level..."* [79]

Perhaps the last word on Punctuated Equilibrium should be given to American scholar Marvin Lubenow who wryly commented: *"It must be the only theory put forth in the history of science which claims to be scientific but then explains why evidence for it cannot be found."* [80] It should be reiterated that even in domestic breeding programs, although marked variations can be achieved, a sheep never grows to the size of a cow nor a cow to the size of an elephant despite hundreds or thousands of generations. There are inbuilt information limits on domestic breeding because of the constraints imposed by microevolutionary change and unless new information can be introduced, it is impossible to go outside the pre-set information boundaries. This explains stasis within species. Once a species is 'optimised' within its fitness landscape for a particular environment, it has nowhere to go except downhill, possibly into extinction.

Common Ancestry/Descent

The third element of Darwin's Theory of Evolution after: 'change over time' and random variation with natural selection was 'Common Descent'. Universal common descent was a central plank of Darwin's hypothesis. However, human beings were not around to document the changes in animal species over the generations, so Darwin had to infer what had happened in the past, from the evidence that he could collect. One notable observation is the similarity in body plans, such as the anatomy of the forelimbs of various mammals, which is called *homology*. Darwin considered that the similarity in the structure of mammalian forelimbs was unlikely to have occurred by chance, so Darwin suggested that this

76. Meyer, S.C. et al: (2009) p.33.

77. Valantine, J. & Erwin, D.: Interpreting Great Developmental Experiments: The Fossil Record. p.96 *in* Raff, R.A. & E.C. eds.: *Development as an Evolutionary Process.* Alan R. Liss Inc., NY (1985) *in* Sarfati, J.: (2010) p.117.

78. Meyer, S.C. et al: (2009) p.38 Ref. 35.

79. Gould, S.J.: *The Structure of Evolutionary Theory.* (2002) p.749 *in* Meyer, S.C. et al: (2009) p.34.

80. Lubenow, M.L.: (2007) p.182.

was evidence that they had descended from a common ancestor. Jonathan Wells notes: *"Darwin considered homology important evidence for evolution, listing it among the facts which 'proclaim so plainly, that the innumerable species, genera and families, with which this world is peopled, are all descended, each within its own class or group, from common parents.' "* [81] Wells points out that as a result: *"Almost every biological textbook uses vertebrate limbs to illustrate, homology, and claim that homology is evidence for common ancestry. But most of those textbooks also define homology in terms of common ancestry. They thereby fall into the same vicious circle that biologists and philosophers have been criticising for over half a century...Such circular reasoning lulls students into sloppy and uncritical thinking."* [82] From Figure 5.4 it can be seen that there is a striking similarity and Darwin correctly identified the similarity in the *information* that is shared between these species. This is an action belonging to Tier 2 in Figure 1.5 and is perfectly valid science but he then attributed Tier 3 and 4 causation, which he had to *assume*, because of the paucity of the evidence.

Darwin had limited evidence from which to deduce his Theory of Evolution. However, over the years, more evidence (information) has been forthcoming. Today, evidence for homology is sought from:

a. The fossil record — such as horses;
b. Comparative anatomy — the similarities observed amongst living species;
c. Embryogenesis — Ernst Haeckel's 'first stage';[83]
d. Molecular similarities;
e. DNA comparisons.

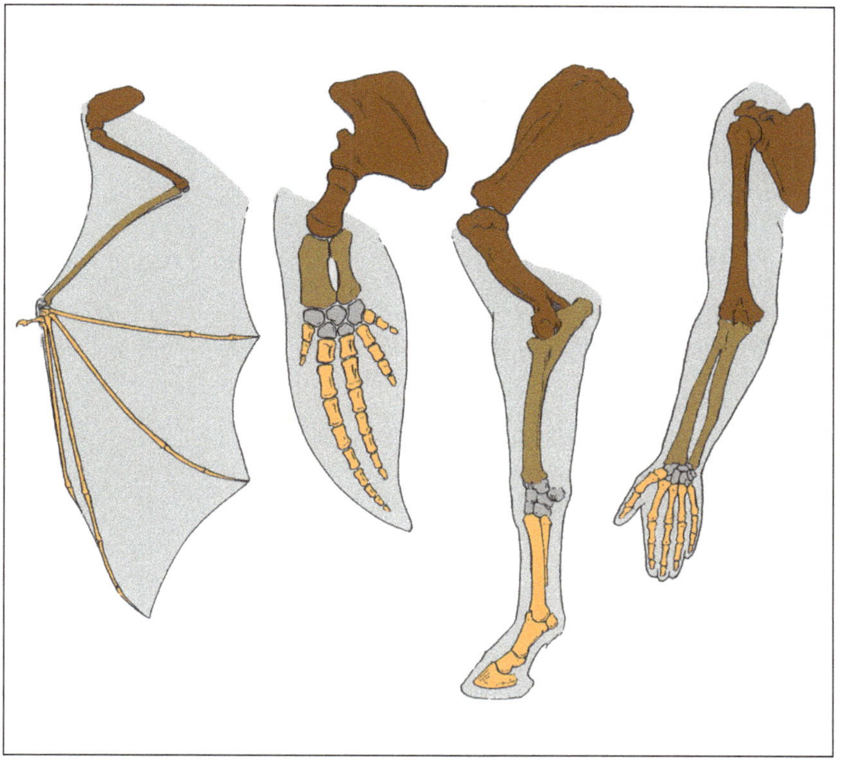

Figure 5.4: Homologous Bony Structures in vertebrate Limbs
Left to right: bat, porpoise, horse and human being.[84] *[Copyright Jody F. Sjogren (2000). Used with permission.]*

81. Wells, J.: (2002) p.61.
82. *Ibid.* pp.77–8.
83. Wells, J.: (2002) pp.82–6.
84. Jody F. Sjogren (2000).

The value of each of these is directly proportional to the information content they retain regarding their association with life on Earth.

All the above evidence, however, relies on assumptions about ancestry, as there is no reliable way to track the purported ancestries into prehistory. The usefulness of each piece of evidence depends on the amount of digital information that can be extracted from it; for example, DNA is more precise than molecular information, which is more accurate than the embryological evidence, which in turn, is better than comparative anatomy, which is superior to the fossil record. This allows the evidence to be weighted in order of reliability; however, Darwin was not privy to DNA or molecular information. So Darwin had the three sources with the least intrinsic information from which to derive his theory. Darwin saw embryology as: *"by far the strongest single class of facts in favour of"* his Theory of Common Descent[85] and of all the sources available to him embryology did have the most intrinsic information, but it is not supportive at all.[86] Darwin was not an embryologist and was deceived by Ernst Haeckel's embryology. Due to what Rudolf Raff, Professor of Biology at Indiana University, describes as the 'hourglass effect', Jonathan Wells writes: *"Haeckel's drawings are misleading in three ways:*

1. *They include only those classes and orders that come closest to fitting Haeckel's theory;*
2. *they distort the embryos they purport to show; and*
3. *most seriously, they entirely omit earlier stages in which vertebrate embryos look very different."* [87] See Figure 5.5.

Disquiet regarding the value of embryology as supporting evidence for Darwinian Theory began in the 1800s. Meyer et al wrote: *"In 1894, Adam Sedgwick, an embryologist at Cambridge University, challenged Darwin's two claims: 1) that vertebrate embryos were more alike than the vertebrate adults, and 2) that the younger the embryos, the greater the resemblance. This was 'clearly not' the case, Sedgwick noted."*

"Even the embryos of 'closely allied animals', such as chickens and ducks, display specific differences very early in development. 'I can distinguish a [chicken] and a duck embryo on the second day,' he wrote. Comparing the embryos of a chicken and a shark, he continued, 'There is no stage of development in which the unaided eye would fail to distinguish them with ease.' He added that these early differences come as no surprise, saying, 'Every embryologist knows that they exist and could bring forward innumerable instances of them.' [88] This deception was verified over a century later. Meyer et al continue: *"In 1997, an international team of scientists, led by the embryologist Michael Richardson, compared Haeckel's drawings to photographs of actual embryos at various developmental stages. They found that Haeckel had distorted the evidence at every turn, leading Richardson to tell the journal* Science *that 'it looks like it's turning out to be one of the most famous fakes in biology.' "* [89] At this stage it needs to be pointed out that Haeckel's deceit does not invalidate the Theory of Common Descent but neither does it substantiate embryology as a support for Darwinian Theory either.

Comparative anatomy was the next most significant source of evidence available in 1859. Darwin considered homology to be convincing evidence for common descent but the important question is: is the *common information* which specifies these similarities in body plans derived from a common ancestor? Certain designs are more successful than others, which, as Darwin pointed out,

85. Meyer, S.C. et al: (2009) p.66.
86. Wells, J.: (2002) p.99.
87. *Ibid.* p.102.
88. Meyer, S.C. et al: (2009) p.68.
89. *Ibid.* p.69.

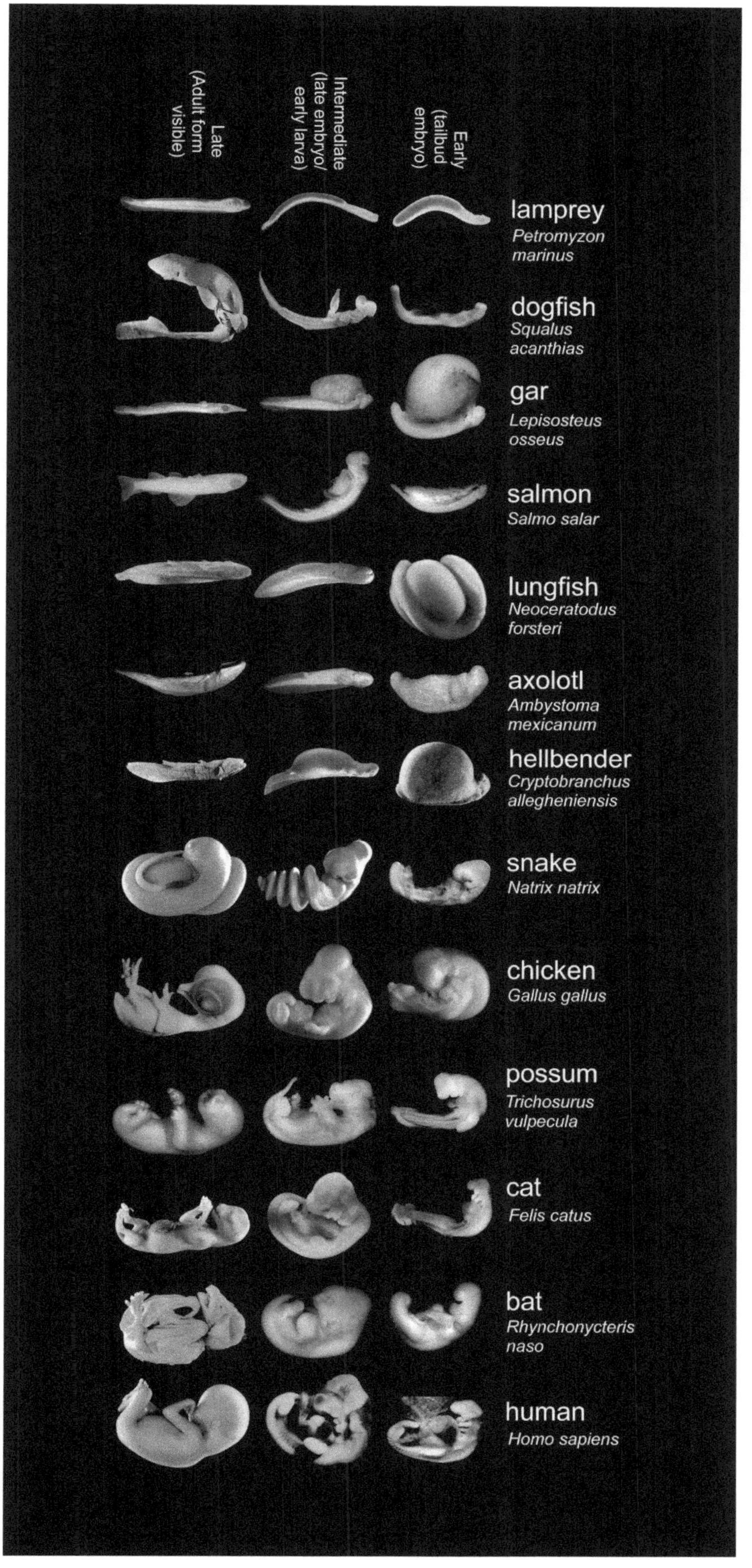

Figure 5.5: Richardson's photos of embryonic development
The pictures speak for themselves. Haeckel's alleged similarities are clearly fabricated. *[Illustration courtesy of Michael Richardson:* Richardson, M.K., Hanken, J., Selwood, L., Wright, G.M., Richards, R.J., Pieau C. & Raynaud, A. *Haeckel, embryos, and evolution. Science. 280(5366):983, 985–6. (1998) Reproduced with permission.]*

would therefore be expected to be found more commonly in nature. In Neo-Darwinian Theory, the process that causes the emergence of these similar features in unrelated species is termed: 'convergent evolution'. In a similar way, we find that most cars have four wheels for optimal balance; cars do not *usually* have five, six or seven wheels. Some simpler models trade wheels for stability and economy, so three-wheeled cars, two-wheeled bikes and one-wheeled unicycles exist but they come with fewer bells and whistles. Brian Goodwin (1931–2009), formerly of the British Open University suggests that homology: *"reflects constraints imposed upon the structure of organisms by the laws of nature. Goodwin contends that the laws of nature dictate that a liquid, for example, has only a limited number of shapes it can take - a spiralling funnel when going down the drain, a droplet when it falls, and so on. In the same way, says Goodwin, the laws of nature ensure that only a certain number of anatomical patterns are possible. Therefore we should expect to see similarities in the anatomical structure of even different types of organisms."* [90] Thus, in living things we repeatedly observe: four limbs, five digits, two eyes and two ears, which appears to represent the optimisation of the design space in animals, in the same way that four wheels and one engine optimise the design space for a car. However, a more reliable way to determine *common* information is through the genetic code, which is being sequenced for many species at this time. Even then, though, there are problems.[91]

Richard Dawkins writes: *"...today we are pretty certain that all living creatures on this planet are descended from a single ancestor. The evidence...is that the genetic code is universal, all but identical across animals, plants, fungi, bacteria, archaea and viruses."* [92] Common DNA sequences point to a common information source but that does not necessarily prove common ancestry or a degree of relatedness. This is the same problem Darwin had with homology. Dawkins correctly observes shared information in Tier 2 but then extrapolates causation in Tiers 3 and 4. As Brian Goodwin and the French anatomist, George Cuvier (1769–1832) suggest, it may just reflect design constraints created by the laws of the Universe. To optimise the design space, certain patterns will be more successful than other patterns as observed previously. Meyer et al demonstrate how the mammalian forelimbs, with one upper arm bone and two forearm bones, are more functional than the opposite arrangement with two bones closer to the body and one forearm bone.[93] These functional constraints created by the laws of nature can also be used as arguments to support convergent evolution although it does not *prove* that convergent evolution occurred either. Meyer et al wrote: *"Some scientists, such as Cambridge University palaeontologist Simon Conway Morris, think that convergence may point to 'directionality' in evolution, where the 'solutions' to the functional problems faced by organisms were 'effectively predetermined from the Big Bang.'"* [94] Conway Morris continues: *"This is because the number of potential 'blind alleys' [for evolution] is so enormous that in principle all the time since the beginning of the universe would be insufficient to find the one in a trillion-trillion solutions that actually work."* [95] Of course, as we discovered in Chapter 3 with Richard Dawkins and Bernd-Olaf Küppers' computer models of evolution, if there is 'directionality' then there must be pre-specified information at or before the 'Big Bang' and the process cannot be Neo-Darwinian in origin or process.

The other problem that Dawkins faces with his previous statement regarding the universality of the genetic code is that since 1985, at least eighteen different genetic codes have been described in various species and Darwinian common descent argues that once one code is set in place, it would be impossible to change, as vast numbers of proteins would end up as gibberish and the

90. Meyer, S.C. et al: (2009) p.43.

91. Seiglie, M.: (2009) pp.8–13.

92. Dawkins, R.: (2009) pp.408–9.

93. Meyer, S.C. et al: pp.46-7 & p.50 Ref 3 & 4.

94. *Ibid.* p.48.

95. Morris, S.C.: *Life's Solution.* Cambridge Univ. Press, Cambridge (2003) p.309 *in* Meyer, S.C. et al: (2009) p.48 Footnote.

organism would undoubtedly die. No way around this problem has been found to date, except to postulate multiple origins for the different genetic codes. Hubert Yockey summarised the difficulty when he wrote: *"The pervasiveness of the standard genetic code has often been cited as evidence of a single origin of life event (Mayr 1982). The other side of the coin is that the existence of several different genetic codes is evidence of a number of independent origins of life events."* [96] And yet all of the origins appear to be of great antiquity. There does not appear to be evidence of on-going, de novo development of new genetic codes that would suggest that Neo-Darwinian Macroevolution is at work.

If we now look at gene frequency as a measure of relatedness, an article in *Nature* in 2005 found that the chimpanzee genome shares more than 98 percent of its DNA code with that of human beings.[97] This similarity has been cited as evidence of a very close relationship between chimpanzees and humans, thus 'proving' that common descent has occurred but recent research stemming from the ENCODE project and other gene sequencing, suggests that this genetic similarity exists across a wide range of living organisms and that it is the RNA from the epigenome which is the discriminating factor that makes the difference in final body plans. As children, we had Lego blocks and from one box of Lego we could make hundreds of different structures but the basic number of bricks was rather limited. If I was to make a human being and a chimpanzee out of Lego, you would not be surprised if 98% of the bricks were the same in both models. It was the way in which they are combined that made the difference. This principle would seem to apply in life as well. Thus the similarity of the building materials neither proves nor disproves that one species evolved into another. As more and more DNA is sequenced across the various species, this relationship is likely to be better understood.

There is one more 'proof' which Darwin cited as evidence for his hypothesis of common descent in *The Origin of Species*, which is the geographical distribution of species on Earth. This is now referred to as: *biogeography*. The Galapagos finches have earned a place in history through Darwin's work but Darwin also collected information on the other fauna and flora of the Galapagos and the way that they varied from South American species and the rest of the World. The limited range of marsupials and monotremes, principally within Australia and to a lesser extent, South America, is often cited as proof of evolution within the southern supercontinent of Gondwanaland but opossums exist in North America today and I was taught at school that this is evidence that placental mammals out-competed the marsupials and monotremes, which became extinct everywhere else in the World and only survived in Australia due to their isolation. The discovery of a marsupial fossil in China in 2007, which predates any known in the southern hemisphere, would appear to support this theory. Once again, biogeography neither proves nor disproves universal common descent (nor Macroevolution) but does support Microevolution.[98]

So, how can the theory and the evidence be rationalised? Dawkins in *The Blind Watchmaker* writes: *"Darwinism is the only known theory that is in principle capable of explaining certain aspects of life."* [99] The problem is that for science it is *evidence* that counts, not *"principle"*. It is one thing to accept and espouse the principles and another to deliver the evidence. This is where the confusion and the controversy become evident. So how can it be resolved? The answer, as it is science that we are examining, is *empirical evidence*. So can information help to clarify the conflict and what evidence do we have?

96. Yockey, H.P.: *Information Theory and Molecular Biology*. Cambridge University Press (1992) pp.202–3 *in* Meyer, S.C. et al: (2009) p.62 Ref. 15.

97. Chimpanzee Sequencing and Analysis Consortium: Initial Sequence of the Chimpanzee Genome and Comparison with the Human Genome. *Nature* 437:69–87 (2005) *in* Fowler, T.B. & Kuebler, D.: (2007) p.102.

98. Ni, X. et al.: A new Early Oligocene peradectine marsupial (Mammalia) from the Burqin region of Xinjang, China. *Naturwissenschaften* 94:237–241 (2007) *in* Meyer, S.C. et al: (2009) p.78.

99. Dawkins, R.: *The Blind Watchmaker*. W.W.Norton, NY (1986) p.287 *in* Wells, J.: (2002) p.205.

The ultimate question in the debate on descent is: can similarity *prove* relatedness and if so, how 'similar' is sufficient? As noted above, DNA is the richest and most accurate source of information so what does the DNA evidence suggest? Common DNA sequences suggest a common information source and Darwinian evolution predicts that related species will have similar DNA code but as seen earlier in this chapter, the DNA 'tree' is muddled up. Family tree structures are contradictory, with common sequences in various species but there is no logical pattern to trace a line of descent through the DNA evidence.[100] Professor Katz found common DNA sequences in sometimes completely unrelated species, rather than the direct line of descent that Darwin predicted. Microbiologist, Carl Woese and physicist Nigel Goldenfield, both of the University of Illinois, have proposed a mechanism called 'lateral gene transfer' (LGT) to explain these findings but the sharing of information does not explain its origin and does not aid the Darwinian cause in an information audit.[101] The haemoglobin molecule is a good example. Haemoglobin is a complex oxygen carrying molecule found in all vertebrates, which includes human beings, but haemoglobin is also found in earthworms, starfish, molluscs, some insects and plants and even some bacteria; it neither proves nor disproves common descent.[102]

The next point to note is that, although the term: 'common descent' has become cemented in our language since Darwin published *The Descent of Man* in 1871, in information terms it would need to be the *ascent of information*. From Figure 5.1, it can be seen that 'common descent' is a misnomer as it originates from the left-hand side of the diagram and requires the massive generation of complex information, not the simple process implied by Darwinian Evolution, where we as human beings have descended from our grandparents, as seen on the right-hand side of Figure 5.1, which is microevolutionary. Once again, it is clear that there is no evidence to support Macroevolution but a significant amount of evidence to support Microevolution.

Information Change

Information changes in the collective genome of all life on Earth reveals interesting differences between the competing theories of origins when compared to the number of species discovered in the fossil record. Figure 5.6 graphs the information from Figure 5.3 over time.

Note that for more than three quarters of the estimated history of the Earth, only single-celled organisms (if anything can be found at all) are present in the fossil record, with only bacteria and algae being found in Pre-Cambrian rocks. If the generally accepted history of the Earth was compared to a 24 hour day, then the first 18 hours would have only bacteria, algae and other single celled organisms. After 18 hours, some sponges, jellyfish and comb jellies appear in the fossil record until the Cambrian Explosion when within one minute of our compressed timeline, between twenty and thirty five of the World's approximately forty phyla appear almost like magic![103,104]

The Neo-Darwinian model has a number of problems with the Cambrian Explosion which include:

1. **Numbers:** The number of new species which appeared in a very short period of time (a massive new information requirement);

100. Katz, L.A.: The Tangled Webb: Gene Genealogies and the Origin of Eukaryotes. *The Am. Naturalist* 154: 137–145 (1999).

101. Buchanan, M.: Horizontal and Vertical: The Evolution of Evolution. *New Scientist* 205(2744) (2010) *in* Sarfati, J.: (2010) p.93.

102. Blanchard, J.: (2002) p.14.

103. Strobel, L. & Vogel, J.: (2004) p.32.

104. J.W. Valentine et al maintain that all the world's phyla had appeared by the end of the Cambrian Explosion (1999)—See Valantine, J.W. et al.: Fossils, Molecules and Embryos: New Perspectives on the Cambrian Explosion. *Development* 126 (1999) *in* Strobel, L. & Vogel, J.: (2004) p.32.

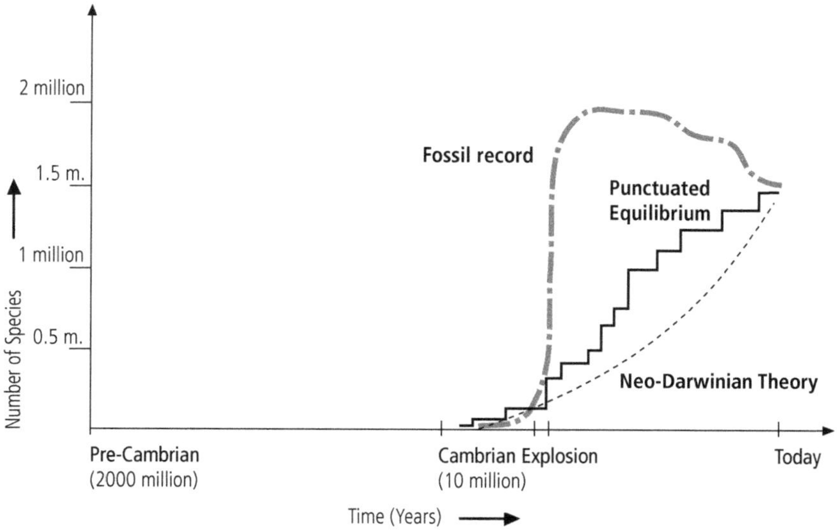

Figure 5.6: Information in the Genome of Life over Time Neo-Darwinian Theory and Punctuated Equilibrium predict very different patterns of information acquisition over time but neither has much resemblance to the pattern of information acquisition observed in the fossil record.[105,106,107]

2. **Variety:** A vast increase in the variety of species (requiring more new information for the new body plans);
3. **Complexity:** The sudden increase in complexity such as the compound eye (needs yet more new information);
4. **Rapidity:** The rapidity of formation (insufficient time to generate the new information);
5. **Mechanisms:** A lack of intermediates (no Neo-Darwinian mechanism to explain the new species).[108]

And these are not the only problems that the Neo-Darwinian Synthesis has with the Cambrian Explosion. The common perception of fossils is of hard bony materials that have been petrified but there are other types of fossils that can provide us with useful information too. Curator of the University of California Museum of Palaeontology, Emeritus Professor James Valentine described the significance of trace fossils in the Cambrian and Precambrian rocks in his book: *On the Origin of Phyla* in 2004. Meyer et al explain: *"Trace fossils provide additional evidence that the Cambrian Explosion was real. When an animal burrows through sediment, it leaves tracks or burrows behind. These tracks can be fossilised. We call these fossilised animal tracks* trace fossils. *Both soft-and hard-bodied animals could leave trace fossils. Here's the point. If lots of soft-bodied animals existed before the Cambrian, then we should find lots of trace fossils. But we don't. Pre-Cambrian sedimentary rock records very little activity. However, at the beginning of the Cambrian Explosion, we see a dramatic increase in trace fossils all over the world."* [109]

The nature of the Theory of Evolution, which advances through the 'survival of the fittest', requires that approximately 95–99 percent of the species that have ever lived to have become extinct. The number of fossil species actually found is around 250,000 while the total number of living species at present is estimated to be around 3 million. This suggests that approximately 8 percent of all species discovered are extinct. Thus, if Neo-Darwinian Theory is true, either there are huge numbers of fossils that are yet to be discovered or the majority of organisms became extinct without leaving a trace.[110] Another possibility is that these observations may point to the need to revise the Theory of Evolution as

105. Wells, J.: (2002) p.43.

106. Bendewald, J. & Sherwin, F.: (2004) p.122

107. Lennox, J.C.: (2007) pp.111–2.

108. Bendewald, J. & Sherwin, F.: (2004) p.125.

109. Meyer, S.C. et al: (2009) p.31.

110. Sarfati, J.: (2008) pp.223–4.

the information within species is more stable than was first imagined. It is also interesting to note that since the Cambrian Explosion, the total number of life forms is decreasing through extinction but if the Neo-Darwinian Theory was true then this number should be increasing all the time. Examine Figure 5.6; from this diagram, it is patently clear that neither the Neo-Darwinian Theory nor Punctuated Equilibrium come even close to explaining the observed fossil record.

Extinctions

Discontinuities are inherent to our world. Earthquakes, tsunamis, volcanic eruptions, asteroid collisions and climate change have all caused alterations to the Earth, resulting in relatively rapid changes (in the geological timeframe) to the environment. To survive, organisms must be able to adapt to these changes. The fact that so many species have survived for so long demonstrates the durability and adaptability of living creatures but even so, there have been five mass extinctions in the fossil record, including the Permian extinction 250 million years ago and the Cretaceous extinction 65 million years ago, when the large dinosaurs disappeared.[111]

In general, the larger an animal and hence the more complex it is, the more information it requires and hence the longer it needs to grow to adulthood. Therefore, the individuals within the reptilian or mammalian species need to live longer so that they can mature and reproduce. So if a sudden (or relatively sudden) change occurs in the climate, the longer the organism's reproductive cycle, the more vulnerable it will be to changes in the environment, as microevolutionary change takes time. Thus, from an information viewpoint, it would be anticipated that the largest animals would be most likely to become extinct first, which is what is observed in the fossil record. Only some of the smaller 'dinosaurs', such as the iguanas, crocodiles, alligators, turtles, tortoises and lizards, still survive today. So how have the whales, especially the blue whale, been able to survive? Being a warm-blooded mammal and living in the sea, it would appear that they are less affected by climate change due to the oceans minimising rapid changes in their local environment.

The other problem with the more complex genomes is: the larger the DNA sequence, the more chance there is for the gradual accumulation of mutations over numerous generations that will compromise its functioning, meaning that there is less viability and therefore less competence within the genome of larger species to cope with relatively rapid change. It could logically be surmised therefore, that the longevity of individuals within the more complex species would gradually decrease as mutational loads accumulate. This narrowing of the gene pool has been observed in the Tasmanian devil, where the only remaining population is now confined to the Australian island of Tasmania and a virus which causes facial tumours is threatening the entire species. Thus the complexity and age of a species can compromise their genetic information and hence be a risk factor for their survival.

Often survival depends upon optimisation of the different component characteristics of the organism. For example, if a rhinoceros is too lightly armoured, it may be devoured by a lion but if it is too robustly built, it may starve in a drought, as it is now too large to adequately nourish itself.[112] This is yet another example of optimisation within each species's design space. The more

111. Sarfati, J.: *Refuting Evolution 2*. Answers in Genesis, AcaciaRidge, Qld, Australia (2002) pp.145–6.

112. Sarfati, J.: (2008) p.218.

specialised or complex an organism is, that is the higher the mountain it sits on in its design space, the greater are the consequences if changing environmental conditions threaten its essential functioning, unless its extra informational complexity enhances adaptability to the environmental change experienced. An example of this would be the fish which have molecules in their bloodstream that act like antifreeze. If the climate gets colder, the extra information that they possess, which allows them to produce natural antifreeze, enhances their chance of survival rather than diminishing it.

Sexual Reproduction (Meiosis)

There is one other function that is relevant to our first species and that is reproduction. Reproducibility occurs in two forms: mitosis and meiosis, both of which require exquisite precision involving over 50 components for the process to occur successfully.[113] Organisms that possess a nucleus (eukaryotes) such as plants, animals and people, reproduce sexually, which requires the *simultaneous development of male and female gametes in one generation* or the species dies out. How might this have occurred? We have already examined mitosis in Chapter 4, where single-celled organisms can reproduce by splitting their genome and cytoplasm to make two daughter organisms. You will recall that Darwinian evolution could not adequately explain how an organism could develop the information to control and execute this function, however, let us assume that mitotic reproduction has been achieved through some yet unknown process, and imagine that through redundancy a bacterium (prokaryote), which reproduces via the simpler, asexual process of mitosis could stumble on the process of hermaphrodisation, whereby each organism in a species can be both male and female at different times in their life cycle. The problem with this is: the organism needs to simultaneously develop male and female reproductive organs, which would require hundreds if not thousands of new complementary biochemical and developmental steps. The complexity of the information required to specify this process is immense and it cannot be broken up into small steps to satisfy Darwinian evolution because it is impossible to only partially reproduce successfully. The next generation is an all or nothing phenomenon and partial reproduction means extinction.

As hermaphrodites, it has been argued, organisms continued to reproduce asexually (by a process called 'parthenogenesis') as they gradually developed their exclusively male and female reproductive systems but the cost of carrying the extra (initially useless) genetic code and machinery means they would be selected against on Darwinian grounds alone. Parthenogenesis is also rare amongst vertebrates and appears to be a degenerative form of sexual reproduction rather than the other way around. In *The Spectrum of life* the authors explain: "*Some scientists think most surviving parthenogenic species have only evolved in the past 20,000 years...Once organisms begin to reproduce in this way the variation that occurs in their population, is greatly reduced. Sexual reproduction increases variation in the population. If there is less variation within a population, there is less chance of having individuals with characteristics suited to a new environment. As the environment changes the species is less likely to survive. In the short term, vertebrates and insects that reproduce by parthenogenesis may be advantaged because they need not waste resources to find a mate...in the long term sexually reproducing organisms may be advantaged*

113. Bendewald, J. & Sherwin, F.: (2004) p.83.

if their environment changes — the history of life on Earth has been one of change." [114] The other problem is the impossibility of retro-writing both the genome and epigenome, to not only develop the sex cells (gametes) with which to reproduce, a process that requires only half the total number of genes per cell but also the organs to store the gametes prior to copulation plus the software to control the entire elaborate process as seen in Chapter 4. After all these requirements are taken into account, the only conclusion that one can confidently draw, is that although hermaphrodites are another enticing thought experiment, this approach does not work from an informational frame of reference.

Further to these problems, Anglo-American Professor of Anthropology, Robin Fox observes: *"...it remains a constant theoretical problem how sexual reproduction can have arisen, since any favourable mutation in an asexually reproducing organism can be immediately and rapidly replicated, while its sexual competitor must dilute the next-generation effect through breeding. Even inbreeding will not help for sexual competitors, since it is bound to be slower than in the asexual, and also will produce lethal homozygotes."* [115] Thus sexual reproduction enhances diversity in the gene pool and dilutes new traits if they should occur; and the gene pool is vast. Sarfati quotes the Professor of Biological Sciences at the University of California and past President of the American Association for the Advancement of Science (AAAS) *"...Francisco Ayala* [who] *points out that humans today have an 'average heterozygosity of 6.7 percent.' This means that for every thousand gene pairs coding for any trait, 67 of the pairs have different alleles, meaning 1,675 heterozygous loci overall. Thus any single human could produce a vast number of different possible sperm or egg cells –* 2^{1675} *(or* 10^{504}*). The number of atoms in the whole known universe is 'only'* 10^{80}*, extremely tiny by comparison."* [116] This vast choice of alleles ensures the dilution of any new deleterious mutations within a population so that they will not damage the gene pool but conversely, beneficial mutations will not be readily established either.

The other even harder problem with sexual reproduction is the intricacy of the process of meiosis, as the first stage in sexual reproduction involves the simultaneous development of male and female gametes (sex cells). The cells have to split, to reduce their genome to half their usual number and then fuse with a complementary sex cell to re-form a new fully competent cell with the correct number of chromosomes. The male and female gametes are not identical, so each male or female sex cell depends on another entity to provide the perfect reciprocal gamete for successful reproduction. This is a high risk strategy. It would be less risky if two complete cells fused first, then split away to leave just one genome in the next generation but that does not happen. No natural process is known that will induce the cell to halve its genome *prior* to sexual fusion, rather than just divide by mitosis as described by Fox above? The organism would need to be directed to take this beneficial yet very risky reproductive detour, which again requires pre-specified information.

Irreducible Complexity

Thus we have seen that living organisms require a vast amount of inter-related information which, when interlinked, is called complexity. Complexity requires meaningful or structured information content. So when a complex entity is made, meaningful information must be generated and incorporated into that entity. Further, the degree of complexity is proportional to the amount of meaningful

114. Aubisson, P. et al: *Biology in Context – The Spectrum of Life.* (2nd Ed.) Oxford Univ. Press, Melbourne (2004) p.201.

115. Fox, R.: "The Conditions of Sexual Evolution" *in* Aries, P. & Bejin, A.: (Eds) *Western Sexuality: Practice & Precept in Past and Present Times.*: Basil Blackwell, Oxford (1985) p.1.

116. Sarfati, J.: (2010) p.44.

information incorporated. If that entity is later deconstructed, what happens to the information? For example when the great library at Alexandria in Egypt was destroyed by fire in 48 BC, what happened to the information:

1. In the scrolls and manuscripts?
2. In the fabric of the building?
3. In the minds of the librarians and keepers of the information?

Much of it was probably lost but some of it may have been able to be transferred, for example: any surviving manuscripts to other libraries, librarians to other institutions, and maybe a painting of the building and/or plans written down to store the information regarding the fabric of the building. A similar analogy could be drawn regarding the deconstruction of a social or societal organisation, such as a sporting club or a government. Most of the time when this type of entity is deconstructed, the incorporated information (sometimes called 'social capital') is lost. However, in deconstruction, is there a critical extent to which an entity can be deconstructed before the entity loses its identity? It would appear that this point is reached when it loses function. Michael Behe has described this condition, which he has called: *irreducible complexity*. Thus irreducible complexity is reached when information levels fall to the extent that the intended function is lost and the mousetrap no longer catches mice.

Complexity and Design Constraints

The next important concept to appreciate is that not all permutations of information are viable due to the immutable laws of nature. Blue whales are never going to crawl up on land, much less fly, because their huge mass would cause them to suffocate (due to gravity). So, since only certain combinations of data result in useful or meaningful information solutions, and as we have previously seen that when only one or two changes occur through random mutation the resulting organism is often unviable and either never grows or dies in infancy, such as in the majority of human trisomies,[117] multiple changes are usually required to go from one viability peak to another in a fitness landscape. These are termed 'design constraints', where not all designs are viable. So, if Neo-Darwinian evolutionary processes change only one or two bases in a genome, this minimal change may not be viable, as a relatively large gap often needs to be crossed to reach the next survivable peak. See the fitness landscape in Figure 3.12.

Redundancy

Complexity often involves redundancy which is our next topic. Redundancy in biology comes in two different forms. The first usage is in the context of 'junk DNA', where no immediate function has been identified for a piece of DNA and the presumption is that the 'redundant' DNA code is the equivalent of a DNA fossil. Today, less than 10 percent of genetic code is considered to be dispensable but as little as fifteen years ago, more than 90 percent of DNA was thought to be redundant. However, as time has gone on, more and more DNA code has been discovered to be required for control sequences and so it may be premature to declare any DNA to be redundant in the light of on-going discoveries. It is more accurate to say that a certain proportion has no known function at this stage.

117. A trisomy is where one chromosome is duplicated inappropriately so that the new organism has three copies of the chromosome instead of the normal two. Down's syndrome is also called: 'Trisomy 21'.

The other connotation for the word redundancy is: a trait which is held in reserve, so that if the first piece of machinery fails, there is a ready-made backup that is immediately available to fill the gap. This way, the whole system does not fail just because one sub-system malfunctioned, as discussed in Chapter 2. Essential services such as banks build redundancy into their computer systems and NASA built redundancy into their space shuttles, with multiple computers backing each other up. This type of redundancy builds resilience into systems and resilience is characteristic of ecosystems so that if climactic conditions change, such as global warming or cooling, the whole system will not fail.[118] This redundancy needs to be reflected in the DNA which codes for it.

Dutch geneticist, Dr Peter Borger, identifies "*three classes of DNA elements: essential, non-essential and redundant. When essential elements mutate... the information carrier as a whole is immediately subject to a severe reproductive disadvantage. In the worst case the mutation is incompatible with life...Essential DNA elements can be defined as biological information that is unable to evolve. Non-essential genes are genes that are allowed to mutate and may thus contribute to allelic variations. As they produce non-lethal phenotypes, they contribute to the variation observed in populations. Classic Mendelian genetics is largely due to variation in the non-essential genes. Variation in non-essential genes is what geneticists call alleles.*"[119] Classic examples of these alleles are eye colour and hair colour in human beings. Borger continues: "*The third class of...genetic elements are correct genes that underlie genetic redundancy. They make up a special class of non-essential genes and have only recently been discovered. That is because their existence cannot be deduced from genetic experiments: they do not contribute to a detectable phenotype. Their peculiarity is that redundant genes may be completely lost from the genome without any effect on reproductive success. That redundant genetic elements make up a major part of the genome of all organisms became evident when biologists interested in gene function developed gene knockout strategies; with the remarkable observation that many knockouts do not have a phenotype.*" [119] Borger points out: "*In* Bacillus subtilis *only 270 of the 4100 genes are essential, and in* Escherichia coli *this is a meagre 303 out of a total of almost 4300 genes. Genetic redundancy is present everywhere...organisms are full of genetic tools that are handy but not essential for survival.*" [120] Borger compares the information within the genome to a Swiss Army knife: "*A Swiss knife contains many functions which are not immediately necessary in a particular environment; but some of them are extremely handy in the mountains, others in the woods, still others are made for opening bottles and cans, or tools for making a fire. Depending on where you are, you may require different sets of functions. Similarly, depending on where the organism lives, it demands different functions (i.e. protein-coding genes and their protein products) from its genome. The environment then determines which part of the non-essential genome is under constraint and it is only this part that will be conserved.*"[121] Thus, no single individual within a population can express all the genetic information within their genome but in addition, no one individual can carry all the possible variations within a species. This information is spread over many individuals, which is why a viable species must have many tens or hundreds of individuals for long-term survival.

Resilient systems have connectivity (with other elements in their network) and diversity (redundancy) through which they derive strength but this form of redundancy is hard to explain in terms of the Neo-Darwinian Evolutionary Model, which distinguishes between traits as a selection between: 'either/or' rather than: 'and/both'. As the Neo-Darwinian Model cannot explain this type

118. Duncan, A.: Graduation ceremony keynote address, University of New England, Armidale, NSW, Australia, 8th Oct., 2010.

119. Borger, P.: Evidence for the design of life: part 2 - Baranomes. *Journal of Creation* 22(3): 71–2 (2008).

120. *Ibid.* pp.68–9.

121. *Ibid.* pp.70–1

of redundancy, some form of pre-specified information (as in the space shuttle computers) would appear to be a better model for explaining this observation of inbuilt redundancy.

Inefficient Design

The charge of 'poor anatomical design' is often levelled at mammals and, in particular, the human body: the retina of the human eye is often said to be 'back to front', the left recurrent laryngeal nerve (one of Dawkins' favourites), as taking an unnecessarily circuitous route and the left vas deferens, as travelling needlessly over the left renal artery and vein before descending to its final destination but a charge of 'poor' or 'unintelligent' design, which is called *dysteleology*, depends upon which criteria one uses to score the design; that is, one's frame of reference. If the only acceptable design is: the 'simplest' or the 'most symmetrical' or the most 'aesthetically pleasing', then the result may be different to: the 'most functional' or 'developmentally sustainable'. The human testes need to be cooler than core body temperature for the production of fertile sperm (so men with undescended testes are infertile) but because the human sex cells for the next generation are set aside soon after the formation of the embryo, the future sex cells (male and female) begin in the centre of the foetus. In males they must then migrate down to the scrotum during the time when the foetus is in his mother's womb. The path that these sex cells travel leaves a loop of vas deferens (their collecting tube) over the artery to the left kidney. A similar developmental process leaves a loop of the left recurrent laryngeal nerve around the major vessels of the upper chest. Yes, the nerve takes a longer route than it might otherwise but maybe the developmental sequence takes precedence over the material route of the anatomical structures. Sarfati quotes *Gray's Anatomy* to point out that the same nerve innervates other tissues in the chest, such as the blood vessels around the heart, as well as other structures in the neck, which may place competing demands on the embryological processes that are required.[122] Dawkins' stance implies that he knows everything that there is to know about the function and embryological development of these structures. In this case, it would appear that the competing requirements to innervate the vessels in the chest and other neck structures supplants the use of the most direct route, which explains why this is such a widely conserved finding. Surely if this truly was 'poor design' and Darwinian Evolution was true, there would be a progressive simplification of these circuitous routes but as far as I am aware, these anatomical features are universal amongst human beings, suggesting that there is some value in conserving them.

Perhaps the most extraordinary statement made by Richard Dawkins, after observing an autopsy of a giraffe is: *"But the overwhelming impression you get from surveying any part of the innards of a large animal is that it is a mess!"* [123] Having studied medicine for over thirty years, assisted at many operations and having given even more anaesthetics, I am continually awestruck at the way the human body is so exquisitely assembled. For people who are used to the symmetry of classical Greek architecture, the human body may be viewed as lacking aesthetics or 'good design' but the Parthenon was never required to eat, sleep, breathe and reproduce, let alone walk and the elasticised, shrink wrapping provided by our skin around a functional unit, which starts as a single cell and can live for more than 100 years is unbelievably remarkable! You would not expect your newly purchased motor vehicle or computer to have anywhere near the same

122. Sarfati, J.: (2010) p.277.

123. Dawkins, R.: (2009) p.371.

performance. The other point is that Dawkins was inspecting a post-mortem specimen, where maceration may have been a factor but this is not the case in healthy living creatures.[124] No matter how it is achieved, all design is a compromise to achieve the best *overall* result. One of the lessons from the history of biology is that what we observe in nature usually turns out to be the best design when we have properly understood *all* the facts. In the vertebrate eye (which includes the human eye), the photoreceptors are located behind two other layers, which at first glance appears to be an inefficient way to introduce light into the system.[125] It is now realised however, that the existing architecture optimises the competing needs for vision, nutrition of the highly active retinal cells and the connection of their output to the brain. The human eye can detect a single photon of light, so any increase in sensitivity cannot improve its function, as the human eye is already maximally sensitive. The bottom line is that the human eye is so much more sophisticated than anything that human beings can build, that even the alleged 'clunky' design of the vertebrate eye is vastly superior to our most sophisticated cameras, so maybe the assessment tool that has been used to rate the competency of the human eye needs to be revised. (See Tables 4.4 and 4.5)

Many other examples of dysteleology could be discussed but I will examine the human pharynx as it includes the important subject of speech and language. The human voice box or larynx has often been cited as suffering from poor design as every year a significant number of people die from choking or pneumonia from food entering the airway and causing infection.[126,127] However, Associate-Professor Gerry Bergman of Medical University, Ohio, writes: "*The pharynx serves as a single passage for three systems — the respiratory, digestive and communicative — for many very good reasons. A major one is, unlike other primates, our airway and oesophagus intersect. This can cause choking, but allows speech...The pharynx structure permits the mouth and nose to alternate as breathing ports — a feature that is critical whenever the nose is plugged, as when suffering from a cold, or the mouth is blocked, such as when it contains food...The tongue, teeth, palate, mandible and cheeks are all necessary for manipulating food, but they are also required for speech. These structures, called articulators, have critical functions in the formation of vowel sounds...The pharynx design allows both simultaneous eating and breathing with greater efficiency and less body bulk than if we had two separate unconnected channels.*" [128] Other integrated functions include: coughing and sneezing, (which are necessary for clearing foreign bodies, such as dust, pollen and excess mucus) and tasting, (which requires the nasal olfactory areas), as discussed in the previous chapter. Without our sense of smell, food would taste bland and uninteresting, as when we have a blocked nose from a heavy cold. This integrated design also permits the gulping of large amounts of air, which facilitates swimming underwater.

More importantly, however, Bergman relates: "*Humans, unlike apes, have a 'descended larynx' (meaning it is located much farther down the trachea towards the lungs than it is for all other primates)...The 'long larynx' in humans is what allows humans to choke but also allows humans to use language. Furthermore, so far as is known, this design feature exists in one primate only —* Homo sapiens. *This design allows speech in humans — the only life form on earth that has complex language...*" [128]Addressing the incidence of choking, Bergman writes: "*Humans swallow about 1000 times a day or 27,375,000 times in an average lifespan. Given this fact life-threatening choking events* [are] *actually a comparatively very rare event when compared with the number of swallowing events during a lifetime.*" [129] He goes on to point out that the majority of choking occurs in very young children, a

124. *Ibid.* pp.356–363.

125. Fowler, T.B. & Kuebler, D.: (2007) p.81.

126. Atrain, S.: *Unintelligent Design. in* Brockman, J.: *Intelligent Thought.* (Ed.) Vantage, NY (2006) pp.129–130.

127. Skybreak, A.: *The Science of Evolution and the Myth of Creationism: Knowing What's Real and Why it Matters.* Insight Press, Chicago, IL (2006) p.109.

128. Bergman, J.: Is the human pharynx poorly designed? *Journal of Creation* 22(1): 41–2 (2008).

129. *Ibid.* p.43.

significant number of whom have placed inappropriate objects in their mouth or are developmentally delayed and have not chewed their food properly. Choking is also seen more often in the elderly, who suffer from a debilitating illness, such as hemiplegia (partial paralysis) following a stroke. Choking is also increased in people who have imbibed excessive amounts of alcohol or who are eating too quickly or are otherwise distracted. All of these examples represent inappropriate use of the pharynx. Choking is a rare event when the pharynx is used appropriately. Thus, once again, the human pharynx is a compromise between competing needs, which from one point of view may be considered to be 'poorly designed' but the alternative would be to sacrifice the unique attribute of human speech, which sets man apart from all the other life on Earth. I consider that the human larynx exhibits exceptional design characteristics with its multiple functions and economy of construction, which uniquely locates human beings within their design space.

Redundant Organs

Vestigial organs are organs which have been considered to have passed their evolutionary use-by date but not yet eliminated by natural selection. They are considered to either have no useful function or vastly reduced function and a list of 86 human 'vestiges' was assembled by the German anatomist, Robert Wiedersheim (1848–1923), in 1893.[130] As our knowledge of anatomy and physiology has increased over the intervening years, a number of the 'vestiges' on Wiedersheim's original list have been discovered to have vital functions, such as the pituitary gland, which is the master hormone coordinator, and the pineal gland, which co-ordinates sleep-wake cycles through the secretion of melatonin. Both of these glands are located at the base of the brain. Over time, the list of vestiges shrank but has since expanded again. Whether a structure or organ should be on the 'vestigial list' is largely up to the person who defines it. The following table lists just a few of the more contentious organs or structures. In 2008, *New Scientist* magazine listed their top five human vestiges which are: goose bumps, wisdom teeth, the Vomeronasal Organ, Darwin's point and the tail bone or coccyx. This list is an interesting mixture of examples that serve to illustrate a number of points which will be discussed in more detail below. However, it is impossible to refer to vestigial organs without first discussing the human appendix, which has always attracted immense discussion. *New Scientist* summarised the debate very succinctly when they wrote: "*The appendix [is] a small worm-like pouch that protrudes from the caecum of the large intestine. It was long thought to be a shrivelled-up remnant of some larger digestive organ — primarily because it is a lot less prominent than its counterpart in rabbits, with which it had previously been compared. In 1980, GBD Scott of London's Royal Free Hospital compared the appendix in different primate species and found the human appendix to be among the largest and most structurally distinct from the caecum. 'It develops progressively in the higher primates to culminate in the fully developed organ seen in the gorilla and man,' he wrote. Scott concluded that the appendix is far from functionless in apes and humans, a view supported by other research. A study published* [in 2007] *in the Journal of Theoretical Biology (249:826) by a group at Duke University School of Medicine, North Carolina found that the human appendix acts as a 'safe house' for helpful, commensal bacteria, providing them with a place to grow and, if necessary, enabling them to re-inoculate the gut should it lose its normal microbial inhabitants — for example, as a result of illness.*"[130]

130. Spinney, L.: Five things humans no longer need. *New Scientist*. 19 May 2008. http://www.newscientist.com/article/dn13927-five-things-humans-no-longer-need.html

The first relevant observation is the fact that when people have their appendix removed, they do not appear to be any less healthy than people who retain them. This is one of the principal reasons why the appendix is cited as a vestigial organ in human beings. One response to this observation is that appendectomy occurs mainly in the developed world, where severe frequent gastrointestinal disease, such as cholera, is generally not a problem. The second interesting observation is that the appendix is universal in human beings. *New scientist* notes that: " *as many as 35% of people have no wisdom teeth at all, suggesting that we may be on an evolutionary trajectory to losing them all together.*" This conclusion is probably true and reflects Microevolution, where information is being lost from the human genome. Wisdom teeth are not critical for life and as such there is evidence that they are being lost as the human jaw appears to be getting gradually smaller. If the same is also true for the appendix, then why is there not a growing percentage of the population which does not have one? Randolph Nesse of the University of Michigan and George Williams, of the State University of New York at Stony Brook, suggested that a smaller, thinner appendix may block more easily and hence be selected against and *not* eliminated.[131] However, having assisted at a considerable number of appendectomies myself, I cannot understand how this would preclude a relatively wide-lumened appendix from progressively shortening itself out of existence over a number of generations, if it was truly a redundant organ. Its universality would seem to verify its necessity, even if we do not fully understand the reasons at this time.

And now for the rest of the 'top five'; *New Scientist* expresses the goose bump argument as follows: "*Though goose bumps are a reflex rather than a permanent anatomical structure, they are widely considered to be vestigial in humans. The pilomotor reflex occurs when the tiny muscle at the base of the hair follicle contracts, pulling the hair upright. In birds or mammals with feathers, fur or spines, this creates a layer of insulating warm air in a cold snap, or a reason for a predator to think twice before attacking. But human hair is so puny that it is incapable of either of these functions. However, they have taken on a minor new role in humans and have become linked with emotional responses – notably fear, rage or pleasure. This may also heighten emotional reactions: there is some evidence, for instance, that a music-induced frisson causes activity in the brain that is associated with pleasure.*" [130] For this music-induced response, there must be a neuro-endocrine connection, which points to information (a connection) that has not yet been described and hence a simplistic comparison with birds and mammals would seem to under estimate this situation, in the same way that the comparison of the human and rabbit appendices also fell short of the mark. Apart from this observation however, and despite what Richard Dawkins contends, goose bumps (or more correctly the pilomotor reflex) in human beings still performs the same functions as in other mammals, namely: warmth, fright and emotion.[132] Many Europeans and people in hot climes have largely lost their body hair but a number of races still retain large amounts of general body hair. At a party I attended as a university student, one male individual of olive complexion had so much long thick black curly body hair that the skin of his trunk was scarcely visible when he removed his shirt (which he did to impress the other partygoers, so magnificent was his 'coat' of hair). The world has been in a warmer climate cycle since the mini 'ice age' of the 15th and 16th-Centuries and no doubt there has been no selection pressure for human beings to have a thick 'hairy coat'; but there is no doubt that some individuals in the population still carry the genes that gave a thicker covering of

131. Nesse, R.M. & Williams, G.C.: Evolution and the Origins of Disease. *Scientific American* 279(5): 86–93 (1998).

132. Dawkins, R.: (2009) p.340.

natural hair. I would like to know which country the male partygoer came from; my impression is south-eastern Europe, such as the Balkan States but I am not sure. I also know from my personal experience, that when I am cold, even though I have only a fraction of the body hair that the fellow described above exhibited, the pilomotor reflex serves to keep me warmer than if I had none. So hair loss is a microevolutionary change and represents information loss which, as discussed previously, can happen rather quickly; certainly within a few hundred years.

The Vomeronasal Organ debate appears to lack essential information also. *New Scientist* wrote: *"Rodents and other mammals secrete chemical signals called pheromones that carry information about their gender or reproductive state, and influence the behaviour of others. Pheromones are detected by a specialised sensory system, the vomeronasal organ (VNO), which consists of a pair of structures that nestle in the nasal lining or the roof of the mouth. Although most adult humans have something resembling a VNO in their nose, neuroscientist Michael Meredith of Florida State University dismisses it as a remnant. 'If you look at the anatomy of the structure, you don't see any cells that look like the sensory cells in other mammalian VNO's,' he says. 'You don't see any nerve fibres connecting the organ to the brain.' So what about the puzzling evidence that humans respond to some pheromones? Perhaps it is possible that we may still secrete them to influence the behaviour of others without using a VNO to detect them."* [130] At the very least, the human VNO (if there ever was one) represents a loss of information and is consistent with Microevolution. However, the fact that humans can still respond to pheromones suggests that there is an alternative detection system that does not require a nerve supply but communicates by some other mechanism, possibly hormones via the bloodstream. Faith Hickman Brynie writes: *"The VNO debate remains unresolved still...Nevertheless, the argument moved away from the VNO in 2006 when Stephen Liberles and Linda Buck reported on TAARs [trace amine-associated receptors]. TAARs are receptors in olfactory cells, but you can't use them to smell the roses."* [133] An article in the Journal of Heredity in 2008 explains: *"Recently, a new gene family of G protein–coupled receptors (GPCRs), denoted trace amine–associated receptors (TAARs), have been identified in both mammalian and fish genomes (Borowsky et al. 2001; Gloriam et al. 2005)...recent studies point to TAARs being mainly involved in olfaction (Liberles and Buck 2006)...It has...been suggested that this new class of chemosensory TAAR receptors is associated with the detection of social cues and thus may be highly relevant for the study of behaviors associated with individual recognition and/or mate choice (Liberles and Buck 2006)."* [134] Thus, the VNO is not universal in the animal kingdom and human beings may never have possessed one but even if they once did and it has since degenerated, the presence or absence of a VNO does not support Neo-Darwinian Macroevolution.

The coccyx or human 'tailbone' consists of four tiny vertebrae at the base of the spine to which ligaments attach to anchor the human pelvic floor and anus. Because the bones are so small in comparison to the tails of monkeys, they have been assumed to be vestigial but once again, they are universal in human beings and hence appear to be essential, as the anus is not optional unless you have a colostomy fashioned by a surgeon for the treatment of bowel cancer or inflammatory bowel disease. This example once again serves to illustrate that direct comparisons between species, even if the organs appear to share similar functions or their owners are considered to be 'closely related' in an evolutionary sense, can be very misleading and can result in misleading conclusions.

Darwin's point is a minor malformation on the ear and is an autosomal

133. Brynie, F.H.: *Brain Sense: The Science of the Senses and How We Process the World Around Us.* Amacom Books, NY (2009) p.60.

134. Mueller, J.C., Steiger, S., Fidler, A.E. & Kempenaers, B.: *Biogenic Trace Amine–Associated Receptors (TAARs) Are Encoded in Avian Genomes: Evidence and Possible Implications.* Oxford Journals Life Sciences Journal of Heredity 99(2): 174–6 (2008).

dominant genetic mutation. As such, it is a corruption of the original information and further evidence of microevolutionary change. Other organs that have been named as vestigial include: the tonsils, adenoids and the thymus gland which were all considered vestigial at one stage, but are now known to be important for the human immune system. Once again, they can all be removed from adults without obviously damaging the health of the individual. However, much of the finer workings of the immune system are still not completely understood and it appears that their most important function is in the development of the immune system in early childhood and that by the time they are removed, their principal function has already been completed. This particularly applies to the thymus gland, which shrinks after eight years of age.

Another 'vestigial organ' mentioned by Richard Dawkins is: "*...the pouch of that iconic Australian animal the koala, which — not a great idea in an animal that spends its time clinging to tree trunks — opens downwards, instead of upwards as in a kangaroo.*" [135] Having observed many koalas since my childhood, it occurs to me that the pouch is backwards for exactly the same reason that the wombat and the marsupial mole have their pouch opening backwards. As the koala drags its abdomen up the eucalypt trees, the pouch would fill with a lot of unwanted bark if it opened upwards as in the kangaroo. It might be supposed therefore that the baby koalas would fall out when the mother climbs trees, however anyone who has observed koalas, knows that 99 percent of the time they move extremely slowly (they can move much faster if chased by a bushfire). Initially the baby koala is only a few millimetres long when it enters the pouch and attaches firmly to a teat. Jonathan Sarfati also notes that the: "*...pouch has a sphincter in the backwardly facing opening to hold the young ones in.*" [136] When the baby koala is too big to ride in mother's pouch, it clings to her back using its claws until it is old enough to live independently. For these reasons I do not consider the koala's pouch an example of a vestigial organ or 'poor' design but one with excellent functioning.

	Feature	Function
1	Appendix	'Safe house' for beneficial bacteria
2	Tonsils	Immunological protection
3	Goose Bumps	Heightens emotional responses in human beings.
4	Wisdom teeth	Gradually disappearing as the human jaw become smaller due to microevolutionary change.
5	Vomeronasal Organ	Possibly microevolutionary information loss but may have a completely different function in human beings that is yet to be understood.
6	Darwin's point	Autosomal dominant congenital defect (mutation).
7	Coccyx	Required for muscle attachments for the pelvic floor and defecation.

Table 5.1: Some structures that have been named as Vestigial Organs in Human beings

Most existing vestigial organ arguments centre round one of the issues identified above:

1. They are degenerative structures where information has been lost (microevolutionary);
2. They are corrupted structures where information has been mutated;
3. They are universal structures where the function is not yet properly understood;
4. Invalid comparisons have been made between other animals and

135. Dawkins, R.: (2009) p.369.
136. Sarfati, J.:(2010) p.283.

1	Teeth in baleen whale embryos[137]	It has been suggested that the teeth act as a framework for the subsequent formation of the baleen plates.[138] The teeth may derive from a common whale body plan with differential expression by control genes. They indicate additional information but they do not prove (nor disprove) Darwinian Macroevolution.
2	Extra digit in developing chickens	This may be a degenerative error of the epigenome. It represents information expression but not proof of Darwinian Macroevolution.
3	Non-functional eyes in cave dwellers	The eyes are an infection risk and an expensive accessory, so where there is no survival advantage, microevolutionary processes select against them.[137]
4	Non-functional legs on snakes or legless lizards	These are degenerative forms of previously functional legs which indicate information loss and are hence microevolutionary.
5	Tiny bones on the whale pelvic girdle.[139]	The bones differ in males and females and *"help anchor the reproductive organs"* [140]
6	Degenerative wings on birds and insects	This represents information loss and is another example of Microevolution.
7	Koala's pouch opens 'backwards'.[135]	A backward-opening pouch does not collect unwanted debris (also in wombats); it has a sphincter to retain the young.[136]

Table 5.2: Some structures that have been named as Vestigial Organs in Animals

human beings (for example comparing the human coccyx to a monkey's tail is not valid if human beings did not live in the trees like the lemurs and monkeys do, as is conjectured by Neo-Darwinian Theory).

Macro-Communication (inter-organismal)

All of the more complex species have some form of communication which aids their survival as a species. Even simple bacteria and moulds have forms of communication, which allows them to band together to improve their survival, such as to coordinate their attacks against a host as evidenced by the following report in *Cell*: *"...Howard Hughes Medical Centre Institute researchers have shown that bacterial communication can be intercepted and chemically manipulated to control group behaviour, such as virulence or the production of a useful product."* [141] However, this is not the first time this type of group behaviour has been described in microorganisms.

Brian Goodwin, formerly Professor at the Open University at Milton Keynes, UK noted that slime moulds communicate via cyclic-AMP (cAMP) to form fruiting bodies. Fowler and Kuebler write: *"When the individual amoeboid cells sense a lack of nutrients in the environment, they begin to secrete a signalling molecule, cAMP to notify the neighbouring cells that it is time to form an aggregate. The cAMP molecule is propagated from cell to cell in the form of a wave as each cell responds to the cAMP signal by making more of it."* [142] Although the cells are individual and discrete, they organise together to form a patterned fruiting body. Goodwin cited the intricate patterns as a self-organising system. He wrote: *"There is no plan, no blueprint, no instructions about the pattern that emerges. What exists in the field is a set of relationships among the components of the system."* [143] But what determines the *"relationships among the components"*? Surely this is pre-specified, intrinsic information in the slime mould genome which tells it to release cAMP when the food source dwindles and where did that information come from? 'Self-organising' systems only have the *appearance* of being so. When the system is fully investigated, pre-specified information will be found to lie behind it.[144] Fowler and Kuebler write: *"They also need a mechanism for secreting cAMP, receptors for cAMP, a mechanism for secreting phosphodiesterase, and a mechanism to desensitise themselves to cAMP levels in the fluid around them immediately after exposure. How all of these field components came to be aligned to interact properly is not clear...The*

137. Fowler, T.B. & Kuebler, D.: (2007) pp.82–83.

138. Sarfati, J.: (2002) p.210.

139. Dawkins, R.: (2009) p.342.

140. Sarfati, J.: (2010) p.262.

141. Bassler, B.L.: *Scientists Manipulate Group Behaviour in Bacteria.* Howard Hughes Medical Institute, 8 August 2008 *at* http://www.hhmi.org/news/bassler20080807.html

142. Fowler, T.B. & Kuebler, D.: (2007) pp.313–4.

143. Goodwin, B. *How the Leopard Changed Its Spots: The Evolution of Complexity.* Charles Scribner's Sons, NY (1994) pp.51–2 *in* Fowler, T.B. & Kuebler, D.: (2007) p.314.

144. Fowler, T.B. & Kuebler, D.: (2007) p.315.

gradual accumulation of the components necessary to build this dynamic field appears to require some type of selection. But it is hard to see how receptors for cAMP would be useful if there were no mechanism in place for secreting it during times of starvation and vice versa." [144] It is apparent once again that additional, pre-specified information is required to build and operate the 'self-organising' system.

This external message results in information travelling from cell to cell via a messenger (the cAMP), like a postman carrying a letter, to alert the cell to adopt an advantageous behaviour. The cAMP in this case is a carrier of the message, which says: 'Food is running out; form fruiting bodies now'. The carrier therefore has the information that constitutes the message imbedded in it. As such, the carrier is the physical entity which carries the nonphysical information and so it is actually a physical symbol of the otherwise undetectable information, which is the message. It is important to note however that the cAMP is only a symbol of the message. It has this meaning for slime mould but other species use cAMP for other functions. The 'form fruiting bodies' message is not built into the message carrier in the way that shape is built into a sequence of amino acids in a protein. The interpretation or meaning is determined by the process that encoded the information into the slime mould which produced the cAMP and the second slime mould which received and decoded the message; the message is extrinsic to the carrier of the message. Edgar Andrews writes: *"In linguistics and language theory, a 'symbol' is defined as something that has meaning in isolation and in the absence of an object. However,* [one can] *employ it in the broader sense used in information theory, where it can include not only meaningful written signs but also recognisable sounds, scents, actions and so on."* [145] Further, Andrews states that if there is to be anything which requires meaningful content, there must be a meaningful signal which is called a 'code'. Andrews continues: *"A code is simply a kind of alphabet — something in which symbols are assigned agreed meanings or functions...Many — perhaps all — animals communicate using codes, which range from simple alarm cries or a hen's call to her chicks, to the complex array of whistles by which dolphins keep in touch. Honeybees perform a 'dance' to tell the hive where nectar is to be found, the 'steps' in the dance indicating both the direction and distance of the hoard. This too is a code. Animal scents may also provide a simple communication system between members of a species. Ant colonies, for example, use pheromones to pass information around. While none of these animal codes amount to a language in the conventional sense, they all use signs or symbols to pass useful information from one creature to another."* [145] So whenever we encounter codes, we know there must be embedded meaning within them to make them effective. And in a very simple form, the cAMP of the slime moulds is a type of very simple language. Edgar Andrews notes: *"The development of a language of any kind requires that the information sender has 'intention' and the receiver has 'comprehension.' "* [146] We do indeed observe this in the slime mould described above. The 'intention' of the mould which sent the cAMP is to alert the colony to create fruiting bodies and the receivers must have some form of comprehension (in this case cAMP receptors) or they would not respond and produce the fruiting bodies. So the information is effectively invisible unless a material vector (the cAMP) or a changed outcome (the production of fruiting bodies) is observed. If the recipient of the message had suffered a mutation and could not decode the message though, there may have been information in the vector but no change will be observed in the recipient mould. So the information within the vector is only one quarter of any communications system. For effective communication there must be:

145. Andrews, E.: (2009) p.176.
146. *Ibid.* p.188.

1. Information which is the (non-material) reason for the communication;
2. An initiator which combines the non-material information with the material carrier to transmit the communication, (such as the initial slime mould which can make cAMP);
3. A vector to transmit the information within its carrier (such as the cAMP which is transported within a fluid medium). The vector and the carrier must be specified by yet more information, as well as a mutually understandable code, which itself requires another layer of information. And finally,
4. A second party which can recognise the code with its information embedded in the vector (such as the cAMP receptors, which again requires information to form and operate them).

Thus communication is a four step information-rich process. As noted above, many species have some form of communication between their members and some mammals such as whales appear to have quite sophisticated levels of communication, which allows individuals within a species, such as killer whales, to co-ordinate their hunting. There are many reports from Twofold Bay on the south coast of New South Wales, Australia about one particular whale dubbed by the whalers as: "Old Tom", who was the leader of a pod of orcas. The report reads: "*The last whale was taken in Twofold Bay 81 years ago* [in 1928] *but prior to that a pack of killer whales, including old Tom, would herd the baleen whales into the expansive bay. They would then alert the whalers by thrashing the water with their tails and then sit back and wait for the whales to be harpooned. Once the whales had been struck, some of the killers would roll over the whale* [to submerge its] *blow-hole to stop it breathing, while others would swim below to prevent a hit whale from escaping with minor injuries. Once a whale was dead, the killers would help themselves to the tongue and lips of the dead mammal, leaving the rest for the whalers.*" [147] This report clearly illustrates not only the communication within the killer whale species but also between the orcas and human beings for their mutual benefit. This relationship between the human beings and the orcas had persisted for more than 100 years.[148]

The ultimate level of communication however is human language which far surpasses the abilities of any other living species. Much of the development of our modern human society and especially language development has to be surmised from archaeology and written history; however, a rare opportunity occurred in 1980 in Nicaragua when 500 deaf children were brought together for the first time in schools for the deaf. The children had previously been scattered in small communities across the country and had not been exposed to formal education or established sign language. However, once they had come together in the special schools, Peter Radetsky reports that it did not take very long for a sign language to develop. Although it was rudimentary at first, with time it developed recognisable rules of grammar and syntax. Judy Kegl, a behavioural neuroscientist at Rutgers, commented that this was: "*the first documented case of the birth of a language...There is nothing that they could have used as a model. It's clear evidence of an innate language capacity.*" [149] These observations and comments reinforce the general observation that all human beings around the world learn a language and they usually develop it before the age of three years. This universal ability to communicate through sophisticated language suggests that the human brain is either hardwired or pre-programmed to develop it. If that is indeed the case, then the pre-programming

147. The legend of Old Tom. *The Magnet*, Eden, NSW, Australia (29 October 2009) p.9.

148. The relationship only ended after the cessation of whaling in Twofold Bay. Old Tom's carcass was washed ashore on a local beach on 18th September, 1930. It is presumed that Old Tom died of old age. After Old Tom died, the killer whales disappeared from Twofold Bay. See Meacham, S.: The killer Whales of Eden. *Australian Medicine.* (Nov., 2010) p34.

149. Radetsky, P.: Silence, signs and wonder. *Discover Magazine.* 15(8): 60–68, August 1994 *in* May, K.: Born to communicate. *Creation Magazine* 31(2):41 (2009).

required information which must have been generated at some time in our past history. The next obvious question is: where is it programmed and how is that done? We are not in a position to answer these questions yet so, at this stage, let it suffice that we recognise that such a language facility exists and we will investigate this topic further in Chapter 10.

The Selfish Gene Theory

Richard Dawkins proposed the *Selfish Gene Theory* in 1976,[150] in which he stated that genes are the 'masterminds' of all living organisms and control everything that the organism does. The genes aim to replicate themselves through the body which they specify, which is their vehicle for survival. Traditionally, biologists have considered that the body, which the genes specify, replicates itself through the mechanism of its constituent genes. Dawkins' argument is: in our most basic form, we are our genes. However, it can be seen that the genome is essentially a string of information recorded on a DNA molecule that is divided into functional sequences, which are known as genes. This does *not* suggest some higher, hidden function for genes and Richard Dawkins would be the first to point out that genes do not have a mind to be able to think or act in a 'selfish' way. They can only act the way that they are pre-programmed to act, by their constituent information. It is therefore the information, which has previously been stored in the DNA that forms the gene that causes the genes to act 'selfishly'. Dawkins' concept might therefore be more appropriately described as: *the selfish information theory*.

The next problem is that it is not the individual genes that possess sufficient information to replicate its own body but the integrated whole. Taking a part of the genome and expecting it to replicate itself will not work. There is a minimum amount of information that is required for successful replication to occur. This amount of information is the whole genome or 100%. So we are back to the concept that the whole-body replicates itself; individual genes are insufficient. It is therefore the pre-programmed information that is the star in the story. From this analysis, it can be seen that the true identity of each living organism, including human beings, is ultimately not their body, nor their genes (which are merely the vessels or carriers of information) but the information itself, which is programmed into their genes. Thus, the question which will not go away is: where did the information come from? This is discussed in detail in Chapter 12.

The Minimum Information requirement for the First Species

To summarise our investigations into the formation of the first species, the diagram illustrating the information requirements for the first organism has increased further and it can be seen that the information required is now formidable. See Figure 5.7.

Conclusion

In summarising our research to date, we have discovered that Macroevolution requires the introduction of new information, whereas Microevolution does not. So Microevolution can explain the variation in Galapagos finch beaks but not how the finches (birds) appeared in the first place (which is macroevolutionary).

150. Dawkins, R.: *The Selfish Gene*. Oxford University Press (1976).

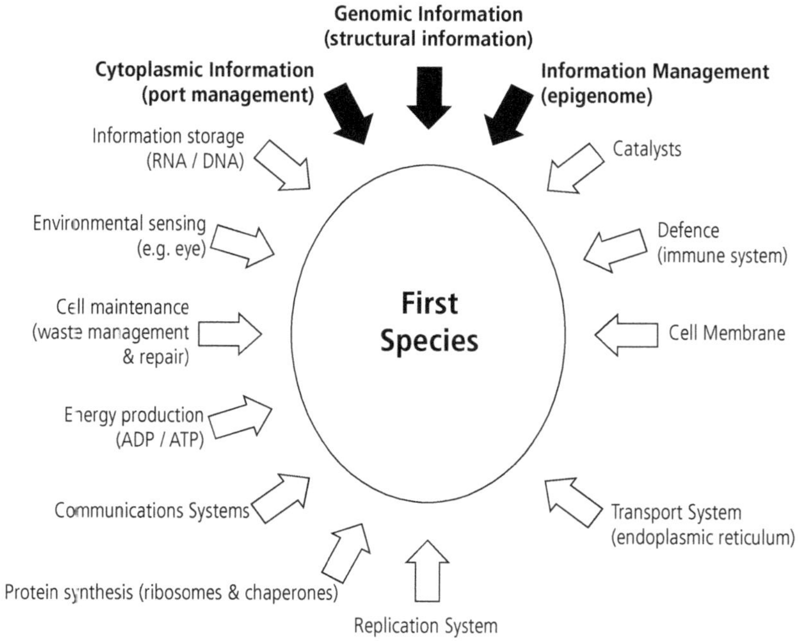

Figure 5.7: The Minimum Information requirement for the First Species
This illustration depicts the minimal information requirements for the first species. Any suggestion that this degree of integrated complexity could have appeared without some *"guiding mechanism"* as Richard Dawkins expresses it, seems absurd. The problem for Neo-Darwinian theorists is that random variation with natural selection cannot explain the information content.

Similarly, Microevolution can explain the resistance of bacteria to antibiotics but not the fact that they exist at all and certainly it cannot demonstrate that finches evolve from simpler forms, such as bacteria. It is becoming clear that what we were taught in Biology at school and university, despite being espoused by some of the most brilliant minds of the 20th-Century, needs further explanation but as Kenneth Miller, Professor of Biology at Brown University, Rhode Island, USA wrote: *"...appearances can deceive even eminent scientists."* [151] And yet, after all of the examples cited already, there are still more really difficult macroevolutionary questions, such as how did cold-blooded animals switch their systems to being warm-blooded in the Neo-Darwinian transition from reptiles to birds and mammals?

Edgar Andrews summarised it well when he wrote: *"...the essence of physical life* [has] *turned out to be organised information — something, moreover, that cannot be stored, transmitted or put to work without the use of communication or 'language'."*[152] Language requires yet more information which will be investigated further in Chapter 10.

Thus, the further we look in biological systems, the more information there is to find. It is essential not only for living our lives but for our physical fabric as well. However, it does not end there. We have already seen that non-material activities such as communication are also information dependant and further non-material human attributes will be investigated in Chapter 10. But before we can broach those subjects, it is necessary to turn our attention to the biosphere and to search further, peering into the Solar System and deep into our known Universe.

151. Miller, K.: What Does It Mean To Be One Of Us? *Life Magazine* (Nov., 1996) pp.38–56 *in* Wells, J.: (2002) p.104.

152. Andrews, E.: (2009) pp.194–5.

6

Information, the Biosphere and the Universe

Philosophy is written in this grand book — I mean, the universe — which stands continually open to our gaze, but it cannot be understood unless we first learn to comprehend the language in which it is written.[1]

Galileo, 1623

As the 20th-Century progressed, it became increasingly apparent that the complexity of the Universe was far greater than had previously been envisaged. The complexity of life here on Earth, involving the intricacies of DNA, the internal architecture of the cell and the interdependence of the neuro-endocrine systems is immense and the information associated with the biosphere is no less amazing. Since the time of Sir Isaac Newton, the clockwork universe had been assumed to be a relatively straight forward affair but the interconnectivity of systems, creates nonlinear results, so that our intuition or 'common sense' is sometimes tricked.[2] In complex systems, such as the world economy and living ecosystems, the whole is more than just the sum of its material parts due to the effect of added information.[3] In this chapter we will be expanding our search into the Earth's biosphere and thence to our solar system, our galaxy, The Milky Way and on to the Universe.

Lord Rees, President of the Royal Society, in his book: *Just Six Numbers* reflects on the extraordinary precision of what he describes as the six mathematical constants that define the form of the Universe and suggests that if any of them was changed even slightly, the Universe and the human race could not exist.[4] This means that the probability that our universe occurred by chance is very small and yet all the constants fit together so that human beings, our biosphere and our universe form one interconnected complex entity. How can such complexity exist, balanced on a knife's edge as it is? Normally complicated things that are unplanned end up as a jumbled mess. Why does the immense complexity of the Universe work? It is because information is embedded in the laws of the Universe, which allows the many inter-meshing systems to work like the cogs of the proverbial Swiss watch (pre-microchips of course) or the software in a well written computer program which allow it to run flawlessly.

In Europe, our understanding of the Universe was erroneous until the telescope provided the data that allowed Galileo and Kepler to more accurately describe the motions of the planets, which then allowed Newton to formulate universal laws. This allowed the apparent randomness of the movements of the planets to be explained logically and mathematically; the inaccuracies within the

1. Gonzalez, G. & Richards, J.W.: *The Privileged Planet*. Regnery Publishing, Washington DC (2004) p.331.

2. Davidson, J.D. & Rees-Mogg, W.: *The Great Reckoning: How the World will change before the Year 2000*. Pan Books, London (1994) pp.245–260.

3. *Ibid.* p.283.

4. Rees, M.: *Just Six Numbers: The Deep Forces that Shape the Universe*. Weidenfeld & Nicolson, London (1999).

previous Greek model of planetary motion had disguised these mathematical laws. Thus a lot of what appears to be random or inexplicable within our Universe may yet be found to have structure and order but as yet we have not found the appropriate 'telescope' to allow the accurate data to be collated and thus the laws to be discovered. In the absence of obvious patterns, we often assume that there are no laws but maybe the nonlinear observations that have been made in the past conceal a more complex system, with new laws or principles still waiting to be described.[5] One such recent discovery is the Anthropic Principle.

The Anthropic Principle was first described by the University of Cambridge astrophysicist and cosmologist, Brandon Carter in a 1973 paper delivered at a conference held in Poland to celebrate the 500th birthday of the father of modern astronomy, Nicolaus Copernicus. The Anthropic Principle is an attempt to explain Lord Rees' six numbers; the observed fact that the fundamental constants of physics and chemistry are 'just right' or 'fine-tuned' to allow the Universe and life to exist. The Anthropic Principle says that the seemingly arbitrary and unrelated constants in physics have one strange thing in common - they are precisely the values that are required to produce a universe capable of sustaining life and in particular, human life.[6] The Universe gives the appearance that it was designed or finely tuned to support life on Earth. It is another example of Paley's watch. Apart from the six constants highlighted by Lord Rees, physics textbooks are littered with precisely defined relationships that are necessary for our universe to exist the way it is; the following examples give just a tiny taste of the precision required for our universe to exist in a state capable of supporting life. More examples are summarised in Tables 6.2 to 6.4:

- The nuclear weak force is 1028 times the strength of gravity. Had the weak force been slightly weaker, all the hydrogen in the Universe would have been turned to helium (making the formation of water impossible);
- If the strong nuclear force was only two per cent stronger, it would have prevented the formation of protons, leaving our universe devoid of atoms. Decreasing it by five per cent would have created a universe without stars;
- If the difference in mass between a proton and a neutron were not exactly as it is, roughly twice the mass of an electron, then all neutrons would have become protons or vice versa; chemistry as we know it and hence life could not exist.

On 7th September 2008, *New Scientist* published an article by Stuart Clarke entitled: "*Unknown Earth: Our Planet's Seven Biggest Mysteries.*" In this article Clarke asks the question: "*How come earth got all the good stuff?...We know that its distance from the Sun provides the right amount of heat and the light to make the planet habitable...Without the unique mix of carbon, hydrogen, nitrogen, oxygen, phosphorus and sulphur that makes up living things, and without liquid water, life as we know it could not have evolved...Another big unknown is how Earth acquired its life-giving water supply. Being so close to the sun, it was probably too hot for water to simply condense out of the gas cloud as the planet formed.*" [7]

The fact that we are living and can observe the Universe, implies that the fundamental constants must be 'just right' to produce life. There is an element of circular reasoning here, because if the constants were not 'just right', we would

5. Davidson, J.D. & Rees-Mogg, W.: (1994) pp.285–6.

6. Gonzalez, G. & Richards, J.W.: *The Privileged Planet.* (2004) p.136.

7. Clark, S.: Unknown Earth: How come Earth got all the good stuff? *New Scientist,* 199(2675): 29–30 (2008) in Schroeder, J.R.: Creation or Evolution: Which is More Believable? The Good News 14(6):6 (2009).

not be here to observe the Universe. However, the fact is that the Universe does not seem to be a random or chance event. We can postulate a multiple universe scenario, in which only one or some universes produce life, but we cannot validate that scientifically because we only live in one of those universes. However, if we are to understand the implications of a Multiverse we first need to have a basic understanding of how our universe works, which is the study of cosmology.

The Big Bang

So what facts do we know about the formation of the one universe which we can study? Lambert Dolphin, senior space physicist at Stanford Research wrote that the Universe: *"...began with a burst of intense light...the 'Big Bang' hypothesis, which is not a proven fact, holds that all the matter in the universe was in the beginning contained in a small 'baseball' compressed so tightly that all matter was converted into energy. In such a state all atomic structure should be squeezed out of normal existence and particles transformed into radiation energy."* [8] Since Dolphin wrote this, cosmologists have traced the Universe back further and propose that it began as a single point smaller than the full-stop at the end of this sentence. The current theory on the formation of the Universe envisions that all matter and energy within the Universe appeared from nothing in the first second. Paul Davies stated that the Big Bang: *"represents the instantaneous suspension of physical laws, the sudden abrupt flash of lawlessness that allowed something to come out of nothing. It represents a true miracle..."* [9] What generated all that energy and matter out of nothing and once it was generated, what caused it to adopt one particular behaviour rather than another? That is, where did the laws of the Universe come from and why would they have been suspended at the beginning?

The Universe was then thought to have expanded very rapidly for a brief period called an 'inflationary' period but how could 'inflation' happen when there was nothing to cause it and why should the normal laws of the Universe have been suspended unless something caused them to be suspended? And science is not supposed to believe in miracles because they are not testable, so why is one invoked now? The only way to explain it is to infer that some process 'nudged' the infant universe into its current configuration, along with its laws and precise settings, in the same way that the epigenome guides the human embryo to grow into a mature adult. In other words, there must have been information which came with the matter and energy that described the way the universal laws were to behave and hence explain the resultant form of the Universe. Einstein's Law of Special Relativity states that matter and energy can neither be created nor destroyed but only converted from one to another ($E=mc^2$). So it would be more logical to hypothesise that the matter and energy came from another place rather than out of nothing and as we shall see in Chapter 8, Einstein's equations suggest that the opposite of a black hole event, a 'white hole' is theoretically possible and might have ushered the matter/energy and information into our universe at its inception. Then, rather than a chaotic, random explosion, the information introduced into our infant universe could have caused it to form as we know it today. There are constraints however, as white holes cannot exist in the presence of matter and the next obvious question is: "Where did the white hole event originate?" These and other questions will be addressed in detail in Chapter 8.

Another unexpected finding is that the rate of formation of the Universe must be tightly controlled for life to be possible. Astrophysicist, Hugh Ross wrote

8. Dolphin, L.: *Lord of Time and Space*. Good News Publishers, Westchester, Illinois (1974) pp.8–9.

9. Davies, P.: *The Edge of Infinity*, Simon and Schuster, New York (1981) p.161 *in Journal of Creation* 23(3): 124 (2009).

that if the Universe expanded too rapidly then no galaxies or planets would have had time to form but: *"If it expanded too slowly, the whole universe would collapse into a super dense lump before any solar-type stars could form."* [10]

And that is not all. There is another problem which was noted by Fred Hoyle, Hermann Bondi and Thomas Gold in the 1940s, in that there is remarkable uniformity throughout the Universe. Edgar Andrews writes: *"To explain how this* [uniformity] *could happen in an expanding universe, Bondi, Gold and Hoyle suggested in 1946 that matter was being created continually, so that its average density throughout the universe remained the same in spite of the expansion. To achieve the required balance, matter would have to be created at the rate of about one atom of hydrogen for every litre of space every billion years - not something the casual observer is likely to notice."*

"Hoyle and his colleagues were motivated at least in part by a reluctance to accept that the universe had a beginning. For entirely philosophical reasons, they were allergic to the idea of a 'big bang' origin. In fact, when Fred Hoyle coined the term 'big bang' [in 1949] *he intended it as a term of derision. Eventually, the force of evidence persuaded him to abandon 'continuous creation' but he was never entirely happy with the idea that the universe actually began."* [11]

So the notion of a finite universe, which had a beginning, is relatively recent. In fact, it was not until the discovery by Arno Penzias and Robert Wilson in 1965, of low level energy readings that are thought to be a remnant of the energy released at the formation of the Universe — called the 'Cosmic Microwave Background Radiation' (CMBR) — that cosmologists generally became satisfied that there was indeed sufficient evidence to demonstrate a finite universe.[12] George Gamow, Ralph Alpher and Robert Herman had not only predicted this in 1948 but they had fairly accurately estimated its value, at 5° above absolute zero. Andrews writes: *"...it is believed to represent the cooled-down radiation left over from an earlier ultra-hot universe and it convinced most astronomers that the big bang theory was correct. Penzias and Wilson were awarded the 1978 Nobel Prize in Physics for their discovery."* [13] The CMB was subsequently found to be 2.7° Kelvin.[12] This Big Bang model for the formation of the Universe has become known as the 'Standard Model of Cosmology'.[14]

Other evidence enlisted in support of the Big Bang Theory is a phenomenon known as the 'red shift'. In 1922, the Soviet cosmologist Aleksandr Friedman (1888-1925) predicted that light coming from distant stars in an expanding universe should exhibit a Doppler effect, which would create a shift in the frequencies of the starlight towards the red end of the spectrum, which was confirmed by the US astronomer Edwin Hubble in 1929.[15]

A third piece of evidence cited by Jim Al-Khalili, Professor of Theoretical Physics at the University of Surrey, UK is *"...the observed abundances of the light elements. The fact that roughly three quarters of all the atoms in the universe are hydrogen atoms and one quarter helium, the lightest and easiest elements to make, with only a tiny amount of all the other elements, requires a universe that was initially hot and dense but which rapidly cooled as it expanded...The conditions of temperature and pressure had to be just right for these atoms to form...Almost all the other elements had to wait until they could be cooked inside stars. The Big Bang model predicts the correct proportions of hydrogen and helium observed by astronomers."* [16]

Now although the Big Bang Theory has been adopted as the 'Standard Model' and is accepted by the majority of physicists, it is not able to explain all observations and calculations. Edgar Andrews writes: *"A particular problem is that*

10. Ross, H.: *The Creator and the Cosmos*. NavPress, Colorado Springs (1993) *in* Heenan, D.: *The Skeptic's guide to God*. Show and Tell. Sanctuary Cove, Queensland (2008) p.25.

11. Andrews, E.: *Who Made God? Searching for a Theory of Everything*. EP Books, Darlington, England (2009) p.99.

12. Gonzalez, G. & Richards, J.W.: *The Privileged Planet*. (2004) pp.174–5.

13. Andrews, E.: (2009) p.103.

14. *Ibid*. pp.99–100.

15. Al-Khalili, J.: *Black Holes, Worm Holes & Time Machines*. Taylor & Francis Gp, NY (1999) pp.47–56.

16. *Ibid*. pp.56–7.

to explain both the shape and the extreme uniformity of the universe, theoreticians have had to introduce the idea that (in its initial stages) the universe 'inflated' from nothing to 1 metre in diameter in a period of 10^{-35} *seconds - a speed of* 10^{35} *metres per second which is a trifling* 10^{27} *(10 followed by 27 zeros) times faster than the speed of light. That's 1 billion, billion, billion times faster than the fastest thing we know."* [13] These numbers are so extremely large that it is difficult to comprehend them. Hawking and Mlodinow describe it graphically when they write: *"It was as if a coin 1 centimetre in diameter suddenly blew up to ten million times the width of the Milky Way."* [17] Now Einstein's Special Theory of Relativity forbids travel *within* our universe that is faster than the speed of light but Andrews points out that: *"the expansion of the universe we observe is not the flight of galaxies through space, but an expansion of space itself."* [18] Jim Al-Khalili writes: *"...the big bang was not like a supernova explosion with all the matter flying away from a central point. The expansion of the universe is a stretching of space itself, with the matter embedded within space and carried along for the ride."* [19] This is not an easy idea to grasp, so let's try to simplify it.

One of the clearest ways to visualise the growth of our universe was described in 1931 by Cambridge University astronomer, Arthur Eddington (1882–1944).[20] Andrews explains: *"Take a round toy balloon and begin to blow it up. The two-dimensional surface of the balloon represents the three dimensions of space (this means of course that nothing else about the balloon, like the centre of its spherical shape, has any meaning. The surface represents everything there is). Next, attach small circular stickers all over the balloon's surface to represent galaxies, and continue to inflate the balloon. What happens? As the balloon grows in size, the stickers move away from one another uniformly in all directions. If you were a bug sitting on a sticker you would see exactly the same thing whichever sticker you were on — all the other stickers would be receding from you equally in all directions. What is more, the further away a sticker is, the faster it will move away from you...Relativity theory only forbids faster-than-light movement through space; it has no problem with space itself expanding faster than light!"* [21]

So, the Big Bang is not so much a 'Big Bang' as a 'Sudden Searing Seismic Stretch'. Al-Khalili likens it to a sheet of rubber that is stretched further and further and as it is stretched, the galaxies spread out with it;[22] but there are laws that govern this process and hence there must be imbedded information, which as always, needs to be accounted for in our auditing process.

Information in an Expanding Universe

As we have seen, the reigning paradigm at the beginning of the 20th-Century was an eternal, static universe. So, when Einstein deduced his Theory of General Relativity in 1915, he altered his initial equations by adding a constant to reflect this widely accepted belief that the Universe was static; but as Edgar Andrews describes, it did not remain this way.

"Aleksandr Friedman in 1922 and Georges Lemaitre in 1927 independently found solutions to Einstein's equations that described evolutionary as opposed to static models of the universe. This implied a beginning which could be represented mathematically as a 'singularity' in Einstein's equations.[23] *(A singularity is a point where some physical quantity becomes infinite - in this case the density and temperature of the universe, implying a 'hot big bang' beginning.)"* [24] At a singularity

17. Hawking, S. & Mlodinow, L.: *The Grand Design*. Bantam Press, London (2010) p.129.

18. Andrews, E.: *Who Made God? Searching for a Theory of Everything*. (2009) p.104.

19. Al-Khalili, J.: *Black Holes, Worm Holes & Time Machines*. (1999) p.58.

20. Hawking, S. & Mlodinow, L.: (2010) p.125.

21. Andrews, E.: (2009) pp.104–5.

22. Al-Khalili, J.: (1999) pp.52–3.

23. This was an international effort and as well as the Soviet cosmologist Friedman and Belgian cosmologist Lemaitre, the Dutch astronomer William de Sitter also concluded that the Universe was expanding. See Al-Khalili, J.: (2009) p.48.

24. Andrews, E.: (2009) p.101.

the mathematics and hence the theory breaks down. Hawking and Mlodinow expand this description of the Standard Model when they write: "*...although one can think of the big bang picture as a valid description of early times, it is wrong to take the big bang literally, that is to think of Einstein's theory as providing a true picture of the origin of the universe. That is because general relativity predicts there to be a point in time at which the temperature, density and curvature of the universe are all infinite, a situation mathematicians call a singularity. To a physicist this means that Einstein's theory breaks down at that point and therefore cannot be used to predict how the universe began, only how it evolved afterwards. So although we can employ the equations of general relativity and our observations of the heavens to learn about the universe at a very young age, it is not correct to carry the big bang picture all the way back to the beginning.*" [25] Edgar Andrews resumes the narrative: "*Much later, George Gamow, Ralph Alpher and Robert Herman developed the big bang theory of the early universe in a paper titled 'The Origin of Chemical Elements' published in the* Physical Review *on 1 April 1948. In spite of the date it was no 'April fool' spoof and turned out to be one of the most celebrated scientific papers ever.*" [24]

We live in a truly extraordinary universe. There are many things that make it extraordinary and the first and most obvious is its enormous size and dynamics. Our home planet, Earth, spins around our star, the Sun, at 107,000 kilometres per hour and our solar system spins around the centre of our galaxy at 914,000 kilometres per hour.[26] Our home galaxy, the Milky Way is estimated to have 100 billion stars and to cross it at the speed of light (the fastest speed attainable within our Universe according to Einstein's equations) would take 100,000 years. The Milky Way is grouped with about nineteen other neighbouring galaxies in the 'local cluster' but over 1 billion galaxies like our own can be observed, each with 100 billion stars, so the stars in the known universe exceed 10^{23} and to cross the known universe would take 15 billion light years![27] And the amazing fact is, it is still expanding. Paradoxically this vast universe is composed of minute atoms which are so small that we have trouble even detecting them. However, their intricacy allows the construction of the elements which permit life to exist and the most important elements for life are carbon and water.

25. Hawking, S. & Mlodinow, L.: *The Grand Design*. (2010) p.128–9.

26. Gaensler, B.: Extreme Speed. *Cosmos* 41:71 (2011).

27. Dolphin, L.: *Lord of Time and Space*. (1974) p.34.

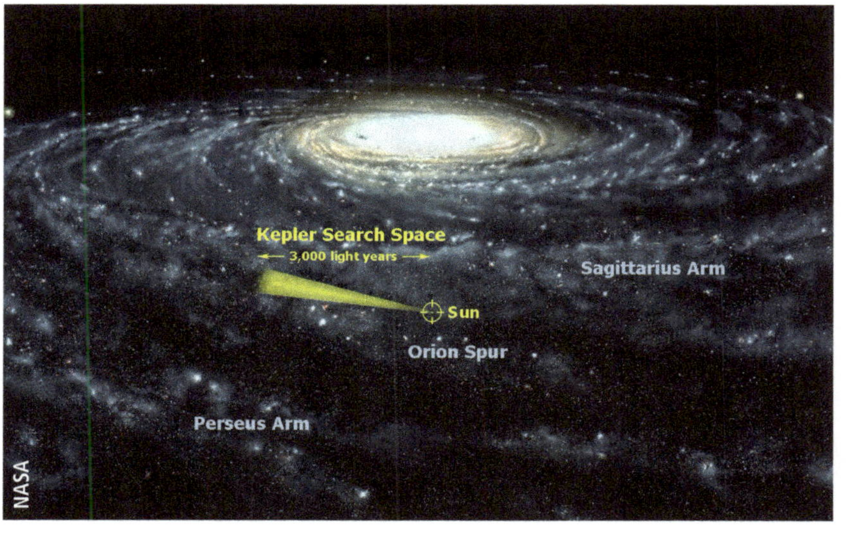

Figure 6.1: The Milky Way
The Sun with our solar system can be seen lying between two of the spiral arms. It is travelling at 914,000 kilometres per hour around the galactic centre.[26] *[Reproduced from Wikimedia Commons© Jon Lomberg.*

Extraordinary Carbon

Carbon is unique among the elements in its ability to form long chains with itself that can then fold into many different configurations. It has four bonds which allow it to form a straight chain with side branches but it is one of the lighter elements and hence its atomic volume is not very large. Other elements in the periodic table, such as silicon, have similar bonding characteristics but the size of their nuclei prevents the formation of the large folded chains that are possible with carbon, which produce the vast array of uniquely shaped proteins. Silicon's inability to form tightly folded molecules due to the size of its nucleus is called 'stearic hindrance' which restricts organic chemistry to carbon alone. Hawking and Mlodinow also note: "*Carbon dioxide, for example, is gaseous at room temperature, and biologically very useful... However, silicon dioxide, quartz, is far more useful in a rock collection than in an organism's lungs.*" [28] Therefore most scientists conclude that, all life in the Universe must be based on carbon.

However, the synthesis of carbon depends upon a remarkable coincidence in the ratio of the strong force to the electromagnetic force, which allows a resonance to occur in the carbon nucleus. British physicist, Paul Davies wrote: "*The Big Bang that kicked off the universe coughed out plenty of hydrogen and helium, but no carbon. So where did the carbon in our bodies come from? The answer was worked out in the 1950s: most of the chemical elements heavier than helium were manufactured in the cores of stars, as the product of nuclear fusion reactions. It is the energy released by these reactions that makes the Sun and stars shine.*

However, while the details of stellar nuclear reactions are fairly straightforward, there is a notable exception: carbon. Most nuclear reactions in stars occur when two atomic nuclei, rushing around at tremendous speed care of the searing temperatures, collide and fuse, forming a heavier element. But carbon cannot be made this way because all the intermediate steps from helium to carbon involve highly unstable nuclei. The solution, spotted by University of Cambridge astronomer Fred Hoyle, is for carbon to form from the simultaneous collision of three helium nuclei.

There is however, a snag. The chances that three helium nuclei will come together at the same moment are tiny. So Hoyle reasoned that a special factor must be at work to boost the rare reaction and lead to our abundance of carbon. If not, then life in general, and Fred Hoyle in particular, would not exist!

Hoyle knew that nuclear reactions can sometimes be greatly amplified by the phenomenon of resonance, similar to the way that an opera singer can shatter a glass by hitting a certain pitch. Carbon nuclei can resonate too, if the masses and energies of the colliding particles that go to form it are just right. Hoyle worked backwards — he knew the particle masses and energies, and he used them to predict the existence of a carbon resonance.

He then pestered Willy Fowler, a nuclear physicist at the California Institute of Technology, to do an experiment to test the prediction. And sure enough, Hoyle was right. Carbon has a resonant state at exactly the right energy to enable stars to manufacture abundant carbon, and thereby seed the universe with this life-encouraging substance.

Hoyle immediately realised just what a close-run thing this mechanism is. Like Baby Bear's porridge in the story of Goldilocks, the energy of the carbon resonance has to be 'just right'. Too high or too low, and the consequences for life would be catastrophic.

So what determines the carbon resonance? Ultimately it depends on the strength

28. Hawking, S. & Mlodinow, L.: *The Grand Design*. (2010) p.157.

of the force that binds protons and neutrons together in the nucleus. That force is one of the unexplained parameters of basic physics — one of the knobs on the Designer Machine if you like. If the strength of the force that determined the carbon resonance was only a fraction stronger or weaker, it is doubtful there would be observers in the universe to worry about the distinct absence of carbon.

Hoyle himself was deeply impressed by this discovery. 'It looks like a put-up job,' he quipped. 'A commonsense interpretation of the facts suggests that a superintellect has monkeyed with physics,' he later wrote. A similar conclusion was reached by the Princeton physicist Freeman Dyson: 'In some sense, the universe knew we were coming.' " [29]

In 2004, Guillermo Gonzalez, Assistant Research Professor in Astrophysics at Iowa State University and US philosopher, Jay Richards, published a book called: *Privileged Planet,* which describes this unexpectedly delicate balance in our universe. Gonzalez and Richards explain: *"A nuclear resonance is a range of energies that greatly increases the chances of interaction between a nucleus and another particle - for example, the capture of a proton or a neutron. And energy resonance in a nucleus will accelerate reactions if the colliding particles have just the right kinetic energy. Resonances tend to be very narrow, so even very slight changes in their location would lead to enormous changes in the reaction rates."*[30] This resonance makes it possible for carbon-12 to reach an excited state of exactly 7.65 MeV at the temperature typical of the centre of stars, which creates a resonance involving helium-4 and beryllium-8 to form carbon-12. This allows the necessary binding to take place during a tiny window of opportunity lasting only 10^{-16} seconds. As explained above, in 1952, Fred Hoyle predicted this previously unknown resonance, simply due to the significant quantity of carbon in the Universe. Gonzalez and Richards go on to explain that further coincidences prevent other resonances converting most of the carbon into oxygen-16 and neon-20. This very unlikely series of coincidences between the energy states of the various elements makes the existence of life an extraordinarily low probability event.[31]

Amazing Water

Water complements carbon perfectly. However, the very nature of water, which is so vital to life, is something of a mystery, a point noted in 1913 by Harvard chemist Lawrence Henderson, one of the forerunners of anthropic reasoning.[32] Unique amongst the molecules, water is lighter in its solid state than its liquid form - hence ice floats! If it did not, the oceans would freeze from the bottom up and Earth would now be covered with solid ice. This property derives from the unique properties of the hydrogen atom. However the properties of water are best showcased when they are studied in Earth's biosphere. Life relies on water being liquid. Water must be in its aqueous phase for molecules to be able to move and react (unlike ice), but dense enough for reactions to take place sufficiently often, that is, fast enough, to support life (unlike steam). It has been estimated that if the Earth was five per cent closer to the Sun there would be a runaway greenhouse effect but if the Earth was twenty per cent further away from the Sun then carbon dioxide would condense in the atmosphere and our planet would freeze. Other unique properties of water are its very high latent heat of vaporisation, which buffers the Earth against large or rapid changes in heat and its extremely high

29. Davies, P.: Life, the universe and everything. *Cosmos* 14 (2007).

30. Gonzalez, G. & Richards, J.W.: *The Privileged Planet.* (2004) p.198.

31. *Ibid.* pp.198–9.

32. *Ibid.* pp.33–4.

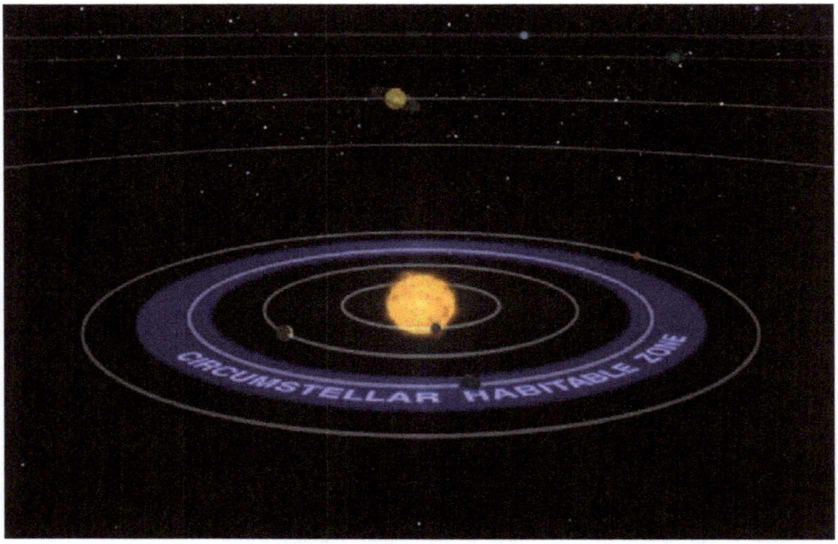

Source: The Privileged Planet DVD (2004)

Figure 6.2: The Circumstellar Habitable Zone or 'Goldilocks Zone' [34] *[Reproduced with permission courtesy of Illustra Media.]*

surface tension in its liquid form, which permits capillary action in soils and plants.[32] The Earth is fortunately the perfect distance from our star, the Sun, for water to remain in its liquid phase all year round. Hawking and Mlodinow write: *"...given any star, scientists define the habitable zone as the narrow region around the star in which temperatures are such that liquid water can exist. The habitable zone is sometimes called the 'Goldilocks zone', because the requirement that liquid water exist means that, like Goldilocks, the development of intelligent life requires that planetary temperatures be 'just right'."* [33]

Without the precisely configured environment that is found on Earth, the water cycle would not operate and life could not exist as we know it. Due to these multiple constraints, it may not be possible for life to exist anywhere other than Earth, unless almost the identical conditions found in our solar system are replicated somewhere else in the Universe.

So why do carbon and water behave the way they do? It is their intrinsic chemical properties, which includes their melting points and vaporisation points, which are controlled by their molecular structures. These, in turn, depend upon the atomic arrangements of their composite elements, which derive from their proton and electron configuration. Tracing the chain back to its simplest form, the proton and electron structures depend upon the laws that govern the microscopic particles within the Universe, which appear to have been operative at or before the beginning of the Universe. These laws are determined by information, which must have been in existence for at least the same period as these laws of the Universe. So there must be information that describes the arrangement of protons, neutrons and electrons that dates from the formation of the Universe. More laws (information) specify the bonds and forces that cause atoms to interact. So behind all matter in the Universe is information, which in Chapter 2 was introduced as 'Intrinsic Information'. The four fundamental forces of the Universe were outlined in Chapter 1 and are summarised in Table 6.1.

Each of these universal forces has a predictable, reproducible action, so within each entity there must be a defining factor that gives it these unique characteristics. This identifying factor, which specifies each force individually, is ultimately dependent on some form of predetermined information, which causes

33. Hawking, S. & Mlodinow, L.: *The Grand Design*. (2010) pp.152–3.

34. Gonzalez, G. & Richards, J.W.: *The Privileded Planet*. (DVD) Illustra Media (2004).

Force	Effect	Energy	Carrier
Gravitation	Determines the orbits of the planets (cosmology)	Tidal power	(Graviton)
Strong (nuclear) force	Holds protons and neutrons together in nuclei, preventing the nuclei from exploding (physics)	Nuclear fusion & the hydrogen bomb	Gluon
Weak (nuclear) force	Determines radioactive decay (physics)	Nuclear fission & the atom bomb	W & Z bosons
Electromagnetic force	Determines the electron shells (chemistry)	Chemical explosions (such as dynamite)	Photon

Table 6.1: The Four Fundamental Forces in the Universe.[35,36]

each force to act in the way that it does. But where did the information come from to create each force in the first place? The answer to this question is not known yet but will be examined further in Chapter 8. Before then we need to learn a little more about what is already known about our Universe.

Fine Tuning

So our material universe is underpinned or specified by intrinsic information. But the *most* unusual feature is not that the information is there, but that it blends together so perfectly to make life possible. Over one hundred criteria have been described that require tightly controlled values if life is to exist on the Earth.[36] Most of the criteria are finely balanced and significantly changing any one of them may result in the extinction of life. Had the terrestrial and universal parameters been different in the first place, life may not have been sustainable. Table 6.2 lists properties of the Earth, while Table 6.3 lists features in the Solar System and galaxies. Table 6.4 lists, subatomic characteristics necessary for life and this is not an exhaustive list but is indicative. The precision of this balancing act is illustrated by astrophysicist Hugh Ross who comments "*...unless the number of electrons is equivalent to the number of protons to an accuracy of one part in 10^{37}, or better, electromagnetic forces in the universe would have so overcome gravitational forces that galaxies, stars and planets never would have been formed.*" [10] This is reiterated by astrophysicist Paul Davies who wrote: "*This* [fine tuning] *is one of a long list of properties in underlying physics that seem to be prerequisites for a universe with life. For example the strengths of the other forces are all important - the masses of the various subatomic particles; if all of these things were even a little bit different then life... certainly life as we know it could not exist.*" [37]

The other planets in our solar system clearly demonstrate the effects that a lack of the fine balance between these multiple factors will deliver: Mercury and Venus are too hot and Venus has a run-away greenhouse effect; Mars is too cold and dry; Europa, the watery moon of Jupiter has a deep ocean that encases it in a thick shell of ice, while the largest planets: Jupiter, Saturn, Uranus and Neptune have too much methane and ammonia to support life. Thus, none of the other planets in our solar system come close to being able to support life as we know it.

Fine tuning is not confined to the Earth alone though. The Earth is dependent on its position in the Solar System, which is dependent on its place in our galaxy, the Milky Way. We have seen how important the 'Goldilocks Zone' is for life on the Earth. However, there are many other factors that are just as important, a number of which are listed in Table 6.3. Despite their distances from the Earth being great, even apparently unrelated facts such as the size of the outer gas giants: Jupiter, Saturn and Uranus, are considered by scientists to be important for the existence of life, as they shield the Earth from disruptive (and potentially

35. Hawking, S. & Mlodinow, L.: *The Grand Design*. (2010) pp.103–4.

36. Gonzalez, G. & Richards, J.W.: *The Privileged Planet*. (2004) pp.197–210.

37. Gonzalez, G. & Richards, J.W.: *The Privileded Planet*. (DVD) Illustra Media (2004).

	Properties of the Earth	Effects if too great	Effects if too small	Notes
1	Distance from the sun	If 20% larger, all water would freeze.[34]	If 5% less, runaway greenhouse effect.[34]	Defines the 'Goldilocks Zone'.[34]
2	Speed of Rotation (Length of a day)	Winds velocities become destructive.[38]	Earth's surface gets too hot by day and too cold by night.[38]	
3	Ratio of Oxygen to Nitrogen	Forest fires would consume the Earth.[38]	Complex life would not be supported.[38,39]	
4	Tilt of the Earth's axis (23½°)	Seasons are disruptive as temperatures vary too much.[39]	There are no seasons or monsoon which results in aridity.[39]	Doubles the arable land compared with other tilts.[39]
5	Size of the Moon (Our Moon is large at ¼ of the size of Earth)	Tides become disruptive.[40]	Loss of stabilisation of Earth's rotation.[40]	It sustains life by aerating and mixing ocean heat.[40]
6	Number of Moons	More could cause axis destabilisation.[38]	Loss of stabilisation of the Earth's axis.[38,39]	
7	Thickness of crust (seismic activity)	Too much oxygen is lost to the crust.[41]	Volcanic activity is too destructive.[41]	Controls plate tectonics.[41]
8	Thunderstorms & Lightening	Destruction of forests by fire.[38]	Soil infertility (lack of nitrogen for plants).[38]	
9	Albedo (reflectivity of Earth)	Seas freeze in a perpetual ice age.[38]	Uncontrollable greenhouse effect.[38]	
10	Ratio of land to water	Aquatic planet (e.g. Jupiter's moon, Europa).[38]	Arid planet (e.g. Mars).[38]	Determines Ocean depths.[38]
11	Diameter of Earth (surface gravity)	Too much methane and ammonia.[39]	Water is lost to space.[39]	
12	Orbital eccentricity	Temperature and climate vary too much for life.[42]	Nil. Orbit is almost circular which creates a very stable climate.[42]	Eccentricity = 0–0.6%[42]
13	Transparency of the atmosphere	Life is destroyed by cosmic rays.[43]	The Earth would freeze.[43]	
14	Ozone Layer	Surface temperature too low.[38]	Sun's UV light sterilises the Earth & surface temperature is too high.[38]	Depends on constant UV from Sun.[44]
15	Salt in oceans	Salt kills all life (e.g. Dead Sea).[43]	Putrid oceans become lifeless.[43]	
16	Water vapour	Greenhouse effect overheats the planet.[43]	No rain.[43]	
17	Magnetic field	Electromagnetic storms too destructive.[45]	Cosmic radiation would kill life.[45]	Core composition (iron/nickel) and it must circulate.[45]
18	CO_2 concentration	Greenhouse effect destroys life.[43]	The seas would freeze.[43]	
19	Water cycle	Soil infertility from erosion & leaching.[43]	Aridity would result.[43]	
20	Thickness of atmosphere	Air pressure, too great for large mammals.[43]	Not enough protection from gamma rays.[43]	

Table 6.2: Properties of the Earth

38. Missler, C.: *The Beyond Collection: Beyond Coincidence.* (DVD) Version 2.0. Koinonia House, Coeur d'Alene, Idaho (2010).

39. Seiglie, M.: Prove There is a God Even Without the Bible. *The Good News* 17(2):15 (2012).

40. Strobel, L. & Vogel, J.: *The Case for a Creator*. Zondervan, Grand Rapids (2004) p.52.

41. Sarfati, J.: *By Design*. Creation Book Publishers (2008) pp. 238–9.

42. Gonzalez, G. & Richards, J.W.: *The Privileged Planet.* (2004) p.182.

43. Missler, C.: *Cosmic Codes: Metacodes.* (CD) Koinonia House, Coeur d'Alene, Idaho (2004).

44. Gonzalez, G. & Richards, J.W.: *The Privileged Planet.* (2004) p.135.

45. Ibid. p.131.

catastrophic) bombardment by meteorites from space. Likewise, the position of our solar system in-between two of the spiral arms of the Milky Way, about two thirds the distance from the galactic centre, is considered crucial by cosmologists, as the radiation from exploding supernovae in the galactic centre or the spiral arms are also likely to damage or eradicate life on Earth.

	Solar System Properties	Effects if too much	Effects if too little	Notes
1	Size of the Sun (which also determines luminosity)	Gravity prevents the Earth's rotation, making it too hot for life.[46]	Goldilocks Zone is too shallow to allow for the eccentricity of Earth's orbit.[9]	
2	Distance of the moon from the Earth	Insufficient Earth orbit stabilisation.[38]	Ocean tides become too disruptive.[38]	
3	Position in Milky Way (the galactic orbit) [10]	The Galactic centre has many ionising events risking sterility.	Solar system won't form in the periphery due to low Metallicity.	Galactic Habitable zone is equivalent to the Goldilocks zone.
4	Wavelength of the sun's light	High energy wave-lengths damage life (e.g. gamma rays) [48]	Low energy wave-lengths are insufficient for photosynthesis.[48]	
5	Galactic orbital speed matches that of the spiral arms	Loss of safety of space between spiral arms.[47]	Entering spiral arms risks supernovae events.[47]	The Solar System travels at 914,000 km/hour [26]
6	Metallicity of the Sun	Planets are too large for intelligent life.[47]	No terrestrial planets.[47]	Metallicity: amount of heavier elements
7	Variability of the Sun	Risks sterilising events on Earth.[49]	Constancy is good.[49]	
8	Size of Earth	Comet impact rate becomes too high, destroying life; also retains too much ammonia & methane[43]	Cools too quickly & molten core is not protective of ionising radiation; also loses oxygen and water.[43]	
9	Number & size of the large outer planets	Disrupts the stable orbits of the planets.[38]	No protection from asteroid impact.[38]	
10	Core spin compared with orbit	Migration patterns of birds disrupted.[50]	Insufficient protection from cosmic rays.[50]	

Table 6.3: Properties of the Solar System

Just as our universe has immense size which can be studied, so it has minute detail which provides the fine structure that constructs the atoms of which we are made. Most people have heard of electrons, protons and neutrons, which make up the atoms that are the building blocks of the Universe but many people will be surprised to discover that these structures are mostly vacant 'space'. Dolphin writes: *"...electrons, protons and neutrons thought in the 1920's to be hard particles, now have added mysteries of electric charge and wave-like character added to their descriptions. Since the vast empty spaces between atoms are a thousand times greater than the space it takes to contain an atom, so-called 'solid matter' is mostly empty space. Were it not for the rapid motions of the molecules of matter, matter would readily interpenetrate and 'solid' objects would not exist. This would mean we would have trouble standing, walking, sitting or supporting anything at all."* [51] One analogy is that if the atom is envisaged as an Olympic sports stadium and a peppercorn placed on the halfway line in the centre of the stadium's playing field represents the proton, then the electron would be a gnat on the upper, outermost terrace for spectators. Dolphin continues: *"One of Einstein's most significant discoveries was that energy and matter are different forms of the same thing. The universe loses almost all of its solidarity and substance! 'Solid' matter is mostly empty space. When one rarely encounters an atom, it is not composed of hard particles or spherical building blocks but electrically charged waves...Since energy and matter are equivalent, one can say that solid matter is the way that space happens to be vibrating in a certain way at a certain time and place."* [52] Table 6.4 lists some of the characteristics of fundamental particles which, once again, need to be finely balanced for life to exist.

Walter Bradley, whom we previously met in Chapter 4, summarised a change in thinking regarding this fine tuning when he wrote: *"More than one hundred similar cosmological coincidences have been identified. Initially there was a*

46. Gonzalez, G. & Richards, J.W.: *The Privileged Planet.* (2004) pp.127–132.
47. Ibid. pp.152–164.
48. Ibid. pp.66–7.
49. Ibid. p.253.
50. Glatzmaier, G.A. & Roberts, P. H.: Rotation and Magnetism of Earth's Inner Core. *Science* 274(5294): 1887–1891 (1996).
51. Dolphin, L.: Lord of Time and Space. (1974) pp.35–6.
52. Ibid. p.36.

	Fundamental Properties	Effects if greater	Effects if smaller	Notes
1	Charge on an electron	Atoms are unstable.[43]	Atoms are unstable.[43]	
2	Gravitational force coupling constant	Universe would have collapsed after the 'Big Bang'.[43]	Matter disperses too rapidly after the 'Big Bang' for the formation of planets, stars and galaxies.[43]	
3	Electromagnetic force coupling constant	Electrons bind too tightly to bond to other atoms and water would not form.[43]	Electrons bind too loosely & water would not form.[43]	The tolerance is 4%.[53]
4	Strong nuclear force coupling constant	Too few light elements (like hydrogen) would form; with none heavier than iron.[43]	Nuclei would not hold together; the Universe would be mostly hydrogen; fewer stable elements.[43]	Holds protons and neutrons together-tolerance, ½ %[58]
5	Weak nuclear force coupling constant	There would be no helium or carbon.[55]	Universe would be all helium & no stars.[55]	
6	Cosmological constant (Anti-gravity force)	Universe would have expanded too rapidly for galaxies, stars & planets to form.[56]	Universe would not have expanded after the 'Big Bang'.[56,57]	This equates to the energy density of empty space.[58]
7	Properties of water	If ice was denser than water the seas would freeze.[43]	There would be no water cycle if water did not vaporise at relatively low temperatures.[43]	
8	Carbon and oxygen nuclei energy levels Resonance (Hoyle)	Most carbon would be converted to oxygen &/or neon[59]	Carbon would not form from beryllium in sufficient amounts[59]	
9	Proton to electron mass ratio	If mass of a proton was 0.2% greater, it decays to a neutron which destabilises atoms.[55]	Molecules do not form and life is impossible.[43]	
10	Proton to neutron mass ratio	Atoms are unstable.	Fusion in stars would cease depriving life of an energy source.[60]	Tolerance 0.1%

Table 6.4: Fundamental Properties of the Universe

tendency to attribute these coincidences to fortuitous accidents, as evidenced by an early quotation from Hoyle (1951, 103–4): 'I have often seen it stated that our situation on earth is providential. The argument goes like this. It is providential that the earth is of the right size and at the right distance from the sun. It is providential that the sun radiates the right kind of heat and light. It is providential that the right chemical substances occur on earth. A long list of this sort of statement could be compiled, and to some people it looks as if there is indeed something very strange and odd about our particular home in the universe. But I think that this outlook arises from a misunderstanding of the situation. Because if everything was not just right we should not be here. We should be somewhere else (or more likely, nowhere).' However, many scientists are no longer willing to attribute these coincidences to chance, as is evidenced by a much more recent quotation from Hoyle: 'Such properties seem to run through the fabric of the natural world like a thread of happy coincidences. But there are so many odd coincidences essential to life that some explanation seems required to account for them' (Hoyle 1983, 220)" [61]

The Four Universal Forces

We have already seen in Table 6.1 that there are four fundamental forces in the Universe, which determine its composition, size and activity. In addition, Tables 6.2 to 6.4 show that there is a high degree of precision required to balance these forces to allow life to occur. We now need to look at these four entities in more detail, as our understanding of the information in the Universe is dependent upon these four forces.

53. Hawking, S. & Mlodinow, L.: *The Grand Design*. (2010) p.159.

54. Gonzalez, G. & Richards, J.W.: *The Privileged Planet*. (2004) p.201.

55. Hawking, S. & Mlodinow, L.: *The Grand Design*. (2010) p.160.

56. Ibid. (2010) p.162.

57. Gonzalez, G. & Richards, J.W.: *The Privileged Planet*. (2004) p.205.

58. Strobel, L. & Vogel, J.: *The Case for a Creator*. (2004) p.48.

59. Lennox, J.C.: *God's Undertaker. Has Science Buried God?* (2007) p.69.

60. Strobel, L. & Vogel, J.: (2004) p.49.

61. Bradley, W.L.:'Nature. Designed or Designoid' *in Mere Creation* Dembski, W.A. (ed): *Mere Creation. Science, Faith and Intelligent Design*. InterVarsity Press (1998) p.40 .

Gravitation

When asked about gravity, many people may be prompted to think of Sir Isaac Newton (1643–1727) watching apples fall whilst sitting in his apple orchard on an English summer's day but there is much more to it than this. Newton published his Law of Gravitation in 1687, which revolutionised science at that time but we now have a far greater understanding than was possible in the 17th-Century.[62] Jim Al-Khalili writes: *"Long before Newton, it was realised that the reason objects fall is because the Earth exerts a force on all things that pulls them towards it. It was also known that the Moon orbits the Earth because the Earth exerts some mysterious force on it stopping it from floating off into space. Newton made the connection between these two phenomena. Attributing the motion of the Moon and the falling apple to one and the same force (gravity) was a bold stroke of genius. Until then it was believed that entirely different laws of nature governed the behaviour of earthly objects (apples) and heavenly bodies (the Moon)."* [63]

"Newton's law of gravity states that any two objects in the Universe will be attracted towards each other by an invisible force" [63] that is proportional to the masses of the objects and inversely proportional to the square of the distance between them. This is not necessarily an intuitive concept if you have not studied physics, so Al-Khalili explains:

"...space should not simply be thought of as 'somewhere to put things', but instead that it has its own geometrical properties. These properties are altered in the presence of mass...Imagine rolling a small ball across a trampoline. It should go across in a straight line. Now what if you stand still in the middle of the trampoline and get someone to roll the ball again? You will have made a dent causing the trampoline's material to bow down a little. If the ball's path takes it close enough to this dent, it will follow the curvature and be bent round to move in a different direction...Viewed from above, it would appear as though you had exerted a mysterious force on the ball causing it to be attracted towards you and away from its original straight path. This is how we imagine matter to curve space around it. The curvature causes other objects to follow a different path to the one they would in the absence of the curvature...Such paths in curved space are called geodesics...a geodesic path is the shortest distance between any two points...A geodesic is only a straight line when the space is flat." [64]

Massive celestial bodies, such as stars, deform the space around them so that smaller bodies like planets or comets, which enter this curved space, experience an alteration to their trajectory as they travel through it. This curvature is what we call the attractive force of gravity.[65] Gravitation is the weakest of the four universal forces but it is a long-range force, therefore it can act at a distance on anything with mass.[35]

"So now we can understand Einstein's interpretation of gravity. [The General Theory of Relativity] *All material bodies warp space around them, by an amount that depends on how massive they are, and this warped space then guides all bodies that are moving in it, making them travel along geodesic paths."* [65] The question that remains unanswered is: what is the entity that makes up the item that we call 'space', which can be 'warped'? If space was *absolutely* nothing then there would be nothing to be warped. Information of some kind is implicated in this process to make the planets change their course to follow a geodesic path. The answer to: "What is space?" is not directly known yet, so we will examine the other universal forces to gain a better understanding of the construction of the fabric of our universe before we return to this conundrum in later chapters.

62. Hawking, S. & Mlodinow, L.: *The Grand Design*. (2010) p.87.

63. Al-Khalili, J.: *Black Holes, Worm Holes & Time Machines*. (1999) pp.23–4.

64. Ibid. pp.32–3.

65. Ibid. p.34.

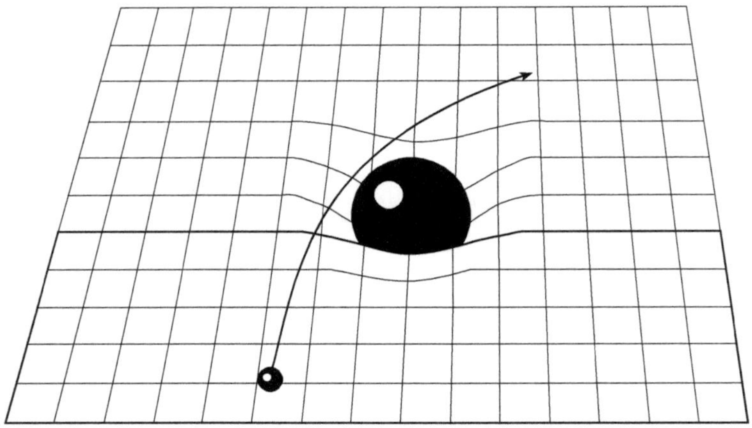

Figure 6.3: Einsteinian Gravitation and the Warping of Space [65] *[Reproduced with permission of Al-Khalili, J.S.: Black Holes, Worm Holes & Time Machines. Taylor & Francis Group, NY (1999) via Copyright Clearance Center.]*

The Strong Nuclear Force

The strong nuclear force acts within the nucleus of each atom to bind the protons and neutrons together to prevent them from flying apart, which is what would be expected to happen when a number of positively charged particles are placed close to each other in an atomic nucleus. Gonzalez and Richards write: *"In such close quarters, it is strong enough to overcome the electromagnetic force and bind the otherwise repulsive, positively charged protons together. It is as short-range as it is strong, extending no further than atomic nuclei."* [54] The energy to produce sunlight derives from the strong nuclear force. If the strong nuclear force was 2 percent weaker than it is, there would be damaging radioactivity in the lighter elements which are necessary for life; this would risk mutations in the DNA at the very least and prove lethal at worst. There would also be fewer heavier elements, for example, iodine, which is also essential for complex life.[54] Hugh Ross wrote that if: *"...the strong nuclear force...[was] 0.3% stronger than it actually is, life would be impossible at any time and any place within the universe."* [10] In Table 6.4 we saw that this was due to too few light elements (like hydrogen and hence water) forming with none heavier than iron.[43] The degree of fine tuning is once again clearly evident.

The Weak Nuclear Force

The weak nuclear force governs the conversion of sub-atomic particles from one to another within the nucleus (for example, protons to neutrons, and vice versa) and also the reactions between other sub-atomic particles, such as neutrinos; hence, it underlies the phenomenon of radioactive decay.[66] This affects the formation of helium from hydrogen, and hence the composition of the Universe. Gonzalez and Richards state that it has been estimated that the 'Big Bang' produced 25 percent helium by mass. More helium would have caused faster burning of the Sun, meaning that it would have expended all its fuel more quickly, which would not have allowed enough time for life bearing planets to form.[66]

The Electromagnetic Force

After gravitation, the electromagnetic force was the next to be described. Hawking

66. Gonzalez, G. & Richards, J.W.: *The Privileged Planet.* (2004) pp.202–3.

and Mlodinow relate: "*The first to realise there was some connection was Danish physicist Hans Christian Orsted. While setting up for a lecture he was to give at the university in 1820, Orsted noticed that the electric current from the battery he was using deflected a nearby compass needle. He soon realised that moving electricity created a magnetic force, and coined the term 'electromagnetism'. A few years later British scientist Michael Faraday reasoned that — expressed in modern terms — if an electric current could cause a magnetic field, a magnetic field should be able to produce an electric current. He demonstrated that effect in 1831. Fourteen years later Faraday also discovered a connection between electromagnetism and light when he showed that intense magnetism can affect the nature of polarised light.*" [67] Hawking and Mlodinow continue: "*...in the 1860s, Scottish physicist James Clerk Maxwell developed Faraday's thinking into a mathematical framework that explained the intimate and mysterious relation among electricity, magnetism and light. The result was a set of equations describing both electric and magnetic forces as manifestations of the same physical entity, the electromagnetic field. Maxwell had unified electricity and magnetism into one force. Moreover, he showed that electromagnetic fields could propagate through space as a wave. The speed of that wave is governed by a number that appeared in his equations, which he calculated from experimental data that had been measured a few years earlier. To his astonishment the speed he calculated equalled the speed of light, which was then known experimentally to an accuracy of 1 percent. He had discovered that light itself is an electromagnetic wave!*" [68] Electromagnetism, like gravity, is a long-range force. It is significantly "*...stronger than gravity, but it acts only on particles with an electric charge, being repulsive between charges of the same sign and attractive between charges of the opposite sign. This means that electric forces between large bodies cancel each other out, but on the scale of atoms and molecules they dominate.*" [35] The electromagnetic force determines the specific energy levels (electron shells) around a nucleus, which consequently determines the behaviour of elements and compounds. This precise arrangement of the electron shells permits some materials to react with others and some not at all, which is the discipline called *chemistry*. Chemistry is divided up into 'organic', which describes the behaviour of molecules that make up living organisms and is based on the interaction of carbon chains, while 'inorganic' encompasses the rest; both are vital for life to exist on Earth.

The Anthropic Principle

Critics of the fine tuning theory of the Universe have described the Weak Anthropic Principle which states (as Hoyle suggested in 1951) that: "*If the universe were not fine-tuned for life, then human beings wouldn't be around to observe it.*" [69] Stated alternatively, this principle may be expressed as: "*...we can expect to observe conditions necessary for our existence as observers.*" [70] This argument proposes that the Universe as we know it today was necessary for the existence of life but design was not required. Advocates of this theory often suggest the existence of a Multiverse, with countless other universes that do not support life; we just happen to live in this one that does. There is no way of detecting these other universes however, so there is no scientific way to prove or disprove this hypothesis. Thus, by the conventions of science, the Multiverse is not 'scientific' but 'philosophical'. In the end, it appears to be another example of one person's opinion against another, with not enough hard evidence to arbitrate between the two. Gonzalez and Richards conclude: "*But by itself the Anthropic Principle is not an explanation. It*

67. Hawking, S. & Mlodinow, L.: *The Grand Design*. (2010) p.88.

68. *Ibid.* pp.90–1.

69. Strobel, L. & Vogel, J.: *The Case for a Creator*. (2004) p.50.

70. Gonzalez, G. & Richards, J.W.: *The Privileged Planet*. (2004) p.266.

simply states a necessary condition *for our observing the universe. It's no explanation for why that universe exists, or is fine-tuned."* [70] However, if we once again resort to looking from an information vantage point, it is absolutely clear that the Universe is filled with information, which had to be specified at some point; which is to say that it required an information generator rather than being the product of random chance.

There are thus a vast number of observations which suggest that the existence of life is dependent upon the various parameters that specify our universe being balanced on a knife-edge. Gonzalez and Richards write: *"In many ways, ours is the optimal galaxy for life: a late-type, metal-rich, spiral galaxy with orderly orbits, and comparatively little danger between spiral arms."* [71] Their book caused a lot of controversy as there was the implication that the Universe could only have occurred if it was made that way on purpose, which is to say that it was designed. The strange fact is that theirs was not a new idea. Sixteen years earlier, in 1988, Paul Davies had written: *"It seems as though someone has fine-tuned nature's numbers to make the universe...The impression of design is overwhelming."* [72] So our Universe, as evidenced by the charge on an electron, organic chemistry, DNA, life, the Earth and our solar system among many other examples is intricately balanced and has the appearance of design. Gonzalez and Richards also conclude that the Universe appears *"finely tuned for life"*; the question is: why should physics, cosmology and biology all be finely tuned for life and exhibit extreme specificity, when we are taught at school and university that the Universe occurred by chance or in other words happened 'by accident'? [73] To answer this question, we need to examine the rules that underpin the universal forces.

The Laws of the Universe

The long list of tightly specified parameters in Tables 6.2 to 6.4 is backed up by an equally impressive collection of elegant mathematical laws which describe the functioning of our Universe. The overwhelming question is: "Why do the laws work and how did they 'organise' in the first place?" Edgar Andrews observes: *"The very existence of science testifies to the fact that the universe is subject to physical laws that have operated from its inception. Unless this was so, all attempts to describe or 'explain' the cosmos* scientifically *would be in vain."* [74] T.H. Huxley alluded to these universal laws in 1868 when he compared life to a game of chess: *"It is a game which has been played for untold ages, every man and woman of us being one of the two players in a game of his or her own. The chessboard is the world, the pieces the phenomena of the universe, the rules of the game are what we call the laws of nature. The player on the other side is hidden from us. All we know is that his play is always fair, just and patient. But, also, that he never overlooks a mistake or makes the smallest allowance for ignorance."* [75] Andrews points out how appropriate the analogy of chess is when he writes *"...when we play chess, the laws determine the moves we can make but not the moves we do make. That is, the laws are not deterministic; they don't impose a particular outcome for the game. In the same way, the laws of nature determine what is and what is not physically possible, but they do not determine what actually occurs within the multitude of available possibilities. Scientists once thought they did but* [this concept which is called]*...scientific determinism was abolished by quantum mechanics."* [76]

So, what are these laws? We need to be clearer about what constitutes these rules that have controlled the formation and operation of the Universe since

71. Gonzalez, G. & Richards, J.W.: *The Privileged Planet.* (2004) p.167.

72. Davies, P.: *The Cosmic Blueprint.* Simon and Schuster, NY (1988) p.203 *in* Lennox, J.C.: (2007) p.70.

73. Lennox, J.C.: *God's Undertaker. Has Science Buried God?* (2007) p.148.

74. Andrews, E.: *Who Made God? Searching for a Theory of Everything.* (2009) p.138.

75. Huxley, T.H.: quoted in 'Professor Huxley's hidden chess player' by Hutton, R.H. *The Spectator.* (11 January, 1868) *in* Andrews, E.: (2009) p.138.

76. Andrews, E.: (2009) p.139.

its inception. Andrews explains: "*The laws of nature constitute unchanging reality whereas the laws of science are our frequently imperfect attempts to describe reality. It is the business of science to discover laws that describe, more or less accurately, how the universe works. Such discoveries not only deepen our understanding of the cosmos but can often be exploited for man's benefit in what we call 'technology'. But neither science nor technology would be possible without the unchanging substratum of natural law — the ultimate physical reality that science strives to probe and understand. As we have seen, Einstein and other leading scientists have stressed this difference, pointing out that beneath our scientific descriptions of the universe there may well lie 'realities' that we do not (and perhaps cannot) comprehend. And it is possible, even inevitable, that sometimes science's laws and models lead us away from these realities rather than towards them. The long-held belief that light propagated through an omnipresent medium called the 'ether' is a historical example. This idea was genuinely 'scientific', being based on the fact that waves generally occur in some medium — ripples on a pond propagate in water and sound waves propagate through air (or some other fluid or solid medium). But until it was proven wrong, this perfectly reasonable scientific model stood in the way of a true understanding of electromagnetic fields and radiation...All that said, the laws of science that we establish by practising science represent, at any time, our best approximations to the underlying realities of nature. In so far as they can be verified by experiment (and that is an important proviso) they enhance our understanding of the cosmos and should not be despised because they are approximations, liable to refinement and correction.*"[77] Thus the 'laws of science' are man's best attempts to describe the 'natural law' which we observe in our universe.

Sir Isaac Newton revolutionised physics and cosmology when he published his universal laws of motion but these laws only describe the behaviour of the Universe. They do not explain how the laws or the Universe formed or what makes the laws work. In information terms, the laws specify the relationships between the foundational data sets of the Universe and hence describe the parameters established by the information which underlay the formation of the Universe but they do not explain the information itself; neither do they explain the laws themselves. Similarly, Albert Einstein when he published the Theory of Relativity and Louis de Broglie, Niels Bohr and others when they described Quantum Mechanics very elegantly described the behaviour of our universe (describing 'what' and to some extent 'how') but none of these theories addressed *why* the Universe behaves as it does. To answer this question we need to seek the information that specified it in the first place. The inescapable conclusion is that information is a pre-requisite for all systems in the Universe. The corollary of these laws is: information must have been a pre-requisite for the formation of the Universe and must have been generated prior to it coming into existence. Paul Davies reiterated in 1989: "*There is for me powerful evidence that there is something going on behind it all...It seems as though somebody has fine-tuned nature's numbers to make the Universe.*"[78]

In the physical universe, this information is located within the entities of matter and energy themselves (which are its carriers) but it also lies within the laws that govern them. This pre-specified information controls the characteristics of each respective entity, such as the constitutional components of gravity or light but it also determines their behaviour. This presents another paradox in that this information describes laws that act within and control our material universe and yet they are non-material (for example Newton's Law of Gravitation). Thus, the non-material laws govern the material universe.

77. *Ibid.* pp. 141–2.

78. Davies, P.: *The Cosmic Blueprint: New Discoveries in Nature's Creative Ability to Order the Universe.* Touchstone Books, NY (1989) p.203 *in* Gonzalez, G. & Richards, J.W.: (2004) p.195.

Although our knowledge of the Universe is still incomplete, there are a number of findings that are completely unexpected if the Universe is simply a random, unplanned event. The first is that mathematical relationships exist between many entities within the Universe. Why should they exist at all? Andrews expressed it poignantly when he wrote: "*The laws that describe the physical cosmos possess a number of striking properties. First of all, they are almost universally expressed in terms of mathematics, that is, they take the form of algebraic equations rather than verbal statements. Of course, we can often put into words what the equations are telling us, but the laws can only be stated precisely and definitely using the language of mathematics. For example, Newton's law of gravity says that the force of attraction between two objects is proportional to the product of their masses divided by the square of the distance between them. But these words do nothing more than describe the equation:*

$$F = G\{m_1 m_2\} / d^2$$

That is, the law itself is a mathematical expression. The whole of nature appears to be built out of mathematical Lego!" [79]

Another slightly more foreign concept to most of us is a property called 'symmetry', which is important in understanding time. Andrews continues: "*A second feature of natural law is one that underlies much of science itself — namely, the concept of symmetry. This is so important that in his book* The Fabric of the Cosmos *Brian Green declares: 'During the last few hundred years there have been many upheavals in science, but the most lasting discoveries have a common characteristic; they've identified features of the natural world that remain unchanged even when subjected to a wide range of manipulations. These unchanging attributes are what physicists call symmetries.*" [80] Most people are familiar with the symmetry of objects. No matter which direction an astronaut takes a photograph of the Earth, it always has the same shape in outline and that outline does not change no matter which way the axis of the Earth is changed; this principle also applies to an orange. In contrast, a pear has a much more limited symmetry (only along its long axis) and so it exhibits 'rotational symmetry' whereas a banana changes its profile no matter which direction it is rotated.

In a similar way, there are properties of the Universe that remain unchanged when some other characteristic or parameter changes. Andrews cites uniformity of the laws of the Universe as belonging to this category. This can be demonstrated on Earth by performing the same scientific experiment (such as Galileo's two canon ball experiment) at the South Pole and the equator (or anywhere in between) and the same result will be obtained because the same physical laws apply (in this case gravity). This is called 'translational' symmetry. Although scientists cannot go there, observations of distant galaxies appear to be consistent with the natural laws we observe on Earth so scientists infer that this principle of translational symmetry, which is observable on Earth, also applies throughout the Universe. This allows science to gain a greater understanding of places we cannot visit, such as distant galaxies. Andrews adds: "*Einstein's general theory of relativity is an outstanding example of this, having as its starting point the assumption that all observers — wherever they might be in the universe and however they are moving relative to other observers — will find that the same laws of nature apply unchanged.*" [81]

79. Andrews, E.: *Who Made God? Searching for a Theory of Everything.* (2009) pp.142–3.

80. *Ibid.* p.144.

81. *Ibid.* p.145.

In later chapters we will be examining the phenomenon of time in greater detail but it is important to note here that time is yet another entity in our universe that exhibits symmetry. Andrews writes: *"Another major symmetry is symmetry with respect to time...almost all the laws of science remain unchanged if time is replaced by negative time - that is, they are mathematically independent of the arrow of time. The only known exceptions are the law of increasing entropy (the second law of thermodynamics) and the behaviour of an obscure quantum particle called a 'kaon' — which seems to know which way the arrow of time is pointing. This 'strange case of the kaon' may have some other explanation that doesn't break time-symmetry, but entropy is highly important since it appears to determine the direction of time. However, time-symmetry is still the overwhelming character displayed by nature's laws."* [81]

It is interesting to note that symmetry exists but the more important question is to ask why does it exist at all? If the Universe were a random event, as many leading scientists maintain, then how and why does symmetry occur? Andrews continues: *"As Brian Green comments:* 'The laws of physics didn't have to operate in this *[symmetrical]* way' *(his emphasis)."* [82] Andrews points out: *"Very few writers offer any explanation for the ultimate origin of natural law. They simply accept it as 'given' - as the unexplained and probably inexplicable substratum of reality that science seeks to explore and without which it could not exist."* [83] In his book: '*God; the failed hypothesis: How science shows that God does not exist*', Victor Stenger does attempt to answer this question. Stenger writes: *"So where did the laws of nature come from? They came from nothing!' They 'follow from the symmetries of the void out of which the universe spontaneously arose'."* [84] I wish the money in my bank account operated under the same principles! In our material world of cause and effect, Stenger's statement does not make sense and this will be discussed in greater detail in later chapters.

Andrews adds clarity to the discussion by defining the different states of 'nothing' that existed at the formation of the Universe. He writes: *"There are several problems with Stenger's bold claim. First of all, he confuses 'nothing' with 'nothing'... Remember that, according to big bang cosmology, the universe began as a tiny object which contained space and time within itself. Without the expanding boundaries of the primordial cosmos there was neither space nor time, so let's call this region external to the cosmos 'void-zero'. According to Stenger, this is 'the void out of which the universe spontaneously arose' — the void who's supposed symmetries, says Stenger, gave birth to the laws of nature."*

"But void-zero has no physical or material properties (such as symmetries) because it has no physical or material existence! By definition it lies outside the physical and material universe...Void-zero is the eternally pre-existent, non-physical framework in which the physical universe began and must, by definition, lie beyond the reach and remit of science. Stenger's claim is thus an unwitting admission that the laws of science had a non-physical...origin."

"But Stenger doesn't intend to make any such admission. He is simply confusing void-zero ('from which the universe spontaneously arose', as he put it) with a totally different void — one which lies entirely within the material universe. The flexible space-time continuum of general relativity is a constituent of the cosmos (its fabric, if you like) and it contains an awful lot of empty space — sometimes called 'the vacuum of outer space'. Let's call this empty space 'void-one' to distinguish it from void-zero. [See Figure 6.4] *It is, of course, characterised by the absence of matter and energy, at least 'as we know it', but there is more to void-one than meets the eye. According to Heisenberg's uncertainty principle, it must undergo quantum fluctuations that*

82. Greene, B.: *The fabric of the cosmos.* Alfred Knopf (2004) p.222 *in* Andrews, E.: (2009) p.145.

83. Andrews, E.: (2009) p.148.

84. Stenger, V.J.: *God, the failed hypothesis: How science shows that God does not exist.* Prometheus Books, NY (2007) p.131 *in* Andrews, E.: (2009) p.150.

lead to the appearance and disappearance of phantom (or virtual) particles — which are created as particle-antiparticle pairs and immediately vanish again by mutual annihilation. More than that, according to some theories, void-one is actually filled with an 'ocean' of Higgs particles. This 'Higgs field' is a kind of cosmic treacle, resisting the movement of known particles and imparting to them the property of mass. But all this relates to empty space within the space-time fabric of the physical universe — a fabric that has, by definition, existed only from the moment of creation. When Doctor Stenger says that the void possesses symmetries, he is talking about void-one. He may be right. After all, if we allow that the vacuum of space is awash with phantom particles and Higgs bosons, it would be a bit mean to say it can't have symmetries as well...So let's suppose for a moment that Stenger is right so far, and that laws of nature originate from the symmetries of void-one. What does that tell us? Since void-one is an integral constituent of the cosmos it tells us that the laws of nature follow naturally from the existence of the cosmos. Put more simply, they are just part of the created physical order. But we already knew that, because science cannot conceive of a physical order which is not governed by rules; laws are built into the very concept of 'physical order'. So the symmetries of void-one (if they exist) do nothing to explain the origin of the laws of nature, being themselves simply an expression or manifestation of those laws. And it's no good saying (as some do) that in the early moments of the big bang there was chaos rather than order, because scientifically speaking even chaos is governed by laws. James Gleick's book Chaos *is subtitled 'Making a new science', while the front cover calls it 'the amazing science of the unpredictable'. I hardly need to add that where there is science there has to be law."* [85]

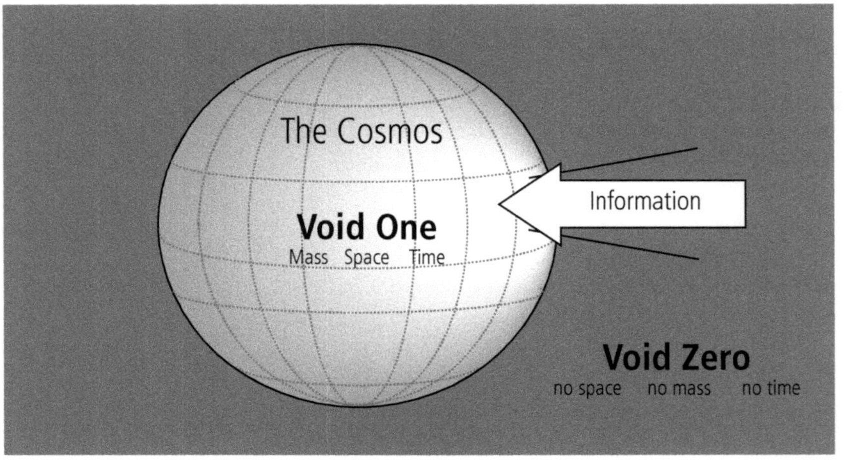

Figure 6.4: The Formation of our Universe
Void One represents our infant universe which required the injection of information from Void Zero, possibly through a transient white hole (see Chapter 8). Void Zero has no mass and therefore no space or time (it is timeless). Void One acquires mass with space and time (as well as information). Note that although the Universe is depicted as a sphere at this early stage in its formation, some theories suggest it was cone-shaped and others suggest 'curved sheets'. The shape is not significant and the information principles apply to whatever shape it took.

Ernan McMullin, U.S. Professor-Emeritus of Philosophy at the University of Notre Dame helps to further clarify the concepts involved when he writes regarding Void One: *"...the term 'vacuum'... presumably means that the relevant space is 'empty'. Empty of what? Presumably of rest mass."* [86] Thus what we consider to be a 'vacuum' is only empty of *physical matter* but not necessarily empty of non-material entities, such as the Laws of the Universe and information.

So rather than decreeing that the Laws of the Universe arose out of nothing, a far more plausible supposition is that pre-defined information was introduced from 'Void-Zero' to create 'Void-One', at its formation, to sculpt our physical universe, as in Figure 6.4. The options as to how this may have occurred are addressed in Chapters 8 and 9. However, at this stage our investigations

85. Andrews, E.: *Who Made God? Searching for a Theory of Everything.* (2009) pp.150–2.

86. McMullin, E.: *From matter to materialism ... and (almost) back.* p.34 *in* Davies, P. and Gregersen, N.H. (eds.): *Information and the Nature of Reality: from Physics to Metaphysics.* Cambridge University Press (2010).

suggest that there is a non-physical entity, for which information fits the job description, which stretches back to the beginning of time itself that is the mechanism underlying all of life and the physical universe.

Did the Earth and Universe Evolve?

The question as to whether the Universe evolved depends upon which definition of 'evolution' is used. If evolve means simply 'to change', then the answer is 'yes', but if the meaning of evolve is 'to create itself from nothing' then the answer is 'no'. The formation of simple structures such as rocky asteroids or a comet made of ices which require limited amounts of information to specify and control may be *imagined* but as we have seen, the spontaneous development of complex systems with multiple levels of information control such as the laws of the Universe or the myriad of precisely defined constants is impossible because the generation or existence of information is a pre-requisite and our physical universe lacks an information generating mechanism.

It is obvious from the discussion to date that there is more information in the Universe than is understood at present (for example Higgs fields). There is also information which we know is not well understood such as invisible or 'dark' matter. In general, Western science teaches that invisible things do not exist but just as we can detect the wind and radio waves, there is significant scientific evidence to support the existence of dark matter. Jim Al-Khalili writes:

"...the density of matter visible to us (that is the number of galaxies in a given volume of space) is about one per cent of the critical value required for a flat [stable] universe. Oh, oh, we have a problem! Where could the other ninety nine per cent be? Part of this missing mass of the universe is believed to be made up of something known enigmatically as dark matter. It is believed that there is probably between ten and forty times as much matter, or 'stuff', out there in space than we can actually see. This is not because it is so far away or hidden behind other objects, but because it is literally invisible...galaxies appear to weigh a lot more than the sum of stars and other visible objects they contain and must also be made up of a cloud of invisible material that extends beyond the visible stars. This strange conclusion is reached from two quite independent routes. The first is by studying the way stars on the outer rims of galaxies orbit the centre." [87]

From the material in the Universe that is visible to us, it has been calculated that galaxies should revolve more slowly than they do, so some entity must compel them to move as fast as they do and greater mass would explain this. Al-Khalili continues: *"Another indication that galaxies are more massive than they appear is found by directly measuring their weight! This is done using Einstein's idea that the gravity of a massive object warps space around it...the general theory of relativity was first tested experimentally by observing the deflection of the light from distant stars as it passed close enough to the Sun's gravitational field. In the same way, a galaxy will deflect the light from a more distant galaxy when it is in the line of sight between the more distant one and us. The amount by which the light is bent tells us how much mass the nearer galaxy has. Again we find that galaxies contain much more matter than just the visible stuff. Until recently it was not clear what this dark matter could comprise of...We are now certain not only that dark matter exists but that most of it must be made up of a new kind of substance we have yet to discover.*

Experiments carried out in Japan and announced in mid-1998 have suggested that part of it is probably made up of elementary particles called neutrinos. The

87. Al-Khalili, J.: *Black Holes, Worm Holes & Time Machines.* (1999) p.68.

trouble with these tiny entities, which were predicted theoretically in the early 1930s and discovered in laboratory experiments in 1956, was that no one had been sure that they actually weighed anything at all. They are extremely elusive since they travel through solid matter, including our measuring devices, as though it was not there. In fact, billions of neutrinos, mainly produced in the Sun, are at this moment streaming through your body without you knowing it...they do indeed have a very tiny mass which is enough to account for part of the invisible matter of galaxies, due to their sheer number...Neutrinos would make up what is called 'hot dark matter' because they zip around at high speeds. We are now confident that there must be more, probably in the form of slower moving heavy particles that would make up 'cold dark matter'...So, the best current estimates for all the matter in the Universe (both visible and dark matter) only account for about a third of the density required to make [a stable]*...Universe."* [88]

Thus, although the Universe is still incompletely understood, there is no evidence that it built itself up from nothing. In fact, the only evidence available suggests the contrary view; that the way in which the Universe 'evolves' is through decay. John Lennox writes: *'The Second Law of Thermodynamics describes the natural tendency of closed systems to degenerate, to lose information, order and complexity; that is to increase their entropy.'* [89] Dolphin adds the proviso: *"It is possible in accord with the Second Law for entropy to decrease in a small region of space, provided the entire system experiences an overall entropy increase. An example of this situation is the growth of a man from embryo to adult. The genetic material and the enzyme motors of the living cells in man draw external energy from the environment around man to increase the complexity of the order and structure of a man's body, but a man's body is grown by nature at the expense of increasing the total entropy of the universe. The Second Law wins out more fully in the long run, for at the death of the body the complex structures of the body are broken down to dust. The life of each man raises the total entropy of the universe."* [90]

Thus, it can be seen that although life can temporarily oppose the Law of Entropy, this is only a short-term effect and the notion that an entity devoid of organised information, could have had its beginning in an uncontrolled explosion at the 'Big Bang', yet evolve into the complex universe we know today is once again, wishful thinking.

Conclusion

Hawking and Mlodinow elegantly summarise many of the concepts in this chapter when they write: *"The laws of nature form a system that is extremely fine-tuned, and very little in physical law can be altered without destroying the possibility of the development of life as we know it. Were it not for a series of startling coincidences in the precise details of physical law, it seems, humans and similar life-forms would never have come into being."* [91] By way of an explanation for this extraordinarily finely balanced universe, Paul Davies observes: "*Galileo had proclaimed, 'The great book of Nature can be read only by those who know the language in which it is written, and this language is mathematics'* (Drake 1957)" [92] which accounts for the precision noted by Hawking and Mlodinow. But why should a randomly created universe exhibit precision, beauty and order at all? The Standard Model cannot explain it.

John Lennox writes: *'...as Sherlock Holmes might have told us, if chance and necessity, either separately or together, are not capable of biogenesis, then we must consider the possibility that a third factor was involved. That third possibility*

88. *Ibid.* pp.68–70.

89. Lennox, J.C.: *God's Undertaker. Has Science Buried God?* (2007) p.121.

90. Dolphin, L.: *Lord of Time and Space.* (1974) p.71.

91 Hawking, S. & Mlodinow, L.: *The Grand Design.* (2010) p.161.

92. Drake, S.: Discoveries and Opinions of Galileo. Doubleday-Anchor, NY (1957) *in* Davies, P.: Universe from Bit. pp.65-91 *in* Davies, P. and Gregersen, N.H. (eds.): *Information and the Nature of Reality: from Physics to Metaphysics.* Cambridge University Press. (2010) p.67 .

is the input of information.' [93] And in the same way that this statement applies to biogenesis, it equally applies to the whole of the Universe as well.

In the previous chapter it was observed that the information in living systems cannot be explained by random mutation with natural selection. In this chapter it has been demonstrated that the complexity and intricacy of the Universe, with its embedded laws determined by information cannot be explained by random processes either. However, if the formation of the Universe was not random, what options do we have? In the next chapter we will begin to investigate some alternatives.

93. Lennox, J.C.: *God's Undertaker. Has Science Buried God?* (2007) p.148.

7

The Biological Information Hypothesis

What lies at the heart of every living thing is not a fire, warm breath, nor a "spark of life". It is information, words, instructions... Think of 1 billion discrete digital characters... If you want to understand life think about digital technology.[1]

Richard Dawkins

Universal laws in chemistry and physics have been recognised for over 100 years but why are there not an equivalent number of universally recognised laws in biology? In the last few years unifying laws in biology have begun to be described, such as the Conservation Laws in Biology[2] and Universal Scaling Laws.[3] Robert Laughlin a physicist and Nobel laureate offered one explanation when he wrote: *"Much of present-day biological knowledge is ideological. A key symptom of ideological thinking is the explanation that has no implications and cannot be tested. I call such logical dead ends anti-theories because they have exactly the opposite effect of real theories: they stop thinking rather than stimulate it. Evolution by natural selection, for instance, which Darwin conceived as a great theory has lately come to function as an anti-theory called upon to cover up embarrassing experimental shortcomings and legitimise findings that are at best questionable and at worst not even wrong. Your protein defies the laws of mass action — evolution did it! Your complicated mess of chemical reactions turns into a chicken — evolution! The human brain works on logical principles no computer can emulate? Evolution is the cause!"* [4]

So how can a biologist answer these accusations? The first thing to do is to take the criticisms on board and try to answer them directly. One important observation comes from Andrew Briggs, Professor of Nanomaterials at the University of Oxford and Director of the Quantum Information Processing Interdisciplinary Research Collaboration, who gives us a lead when he writes: *"As you make things very small, so discrete quantum states emerge, and these can be studied and manipulated. This has led to the emerging discipline of quantum nanotechnology."*[5] Briggs quotes David Deutsch, also of the University of Oxford, who wrote a paper in 1985 which extended Alan Turing's 1936 paper on universal computing machines. Deutsch wrote that: "*Turing's insight* [was that]: *'Every 'function which would naturally be regarded as computable' can be computed by the universal Turing machine.' David Deutsch...criticised formulations such as Turing's as being vague...he re-expressed Turing's formulation as a physical statement: "Every finitely realisable physical system can be perfectly simulated by a universal model computing machine operating by finite means." Briggs comments: "This may sound like mere*

1. Dawkins, R.: *The Blind Watchmaker.* Longman, London (1986) p.112 *in* Lennox, J.C.: *God's Undertaker. Has Science Buried God?* Lion Books, Oxford (2007) p.127.

2. Azbel, M.A. Exact Conservation Laws and Exclusion Principle of Biological Evolution. *American Physical Society, Annual APS March Meeting* (2003).

3. West, G.B., Brown, J.H. & Enquist, B.J.: Life's Universal Scaling Laws. *Physics Today* 57(9):36–42 (2004).

4. Laughlin, R.: A Different Universe: Reinventing Physics from the Bottom Down. Basic Books, NY (2005) pp.168–9 *in* Lennox, J.C.: (2007) p.149.

5. Briggs, A.: And Information Became Physical. *in* Berry, R.J. (Ed.): *Real Scientists Real Faith.* Monarch Books, Oxford (2009) pp.73–4.

philosophical nitpicking, but it turns out to have immensely practical consequences. If information is to be subject to computation in a physical system, then it had better be regarded as quantum, because at the relevant level physical systems are quantum. And if it is quantum, then whatever you can do with a quantum object you can do with information." [6] Therefore, if information is digital and biology is defined by information, does that make biology digital? It has already been established that life runs on a physical platform which contains information, so if David Deutsch is correct, then the information in life is quantised and so it must be digital. The fact that life is digital and not a continuum implies that not all physical states in life are possible but only the ones that some quantising process or law allows. But what is this quantising process and how does it work? If this process can be uncovered then it might help to respond to some of Robert Laughlin's criticisms.

Information in living things

In Chapters 1 and 2 we explored the idea that information is a discrete entity that is different from every other entity in the Universe; we also looked at its place in biological systems. The question was asked whether the information was real or whether it only had the appearance of being real and hence only a useful theoretical concept. One way to test this idea is to take DNA (which contains information) and rearrange it so that the DNA now contains the same building blocks but in a different order, which will almost certainly create a nonsense string of nucleotides containing no *useful* information. This is referred to as 'noise' rather than information. Paul Davies wrote: *"Communication theory - or information theory, as it is known today — says that noise destroys information, and that the reverse process, the creation of information by noise, would seem to be a miracle. A message emerging on its own from radio static would be as surprising as the tide making footprints on the beach. We are back with the same old problem: the second law of thermodynamics insists that information can no more spring into being spontaneously than heat can flow from cold to hot."* [7] The point is reiterated by Dolphin who wrote: *"The implications of the Second Law...are even stronger when one asks if inorganic matter can be synthesised into organic matter by means of the input of external energy alone. The answer to this question is that external programming information, genetic material or outside intelligence is also necessary to assemble inorganic material into organic molecules of the sophisticated type found in all living cells. Once organic matter exists, external energy such as sunlight alone is sufficient to allow the organic material to duplicate itself but not to ascend further in the direction of increasing complexity. The action of sunlight or influence of lightening in a primordial sea of organic matter cannot lead to the production of complex organic molecules found in living cells without violating the Second Law unless outside programming information or instructions are given."* [8] The same principles apply to physics and the information stored in the configuration of elements and the physical laws of the Universe as discussed in Chapter 6. To test this, try changing the Strong or Weak Nuclear forces or the number of electrons in the outer shell of an element such as carbon. Thus, we have seen that there is precisely arranged information in both living things and in the Universe. Further, it would appear that both physical and biological digital information systems must be interwoven for life to exist. By way of overview, Table 7.1 summarises the requirements for a cell to produce a single protein.

From the preceding discussion, it can be seen that cells are biological information processing systems similar to microchips.[9] They take inputs of raw

6. Briggs, A.: And Information Became Physical, *in* Berry, R.J. (Ed.): (2009) pp.74-5.

7. Davies, P.: *The Fifth Miracle.* Simon & Schuster, NY (1999) pp.56–7 *in Journal of Creation* 23(3):126 (2009).

8. Dolphin, L.: *Lord of Time and Space.* Good News Publishers, Westchester, Illinois (1974) pp.71–2.

9. Missler, C. & Meyer, S.: *In the beginning there was... information.* (CD) Koinonia House (2000).

	Requirement	Notes
1	DNA sequence	A storage system for extrinsic information specifying what to build.
2	Epigenome	Instructions defining how to apply the information in the DNA.
3	The Strong Force	Intrinsic information that specifies how atoms bind to each other which is defined by the laws of the Universe.
4	The Weak Force	Intrinsic information specifying how molecules bind to each other.
5	Cell cytoplasm	A protected environment in which to construct the protein.
6	Amino acids	Raw materials from which to build proteins (left-handed only).
7	Ribosomes & chaperones	Machines to construct and precisely fold each protein.

Table 7.1: Summary of the Multi-layered Digital Information required to make a single Protein

materials and energy and use information processing to convert them into outputs of greater value, such as proteins. This digital processing ability has enormous implications.

Biological Law of Scaling

In Chapter 1, when discussing common descent in the context of the Neo-Darwinian synthesis, it was shown that related body plans (described as *homologous* in Neo-Darwinian terms) must share information. Richard Dawkins in: *The Greatest Show on Earth* describes the work of a Scottish zoologist and mathematician, D'Arcy Thompson (1860–1948), who published a book in 1917 entitled: *On Growth and Form,* in which he demonstrated how the form of one organism, such as a crab, could be mathematically manipulated to create five other crab species.[10,11]

Dawkins cites this comparative similarity as evidence for common descent; however it was noted in Chapter 1 that this assertion is circumstantial rather than proven. It was established in Chapter 5 that Microevolution produces significant genomic variation within a single species, as individuals change over time (such as the 200 breeds of dog within the canine family). Thompson also demonstrated a relationship between humans, chimpanzees and baboons indicating shared information, however, when Dawkins attempted to show the transformation from *Australopithecus* through intermediates to *Homo sapiens* he was unsuccessful. Dawkins concludes: *"In practice it proved very difficult to do anything like that."* [12] See Figure 7.2.

These examples 'morph' from one into another in the same way that the crab varieties did. This implies that they share a degree of mutual information too.

This evidence suggests that transformations within species do occur due to shared information but transformations across species only occur when there is sufficient common information. If shared information was not involved in defining the different body plans then there should be no relationship demonstrable. The ability to mathematical morph one crab body plan into another is not inconsistent with Richard Dawkins' assertion that there is a relationship because of common descent, however this is circumstantial evidence, not proof; pre-specified information for any other reason is just as valid an explanation. The most surprising finding is that the relationship is mathematically scalable! If it is mathematically scalable then the relationship is *digital,* which is completely unexpected and implies that there is more than a superficial similarity. As observed in Chapter 1, this confirms systematised information content but it does not explain why. Slightly disturbingly, this scaling phenomenon may be far

10. Dawkins, R.: *The Greatest Show on Earth.* Bantam Press, London (2009) pp.308–315.

11. Sarfati, J.: *The Greatest hoax on Earth? Refuting Dawkins on Evolution.* Creation Book Publishers, Atlanta , Georgia (2010) p.91.

12. *Dawkins, R.: The Greatest Show on Earth.* (2009) p.314.

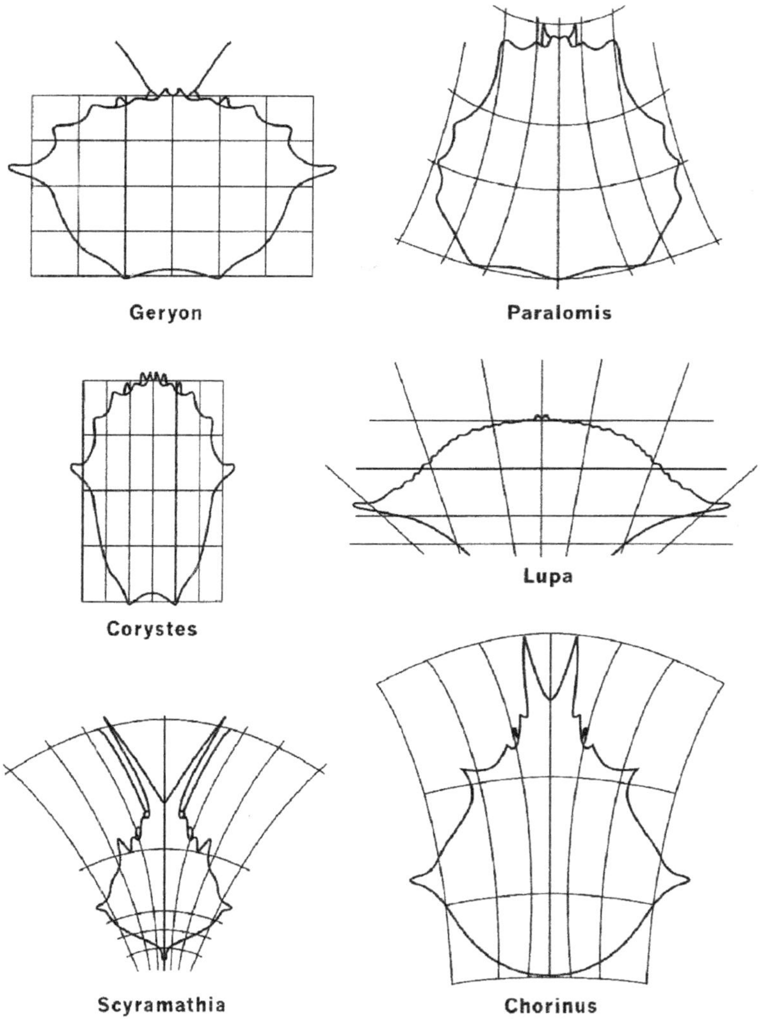

Figure 7.1: D'Arcy Thompson's crab 'transformations' [13]
The ability to geometrically manipulate one variety into another implies an underlying (hidden) mathematical relationship between the different members of the species, which suggests that there is a high degree of shared information.

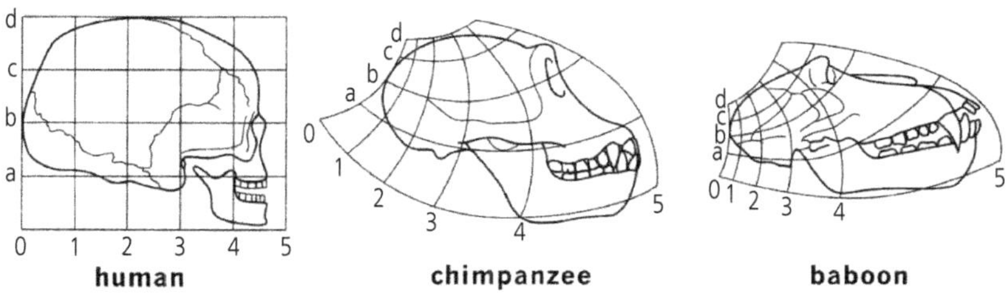

Figure 7.2: D'Arcy Thompson's skull 'transformations' [14]
These examples 'morph' from one into another in the same way that the crab varieties did. This implies that they share a degree of mutual information too.

greater than has been previously imagined. There was a paper published in *Physics Today* in 2004, describing Universal Scaling Laws in biology where mathematical relationships were again drawn between diverse biological characteristics. Geoffrey West et al. wrote: "*...metabolic rate scales as the three-quarter power of mass over 27 orders of magnitude ranging from molecular and intra-cellular levels up to the largest multicellular organisms. Similarly, time-scales (such as life spans and growth-rates) and sizes (such as bacterial genome lengths, tree heights and mitochondrial densities)*

13. Thompson, D.W.: *On Growth and Form*. Cambridge Univ. Press (1917) pp. 770–1 at http://archive.org/details/ongrowthform1917thom..

14. *Ibid*. pp.770–771.

scale with exponents which are typically simple powers of one quarter. This suggests that fundamental universal principles underlie much of the course-grained structure and organisation of living systems...These...lead to a general qualitative, predictive theory that captures the essential features of many diverse biological systems, including circulatory systems of animals, plant vascular systems, growth, cancer, ageing and mortality, sleep, cell size and genome lengths, mitochondrial densities, DNA nucleotide substitution rates and the concept of a universal molecular clock." [3] We have always accepted that there was a mathematical basis to physics and to a lesser extent chemistry (such as molar relationships in chemical equations) but never to biology, so this is rather startling. There are two things which result from the realisation that there is a digital basis to at least some aspects of biology. The first is that there must be a deeper order that specifies the digital relationships in biology and the second relates to a digital framework within the Universe itself which we will investigate further in Chapters 8 and 9.

Thus, the factor that links everything together from the smallest virus to the largest galaxy is digitally coded information. To paraphrase the immortal words of Douglas Adams: information becomes our key to unlocking, *"life, the universe and everything"*. The answer turns out not to be "42" but information.[15] In terms of the observation that a deeper, digital relationship exists between species, both Dawkins and West suggest that this is due to their shared inheritance from a common ancestor but specified design, as proposed by advocates of 'Intelligent Design' and some biblical scholars could equally explain the same relationship. The Information Paradigm can only say that shared information is present and we must look to some other source to try to establish if a common descent or pre-specified information explanation is more successful at explaining the observations.

Information Stasis

Relentless change is the hallmark and essential core of Neo-Darwinian evolution as organisms gradually and progressively modify their genetic information, however, geneticist Michael Denton observed exactly the opposite; there was stasis when different biochemical forms were compared with each other. He wrote: *"Molecular biology has also shown that the basic design of the cell system is essentially the same in all living systems on earth from bacteria to mammals. In all organisms the roles of DNA, mRNA and protein are identical. The meaning of the genetic code is also virtually identical in all cells. The size, structure and component design of the protein synthetic machinery is practically the same in all cells. In terms of their basic biochemical design, therefore no living system can be thought of as being primitive or ancestral with respect to any other system, nor is there the slightest empirical hint of an evolutionary sequence among all the incredibly diverse cells on earth. For those who hoped that molecular biology might bridge the gulf between chemistry and biochemistry, the revelation was profoundly disappointing. In the words of Monod: '...we have no idea what the structure of a primitive cell might have been. The simplest living system known to us, the bacterial cell...in...its overall chemical plan is the same as that of all other living beings. It employs the same genetic code and the same mechanism of translation as do, for example, human cells. Thus the simplest cells available to us for study had nothing 'primitive' about them...no vestiges of truly primitive structures are discernible.' "* [16] Edgar Andrews echoes this point when he writes: *"... environmentalist David Swift points out that the functional sites in proteins, and thus*

15. Adams, D.: *The Hitchhikers Guide to the Galaxy*. Pan books, UK (1979).

16. Denton, M.: *Evolution: A Theory in Crisis*. Adler & Adler, Chevy Chase, Maryland (1986) p.250.

the shapes into which proteins fold, are virtually identical (for a given protein type, e.g. cytochrome c) in all living organisms; that is, they show no evidence of having evolved. He writes: 'crystallographic studies have shown that the 3D structures of eukaryotic cytochrome c are scarcely distinguishable from each other, including those from higher plants to higher animals, even though they are presumed to be separated by many hundreds of millions of years of evolution.' [17]

These observations are at odds with most biology textbooks printed in the last 50 years. When I was in medical school, the malarial parasite was cited as the organism which demonstrated *par excellence* the power of evolution. It not only demonstrated the development of resistance to anti-malarial medications (antibiotics), such as chloroquine but also how one species will change in an effort to evade the competitive advantage of another (as in the sickle cell mutation in human beings). A single point mutation changing one amino acid on the β-chain of the haemoglobin molecule, which transports oxygen in every human red blood cell, can cause the red blood cells to change their shape from biconcave disks to sickle-shaped cells. This significantly reduces the virulence of the malarial parasite by making it more difficult for the microorganism to latch on and infect the red blood cell. Because this change is in the human DNA, the condition is inheritable by the next generation but there is only one small problem, in that the human carrier has reduced oxygen carrying capacity due to the affected red blood cells. This inconvenience is not as bad as dying from malaria but it can leave the afflicted person mildly anaemic and so the condition is termed: 'sickle cell trait'. The condition is usually advantageous if the affected individual lives in an area infested with malaria-carrying mosquitoes, which transmit the malarial parasite when they bite people. However, there is an advantage only if the person possessing the sickle cell gene has only *one copy* of the mutated gene. People who carry only one copy of the gene are said to be *heterozygous* for the condition. If a child should inherit two copies of the sickle cell gene (one from *each* parent), they are said to be *homozygous* for the condition and they suffer severe anaemia, which results in death and this condition is called: 'sickle cell anaemia' or 'sickle cell disease'. One of my jobs as a hospital resident doctor was to transfuse an eight-year-old boy whose parents were first cousins from Lebanon and every second month he came into the Alexandra Hospital for Children for a blood transfusion, which not only kept him alive but allowed him to live a normal life.

Therefore, in the case of sickle cell inheritance, one copy of the mutated gene may be beneficial but two copies are not better but lethal from both a personal and an evolutionary point of view. This limits the level to which the gene can rise within the population as, if everyone became homozygous for the condition, the whole population would die out. There is also a disadvantage if the individual moves to a non-malarial area such as: the USA, Canada, Europe, Australia or New Zealand, where even being homozygous is a disadvantage rather than an advantage. This is a very complex topic which is explained by Michael Behe in his book: *The Edge of Evolution* and there is not room to discuss it all here but it is interesting to note that there is a subgroup in New York City (ironically a non-malarial area), where a balancing second mutation has occurred, which to some extent reduces the disability of the sickle cell mutation. The individual now has two mutations, which improves their location in their design-space from when they had only one but they are not as healthy as individuals that lack the sickle cell trait altogether and the information in their genome continues to be depleted, not enhanced. Since supremacy of the sickle cell gene would result in extinction

17. Andrews, E.: *Who Made God? Searching for a Theory of Everything*. EP Books, Darlington, England (2009) pp.238–9.

of the species, this mutation cannot be macroevolutionary in the Darwinian sense.[18] These changes *adapt* an individual to be more competitive in a malarial micro-environment and hence the mutations are microevolutionary; they do *not* demonstrate Macroevolution.

Edgar Andrews amplifies this when he writes: "*...no known 'beneficial' mutation does anything but produce a new strain or variety within the same species, and the changes are trivial in the sense that no new biological structures are ever formed. For example, as Behe points out, such mutations have never been found to create new protein-to-protein binding sites — as would be necessary for the formation of novel molecular machinery within the cell...The cell's molecular machines (like the ribosome we considered earlier) always require different proteins to act together in 'gangs' of six, seven or more different molecules. For this coordination to occur, the different proteins must bind together (must link arms, as it were) to form working structures. For a novel working structure to arise in a cell, therefore, new binding sites must be created on several different proteins, and these binding sites must match to allow the molecules to stick together.*" [19] If evolution cannot develop even one new receptor site on one protein, it is never going to build nano-machines, let alone new cells or organisms. This is why information stasis is observed throughout microbiology.

Stasis in the Fossil Record

Turning to the sphere of palaeontology, Niles Eldredge, also observed stasis in the fossil record: '*...in the 60s as I tried in vain to document examples of the kind of slow directional change we all thought ought to be there ever since Darwin told us that natural selection should leave precisely such a tell-tale signal...I found instead that once species appear in the fossil record they tend not to change very much at all. Species remain imperturbably, implacably resistant to change as a matter of course-often for millions of years.*' [20]

Stephen Jay Gould also noted the phenomenon of stasis within species. He wrote: "*The history of most fossil species includes two features particularly inconsistent with the idea that they gradually evolved:*

1. *Stasis. Most species exhibit no directional change during their tenure on earth. They appear in the fossil record looking pretty much the same as when they disappear; morphological change is usually limited and directionless.*
2. *Sudden appearance. In any local area a species does not arise gradually by the steady transformation of its ancestors; it appears all at once and 'fully formed.*'" [21]

Then there are the 'living fossils' noted previously, such as the coelacanth, ginkgo, comb jellies and the Nautilus shell which have not changed since they first appeared in the fossil record.[22]

These observations would suggest that the major information generators have stopped producing new information to form new body plans and from Denton's observations this would appear to have been about the time of the Cambrian explosion. Since that time there has been some genetic drift but no significant information gains. To state this in another way, the existing information

18. Behe, M.J.: *The Edge of Evolution*. Simon & Schuster, New York (2007) pp.22–34 & pp.111–2.

19. Andrews, E.: *Who Made God? Searching for a Theory of Everything*. EP Books, Darlington (2009) p.241.

20. Eldredge, N..: *Reinventing Darwin*. Phoenix, NY (1996) p.3 *in* Lennox, J.C.: (2007) pp.112–3.

21. Gould, S.J.: The Episodic Nature of Evolutionary Change *in The Panda's Thumb: More Reflections in Natural History*. W.W. Norton & Co, New York (1985) *in* Lennox, J.C.: (2007) p.111.

22. Meyer, S.C. et al: *Explore Evolution*. Hill House Publishers, London (2009) p.25.

in the genetic economy has been traded or churned but the total information supply has not increased.

American biophysicist, Werner Loewenstein, whose research was in cell communication and biological information transfer, wrote regarding the genetic code: *"This genetic lexicon goes back a long, long way. Not an iota seems to have changed over two billion years; all living beings on earth, from bacteria to humans, use the same sixty-four word code."* [23] Again, Loewenstein's observations suggest that the information generators that produced the DNA codes stopped generating new information over two billion years ago. The information had to have been introduced prior to this and then remained essentially unchanged since that time. This poses a problem not only for gradual evolution but also the theory of Punctuated Equilibrium. The information generator(s) was active and then went silent.

Andrews sums up the modern history of Evolution when he writes: *"No new species* [have] *emerged and no genuine evolution occurred...there is nothing here that can account for macro-evolution – meaning the creation of new organs, a rise in biological complexity or the bridging of genuine gaps between different types of organism. Evolution simply assumes that many micro-evolution steps can be extrapolated to account for macro-evolution but this claim has no scientific justification."* [24,25]

Bacterial Resistance to Antibiotics

So, if stasis is a fact, how do we account for bacterial resistance to antibiotics, which certainly has appeared within our own lifetimes? This is a complicated topic, which requires a detailed investigation as it is illustrative of many information processes. In 1928, when the British physician Alexander Fleming (1881–1955) discovered the first antibiotic, which he called 'penicillin', there appeared to be no bacterial resistance. The penicillin was a natural substance produced by the mould *Penicillium*, which accidentally contaminated Fleming's agar culture plates.[26] Initially, production of penicillin was very slow but during World War II, Howard Florey (1898–1968) and Ernst Chain (1906–1979) developed a method to mass produce it. When my father graduated in medicine in 1942, he described how penicillin rapidly cleared all bacterial infections but by the late 1960s, bacteria were noticeably acquiring resistance. To understand the process, we need a comprehensive knowledge of how each antibiotic works at the microbiological level. Bacteria, unlike eukaryotic cells such as in human beings, possess a cell wall, as well as a number of other differences, which antibiotics can exploit without harming the host. Penicillin (which contains the beta-lactam ring) interferes with the manufacture of the bacterial cell wall, which results in the death of the bacterium, usually without harming the person being treated.

Since the 1940s, a significant number of bacteria have acquired a gene which produces the enzyme 'penicillinase', a β-lactamase enzyme, which cleaves penicillin and renders it harmless to the bacterium. The question is: how did bacteria acquire it? The complexity of the gene and enzyme product suggests that many steps would have been required to develop it so it is beyond microevolutionary processes, yet to develop it by Neo-Darwinian macroevolutionary mechanisms would have required thousands of years. So how can we explain its appearance? One needs to remember that penicillin was not a synthetic product but was manufactured by the mould, *Penicillium*.[27] Biochemist,

23. Loewenstein, W.: *The Touchstone of life*. Penguin Books, London (2000) p.64 *in* Lennox, J.C.: (2007) p.136.

24. Andrews, E.: *Who Made God? Searching for a Theory of Everything*. (2009) p.238.

25. Andrews is referring to industrial melanism in the British peppered moth population but the comments equally apply to all Macroevolution.

26. Meyer, S.C. et al: *Explore Evolution*. (2009) p.99.

27. Fowler, T.B. & Kuebler, D.: *The Evolution Controversy*. Baker Academic, Grand Rapids (2007) pp.174–5.

Jonathan Sarfati writes: "*A control gene regulates the production of an enzyme that destroys the antibiotic, e.g. penicillinase that destroys penicillin. (The information to make this complex enzyme was already in the gene pool prior to the discovery of penicillin...A mutation* disabling *this gene destroys the regulation of the production, so far more enzyme is produced. Such a bacterium can cope with more antibiotic than others can, but in the wild, it would be less fit than normal because it's wasting valuable resources producing more enzyme than is needed.*" [28]

Fowler and Kuebler further explain: "*To overcome resistance, new classes of β-lactam antibiotics have been developed to better fight infections caused by these bacteria. The third generation of β-lactam antibiotics were successful in dealing with many strains that had the β-lactamase enzyme, because the enzyme was unable to efficiently deal with these newer antibiotics. Since their introduction in the late 1970s and early 1980s, though, a number of modifications have occurred in the bacteria's β-lactamase enzyme that have expanded its binding site to better fit these newer antibiotics.*" [27] The method of acquisition of the pre-existing penicillinase-producing DNA is usually by the transfer of small rings of DNA called plasmids. This process is variously known as *lateral* or *horizontal gene transfer* and can occur rapidly, even between different species of bacteria. We do not know where the penicillinase gene came from originally but its presence neither confirms nor denies Neo-Darwinian evolution because it is unknown where or how the information was initially generated. However, our analysis of information so far would suggest that it is beyond the information generating capacity of Neo-Darwinian evolution as we presently understand it.

Not all is lost however. Some new light has been shed by an article in *Nature* in September, 2011 which shows that the genes for antibiotic resistance have great antiquity: "*...analyses of rigorously authenticated ancient DNA from 30,000-year-old Beringian permafrost sediments* [confirm] *the identification of a highly diverse collection of genes encoding resistance to β-lactam, tetracycline and glycopeptide antibiotics. Structure and function studies on the complete vancomycin resistance element VanA confirmed its similarity to modern variants. These results show conclusively that antibiotic resistance is a natural phenomenon that predates the modern selective pressure of clinical antibiotic use.*" [29]

So the origin of penicillin, tetracycline and vancomycin resistance is ancient and unclear but what about the other antibiotics? Different antibiotics attack bacteria in different ways. Some antibiotics affect DNA replication, others disrupt RNA synthesis and still others attack protein synthesis, such as at the ribosomes or polymerase stage (like penicillin). There are a number of other mechanisms by which bacteria have acquired antibiotic resistance. Describing a second mechanism, Sarfati continues: "*A* pump *in the cell wall takes in the antibiotic. A mutation* disabling *this pump will prevent the bacterium pumping in its own executioner. But in the wild, a bacterium with a disabled pump will be less fit than other bacteria because the pump also brings nutrients, etc., into the cell.*" [28] Describing a third mechanism, Sarfati writes that where: "*An enzyme is highly specialised to break down one specific type of chemical very well, and hardly affects other chemicals,* [a] *mutation* [can] reduce its specificity, *i.e. it no longer does its main job so well, and affects other chemicals to some extent too. Normally, a biological system with such a mutation would not function as well, and reduced specificity is reduced information* by definition. *But sometimes the other affected chemicals happen to be antibiotics, so this type of mutation confers resistance. Resistance to the antibiotic streptomycin can be caused by an information-losing mutation that degrades the surface of a bacterium's*

28. Sarfati, J.: *The Greatest hoax on Earth? Refuting Dawkins on Evolution.* (2010) p.69.

29. D'Costa, V.M. et al: Antibiotic Resistance is Ancient. *Nature* 477:457–461(2011).

ribosome, where decoding of DNA information to proteins occurs. This reduces *the binding ability of the drug to the ribosome, preventing it from ruining its operation."* [28]

As bacteria have acquired mechanisms for evading antibiotics over the last sixty years, human beings have developed new antibiotics with different mechanisms for incapacitating the bacteria and so a number of forms of antibiotic resistance have developed. Another form of bacterial resistance is acquired through genetic mutation which is simply a copying error. We have previously seen that the mutation rate is somewhere between one mutation in 100 million to 1 billion bacterial cell divisions but bacteria can replicate in only a few hours and exponential growth under appropriate conditions can achieve this number of bacteria in only two and a half days. The bacteria can neutralise the effect of the antibiotics if they can (randomly of course) mutate the active site that is attacked by the antibiotic, so that the antibiotic cannot interfere or at least not attack as effectively as before.[30] The mutation needs to alter the antibiotic binding site but not incapacitate the ribosome or active protein, so that the bacterium can continue to function and replicate, albeit less efficiently. Now it needs to be remembered that each of these antibiotic binding sites is at a critical point in the bacterial metabolic system, otherwise the antibiotic would not have worked in the first place, so if the bacterium was in an optimal place in its design space and mutations start to tinker with the bacterium's vital machinery, it will (at the very least) find itself at a slightly less optimal place in its design space after mutating some of its core machinery but at least it is alive and not wiped out by the antibiotic like its un-mutated brothers. This loss of efficiency suffered by the bacterium

Table 7.2: Types of Bacterial Resistance

Mechanism	Factor conferring resistance	Example
1. Control gene mutation	Mutations in the control gene for penicillinase production leads to an increase in penicillinase production.	Penicillin [28]
2. Influx pumps	Mutations which reduce transport proteins/channels can decrease the import of antibiotics, stopping the cell importing the antibiotic (e.g. Klebsiella) [31]	Kanamycin[32]
3. Efflux pumps	Mutations of the regulation process can increase the synthesis of efflux pumps, pumping more antibiotic out of cells.[32]	Tetracycline,[33] Isoniazid[34]
4. Enzyme mutations	Mutations which reduce enzyme specificity may allow an enzyme to act more broadly, meaning that it can now break down antibiotics. [32]	
5. Ribosomal mutations	Mutations to the surface of ribosomes may reduce antibiotic binding.	Streptomycin [28]
6. Antibiotic degrading enzymes	Cellular genes may produce enzymes (such as penicillinase) to degrade antibiotics. [27]	Penicillin G, Cephalosporin[32]
7. Genetic deletions	Small genetic deletions can block gene expression. [32]	Erythromycin, Chloramphenicol[32]
8. Membrane disruption	The *Plasmodium falciparum* chloroquine resistant trait (PfCRT) is a protein of 424 amino acids that makes the storage vacuoles leaky so that chloroquine does not build up to the same extent.[35]	Chloroquine[35]
9. Enzyme mutation	Enzyme mutations can make the binding of antibiotics ineffective (eg malaria).[36]	Sulfadoxine with pyrimidine, [36] Quinolones [37]

30. Meyer, S.C. et al: *Explore Evolution*. (2009) p.100.

31. Chevalier, J., Pages, J.M., Eyraud, A. and Mallea, M., Membrane permeability modifications are involved in antibiotic resistance in Klebsiella pneumoniae, Biochemical and Biophysical Research Communications 274(2):496–9, (2000) in Bergman, J.: Does the acquisition of antibiotic and pesticide resistance provide evidence for evolution? *Technical Journal* 17(1):26–32 (2003).

32. Fowler, T.B. & Kuebler, D.: *The Evolution Controversy*. Baker Academic, Grand Rapids (2007) p.94.

33. Speer, B.S., Shoemaker, N.B. and Salyers, A.A: Bacterial resistance to tetracycline: mechanisms, transfer, and clinical significance. *Clin Microbiol Rev*. 5(4):387–399 (1992).

34. Machado, D., Couto, I., Perdigão, J., Rodrigues, L., Portugal, I. et al.: Contribution of Efflux to the Emergence of Isoniazid and Multidrug Resistance in *Mycobacterium tuberculosis*. PLoS ONE 7(4): e34538 (2012).

35. Behe, M.J.: *The Edge of Evolution*. Simon & Schuster, New York (2007) pp.48–9

36. Fowler, T.B. & Kuebler, D.: *The Evolution Controversy*. Baker Academic, Grand Rapids (2007) p.259.

37. Wolfson, J.S. & Hooper, D.C.: Bacterial resistance to quinolones: mechanisms and clinical importance. *Reviews of Infectious Diseases* 11 (Supplement 5):S960–8 (1989).

is called the 'fitness cost' of the new mutation. This is yet another example of microevolutionary change — small changes *within* a species.

Doctors try to overcome any post-mutation decrease in binding affinity of the antibiotic by using a higher dosage of antibiotic administered to the patient, so that if the antibiotic only binds weakly to the active site and therefore is dislodged easily, the higher concentration of antibiotic might occupy the active site for longer because it is in higher concentrations in the bacterium. All bacteria appear to be able to bear the fitness cost of one or two mutations, especially if the second mutation compensates for some of the cost of the first mutation but if there are multiple mutations, then the fitness costs add up. This is not usually a problem in the continuing presence of the antibiotics for which the bacteria carry resistance but if the antibiotics are removed, then the other (non-resistant) bacteria which do not have to bear the antibiotic resistance fitness cost will prove more efficient and can out-compete the resistant strains of bacteria. This explains why periodically rotating antibiotics is more effective than continuing with the same one for long periods.

As previously mentioned, the empirical practice of sending patients with multiple antibiotic resistances, such as multi-resistant *Staphylococcus aureus* (MRSA), home from hospital where there are no antibiotics and usually ample wild strains of bacteria lacking the resistant genes, can also be effective. This was discovered by accident. When there were no other treatments left, the patients' antibiotics were stopped, as it was a waste of time and money to continue with them and the patients were sent home to die. The unexpected result was that sometimes the patients got better and eliminated the MRSA from their bodies. The success of this practice demonstrates how mutations decrease the robustness of the mutated strains of bacteria. The Information Model predicts that each mutation risks reducing the global competency of the information in the bacterial genome, thereby decreasing the bacteria's repertoire of alternatives when challenged with new environments. This leaves the bacteria less universally competent, narrowing the conditions under which the bacteria can survive. In the same way that a small population on an isolated island may have a limited gene pool, a mutated bacterium is more at risk of dying by being too specialised and not being able to cope if the physical environment changes again. So antibiotic resistance illustrates how microevolutionary change can cause the relatively rapid changes in bacteria that we observe in our hospitals. In contrast, macroevolutionary change has never been observed through antibiotic resistance, a fact which was corroborated by British microbiologist, Alan Linton who wrote: "*Throughout 150 years of the science of bacteriology, there is no evidence that one species of bacteria has changed into another.*" [38] This is despite the fact that every human being carries one trillion bacteria in their intestines and roughly another trillion bacteria on their skin, so new mutations are not uncommon. Despite these vast numbers, Alan Linton's statement still stands.

In stark contrast, the factors that *counter* the simultaneous change of characteristics in living organisms can be demonstrated in the development of microbial resistance to antibiotics. As discussed previously, it has become quite common for organisms such as tuberculosis (TB), *Staphylococcus aureus* ('Golden staph.') and the malarial parasite (*Plasmodium falciparum*) to become multiply resistant to antibiotics but it has been discovered that if two and especially three antibiotics (in the case of TB) are used simultaneously, this occurs much less frequently (if at all). As discussed above, bacterial resistance results not from the

38. Linton, A.: Scant search for the maker. *Times Higher Education Supplement*. (20 April, 2001) p.29 in Meyer, S.C. et al: (2009) pp.104–5.

generation of entirely new information code but from the incapacitation of some of the normal functioning of the bacterial machinery, such as the immobilisation of pumps. If Neo-Darwinian Evolution was indeed true, it might be envisaged that a sophisticated bacterium would develop a switching mechanism in its epigenome to turn the pumps on and off in the presence or absence of antibiotics. However, this would also require antibiotic detection equipment to know which antibiotic was present so that it would know which evasive action to take: to change the switches on the pumps, alter settings on ribosomes or adjust other machinery and this has not been described to date. This lack of ability to develop switches in the epigenome is predicted by the Law of Coding, as it would require sending information the wrong way up the intracellular information pathways, which cannot happen. (See Figure 3.11)

Despite the lack of epigenetic switching as described above, over time, various microbes have developed mechanisms to evade almost every known antibiotic. As already noted, bacteria can store these alterations to their RNA coding in small circular chains of RNA called plasmids, which can be copied and transferred to other bacteria. Individual bacteria can sequentially acquire resistance to a number of antibiotics in this way to become multiply resistant in a relatively short period of time and some bacteria, such as MRSA and Vancomycin Resistant Enterococci (VRE) have become resistant to all known antibiotics.[39] When I was in medical school the cardinal sin in treating tuberculosis was to fail to use three antibiotics simultaneously as the treatments were typically for eighteen months duration and antibiotic resistance was found to develop rapidly unless specific combinations of three antibiotics were used together and we now know why; because the microbes can develop one or two simultaneous mutations to gain resistance but never three mutations *simultaneously* as explained by Behe.[40] Edgar Andrews cites Behe when he writes: *"...all known 'beneficial' mutations involved loss of, or damage to, original gene function. They are changes that* fortuitously *protect the organism from attack by specific 'enemies' but which otherwise weaken it relative to the original ('wild-type') organism."* [41]

This principle also applies to warfarin tolerance in rats and insecticide resistance. Andrews quotes Swift who commented: *"Thus regarding insecticide resistance, ...:'The mutation that confers resistance...is invariably inferior to the normal 'wild-type' macromolecule. That is, in the absence of the relevant insecticide, the resistant strain fares less well,* [and] *is less fit, than the normal phenotype.'"* [42]

The last point that needs to be discussed here concerns what Richard Dawkins calls: 'the arms race'. Dawkins maintains that if leopards develop new traits over numerous generations that allow them to run faster than the antelope which they predate, the antelope that are not caught are likely to be those that run faster, as the slower ones become the leopard's dinner. As a result, both predator and prey are forced to enhance either their offensive or defensible capabilities in the battle to survive. However, this process is not open-ended as implied by Dawkins and there is a limit to how far this can proceed as it is dependent upon the pre-programmed information in each animal's genome. Referring back to the malarial parasite, Andrews writes: *"There is one further lesson from the* [sickle cell] *mutation; the malaria parasite has failed to evolve an effective 'weapon' against it despite billions of generations and huge parasite populations. So how stands the arms race concept now?"* [43] The explanation is, of course, that this is still change within a species and is microevolutionary not macroevolutionary and the malarial parasite does not possess the DNA code for macroevolutionary modification.

39. Sarfati, J.: *By Design*. Creation Book Publishers (2008) p.18.

40. Behe, M.J.: *The Edge of Evolution*. Simon & Schuster, New York (2007) pp.142–7.

41. Andrews, E.: *Who Made God? Searching for a Theory of Everything*. EP Books, UK (2009) pp.240–1.

42. Swift, D.: *Evolution under the microscope*. Leighton Academic Press, Stirling, UK (2002) p.237 in Andrews, E.: (2009) p.241.

43. Andrews, E.: *Who Made God? Searching for a Theory of Everything*. (2009) p.239.

Before leaving this topic regarding the rapid adaptation by bacteria to novel environments, I would like to discuss one more interesting example. In 1975, Japanese scientists were surprised to discover that certain bacteria could adapt to living on the waste products of nylon manufacture as their only source of carbon and nitrogen. This ability was found to be due to at least three genes which appeared to be unrelated to other DNA in the genome but resided exclusively on plasmids. There were a number of inexplicable findings relating to the bacteria's ability to break down nylon, which included the discovery that one strain of bacteria, *Pseudomonas aeruginosa* could develop the ability to degrade nylon de novo in only nine days, which is far too fast for Neo-Darwinian evolution. Australian biologist, Don Batten wrote: *"It seems clear that plasmids are...features of bacteria that enable adaptation to new food sources or the degradation of toxins. The details of just how they do this remains to be elucidated...This mechanism might be analogous to the way that vertebrates rapidly generate novel effective antibodies with hypermutation in B-cell maturation."* [44] Other possibilities also include the mutation of enzyme binding sites to broaden their specificity, the co-option of detoxifying enzymes that usually have other functions or the possession of pre-specified enzymes with this ability, as reported by D'Costa et al in their article on Ancient Antibiotic Resistance in *Nature*.[29] The conclusion that can be drawn from these findings is that there is a lot more information hidden within living systems, of which we are not yet fully cognizant, let alone able to explain its origin.

Discontinuities

Human timeframes are comparatively short and our ability to perceive change over time is often restricted by our relatively limited memories; days and events appear to blur imperceptibly into each other. As a result we often perceive life and the World as a series of continuities. Life appears to be a continuum, where change 'evolves' slowly. However, during the history of the Earth, episodic events, such as earthquakes, tsunamis, volcanic eruptions, meteorite impacts, hurricanes, floods and firestorms can wreak violent and sudden change and have occurred not infrequently. These discontinuities clearly interrupt the physical status quo but in science, discontinuities can also represent the lack of a logical connection or a lack of cohesion within a scientific argument.[45] In the case of an earthquake, the discontinuity is obvious but in science it is not always that clear. To distinguish a continuum from a scientific discontinuity, it is possible to keep paring down the entity to reach its basic elements until the removal of any further parts will disable the entity. In Chapter 1, we saw that Michael Behe introduced this concept by way of the common household mousetrap as an example of what he termed: 'irreducible complexity'. This mousetrap had five basic components and he asked how it might 'evolve' in a naturalistic setting, as all of its five components are critical for it to function as a mousetrap. Removing one of its components does not catch four-fifths as many mice. It now catches no mice, as it loses that purpose completely. This is a discontinuity, as there has been a disruption to its normal functioning. The question is then: could it have had a previous function which changed when it acquired the next component? Or were all five components put together simultaneously as it did not have a previous function?

Jerry Coyne's response to Michael Behe's example of the mouse trap was to demonstrate a possible previous function by using it (with one less piece) as a tie pin. However, we all know that tiepins did not evolve into mouse traps in the real

44. Batten, D.: The adaptation of bacteria to feeding on nylon waste. *Journal of Creation* 17(3):3–5 (2003) at Creation.com

45. *The New Collins Concise English Dictionary*. William Collins Sons & Co, London & Glasgow (1984).

world, so rather than pursuing this non-biological analogy any further, a real-life situation would be more relevant and useful. As described in Chapter 4, Michael Behe has also championed the irreducible complexity of the flagellar motor and the Darwinian counter has been the bacterial injector of *Yersinia pestis* (the plague bacterium), which shares a number of identical proteins. The Neo-Darwinian argument is that the bacterial injector, with fewer proteins has evolved, protein by protein, into the flagellar motor. For this to have occurred there would have needed to be stepwise increases in intricacy, with functional uses for the protein complex at each stage as it changed roles. As was seen in Chapter 4, there are ten proteins in the injector and forty proteins in the flagellar motor. So there are thirty new proteins required, which we will assume are composed of a minimum of 100 amino acids each or at least 3000 steps to transform the information that specifies an injector into information that describes the flagellar motor. However, there is no experimental evidence to support this scenario. The research by Scott Minnich suggests that the epigenome permits the bacterium to select between one completed structure or the other during its development, depending upon the environmental temperature as explained in Chapter 4.[46]

Thus the management software, which controls the growth of the bacterium, is far more complex than it at first appeared. But if we suppose for a minute that the flagellar evolutionary process had overcome all these problems, could the bacterium have kept the prototype flagellum in a suppressed state as it was being morphed from an injector, during the development phase of the flagellum and only flip the control switch when the flagellum was completely developed? It sounds appealing but it will not work because the Darwinian mechanism needs a working phenotype (prototype) to select for each new stage in the transformation process and neither broken nor unexpressed machinery will be selected via the mechanism of natural selection as there is no survival advantage. There needs to be an absolute minimum of twenty nine intermediate configurations if Punctuated Equilibrium is postulated but possibly a maximum of 2999 intermediate configurations if each amino acid needs to be added one at a time. Most crucially, each stage must construe a survival advantage for the process to be possible and none are known to date. The other problem is that the epigenetic control mechanism must be retrofitted once the two alternative phenotypes have evolved, yet from previous information analysis it has been seen that there is no credible mechanism for this within Neo-Darwinian evolutionary rules. The only interpretation which explains the data is that the flagellar motor and the injector are alternative phenotypes, which the nascent bacterium can select between as it is growing, depending on the environment in which it finds itself as it develops. This means that both phenotypes had to have been pre-specified with their supervising epigenetic DNA.

Information Change

Thus, discontinuities in the evolutionary process are real, which is a problem for Neo-Darwinian Theory. To overcome the discontinuities which have been observed, there needs to be gradual information change. So what would it take to convert the information that describes one type of animal, such as a reptile, into information that describes another kind of animal, such as a mammal? Neo-Darwinian Theory says that a reptile-like progenitor changed into a bird and

46. Minnich, S.A. & Meyer, S.C.: Genetic Analysis of Coordinate Flagellar and Type III Regulatory Circuits in Pathogenic Bacteria. *in* Dembski, W.A.: (Ed.) *Darwin's Nemesis.* Inter-Varsity Press, Downers Grove. Illinois (2006) pp.214–223.

System	Reptiles	Birds	Mammals
Forelimbs	Claws	Wings	Fingers/fins
Feeding	Teeth	Beak	Teeth/baleen
Covering	Scales	Feathers	Fur/hair
Metabolism	Cold-blooded	Warm-blooded	Warm-blooded
Circulation	Three chamber heart	Four chamber heart	Four chamber heart
Respiratory system	Two phase breathing	Four phase breathing	Two phase breathing
Brain development	Hind brain only	Fore-, Mid- & hindbrain	Fore-, Mid- & hindbrain
Reproduction	Egg laying	Egg laying	Uterus

Table 7.3: Examples of Information change between animal types [47]

yet another transformed into a mammal; Table 7.3 summarises a few of these conversions.

It is readily seen that there are many discrete differences between reptiles, birds and mammals. Devising a logical sequence to convert a three chambered reptilian heart into a four chambered mammalian heart or a two phase reptilian respiratory system into a four phase avian respiratory system (which all birds possess) is a daunting challenge for an intelligent human being, let alone blind evolution. In birds for example, the four stage breathing process depends on having three discrete regions in the lungs so that air can flow into the posterior air sacs, then to the anterior air sacks and finally out through a one-way system. On top of these differences, in life, all systems are functionally integrated. When surgeons perform a heart-lung transplant, they install the finished product but when you developed your own heart and lungs you required more than just the end organs; you required the instructions for auto-construction too. Each of the systems in Table 7.3 is itself an integrated subsystem, which requires its own unique embryology to grow with individual epigenomes to control both the embryological development and the functioning of the adult system. So as each animal subsystem evolves into a new form, it must develop its own unique embryology *and* epigenome in parallel with its adult phenotype, which is a problem of almost unimaginable complexity. Finally, if the above requirements appear complicated, it must also be remembered that under Neo-Darwinian rules, each new modification must be functionally equal to or superior to the previous one to provide an advantage for this 'intermediate' or 'missing link' creature to survive or it will not be selected in preference to its siblings. Therefore in summary, every step on the Neo-Darwinian pathway requires simultaneous mutations in each of its genome, its epigenome and the embryological pathway which the organism followed to reach adulthood with each subsequent step being fully functional and in a superior location to their parent's position in their design space and do not forget that the 'inter-species chasms' described in Chapter 5 require thousands of steps to cross. If this process defies human design capabilities, how much less achievable is it for blind evolution?

Assumptions

With so many insurmountable obstacles it is surprising that Neo-Darwinian Macroevolution has so many supporters. In addition, we have seen that Darwinian evolution also has trouble explaining the origin of life, the generation of novel information and hence explaining biological complexity (such as the ATP synthase generator, amongst many other essential requirements). So why have countless

47. Meyer, S.C. et al: *Explore Evolution* Hill House Publishers, London (2009) pp.129–139.

eminent professors and thinkers the world over, for the last 150 years, believed that Darwinian evolution is the answer? The reason lies in the assumptions that are made. Often, the presence of information in living things is attributed to 'evolution' with no attempt to explain how the information was generated. The argument goes something like: "It seems safe to assume that at the formation of the Universe there was no life in the Cosmos but now it is clearly seen that life is flourishing on Earth in all its varied beauty, so the human eye must have formed somehow and for lack of another acceptable explanation, 'evolution' must have done it!" [48] This would not seem unreasonable if there was no evidence to the contrary and until the 1960s or 1970s, there was not a lot of biochemical evidence to prove or disapprove evolution by natural selection but since that time, there has been a rapidly growing body of information that cannot be explained by Neo-Darwinian evolution and a new explanation needs to be sought. The only other explanation to date, that an external entity intervened to place life on the Earth, has been rejected as lacking material evidence and being 'unscientific' (or not natural) but the weight of evidence for and against each theory is shifting.

The other reason that Neo-Darwinian theory has confused so many people is that Darwin was *partially* correct and until now his theory has had the best explanation in purely scientific terms. 'Survival of the fittest' (Microevolution) does actually work and the other aspects of his theory, such as random mutation with natural selection, have the superficial appearance of creating macroevolutionary change. However it has already been shown that random mutation only corrupts working systems and natural selection then removes them, so that they do not contaminate the next generation with damaged genes that are counterproductive. As was shown in Chapter 1, Microevolution depends on information that is already programmed into the DNA and 'survival of the fittest' means that the species with the most successful phenotypes in that habitat at that time survive to pass on their genes. If a warm period is followed by a cold one, such as an ice age, then a different set of genes (such as long fur rather than short) may prove more successful in the subsequent climactic conditions. This adapts organisms to their individual ecological niche and explains why the Galapagos finches vary from island to island and from plentiful season to lean season. This does not involve mutation of the genome however but selection of a phenotype from the huge library of possibilities within the gene pool that is already within the living population and allows those variations to be expressed as the environment changes.

In Chapter 1, Fowler and Kuebler summarised the Neo-Darwinian synthesis as: "*While there is solid evidence that these mechanisms can lead to small-scale change (microevolution), a fact accepted by all schools* [of thought], *there is at present little* direct *evidence that they can explain* all *evolutionary change (macroevolution); what evidence we have is primarily circumstantial.*" [49] (Emphasis in the original.) Although Darwin and his supporters in the Neo-Darwinian Synthesis have argued for the last 150 years, as Fowler and Kuebler wrote: ..."*that the available evidence supports their theory, and no other theory has a better explanation of this evidence, so they infer the truth of their theory by* inference to the best explanation", (emphasis in the original)[49] it fails the information audit, which suggests that the Neo-Darwinian Synthesis only has the *appearance* of explaining the facts and that the circumstantial evidence is just circumstantial. We need a better explanation for the generation of the vast amount of information in biological systems and an external biological information generator is the only

48. Sarfati, J.: *By Design.* Creation Book Publishers (2008) p.35.

49. Fowler, T.B. & Kuebler, D.: *The Evolution Controversy.* Baker Academic, Grand Rapids (2007) p.344.

deduction possible as an 'internal' information generator does not exist. The nature of this external information generator is yet to be scientifically identified but if we are going to find answers as a scientific community, we need to be looking in the right place to deduce the right answers.

Another significant assumption within the Neo-Darwinian Synthesis is Uniformitarianism which implies that our world and the Universe only change slowly over very long periods of time.[50] There are pragmatic reasons for this assumption. As the World is extremely complex, if we can simplify or summarise concepts, it makes the Universe easier for us to understand and analyse. Thus, if we assume that there has been no change unless we find overwhelming evidence to the contrary, such as the extinction of the large dinosaurs, extrapolation becomes possible to infer details that are currently unknown. However, oversimplification can lead to errors too. This may leave our minds closed to more rapid processes or events which may only occur occasionally, such as a significant meteor strike or massive volcanic eruptions that precipitate global cooling. These are changes with which science can struggle. As stated in Chapter 1, Uniformitarianism began in the 19th-Century with Charles Lyell and his textbook on geology. When he looked at the rock strata and the rate of erosion, he surmised that change was very slow but like the process of evolution itself, Lyell needed to extrapolate from the geologic column because no one was there to observe it actually happening. Extrapolation is often necessary when compiling a theory but it does not constitute incontrovertible proof.

Materialism and Non-Materialism

So if the only choice for the appearance of life in the Universe is a fatally flawed yet materialistic (and hence potentially testable) theory of Evolution or a non-material and hence possibly untestable information generator, which way should a scientist turn? Is there no other way? Can the Information Model shed any light on the differences between materialism and non-material entities? Is there any evidence for non-material entities in the Universe? Well, yes, there is. Information and human thought (which is investigated in Chapter 10) are non-material entities, but we are not used to thinking of them in such terms. Materialism is confined to the first four dimensions (width, length, height and time) and is considered to be within the realm of science, whereas any other dimensions (if they exist) are by definition, non-material and assumed to be unmeasurable. Consequently they are considered to belong to the realm of philosophy or religion.[51] Yet there are a number of entities, like information, which lack mass and yet are real and they steadfastly resists scientific analysis; human thought has already been mentioned but human consciousness and the entity of life itself also belong in this category.

Considering this difference between materialism and non-materialism begs the question: 'what is reality?' Oxford biochemist and theologian, Arthur Peacocke (1924–2006) cites the dictum: *"For to be real is to have causal power"* [52] and there is no doubt that information, human thought, human consciousness and life have causal power. Therefore to summarily dismiss all non-material entities, such as those listed above, as not being real due to their lack of material qualities is invalid as it would be denying our human existence. As a result, over time, an increasing number of scientists have been recognising the inconsistencies in a purely materialist paradigm and have begun to ask questions.

Paul Davies wrote in 1999: *"How did stupid atoms spontaneously write their*

50. *Ibid.* p.52.

51. Ward, K.: *Why There Almost Certainly Is a God.* Lion Books, Oxford (2008) p.50.

52. A dictum attributed to S. Alexander by J. Kim (Kim, 1992, pp.134–5) *in* Peacocke, A.: The sciences of complexity: a new theological resource. in Davies, P. & Gregersen, N.H. (Eds.): *Information and the Nature of Reality. From Physics to Metaphysics.* Cambridge Univ. Press, Cambridge (2010) p.254.

own software? And where did the very peculiar form of information needed to get the first living cell up and running come from. Nobody knows. There is no known law of physics able to create information from nothing." [53] In 2003, Davies wrote: "*Nobody knows how a mixture of lifeless chemicals spontaneously organised themselves into the first living cell.*" [54] In an effort to answer this conundrum, he speculates that maybe: "*...some sort of self-organising physical process could raise a physical system above a certain threshold of complexity at which point these new-style 'complexity laws' would start to manifest themselves, bestowing upon the system an unexpected effectiveness to self-organize and self-complexity...Under the bidding of such laws, the system might be rapidly directed towards life.*" [55] In these quotes, Davies appears to be implying that the product is more than just the amalgamation of the tangible raw materials. In Chapter 5 we saw that this occurs because the formation of organised material entities necessarily incorporates additional non-material information.

From these observations, the real problems begin to become apparent. If science is dependent upon the tangibility of what can be measured and/or limited by time, then materialism can be thought of as the study of the finite, whereas that which is timeless and has no beginning or end is the study of the infinite and as already mentioned is usually considered to be part of the study of philosophy or religion. The example that proves the rule is the field of mathematics, which studies the realm of the infinite and the imaginary (such as the square roots of negative numbers which are called 'imaginary numbers' because they cannot physically exist in our four-dimensional universe). And mathematics is completely intangible. Now if materialism is restricted to that which is finite, then non-materialism must include that which is not. Materialism concerns dimensions one to four, and therefore non-materialism (including information, thought, life and consciousness), would appear to reside somewhere else outside these dimensions, which, at this stage, for the purposes of our discussion, we will call the 'fifth dimension'. From an information perspective, materialism and non-materialism are two parts of a greater whole that do not overlap, yet both are necessary to fully understand our universe. Materialists may argue that the non-material entities identified above are simply 'emergent properties' of material ones, but it has already been observed that a non-material entity, such as information, is external to its material vehicle and quantum mechanics says there are other spaces, even possibly other universes, which we cannot physically detect and therefore we cannot physically measure either. Unfortunately, from where science stands today, these

53. Davies, P.: Life force. *New Scientist* 163(2204):27–30 (1999)

54. Davies, P.: Was life on Earth born lucky? *New Scientist* 179(2403):32 (2003).

55. Davies, P.: *The Fifth Miracle.* Simon and Schuster, NY p.259 (1999) in *Journal of Creation* 23(3):125 (2009)

Table 7.4: Some Material and Non-material aspects of familiar Entities
From this table it can be seen that there are more categories in the non-material universe than the material one; we just notice the material ones.

	Components	Computers	Life	The Universe
Non-material	**Information**	Application software (word processing)	Thoughts	Behaviour of the Universe (planetary motion within the solar system)
	Software	Operating systems (Windows 8/OS 10)	Messaging via nerves and neurotransmitters	Laws of the Universe (physics)
		Machine language	Laws of Chemistry	Information defining the fundamental particles
		Thought processes of the programmer	Biological information generator (see Ch. 3)	Physical information generator (See Chapter 8)
Material	**Hardware**	Electronic components	The physical body	Physical matter
		Factory where the computer was made	DNA and RNA molecules	The fabric of the physical Universe

other spaces or universes are effectively non-material as they are unmeasurable. If you are still unconvinced, dark matter and dark energy (which cannot be measured) make up 96 percent of the Universe according to the cosmologists.[56] So from the materialist's definition of science, ninety-six per cent of the Universe may be off-limits to science and scientists! However, in the light of the evidence we observe, and the principles we deduce from that evidence, it is more *scientific* to conclude that both material and non-material properties exist in the Universe and that *both are 'real'*, even though the concept is foreign to us in our four dimensional materialistic frame of reference. Insisting that every entity in the Universe must be tangible is similar to replacing the 'Flat Earth' principle with a 'Shallow Universe' Model! As will be seen in the following chapters, once one's eyes are open to non-material entities, one realises that they have always been there; it is just that we have not been attuned to recognising them.

Requirements of living things

In Figure 5.7, it was seen that the minimum requirements for the first cell included at least thirteen different entities that were *simultaneously* required for life. These thirteen components can be categorised into four different spheres: matter, energy, information and life itself (see Figure 7.3).

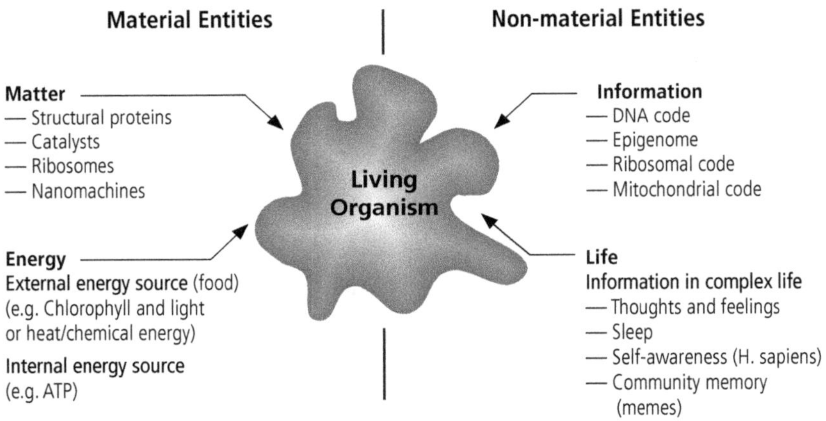

Figure 7.3: Material and Non-material Entities Required for Living Organisms
Living organisms cross the divide between the tangible and intangible. All life requires these components simultaneously to survive, no matter how simple or complex the organism.

Irreducible Complexity

Irreducible complexity, the term introduced by Behe in his book: *Darwin's Black Box*, is interlinked, co-dependent information. Behe wrote: "*By* irreducibly complex *I mean a single system composed of several well-matched, interacting parts that contribute to the basic function, wherein the removal of any one of the parts causes the system to effectively cease functioning*." [57] Expressed in information terminology, a system is irreducibly complex if the loss of any of its information would compromise its intended function.

This understanding allows us to consider: Informationally Competent Systems. An Informationally Competent System (ICS) contains sufficient information to operate independently, without the need for additional information

56. Al-Khalili, J.: *Black Holes, Worm Holes & Time Machines.* Taylor & Francis Gp, NY (1999) p.68.

57. Behe, M.J.: *Darwin's Black Box: The Biochemical Challenge to Evolution.* Free Press, NY (2006) p.39.

input or control. To put it another way, the minimum ICS is the least amount of information that allows the proper, ongoing functioning of an entity. From this definition, it can be seen that viruses are below the ICS threshold, as they cannot reproduce unless they can invade a living cell, whose machinery (and information) they must hijack to replicate themselves.

The Fundamental Laws of Biology

We have come a long way by simply looking at the information required by and contained within living systems and it is nearly time to examine the information in our physical universe in greater depth but before we do, to summarise our discoveries to date, I would like to propose three fundamental laws of biology which can help us in understanding living things.

First law of Biology: Law of Biogenesis

"Only life begets life" [58]

The corollary of this law is that once a living organism dies, it cannot come back to life again in our physical universe by natural physical processes.

As described in Chapter 1, Francesco Redi introduced this concept after publishing his research in 1668, which was confirmed by Louis Pasteur after performing his famous swan-neck flask experiment in 1862 and the Law of Biogenesis is just as true today as it was then. Information alone is not enough for life. The cell needs to overcome the friction caused by the movement of the molecules within each cell so it might equally be thought of as: 'The Biological Law of Friction'. Cells die if the friction forces outweigh the energy input to overcome the friction. Therefore to create life there is a need to overcome friction in each and every cell. How much friction is there? We do not know how to measure it but there is a lot (remember that the viscosity of the cytoplasm has been estimated to be equivalent to the consistency of yoghurt).[59] This is why cyanide is so deadly, because it stops power generation in the mitochondria (the ATP-synthase) and the poisoned cells rapidly run out of energy to power their machinery, so they die.

However, if energy is essential to maintain living things, it is just as necessary for the animation of life in the first place. We have already seen that even the simplest cell is packed with thousands of different molecules and proteins and living cells are frenetically busy places. To create *new* life also requires overcoming the internal friction (initial inertia) of the teeming mass of molecules within the cell and the way life does that is by using another living cell which subsequently divides. Thus Pasteur's observation that: *"only life begets life"* creates a problem for the first cell in a Neo-Darwinian paradigm. To 'jumpstart' the processes in a cell (as one might jumpstart a car with a flat battery by using an electric current from a power source) seems inconceivable, as there are so many intricate chemical reactions involved. Even if all the hundreds of appropriate organic molecules and their catalysts could be artificially put inside a cell membrane, an electric current is (à la Dr Frankenstein) likely to fry the cell and all its contents rather than coax the cell into life. Life uses chemical energy rather than electrical power to function. No credible scientific mechanism has yet been conceived to explain how these preconditions might have been met for the animation of the first cell. Thus a living cell is required as a further precondition before new life can be created. This is why only life begets life.

58. Bendewald, J. & Sherwin, F.: *Evolution Shot Full of Holes*. Evidence Press, Madison, Wisconsin (2004) p.91.

59. Williams, A.: Mutations: evolution's engine becomes evolution's end! *Journal of Creation* 22(2):61 (2008).

Second Law of Biology: Law of Information Primacy

Information and coding precede structure and function

This topic has already been extensively covered in Chapter 3 but to summarise the concept: Proteins, which are the building blocks of cells, are a product of information stored in DNA, which provides the sequence code required to create them and hence the DNA coding must have preceded the production of these proteins. This is a one-way process. Therefore, the DNA, RNA or any other information storage mechanism in the first cell had to have predated the cell it coded for. However, for natural selection to evolve the information stored in the first cell, the cell had to be in existence with physical expression of its traits before the Neo-Darwinian evolutionary processes could sculpt the information storage mechanism. Thus evolution by random variation and natural selection cannot explain the information in the first cell. Information was a pre-requisite.

Third Law of Biology: Law of Genetic Preservation

Random variation with natural selection reverts to the mean unless the environment changes (such as MRSA in the presence of antibiotics).

Changes in Biology trend towards the mean. This is the inevitable conclusion from information stasis and we have observed that in general there is preservation of genetic material over many generations, as observed by Eldridge and Gould in Chapter 5. Without this phenomenon, all DNA would become corrupted within a few generations and all life would have become extinct long ago.[60] The adage that: 'Like promotes like' [61] can be seen in the fossil record, from observations in the wild and in artificial breeding programs over hundreds of years. Bacteria that were described in the late 1800s provide a typical example. One very successful species of bacteria called *Pseudomonas aeruginosa* was first described in 1872 by German doctor and scientist, Joseph Schroeter.[62] P. *aeruginosa* has the ability to replicate within hours and hence over the last 140 years, there have potentially been more generations of this bacterium than the possible number of generations in the ancestry of man and yet *P. aeruginosa* has not changed significantly and certainly has not evolved into identifiably different species. Darwin proposed that gradual small changes would produce new species over many generations but it has never been observed and information flows forbid that it could happen. This fundamental law of biological information is the reason why Macroevolution does not and cannot occur.

The Biological Information Hypothesis

In conclusion, these three biological Laws can be summarised as the *Biological Information Hypothesis* which states: information is the grand conductor of life and in all biological systems, information precedes life. This is the concept of biological information primacy. Many thousands of scientists have expected that with time and increasing technology, the mechanism of random variation with natural selection would eventually be proven to increase complexity in biological systems but microbiological advances have done just the opposite, revealing stunningly complex nano-systems with bewildering arrays of inter-connections and the initial assumption that basic life forms would be 'simple' is just not true. Not only are the

60. Williams, A.: Mutations: evolution's engine becomes evolution's end! *Journal of Creation* 22(2):63–5 (2008).

61. Wells, J.: *Icons of Evolution*. Regnery Publishing, Washington, DC (2002) p.69.

62. *Encyclopaedia Britannica* on-line

simplest life-forms that are presently found on Earth so sophisticated that they cannot be reproduced from scratch, even in a laboratory but a possible prototype that could have initiated life, cannot be devised, even in theory. (No one has a clue how to build a mitochondrion.) In addition to this, the places in the Universe where life can survive, once it has been instigated, are very limited. Most of the Universe is hostile to life. Despite decades of scouring the Universe, not one other planet in the Universe has so far been found to contain any life, let alone be conducive to the spontaneous appearance of intelligent life and the laws of biology suggest that if intelligent life does exist anywhere else in the Universe, it will be exceedingly rare. If it should exist, it will be carbon-based as on Earth. Further, the evidence to date shows that the human brain is the most complex entity in the Universe and that *Homo sapiens* is unique. This will be discussed further in Chapter 12.

Finally, not everything in the Universe can be explained in material terms and information, thought, consciousness and life have non-material elements which need further study and these will be addressed in Chapter 10. However, it is first necessary to dig deeper into the information in our physical universe, which is the subject of the next chapter.

8

The Physical Information Hypothesis

Perhaps the universe is really a frolic of primal information, and material objects a complex secondary manifestation.[1]

Paul Davies

Mathematics, Physics and Information

John Lennox explores the relationship between mathematics and physics by quoting Eugene Wigner (1902–1995), Nobel laureate in physics in 1963, who wrote: *"The enormous usefulness of mathematics in the natural sciences is something bordering on the mysterious, and there is no rational explanation for it...it is an article of faith.' The relationship between mathematics and physics goes very deep and it is very hard to think of it as some random accident. Professor of mathematics Sir Roger Penrose FRS, whose understanding of that relationship is unquestioned, has this to say about it: 'It is hard for me to believe...that such superb theories could have arisen merely by some random natural selection of ideas leaving only the good ones as survivors. The good ones are simply much too good to be the survivors of ideas that have arisen in a random way. There must be, instead, some deep underlying reason for the accord between mathematics and physics."* [2] It is now becoming clear that the reason that there is an accord is because information defines the relationships. Paul Davies writes: *"The traditional relationship between mathematics, physics, and information may be expressed symbolically as follows:*

Mathematics → Physics → Information

According to this orthodox view, mathematical relationships are the most basic aspects of existence. The physical world is an expression of a subset of mathematical relationships, whereas information is a secondary, or derived, concept that characterises certain specific states of matter (such as a switch being either on or off, or an electron spin being up or down). However, an alternative view is gaining in popularity: a view in which information *is regarded as the primary entity from which physical reality is built...Importantly, it is not merely a technical change in perspective, but represents a radical shift in worldview,...The variant I wish to explore here is to place* information *at the base of the explanatory scheme, thus:*

Information → Laws of physics → Matter

1. Davies, P.: Bit before it? *New scientist* 161(2171):3 (1999) *in* Johnson, P.: *Wedge of Truth.* InterVarsity Press, Downers Grove, Illinois (2000) p.56.

2. Lennox, J.C.: *God's Undertaker. Has Science Buried God?* Lion Books, Oxford (2007) p.60.

After all, the laws of physics are informational statements: they tell us something about the way the physical world operates." [3] Thus information (albeit expressed in the terms of mathematics) preceded the physics. As Professor Edgar Andrews explains, the physics is only a description of *what* we observe in the Universe, it cannot explain *why* the Universe is the way it is.[4] It is the underlying information that explains why the physics works.

The Theory of Relativity

If you have a leaning to the biological sciences like me, then understanding our journey to date has been relatively easy and the physicists are the ones who have had to struggle but now we have to delve into the world of physics and I ask that you bear with me as we investigate regions you may have never visited before. We begin this part of our journey on a summer's day, 10th June, 1854, in a lecture at the University of Göttingen when Georg Bernhard Riemann, a German mathematician, used Euclid's geometry to deduce *"that 'forces' were a consequence of a distortion of geometry."* [5] This sounds very esoteric to non-mathematicians but the significance of this was that he was proposing that the physical realm which we inhabit may not be quite as it appears. Dr Chuck Missler explains: *"In one masterful stroke Riemann uncovered the mathematical foundation of geometries of higher dimensional space and thus opened the door to modern physics."* [5] Particle physicist, Professor Michio Kaku relates that Riemann: *"...arrived at this view after imagining a flatworm living in a flat world. Here, there would be no such thing as up or down: only side to side, front and back. But what if one day, a wrinkle appears in the plane it normally inhabits? The worm will have no way of experiencing it. All the worm will know is that a barrier prevents its progress: some strange and invisible force. But Riemann, then made the stunning extrapolation that perhaps the invisible forces encountered in our four dimensional world, like gravity, are also ripples in an unseen, higher dimension."* [6]

Fifty years later, Albert Einstein integrated four-dimensional Riemann geometry to derive his Theory of Relativity. Andrews writes: *"Einstein's special and general theories of relativity, developed in 1905 and 1915 respectively, caused a seismic shift in science's perception of the universe. Until then, space and time were considered to be the 'given' and unchanging stage on which the drama of existence was played out. In Newtonian physics, everything happened in time and space and nothing ever happened to time and space. With Einstein all that changed...He showed that the space-time continuum was not the rigid and unchanging 'stage' (or frame of reference) for everything else, as had previously been supposed, but was itself an actor in the cosmic drama.*[7] Hawking and Mlodinow elaborate: *"Due to the work of Einstein, physicists realised that by demanding that the speed of light be the same in all frames of reference, Maxwell's theory of electricity and magnetism dictates that time cannot be treated as separate from the three dimensions of space. Instead, time and space are intertwined. It is something like adding a fourth direction of future/past to the usual left/right, forward/backward, and up/down. Physicists call this marriage of space and time 'space-time', and because space-time includes a fourth direction, they call it the fourth dimension."* [8] Andrews points out that space-time: *"...was part of the fabric of the universe, capable of change and distortion. Thus gravity ceased to be some mysterious action-at-a-distance and it became instead the result of massive objects causing curvature in space-time. In a relativistic universe, the three dimensions of space and the one dimension of time are all equally flexible. However,*

3. Davies, P.: *Universe from bit. in* Davies, P. and Gregersen, N.H. (eds.): *Information and the Nature of Reality: from Physics to Metaphysics.* Cambridge University Press (2010) p.75.

4. Andrews, E.: *Who Made God? Searching for a Theory of Everything.* EP Books, Darlington, Eng. (2009) p.30.

5. Missler, C.: *Cosmic Codes.* Koinonia House, Coeur d'Alene, Idaho (2004) p.43.

6. Finkel, E.: Fish out of Water. *Cosmos* 2:79 (2005).

7. Andrews, E.: (2009) pp.110–1.

8. Hawking, S. & Mlodinow, L.: *The Grand Design.* Bantam Press, London (2010) p.99.

the equivalence of space and time is only partial. We can move freely backwards and forwards in any of the three dimensions of space, but we can only go forwards in the 'pseudo-space' we call 'time'. For convenience we can (and shall) talk about time itself being unidirectional, or time 'flowing' in one direction...This unique unidirectional nature of time is called 'the arrow of time', a term coined in 1927 by British astronomer Sir Arthur Eddington. Interestingly, almost all the equations and theories of science are symmetrical with respect to time - that is, they hold true whether events are running forwards or backwards in time...[There is] *one glaring exception — a single physical law that is not indifferent to the direction of time. The law in question is the 'second law of thermodynamics', a long-established principle that describes and governs growth in randomness. The law states that while the randomness (technically, the 'entropy') of an isolated system may increase with the passage of time or remain unchanged, it can never decrease."* [9]

Jim Al-Khalili, Professor of Theoretical Physics at the University of Surrey in Guildford, England now takes up the story: *"...special relativity...heralded a revolution in physics. It showed how and why the old notions of space and time had to be ditched and replaced with such a new and unfamiliar set of concepts that, to this day, we still have not been able to shake off the 'old notions'. The space and time that most people still take for granted as 'common sense' were shown to be wrong by Einstein... Newton is rightly acknowledged as having sewn up the whole of classical mechanics with his laws of motion. These describe how objects move and how forces such as gravity affect them by making them speed up, slow down or change direction...*[However] *Einstein's theories of relativity go far deeper than merely stating the laws of motion... Bodies moving freely at constant velocities and in the absence of gravity are described by special relativity. Once the force of gravity is switched on we must turn to general relativity...Newton's law of gravity is only an approximation to the more exact general relativity, but it nevertheless works very well in weak gravitational fields, such as the Earth's. In the same way, Newton's laws of motion are only approximations to special relativity, but the differences now only show up when objects move at very high speeds. For most purposes in everyday life the accuracy of Newtonian mechanics is as much as we need. Even NASA uses Newton's laws to calculate the path a rocket should take to reach the moon,"* [10] It is only when objects approach a significant proportion of the speed of light (which is close to 300,000 km/s) that relativity becomes relevant. To give you some perspective, it takes a commercial airliner about twenty two hours to fly from Sydney to London which is almost halfway round the World but a light beam (that is if we could travel from Sydney to London at the speed of light) takes only one twentieth of a second! Al-Khalili continues: *"Many physicists regard general relativity to be the most beautiful scientific theory ever discovered. Its beauty lies in the simplicity, elegance and richness of its mathematical equations."* [11]

Within ten years of Einstein publishing his landmark theories, Missler explains: *"Theodr Kaluza at the University of Koenigsberg, Germany, would use five-dimensional Riemannian geometry to integrate both gravity and light. Light is now viewed as a vibration in the fifth dimension. Oskar Klein made several improvements, including the calculation of the size of the fifth dimension — the Planck length, 10^{-33} cm — much too small to detect experimentally. One hundred and thirty years after Riemann's famous lecture, physicists would extend the Kaluza-Klein constructs to develop ten-dimensional geometry in their attempt to unite all the laws of the physical universe."* [5]

Thus as a result of Einstein's work on relativity theory, physicists now view time as a physical property which varies with mass, acceleration and gravity.[12] The

9. Andrews, E.: (2009) pp.111–2.

10. Al-Khalili, J.: *Black Holes, Worm Holes & Time Machines.* Taylor & Francis Gp, NY (1999) pp.140–1.

11. *Ibid.* p.217.

12. Missler, C.: (2004) p.45.

converse of this concept is: *that which does not have mass does not experience time.*[13] This has profound implications when we study information, the fundamental forces and the formation of the Universe.

The Self-Organising Universe

There have been many unexpected discoveries in science over the last one hundred years and one of them is that the Universe is finite. When Einstein published his General Theory of Relativity in 1915, he found that the mathematics suggested that the Universe was expanding (or contracting) and hence dynamic and finite, whereas the prevailing view was that the Universe was infinite and unchanging; so Einstein included a factor to adjust for the fluctuation to maintain the status quo of a steady-state universe. Since then it has become generally accepted that our physical universe had a beginning at the 'Big Bang' and it will end in the 'ultimate heat death' when all energy will be equally distributed so that no further work can be done. (Work depends on an energy differential and in the Universe this is effectively heat, so when there is an even temperature throughout the whole Universe, no more work can be done and the Universe will functionally be 'dead'.)[14] Therefore, if our universe had a beginning and will have an end, it is finite in time as well as energy. Further, since the Universe has been observed to be expanding, it is presumed that this expansion had a beginning which has been dubbed the '*Big Bang*', when all matter and energy in the Universe is thought to have been compressed into a space that was less than a pinhead. At that point, pressure and temperature in our embryonic universe would approach infinite values, which physicists call a *singularity*. So the formation of our universe is often referred to as the *initial singularity*, while the end of the Universe is called the *final singularity*. Edgar Andrews defines a singularity as: *"an event or situation in which one or more physical quantities (like temperature or density) become infinite in value."* [15]

Thus we have discovered that our universe is bounded by time but it is also bounded by space. German physicist, Max Planck (1858–1947) discovered that if a material object such as a ruler was repeatedly chopped in half there is a limit to which this process can be repeated. It can be imagined that this process could go on forever producing smaller and smaller lengths but Planck found that the smallest unit of measurement of length in the Universe is 'Planck Length' (= 10^{-33} cm).[16] If we try to measure subatomic particles smaller than this, the matter ceases to have locality. It still exists but it cannot *physically* be measured. (This will be dealt with further in later chapters.) The Universe is also finite in size and has an outer boundary, even though this is very large in human terms. In a similar way, if the same process of endless halving is applied to time, Planck found that there was also a minimum period of time, which is 10^{-43} seconds, which physicists call 'Planck Time'. When discussing quantities in quantum physics (also called 'quantum mechanics') it involves measurements in this size range and physicists often refer to it as: 'Planck scale'. The significance of these discoveries is that space, as well as time, has finite upper and lower limits. The important question is how did it happen that way?

The very existence of the Universe is a mystery. At this stage it is beyond the realm of science to explain *why* the Universe is here (see Table 1.5) but *how* it has come about should be within the ambit of science. One proposal is that it assembled itself. Professor of Chemistry at Oxford, Peter Atkins writes: *"Space-time generates its own dust in the process of its own self-assembly."* He named this

13. *Ibid.* p.369.

14. Dawkins, R.: *The Greatest Show on Earth*. Bantam Press, London (2009) p.414.

15. Andrews, E.: (2009) p.93.

16. Missler, C.: (2004) p.335.

process the *"cosmic bootstrap"* [17] but where did the information required for the assembly instructions originate? We have already seen that assembling biological entities requires information and the same principle applies to our physical universe too. Information was required to specify which chemical bonds are viable and hence, which are not, which properties specify the characteristics of light and which forces will exist in the Universe.

Another viewpoint on the formation of the Universe comes from Arabic teaching which has brought us the Kalam Cosmological Argument which states:

1. Whatever begins to exist has a cause;
2. The Universe had a beginning;
3. Therefore the Universe had a cause.[18]

This is intuitively correct. If I have a computer which I know has not existed forever and therefore had a beginning, then it must have had a cause, which is, that someone manufactured it. So this is true on an intuitive level but what about information? In information terms it could be restated as:

1. Whatever begins to exist requires information;
2. The Universe began to exist;
3. Therefore the Universe required information to begin.

However, not all scientists agree on the causation of the Universe. Paul Davies writes: *"...for me it is much more inspiring to believe that a set of mathematical laws can be so clever as to bring all these things into being."* [19] The problem is that the information implicit in these laws had to come from somewhere or be generated somehow. Even if the laws 'did it', the information is still outstanding and demands an explanation. In effect, the beginning of the Universe was not the *ultimate* beginning because the information that defined our nascent Universe also must have had a beginning; so how was that information generated?

Richard Dawkins' view is: *'The universe is nothing but a collection of atoms in motion, human beings are simply machines for propagating DNA, and propagation of DNA is a self-sustaining process. It is every living object's sole reason for living.'* [20] Dawkins correctly describes DNA replication as a 'self-sustaining process', which it is. Leave it to its own devices and it does naturally replicate (given the right environmental conditions) but this statement overlooks the information stored within the DNA, which dictates what the DNA must do and instructs it *how* to replicate. DNA is self-sustaining once it is in a fully-functional, living system but Dawkins' statement ignores the question of how the system could have formed in the first place. In the same way, inorganic processes will not create complex, organised systems without informational input either.[21] Dawkins maintains that the Universe exists through the combination of chance and necessity. Necessity means that some form of intrinsic information is compelling only one type of outcome to occur. Whereas chance implies that there is no information at all. (See Chapter 2) However this raises the same question once more, which is: where did the information come from to set up the laws which created the *necessity* in the first place?

Thus, in the same way that the Biological Law of Coding states that: *'the function of the product cannot alter the DNA (information) that coded it'* (see Chapter 3), the Physical Law of Coding states that:

17. Atkins, P.: *Creation Revisited.* Harmondsworth, Penguin (1994) p.143 *in* Lennox, J.C.: (2007) p.62.

18. Strobel, L.: *The Case for a Creator.* Zondervan, Grand Rapids, Michigan (2004) pp. 97–109.

19. Davies, P. reported by Clive Cookson: Scientists who glimpsed God. *Financial Times*, April, 29, (1995) p.20 *in* Lennox, J.C.: (2007) p.63.

20. Dawkins, R.: BBC Christmas Lectures Study Guide, London, BBC (1991) *in* Lennox, J.C.: (2007). p.55.

21. Lennox, J.C.: (2007) p.64.

The function of the physical elements in the Universe cannot alter the information that determined their physical properties.

If this were not so, then the constancy of Newton's and Einstein's laws of the Universe would not apply and the physical characteristics of the Universe would be constantly changing.

Stephen Hawking and James Hartle developed a mathematical model to describe the beginning of the Universe and theorised that it came into existence through energy fluctuations in a quantum vacuum.[22] John Lennox explains: *"The expression 'quantum vacuum' can be misleading for someone not familiar with the terminology of physics. For the word 'vacuum' tends to convey the idea that nothing is there at all. A quantum vacuum is a term physicists use for a quantum field in its ground or lowest energy state. It is certainly not 'nothing'."* [23] William Lane Craig, Professor of Philosophy, stated that subatomic particles called: *"'virtual particles'... do not come out of nothing. The quantum vacuum* [constitutes]*...a sea of fluctuating energy, an arena of violent activity that has a rich physical structure and can be described by physical laws. These particles are thought to originate by fluctuations of the energy in the vacuum."* [24] However there is a problem with Hawking and Hartle's solution. Al-Khalili explains that this: *"...obscure yet fascinating, and experimentally proven, effect* [was] *discovered by the Dutch physicist Hendrik Casimir in 1948. It involves a property of what we would consider to be completely empty space. If all the air is pumped out of a chamber then we say that we have a vacuum, meaning that there is no matter inside and hence, I hope you'd agree, zero energy. But down at the quantum level, even the empty vacuum is a busy place...particles and their antimatter partners are continuously popping into existence from nothing before quickly disappearing without a trace. Casimir showed how to harness this process to extract energy from the vacuum even though it has nothing to give."*

"As most of us know only too well, if we borrow money from a bank we must soon pay it back. The rules of quantum mechanics, as expressed within the Heisenberg uncertainty principle, operate in a similar way. But unlike a bank loan where we are free to choose the period over which we make the repayments, the uncertainty principle is rather more strict. It states that energy can be borrowed from the vacuum provided it is paid back very quickly. The more energy that is borrowed, the quicker the debt must be repaid." [25] So people who cite that a fluctuation in the quantum vacuum created the Universe are suggesting that the cancellation of the matter-antimatter particles went wrong and, defying Quantum Mechanics, the entire energy of our universe was extracted from this *"sea of fluctuating energy."* But even if this was possible and all space, mass, energy and time were created this way, there is still no explanation as to where the vast amount of information came from to specify the workings of the Universe and all that is in it, unless there is information embedded in the energy fluctuations. However we do not know of any way that this could occur.

Thus a self-organising universe is beyond the ability of scientific explanation to date. This fact was colourfully expressed by Sir Fred Hoyle in 1981, while discussing the extraordinary properties of the carbon atom, when he summarised the complexity of the Universe this way: *"Would you not say to yourself, 'Some super-calculating intellect must have designed the properties of the carbon atom, otherwise the chance of my finding such an atom through the blind forces of nature would be utterly minuscule.' Of course you would...A common sense interpretation of the facts suggests that a superintellect has monkeyed with physics, as well as with chemistry and biology, and that there are no blind forces worth speaking*

22. Lennox, J.C.: (2007) p.67.

23. *Ibid.* p.183, Ref. 31.

24. Craig, W.L.: *Reasonable Faith.* (Rev. Ed.) Crossway, Wheaton, Illinois (1994) *in* Strobel, L. & Vogel, J.: (2004) p.15.

25. Al-Khalili, J.: (1999) p.212.

about in nature. The numbers one calculates from the facts seem to me so overwhelming as to put this conclusion almost beyond question." [26]

Quantisation

We have already seen that Max Planck was involved in defining Planck Time and Planck Length but more importantly than these, Planck revolutionised physics through his study of black body radiation. He discovered that energy in the Universe came in discrete packets for which he was awarded the Nobel Prize for physics in 1918. These discrete packets are called quanta (singular: quantum). It means that not all energy levels are available as it is only possible to have whole numbers of quanta and it is impossible to have a fraction of a quantum. As time has gone on, more entities in the Universe have been discovered to be quantised. You will recall from Chapter 1 that Edwin Hubble discovered the red shift in the spectra of stars, which suggested that they were moving away from Earth. One of the most surprising and contentious discoveries has been made by Professor Emeritus William Tifft,[27] an astronomer at the University of Arizona, Halton Arp, of the Max Planck Institute for Astrophysics in Germany and others,[28] that the galactic red shifts were quantised, meaning that not all values are possible but only those specified by some (yet unknown) condition of the Universe. If there was no law then any value would be possible and a continuum of red shifts would be anticipated if the red shifts were purely distance related.[27] Quantum Mechanics and the Standard Model explain this quantisation of the red shifts through: *"speculation about quantum fluctuations."* [29] Another explanation is a digital universe predicated upon information. This will be discussed further in Chapter 9.

Quantum Mechanics

Before we broach the topic of Quantum Mechanics I think I need to warn you that even after eighty-five years of research, the best minds in the World are still perplexed by it. Nobel laureate in physics in1965, Richard Feynman (1918–1988) wrote: *"I think I can safely say that nobody understands quantum mechanics"*.[30] Danish physicist and Nobel laureate Niels Bohr was very blunt when he wrote: *"Anyone who is not shocked by Quantum Theory has not understood it."* [31] However before you run off in the opposite direction, I need to point out that it is integral to an understanding of *"life, the Universe and everything"* [32] so we must engage it and wrestle with its concepts even if we still have only a rudimentary understanding of how it works. Jim Al-Khalili explains: *"Despite this it has been the single most successful and important scientific discovery of the twentieth century. Quantum mechanics underpins the whole of modern chemistry and the whole of modern electronics. Without it we would not have been able to understand the structure of crystals, or invent the laser or the silicon chip. Without an understanding of the rules of quantum mechanics there would be no televisions, computers, microwaves, CD players, digital watches and so much more that we take for granted in our technological age."* [33]

Edgar Andrews writes that towards the end of the 19th-Century: *"... it had been widely believed that our physical understanding of the universe was just about complete — there were few riddles left for physics to explain. But then in 1900 the 'quantum theory' burst on the scene, driving a coach and horses through the snug smug world of classical physics and transforming our perception of nature... The basic discovery of quantum theory was that energy can only be transferred in*

26. Hoyle, F.: The Universe: Past and Present Reflections. *Engineering and Science* (Nov 1981) pp.8–12 *in* Sarfati, J.: *By Design*. Creation Book Publishers (2008) p.238.

27. Missler, C.: (2004) p.345.

28. Arp, H., et al: Periodicity of Quasar redshift: *Astron. Astrophys.* 239:33–49 (1990) *in* Journal of Creation 17(2): 94–7 (2003).

29. Fowler, T.B. & Kuebler, D.: *The Evolution Controversy*. Baker Academic, Grand Rapids (2007) p.358.

30. Jamieson, D.: *Accelerated nuclei – probing quantum and geological landscapes* (address to the Australian Academy of Science (21st Feb. 2008) http://www.science.org.au/events/frontiers/frontiers2008/jamieson.html

31. Bohr, N. *in* Gribbin, J.: *In Search of Schrödinger's Cat*. Wildwood House, London (1984) p.5 *in* Kirby, M.D.: Science v Law: The Next Century. *Journal and Proceedings of the Royal Society of New South Wales* 118(1 & 2):51–58 (1985) http://nsw.royalsoc.org.au/journal_archive/kirby.html

32. Adams, D.: *Life, the Universe and Everything* (1982) is the third book in the five-volume *Hitchhiker's Guide to the Galaxy* science fiction series.

33. Al-Khalili, J.: (1999) pp.81–2.

discrete packets called 'quanta'...Quantum theory (or quantum mechanics, as it is more properly called) is a highly successful theory, being borne out experimentally... Quantum theory accounted for many things that classical physics couldn't, such as why hot objects don't radiate an infinite amount of energy; why atoms don't self-destruct by their negatively charged electrons collapsing into their positively charged nuclei; and why light from hot gases is not smeared across the whole rainbow spectrum but is emitted at specific wavelengths (so-called spectral lines, corresponding to distinct colours)." [34] Quantum physicist and Princeton colleague of Einstein, David Bohm wrote: *"The most radical change in the notion of order since Isaac Newton came with Quantum Mechanics. But quantum-mechanical ideas of order contradict co-ordinate order because Heisenberg's uncertainty principle made a detailed ordering of space and time impossible. When you apply quantum theory to general relativity, at very short distances like 10^{-33} cm, the notion of the order of space and time breaks down".* [35] Edgar Andrews explains: *"...the so-called 'uncertainty principle' first propounded by Werner Heisenberg in 1926...means that we can never know everything about even a single particle — let alone all the particles that make up our physical world...once we find our elusive electron...the very act of locating it disturbs it so much that we have no idea what its other properties were before we found it. In other words, once you measure one property of a particle (like its position) you lose all information about some equally important property (like its speed)...Thus the future course of events is unpredictable, not because of our ignorance but on account of the very way the universe is made."* [36]

Stephen Hawking and Leonard Mlodinow point out another fundamental deduction of quantum mechanics when they write: *"Quantum physics tells us that nothing is ever located at a definite point because if it were, the uncertainty in momentum would have to be infinite."* [37] This is consistent with the realisation that at Planck scale, all entities in the Universe have a vibration or wavelike property and are never 'still' as we might imagine that a microscopic particle would be when we observe our macroscopic world. Paul Davies explains: *"Quantum mechanics generally cannot predict* actualities, *only* probabilities. *What it can predict — in principle with perfect accuracy — are wave-function amplitudes, from which the probabilities can be computed."* [38]

As a result, Jim Al-Khalili writes: *"...quantum mechanics ranks even higher than relativity in terms of its importance as a scientific discovery which has affected our everyday lives. The problem is that nobody really understands what it is telling us about the world of the very small which it describes so accurately."* [39] Yet even if they do not understand it, after eighty-five years, it could be imagined that the physicists would have come to some sort of consensus on what it all means but you would be wrong. Al-Khalili continues: *"The interpretation that has held sway for most of the twentieth century is known as the Copenhagen interpretation (as it originated at the institute in Copenhagen where one of the founding fathers of quantum mechanics, Niels Bohr, worked). Its supporters take a very pragmatic view of the whole issue by claiming that we should not worry about trying to understand what is happening down at the scale of atoms which is so far removed from our everyday world...All that matters, say the Copenhagenists, is that quantum mechanics works. The mathematics agrees beautifully with what we see around us in the real world, so why beat ourselves up over it?"* [40]

Although the Copenhagen Model is considered by many to be the 'Standard Interpretation', it has its competitors, the most prominent of which is the 'Many Worlds' interpretation proposed by American physicist Hugh Everett III in the 1950s. Al-Khalili explains: *"According to this view, as soon as a quantum particle, anywhere in the universe, is faced with a choice of two or more options, the*

34. Andrews, E.: (2009) pp.31–2.

35. Bohm, D.: Interview: *On Quantum Mechanics*, (1987) *in* David Bohm Quotes on Quantum Theory *at* http://www.spaceandmotion.com/Physics-David-Bohm-Holographic-Universe.htm

36. Andrews, E.: (2009) p.34.

37. Hawking, S. & Mlodinow, L.: (2010) p.73.

38. Davies, P.: *Universe from bit. in* Davies, P. and Gregersen, N.H. (eds.): (2010) p.81.

39. Al-Khalili, J.: (1999) p.185.

40. *Ibid.* pp.185–6.

whole universe splits into a number of parallel universes equal to the number of options available to the particle. There are, according to this view, an infinite number of universes...There is no experimental evidence whatsoever that parallel universes exist since we cannot make contact with any of the others, but the truth is that it has its good points as a scientific theory and explains away much of the strange behaviour of the quantum world. But at what price? Physicist Paul Davies has remarked that the many-worlds interpretation is cheap on assumptions (a point in its favour) but expensive on universes." [41] David Deutsch of Oxford University has introduced the term: '*The Multiverse*' to describe these multiple parallel universes.

Yet another interpretation of quantum mechanics has been proposed by the American physicist David Bohm, which will be discussed shortly.[42] The number of seemingly bizarre explanations for quantum mechanics may leave you thinking that physicists are a dithery lot who cannot make up their minds, however this conclusion underestimates the scale of the problem. US physicist, Richard Feynman who among other things developed the theory of Quantum Electrodynamics in the 1940's and is widely regarded as one of the most prominent physicists of the 20th-Century wrote: *"It is often stated that of all the theories proposed in* [the Twentieth] *Century, the silliest is Quantum Theory. In fact, some say that the only thing that Quantum Theory has going for it is that it is unquestionably correct."* [43]

To make matters worse though, Bohr's interpretation of quantum mechanics predicted that electrons and other quantum particles only existed when you were looking at them. Does that mean that they do not exist at other times? It makes our usual ideas of reality very tenuous because how can you define what is 'real'? This interpretation disturbed Einstein who wrote: *"I think that a particle must have a separate reality, independent of the measurements. That is, an electron has spin, location, and so forth even when it is not being measured. I like to think that the moon is there even when I'm not looking at it."* [44]

Einstein knew that his mathematics and the Theory of Relativity were correct and so he struggled with Quantum Mechanics which contradicted Relativity Theory even though he was partly responsible for developing Quantum Mechanics. In 1935, Albert Einstein, Boris Podolski and Nathan Rosen devised a 'thought experiment' with the intention of disproving Quantum Mechanics, as the Pauli Exclusion Principle contradicted the Special Theory of Relativity. The experiment describes two quantum particles that were once together so that they are in phase with each other, a state that is described as '*quantum phase entanglement*'. As the two particles move apart from each other in opposite directions, the Pauli Exclusion Principle mandates that the two particles must always have opposite spins, no matter how far apart the particles are; if the spin on one particle changes, the other particle must change also, but in the opposite direction. This communication between the two particles *"does not diminish with distance, cannot be shielded, and travels faster than light."* [45] This has subsequently been dubbed the **E**instein-**P**odolsky-**R**osen experiment or '*EPR paradox*' for short. It is called a '*thought experiment*' because the world of physics had to wait nearly fifty years for the technology to become available to definitively test it and Einstein went to his grave being uncomfortable with Quantum Mechanics, as he was unable to reconcile it with Relativity Theory and the rest of physics.[46]

However in this case, Einstein did not have the last word. Missler writes: *"Bohm remained unperturbed by Einstein's argument. Rather than believing that some kind of faster-than-light communication was taking place, he offered another*

41. Al-Khalili, J.: (1999) p.187.

42. *Ibid.* p.189.

43. Feynman, R.: *in* Missler, C.: *Cosmic Codes.* Koinonia House, Coeur d'Alene, Idaho (2004) p.338.

44. Quoteopia-Einstein. http://www.quoteopia.com/famous.php?quotesby=alberteinstein

45. Herbert, N.: *Bell's Theorem proves Physical Reality must be Non-local. at* <http://mail.cruzio.com/~quanta/bell.html>

46. Einstein, A., Podolsky, B., and Rosen, N.: Can Quantum Mechanical Description of Physical Reality be Considered Complete? *Physical Review* 47:777 (1935).

explanation: 'If subatomic particles do not exist until they are observed, then one could no longer think of them as independent 'things'. Thus Einstein was basing his argument on an error when he viewed twin particles as separate. They were part of an indivisible system, and it was meaningless to think of them otherwise.'"

"In time most physicists sided with Bohr and became content that his interpretation was correct. One factor that contributed to Bohr's following was that quantum physics had proved so spectacularly successful in predicting phenomena, few physicists were willing to even consider the possibility that it might be faulty in some way. The entire industries of lasers, microelectronics, and computers have emerged on the reliability of the predictions of quantum physics." [47]

One of the reasons David Bohm developed his theory of Quantum Mechanics was because of inconsistencies in the existing theories. *"Bohm began his theory with the troubling concern that the two pillars of modern physics, quantum mechanics and relativity theory, actually contradict each other. This contradiction is not just in minor details but it is very fundamental, because quantum mechanics requires reality to be discontinuous, non-causal, and non-local, whereas relativity theory requires reality to be continuous, causal, and local. This discrepancy can be patched up in a few cases using mathematical re-normalisation techniques, but this approach introduces an infinite number of arbitrary features into the theory that, Bohm points out, are reminiscent of the epicycles used to patch up the crumbling theory of Ptolemaic astronomy."* [48]

Bohmian Mechanics is sometimes referred to as the de Broglie-Bohm Theory which *"is a version of quantum theory discovered by Louis de Broglie in 1927 and rediscovered by David Bohm in 1952...*[As usual in] *quantum mechanics...a system of particles is described in part by its wave function, evolving, as usual, according to Schrodinger's equation. However, the wave function provides only a partial description of the system. This description is completed by the specification of the actual positions of the particles. The latter evolve according to the 'guiding equation', which expresses the velocities of the particles in terms of the wave function."*[49] Bohm continues: *"Classical physics says that reality is actually little particles that separate the world into its independent elements. Now I'm proposing the reverse, that the fundamental reality is the enfoldment and unfoldment, and these particles are abstractions from that. We could picture the electron not as a particle that exists continuously but as something coming in and going out and then coming in again. If these various condensations are close together, they approximate a track. The electron itself can never be separated from the whole of space, which is its ground"*.[50] If these ideas seem confusing, do not despair. Using the information concepts which we have already investigated we will be able to make more sense of some of these strange ideas before the end of the chapter. But first we need to understand another interpretation of modern physics called: 'String Theory'.

String Theory

We have discovered the vast gulf that cleaves physics, which divides General Relativity describing the gargantuan world of suns and galaxies and Quantum Mechanics describing the ultra-microscopic world of Plank scale. Al-Khalili summarises it neatly when he writes: *"Quantum mechanics describes the behaviour of the world of the very small, whereas most phenomena that are described within the framework of general relativity tend to involve immense expanses, leviathan masses and titanic forces. Stars, black holes, galaxies, even the whole universe, all rely upon*

47. Missler, C.: *Quantum Physics: The Boundaries of Reality.* (1998) *at* http://www.khouse.org/articles/1998/62/

48. Essay on the Life & Ideas of David Bohm (1917–1992) *at* http://www.spaceandmotion.com/Physics-David-Bohm-Holographic-Universe.htm

49. Bohm, D.: Bohmian Mechanics *at* http://www.spaceandmotion.com/Physics-David-Bohm-Holographic-Universe.htm

50. Bohm, D.: Interview: *On Quantum Physics.* (1987) *in* David Bohm Quotes on Quantum Theory *at* http://www.spaceandmotion.com/Physics-David-Bohm-Holographic-Universe.htm

Einstein's description of gravity and how it affects space and time. They are far removed from the microscopic quantum world...A symbiosis between these two successful descriptions of reality will only be achieved if they can be merged into one unified scheme; an all-encompassing theory of quantum gravity." [51] Einstein was convinced that in the same way that the Special Theory of Relativity in 1905 had described mass and energy very elegantly in the equation: $E = mc^2$, there should also be a 'Theory of Everything', which is now called: 'Unified Field Theory'. It was needed to describe the relationship between the four fundamental forces in the Universe (gravitation, the electro-magnetic force, the strong force and the weak force), which were discussed in Chapter 6. James Clerk Maxwell (1831–1879) had begun this process in 1876 when he demonstrated the relationship between electricity and magnetism and how electromagnetic waves such as light can travel through the 'vacuum' of space.[52] Einstein failed to complete his quest but others have subsequently been more successful. Al-Khalili writes: "*The most mathematically elegant, but at the same time most puzzling, of the candidates for a unified theory that Einstein worked on was due to two mathematicians: a Pole, Theodor Kaluza, and a Swede, Oskar Klein. Kaluza did all the groundwork and, in 1919, sent a paper to Einstein in which he proposed a way of explaining electromagnetic radiation within the framework of general relativity. Kaluza showed that what was needed to achieve this was to write his equations not in 4D but 5D space-time by including an extra space dimension. Although this sounds arbitrary and far removed from what we might consider as reality, it is relatively easy to do mathematically, where we can add as many dimensions as we like. But was this fourth dimension of space that Kaluza was proposing real? We are certainly not aware of it if it is out there. But when he included this extra dimension Kaluza found that light, instead of being an oscillating electromagnetic field through 3D space, was in fact a vibration of this fifth dimension.*"[53] If you are having trouble imagining this, let alone drawing it, do not despair! Professor Al-Khalili has the same problems. He continues: "*So there you go. But don't worry, I don't really understand what this means either. All we can say is that it tries to explain the origin of light at a more fundamental, geometrical level in the same way that Einstein had described gravity as a curvature of 4D space-time. Not only that, but this fifth dimension does not extend in a straight line like the other three dimensions of space but is 'curled up' on itself...The problem of course was that, despite the elegance of the mathematics of Kaluza's theory, there was not a scrap of experimental evidence whatsoever to suggest that this fifth dimension really existed. Even Einstein, while impressed with the way Kaluza had unified light and gravity, was unwilling to believe in the reality of a fifth dimension. After all, he had been rather reluctant to even take on board the idea of four-dimensional space-time to begin with. At least the four dimensions (one of time and three of space) were real. The main reason for Einstein's and others' scepticism was because we never see this extra dimension. This question was answered in 1926 when Oskar Klein suggested that the reason it could not be detected was because it was curled up in a circle so tiny that it was millions of times smaller even than an atom.*" [51] Hawking and Mlodinow add: "*...the extra dimensions are curled up into what is called the internal space, as opposed to the three-dimensional space that we experience in everyday life.*" [54]

This new multi-dimensional theory derived from Kaluza and Klein has come to be known as 'String Theory'. Edgar Andrews writes: "*Many consider the string theory to be the front runner in the Einstein stakes — the race to unify our scientific view of nature.*" [55] Hawking and Mlodinow elaborate: "*According to string*

51. Al-Khalili, J.: (1999) p.238.

52. Andrews, E.: (2009) p.111.

53. Al-Khalili, J.: (1999) pp.241–2.

54. Hawking, S. & Mlodinow, L.: (2010) p.116.

55. Andrews, E.: (2009) p.46.

theory, particles are not points, but patterns of vibration that have length but no height or width — like infinitely thin pieces of string." [56]

Little further progress occurred until in 1968, Italian physicist, Gabriel Veneziano while studying at the European Centre for Theoretical Physics (CERN) in Switzerland found a mathematical formula that described the behaviour of the strong force. Dr Elizabeth Finkel relates: *"The formula, turned out to describe not 'point particles' but vibrating strings."* [57] Andrews continues: *"The theory was born in the late 1960s and early 1970s to explain the behaviour of subatomic particles which experience the strong nuclear force. Yoichiro Nambu (followed by Lenny Susskind and Holger Nielson) realised in 1970 that these strong interactions could be described by a quantum-mechanical model in which particles were not just non-dimensional point objects but one-dimensional strings."* [55]At this point Al-Khalili takes up the story once more: *"After many decades in the wilderness, Kaluza-Klein theory made a comeback in the late 1970s...*[however] *It was not enough for it* [just] *to unify gravity and light. By that time it had been established beyond doubt that all phenomena in nature could, at the most fundamental level, be described by four forces...So the ultimate theory being searched for in the late '70s was one that not only unified gravity with electromagnetism, as Kaluza-Klein theory did, but which also encompassed the two nuclear forces...The reason Kaluza-Klein theory came back into fashion was because of its clever way of being able to unify forces when higher dimensions of space were included in the equations. Of course with four forces to deal with, instead of just two, more than one extra dimension would be required."* [58]

Elizabeth Finkel explains more of the history: *"...in 1984 — after some more mathematical alchemy — John Schwartz at California Institute of Technology and Michael Green at Queen Mary's College in London, found that string theory could also encompass all the other forces."* [57] Missler adds: *"Sometimes these one-dimensional 'strings' vibrating in 10 or 11 dimensions are referred to as "superstrings."* [59] Al-Khalili explains: *"...superstring theory...quickly developed into the most sophisticated, elegant, complicated, powerful and obscure theory ever devised. After all, it was a theory of ten dimensions. If correct, it stated that we lived in a 10-dimensional universe. But now all six extra spatial dimensions would be curled up into a tiny high dimensional sphere that we could never detect, leaving just the four dimensions of space-time. Superstring theory was so named because it suggested that everything is ultimately made of tiny strings which vibrate in 10 dimensions. This may sound crazy but it does achieve the unification of general relativity with quantum mechanics which is, after all, the holy grail of physics."* [60]

Andrews sums up by saying: *"...all the fundamental particles that make up our universe - protons, neutrons, electrons, quarks, mesons, photons, neutrinos etc.; and even the hypothetical 'gravitons' that are supposed to carry the force of gravity - could be regarded as minute 'strings'. That is, instead of thinking of a proton (for example) as a dot it should be regarded as a short string which can vibrate like a violin string...* [T]*he minute strings of string theory can vibrate in different ways to produce...different fundamental particles. So one string vibrating in three different ways can 'be' three different fundamental particles (though not all at once). No wonder particles are so thick on the ground. On this assumption, the theory claims, we can explain all the fundamental particles of nature and the way they interact,...because different vibrational modes correspond to different fundamental particles, the theory neatly explains why there are so many different kinds of particle, and why one kind can decay into one or more other kinds."* [61] Most importantly, the thing that string theory has been able

56. Hawking, S. & Mlodinow, L.: (2010) p.115.

57. Finkel, E.: The Science of String Theory. *Cosmos* **2** (2005) p.77.

58. Al-Khalili, J.: (1999) p.243.

59. Missler, C.: *Cosmic Codes.* (2004) p.342.

60. Al-Khalili, J.: (1999) p.244.

61. Andrews, E.: (2009) pp. 46–7.

to do, which none other has been able to do to date, is that because it predicts the existence of the hypothesized but yet to be discovered theoretical particle that transmits gravitational force (dubbed a *graviton*), it can neatly tie the four fundamental forces of the Universe into one package.[61]

But wait there is more! Now that you know what String Theory is, the next step is '*String Field Theory*'. Finkel continues: "*String field theory is to string theory what the electromagnetic field is to photons: a way of describing how strings act coherently in the real world. In 1974, Michio Kaku* [62] *and Keiji Kikkawa demonstrated that it could work in 10 dimensions. Nearly 20 years later, in 1993, Ed Witten at Princeton University in the US...showed that string theory worked best in 11 dimensions.*" [57] The strings are calculated to be 10^{-33} centimetres in size, which is the theoretical minimum size of particles that possess physical characteristics within the Universe.[55] And finally, if you did not already think that String Theory is as complicated as you can make it, Andrews points out that others now suggest that Superstring Theory predicts that there may be twenty-six dimensions instead of ten or eleven dimensions; it all depends on how you interpret the mathematics! [63] Before we leave String Theory, it needs to be stated that String Theory is not universally accepted by all physicists. Andrews explains: "*Unlike general relativity and quantum mechanics, string theory remains (at the time of writing) a hypothesis - a foundational idea that promises to powerfully unify our perception of the universe but still remains untested in many critical respects.*" [64]

Jim Al Khalili explains: "*It seems that it is just too hard for anyone to fully comprehend. Its mathematics are beyond the current ability of most, if not all, mathematicians. Not only that, but by the early 1990s there were five different versions of superstring theory and no one knew which was the correct version, or indeed whether there was a unique version.*"

"*Then in 1995, a scientist who has been dubbed 'the smartest person on Earth' found an answer, maybe THE answer. His name is Edward Whitten and he works at the Institute for Advanced Study in Princeton, New Jersey (which was where Einstein spent his latter years). Together with a colleague, Paul Townsend of Cambridge University, Whitten believes he has discovered why there are so many versions of superstring theory. The price that has to be paid is relatively cheap under the circumstances. Whitton asks for just one more dimension! With eleven dimensions instead of ten many of the problems of superstring theory go away. Now the tiny strings are replaced by sheets known as membranes and the new theory of Whitten and Townsend is called membrane theory, or M-theory for short.*" [60]

Hawking and Mlodinow acknowledge that M-theory does not equate to Einstein's dream of a single unified model but describe it as the 'M-theory network' which they believe is the basis of a future grand unified theory of everything.[65] Al Khalili then asks: "*But could we just keep on going? What if other versions of M-theory were discovered? Maybe adding a twelfth dimension would cure things. In fact, why not just chuck in another handful of dimensions just to be on the safe side! It turns out that this is not possible. There is something mathematically very special about the ten dimensions of superstring theory and the eleven dimensions of M-theory.*"

"*...One of the main questions to be answered of course is why and how all the extra dimensions get curled round and squeeze down leaving just the four dimensions we see. Current thinking is that this would have happened at the moment of the Big Bang. This implies that there was something before the Big Bang. Maybe our three dimensions of space and one of time were part of a much grander ten-or eleven-dimensional universe in which all the forces of nature were unified into one. The Big*

62. Kaku is Professor of Theoretical Physics at City College, New York.

63. Andrews, E.: (2009) p.26.

64. *Ibid.* p.54.

65. Hawking, S. & Mlodinow, L.: (2010) p.58.

Bang then caused six or seven dimensions of space to be crushed down to a size that we would never be able to access." [66]

So we are left with lots of theoretical physics and a dearth of experimental evidence. Al-Khalili summarises the situation eruditely when he writes: *"The mathematical formalism is right, it is just what the equations mean that is not properly understood, and that is a matter of philosophy not physics. The majority of physicists today believe that despite there being a whole host of different interpretations of quantum mechanics on the market, all of which are equally valid, the underlying mathematical framework is correct and it is purely a matter of personal taste which interpretation an individual subscribes to. Whether you believe the Copenhagen interpretation in which nothing exists until it is observed, or the many-worlds interpretation in which the universe splits into an infinite number of copies, or the Bohmian interpretation in which signals travel faster than light or even, more recently, the transactional interpretation in which signals travel backwards in time, it doesn't matter. No experiment has yet been devised that can discriminate between these rival views."* [67] Thus questions regarding String Theory and M-theory are still unresolved but the apparent 'contradictions' pertaining to the dual theory of light might provide a clue.

The Light Conundrum

While studying physics in high school, we were taught that light is both a particle (with mass) and a wave (without mass) each exclusively and yet simultaneously. This contradiction is known to be true yet it is apparently impossible and is believed by all modern physicists! No wonder I found physics hard; in comparison, medicine was really easy. At school we were told just to accept it, as it had been proven to be correct. (Accepting facts which are impossible can be a challenge for a teenager!) How can we ever hope to make sense of this obvious oxymoron? Well as in many things in physics we have to go back to Isaac Newton. From his prism and other experiments, Newton inferred that: *"light was composed of a stream of tiny particles he called corpuscles."* [68] He noted that sound waves travel around corners but light did not, so he inferred that light was not a wave. English physicist, Thomas Young in the early 1800s found that light did indeed bend when he first performed the double slit experiment to create a diffraction pattern, with light and dark bands on a screen, which are called 'interference fringes'. In 1873 *"... the Scottish physicist James Clerk Maxwell developed a set of equations which showed that all the light we see is a form of electromagnetic radiation,"* [69]

By 1890, scientists thought they had solved all the major problems in physics until Albert Einstein came along in 1905 and described the photoelectric effect and demonstrated that *"light at its most fundamental level is made up of tiny entities called photons."* [70] So how can light be both a particle and a wave, depending on which experiment is performed? Incidentally, this weird duality that applies to light photons was also found to apply to other fundamental entities such as electrons. In 1897, in Cambridge, Joseph John (JJ) Thomson (1856–1940) showed that *electrons* were *particles*, for which he received the Nobel Prize in Physics in 1906. His son, George Paget Thomson (1892–1975) subsequently proved that *electrons* were *waves*, for which he received the Nobel Prize in Physics in 1937! [71] These are experiments that can be conducted by high school physics students: if a single photon of light is directed at an appropriately treated screen such as in a television, then the single packet of light can be seen as a dot on the

66. Al-Khalili, J. (1999) p.245.
67. *Ibid.* p.247.
68. *Ibid.* p.79.
69. *Ibid.* p.80.
70. *Ibid.* p.81.
71. Missler, C.: (2004) p.336.

screen while a stream of electrons can turn a light-weight vane in a vacuum tube, which is consistent with light/electrons being particles but if an identical single photon (or electron) is then directed through parallel slits in a barrier, it would have to pass through one slit or the other if it was a particle but perplexingly it passes through *both simultaneously* and will create a diffraction pattern after passing through the double slit screen, which is a property unique to waves not particles. So where does this leave physics? Both Thomsons, father and son, were correct and deserved their Nobel prizes; it had been proven *beyond doubt* that light is a wave without mass and a particle with mass both at the same time. Thus was born the 'wave/particle duality' of light. It all depends on how you investigate it. This seems impossible but it is true. Ernan McMullin, Professor Emeritus of Notre Dame University, USA summarised the situation when he wrote: *"Roughly speaking, electrons seem to travel as waves but to interact as particles."* [72] In the 1940s, David Bohm was researching plasma physics at Berkeley and later Princeton University as a protégé of Einstein and was intrigued by the contradictions between Einstein's Theory of Relativity and Bohr's Quantum Mechanics which, as already discussed, are both known to be true but explicitly contradictory as well. Bohm summarised it as: *"In relativity, movement is continuous, causally determinate and well defined; while in quantum mechanics it is discontinuous, not causally determinate and not well defined."* [73]

Bohm began his career researching plasma, which is the fourth state of matter after solids, liquids and gases. Plasma is *"an extremely hot and highly ionised state of matter."* [74] Bohm observed that at the subatomic level in plasma: *"the particles stopped behaving like individuals and started behaving as if they were part of a larger and interconnected whole."* [75] Bohm believed that at the subatomic level location ceased to exist (often expressed as having 'non-locality') and he hypothesised that hidden from our view was a deeper order underlying the physical universe. He called this the 'Implicate (folded) Order', which he distinguished from the visible or 'Explicate (unfolded) Order'. Bohm's hypothesis was subsequently researched by Irish physicist, John Stuart Bell (1928–1990) at CERN in Geneva; in 1964, Bell formulated a mathematical approach demonstrating the non-locality.

U.S. physicist, Nick Herbert summarises Bell's Theorem as: *"...despite the fact that Relativity prohibits instantaneous connections, despite the fact that no such connections have ever been observed either in EPR experiments or any other, despite the fact that quantum theory itself predicts no observable instant connections, despite all these considerations from Fact and Theory, the Reality of the EPR particles is such that their initial contact must create an instantaneous [non-local] link between them below the level of appearances. Bell's Theorem (reality must be non-local) is remarkable for several reasons: 1) it is a mathematical proof, not a conjecture or speculation; 2) it is a proof about Reality not Appearances...[and] 3) it is counter-intuitive: why should everywhere local facts need to be supported by a non-local reality?"* [76]

Jim Al-Khalili reconciles Bell's approach with the mutual incompatibility of Bohmian Mechanics and Special Relativity when he writes: *"As a research student my hero was the late John Bell, an Irish theoretical physicist and one of the twentieth century's leading experts on quantum mechanics. He was also, as far as I am concerned at any rate, the voice of reason when it came to the interpretation of quantum mechanics. During the 1920s, the two giants of physics, Niels Bohr and Einstein, had a long running debate about the meaning of the then new theory. Einstein argued that quantum mechanics could not be the last word and that there had to be something*

72. McMullin, E.: *From matter to materialism…and (almost) back. in* Davies, P. and Gregersen, N.H. (Eds.): (2010) p.26.

73. Bohm, D.: *On Quantum Mechanics, Wholeness and the Implicate Order.* (1980) *in* David Bohm Quotes on Quantum Theory *at* http://www.spaceandmotion.com/Physics-David-Bohm-Holographic-Universe.htm

74. Andriessen, M. et al.: *Preliminary Course in Physics 1* (2nd Ed.) Jacaranda, Milton, Qld (2004) p.263.

75. Missler, C.: (2004) p.338.

76. Herbert, N.: *Bell's Theorem Proves Physical Reality must be Non-local. at* http://mail.cruzio.com/~quanta/bell.html

missing, while Bohr claimed that quantum mechanics told us all we could ever know about nature. Bohr was convinced that physical theories do not describe reality directly only what we can know about reality. His version of quantum mechanics became known as the Copenhagen view since that was where his institute was based. Einstein, on the other hand, felt that a good theory had to be ontological in that it described how reality really is. Quantum mechanics should not be any different. It is generally acknowledged that Bohr won that debate and since then generations of physicists have followed the Copenhagen view." [77] Despite this, Bohr's view *"...troubled Bell since he felt that physics should be about trying to understand the deeper meaning of what was going on in nature...I remember once asking Bell a question after a lecture he gave at Queen Mary College London. He had just argued that he was quite a fan of David Bohm's interpretation of quantum mechanics which describes the whole universe as being interconnected on the quantum level so that something happening to an atom here on Earth might instantaneously affect another atom in a different galaxy. This type of connectivity between all the particles in the universe is known as non-locality, or action-at-a-distance, and would require some sort of signalling that must travel faster than light. But surely, I asked Bell, this violates Einstein's special theory of relativity. He replied that he would rather give up special relativity than reality itself, which is literally the price one has to pay if the Copenhagen view is to be believed. You see, according to Bohr nothing even exists in the quantum world until we have measured it and observed it, and since everything is ultimately made up of quantum objects anyway, then nothing (not even the next page of this book) exists until we look at it. Bell maintains that if this were not the case, where would we draw the line between the microscopic world that obeys the quantum rules and the macroscopic world of everyday life?"* [78]

In 1982, physicist Alain Aspect with Jean Dalibard and Gerard Roger at the Institute of Theoretical and Applied Optics in Paris conducted the Two Particle Experiment which confirmed non-locality and definitively answered the EPR question. Missler writes: *"They produced a series of twin photons by heating calcium atoms with lasers, allowed each photon to travel in opposite directions through 6.5 metres of pipe and pass through special filters that directed them towards one of two possible polarisation analysers. It took each filter 10 nanoseconds to switch between one analyser or the other, about 30 nanoseconds less than it took light to travel the entire 13 metres separating each set of photons. In this way Aspect and his colleagues were able to rule out any possibility of the photons communicating by any known physical process."*

"The experiment succeeded. Just as quantum theory predicted, each photon was still able to correlate its angle of polarisation with that of its twin. This meant that either Einstein's ban against faster-than-light communications was being violated or the two photons were nonlocally connected." [79] Author Michael Talbot (1953–1992) explained: *"The problem with this feat is that it violates Einstein's long-held tenet that no communication can travel faster than the speed of light. Since travelling faster than the speed of light is tantamount to breaking the time barrier, this daunting prospect has caused some physicists to try to come up with elaborate ways to explain away Aspect's findings."* [80]

So in summary, the four leading views described by Al-Khalili all have problems:

1. The Copenhagen interpretation makes us believe in unreality (the next page of this book really does exist!)
2. A Multiverse results from the infinite splitting of our universe into new

77. Al-Khalili, J.: (1999) pp.247–8.

78. *Ibid.* p.249.

79. Missler (2004) pp.339–340.

80. Talbot, M.: *The Universe as a Hologram...does objective reality really exist...or is the universe a phantasm?* (1991) http://www.universenews.net/the-universe/the-universe-as-a-hologram-by-michael-talbot-state-of-earth/ See also Talbot, M.: *The Holographic Universe.* Harper & Collins (1991).

universes every time a choice occurs between two alternatives anywhere in the Universe. (Creating yet more universes out of nothing!)

3. Bohmian mechanics implies signals travel faster than light contradicting Special Relativity (which is proven to be correct!)
4. The transactional interpretation describes *"signals travelling backwards in time"* (which is an impossibility in our space-time domain due to the Second Law of Thermodynamics!)[77]

How can we decide? Are all physicists mad? Well, before we call for the men in white coats to take all the physicists to the mental institution we need to return to Chapter 1. Is there any experimental evidence as per Figure 1.4? The truth is that there is not a lot; all we have is Alain Aspect's Two Particle Experiment which would seem to vindicate Bohm's non-locality and option 3, while violating Einstein's Special Relativity which is not allowed. So is there any other way around Special Relativity and option 3? Well, possibly but this time we need to learn about holograms.

Holograms

Although Albert Einstein intimated that lasers were possible in 1917, the first working laser was not demonstrated until 1960, when physicist Theodore Maiman, working at the Hughes Research Laboratories in California used a ruby crystal to produce coherent pulses of light.[81] In 1962, lasers were first used to produce holograms and another fascinating feature of our universe was discovered. A true hologram is like a three-dimensional photograph but *with a difference*. It is a recorded image in which the information that describes the object is distributed or disseminated throughout the medium that stores it. The recording medium that most people are familiar with is a sheet of photographic film. A hologram is created through a type of lens-less photography, by shining a laser beam at an object, such as a rose, which simultaneously illuminates a sheet of photographic film. Dr Chuck Missler continues: *"The film then records the interference between the light waves hitting it directly and the light waves reflected from the object. It is in effect a frequency record rather than a spatial image."* [82] The spatial image is what we see in a conventional photograph, which is clearly identifiable in natural light. In contrast, the holographic film, when developed and viewed in natural light, looks like a blurry, cloudy photographic accident where nothing in particular is distinguishable, as if the camera had been jerked suddenly at the instant the shutter was activated. However, if the holographic film is illuminated with laser light of the same frequency that was used to create it, then the image comes into crisp focus. Missler writes: *"...the result is astonishing: it appears as a window into the three-dimensional space containing the image of the original object! As one moves their eye, one can see around corners;...(the hologram is a Fourier transform of the image space: it is a recording of the frequency information rather than the space-time information recorded by conventional photographic techniques.) The hologram exhibits some very profound properties beyond the three-dimensionality of the image. In fact, it is one of the most profound means to distribute information throughout a given media. All of the information it contains is distributed over the entire image surface. One can remove a portion of the* [photographic sheet on which the] *hologram* [is recorded] *without losing the image! Drill a hole in the hologram, and one can still view the entire object by simply moving ones eye to a more convenient angle. (Some resolution, or*

81. Cowen, C.: The Light Fantastic. *Cosmos* 34:59 (2010).

82. Missler, C.: Exploring the Mind of Man: The Human Brain. p.1 *at* http://www.khouse.org/articles/1999/229/

sharpness, will be lost however.) Cut the film into pieces, and each piece contains the complete image." [83] As a result, holograms have been described as: *"the whole in every part".*

Michael Talbot takes up the story: *"The 'whole in every part' nature of a hologram provides us with an entirely new way of understanding organisation and order. For most of its history, Western science has laboured under the bias that the best way to understand a physical phenomenon, whether a frog or an atom, is to dissect it and study its respective parts. A hologram teaches us that some things in the universe may not lend themselves to this approach. If we try to take apart something constructed holographically, we will not get the pieces of which it is made, we will only get smaller wholes. This insight suggested to Bohm another way of understanding Aspect's discovery. Bohm believes the reason subatomic particles are able to remain in contact with one another regardless of the distance separating them is not because they are sending some sort of mysterious signal back and forth, but because their separateness is an illusion. He argues that at some deeper level of reality such particles are not individual entities, but are actually extensions of the same fundamental something."* [80] The difficulty in visualising how a holographic universe would fit together can be gauged by Hawking and Mlodinow comment: *"...if a theory called the holographic principle proves correct, we and our four-dimensional world may be shadows on the boundary of a larger, five-dimensional space-time."* [84]

Further, it was discovered that Fourier transforms can be used to mathematically describe the holographic image (information) which is dispersed over the photographic plate. Missler explains: "[A] *hologram is a* virtual *image, an image which appears to be where it is not. It requires no more real space than the three-dimensional image you see of yourself in a mirror."* [85] The most unusual thing about a hologram is that the information in the image is not in one particular place but it is spread over the entire plate. Therefore, as explained above, if the photographic plate was cut in half, there are not two halves of one picture but two complete pictures, although each has less resolution than the original image because half the original information lies in each of the smaller images. In this example of the hologram, it can be seen that information is dispersed evenly over the film. In a sense, the information that makes up the holographic image has *non-locality* with respect to the photographic plate on which it is recorded.

Michael Talbot continues: *"To enable people to better visualise what he means, Bohm offers the following illustration. Imagine an aquarium containing a fish. Imagine also that you are unable to see the aquarium directly and your knowledge about it and what it contains comes from two television cameras, one directed at the aquarium's front and the other directed at its side. As you stare at the two television monitors, you might assume that the fish on each of the screens are separate entities. After all, because the cameras are set at different angles, each of the images will be slightly different. But as you continue to watch the two fish, you will eventually become aware that there is a certain relationship between them. When one turns, the other also makes a slightly different but corresponding turn; when one faces the front, the other always faces toward the side. If you remain unaware of the full scope of the situation, you might even conclude that the [two] fish must be instantaneously communicating with one another, but this is clearly not the case. This, says Bohm, is precisely what is going on between the subatomic particles in Aspect's experiment. According to Bohm, the apparent faster than light connection between subatomic particles is really telling us that there is a deeper level of reality we are not privy to, a more complex dimension beyond our own that is analogous to the aquarium. And, he adds, we view objects*

83. *Ibid.* p.2.

84. Hawking, S. & Mlodinow, L.: (2010) p.44.

85. Missler, C.: (2004) p.330.

such as subatomic particles as separate from one another because we are seeing only a portion of their reality. Such particles are not separate 'parts', but facets of a deeper and more underlying unity that is ultimately as holographic and indivisible as the previously mentioned rose." [80]

When David Bohm discovered holograms, he recognised that they might be a model for his Implicate/Explicate Orders. In a hologram, the objects we see in our visible/tangible universe are Fourier transform projections from the hidden digital information that is recorded on the photographic plate. In a similar way, Bohm suspected that hidden digital information underlies the Universe in an as yet imperceptible domain, which Hawking and Mlodinow implied above may be a curled fifth dimension in Space-Time. Some of these variably tangible, yet intangible entities, with the ways that information may present itself in our universe, are summarised in Table 8.1 below:

Bohm's Explicate and Implicate Orders

Entity	The Material	The Non-material
Light	Physical particle	Intangible vibration
Physics	Relativity Theory	Quantum Mechanics
Dimensions in String/M-theory	Four measurable	Six/seven curled (unmeasurable)
Bohm's hypothesis	Explicate Order	Implicate Order
Hologram	Physical object	Holographic image

Table 8.1: Entities that possess both Material and Non-material elements
Explanation of these entities is covered in Figures 8.1, 8.5 and the accompanying text.

The Explicate/Implicate Orders can be thought of as being like a piece of paper covered in iron filings where the iron filings can be controlled by an unseen magnet underneath the piece of paper. The Explicate Order is above the surface of the paper and can be thought of as the tangible world in which there is matter which can be touched and felt. The Implicate Order is like the space below the piece of paper, except that it is 'curled' in the terminology of physics, so it cannot be measured directly. If we could detect or measure it however, Bohm suggests that we would find it was one dimensional with only vibrating waveforms, somewhat analogous to magnetism or light yet not the same. These vibrating waveforms (think 'digital information') control the material world in a similar way that the hidden magnet controlled the iron filings. In the case of the iron filings, the control mechanism is magnetism but there must be some other linkage between the information waveforms in the Implicate Order and the matter in the Explicate Order. The most difficult concept to understand is that the Implicate and Explicate Orders run in parallel or more accurately, they are co-mingled; they always act together like Siamese twins. If you have trouble in imagining this do not worry. Thinking in five dimensions is not intuitive or easy but read on.

David Bohm explains: *"The present state of theoretical physics implies that empty space has all this energy and that matter is a slight increase of that energy and therefore matter is like a small ripple on this tremendous ocean of energy, having some relative stability and being manifest. Therefore my suggestion is that this implicate order implies a reality immensely beyond what we call matter. Matter itself is merely a ripple in this background."* [86] New York based science writer, Richard Panek, explains: *"Scientists reached a consensus in the 1970s that there was more to the universe than meets the eye. In computer simulations of our galaxy, the Milky Way, theorists found*

86. Bohm, D.: *Dialogues with Scientists and Stages: the Search for the Unity.* in Webber, R.: *Quotable Spirit.* p.218.

that the centre would not hold - based on what we can see of it, our galaxy doesn't have enough mass to keep everything in place. As it rotates, it should disintegrate, shedding stars and gas in every direction. Either a spiral galaxy such as the Milky Way violates the laws of gravity, or the light emanating from it — from the vast glowing clouds of gas and the myriad stars — is an inaccurate indication of the galaxy's mass. But what if some portion of a galaxy's mass didn't radiate light? If spiral galaxies contained enough of such mystery mass, then they might well be obeying the laws of gravity. Astronomers dubbed the invisible mass 'dark matter.'" [87] Dark matter was first proposed by Swiss-American astronomer, Fritz Zwicky in 1933; he referred to it as the "*missing mass*". Dark energy was named by University of Chicago cosmologist, Michael S. Turner in 1998.[88,89] Astrophysicist John Hartnett elaborated on this topic further when he wrote: "*The 'standard model' now seems to demand that the universe is about 5% ordinary matter, which is observed through telescopes; 22.5% is dark matter, which is not observed; and the remaining is a mysterious dark energy, 72.5%.*" [90] The problem has been how to study that which can neither be seen nor detected physically. Panek continues: "*One way to study it is to measure so-called baryon acoustic oscillations. When the universe was still in its infancy, a mere 379,000 years old, it cooled sufficiently for baryons (particles made from protons and neutrons) to separate from photons (packets of light). This separation left behind an imprint — called the cosmic microwave background — that can still be detected today. It includes sound waves (acoustic oscillations) that coursed through the infant universe. The peaks of those oscillations represent regions that were slightly denser than the rest of the universe. And because matter attracts matter through gravity, those regions grew even denser as the universe aged, coalescing first into galaxies and then into clusters of galaxies. If astronomers compare the original cosmic microwave background oscillations with the distribution of galaxies at different stages of the universe's history, they can measure the rate of the universe's expansion.*" [91]

It can therefore be seen that there are both detectable and undetectable elements within our universe. The physical elements are relatively well understood but the undetectable entities such as dark energy and dark matter have continued to evade scientists. They add to the growing list of entities that are difficult to explain in purely physical terms such as information, life itself, thought and awareness but fit well with Bohm's Implicate Order. Figure 8.1 [92] categorises some of these entities.

87. Panek, R.: Invisible Force. *Cosmos* 35:70 (2010).

88. Catchpole, H.: Phantom Empire. *Cosmos* 34:40 (2010).

89. Encyclopaedia Britannica *at* eb.com

90. Hartnett, J.: Cosmologists Can't Agree and Are Still In Doubt! *Journal of Creation* 16(3):21–6 (2002).

91. Panek, R.: Invisible Force. *Cosmos* 35:73–4 (2010).

92. Gosler, A.: Surprise and the Value of Life. *in Berry, R.J.: (Ed.) Real Scientists Real Faith.* Monarch Books, Oxford (2009) p.185.

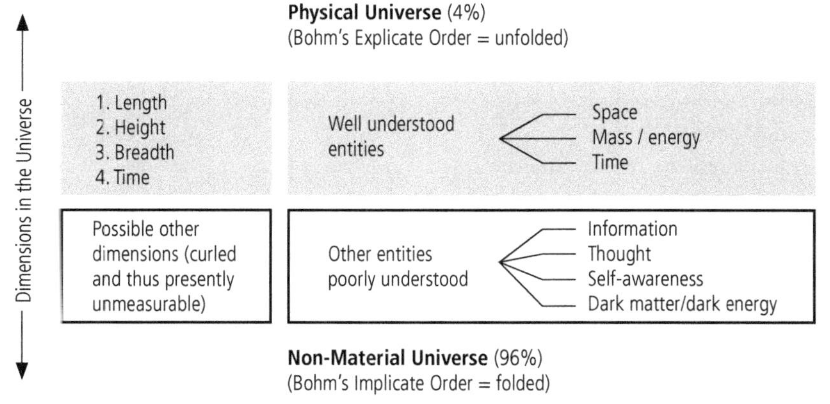

Figure 8.1: Our Information dependent Physical and Non-Material Universes
Andrew Gosler of Oxford University has described a hierarchy of significant qualities dependent on information. They are: existence, life, complex life, conscious life, self-aware life and self-contemplating life.[92]

Fourier Transforms and the Universe

Fourier transforms were mentioned previously as a mathematical tool that describes the relationship between physical and holographic images and a basic understanding of Fourier transforms will aid in understanding the wave/particle duality of light. Remember that a Fourier transform is the representation of an image through the summation of different waveforms of varying shapes, sizes and frequencies. As a result, a Fourier Transform is able to decompose or separate a signal into individual frequencies.[93]

Randall Munroe

Figure 8.2

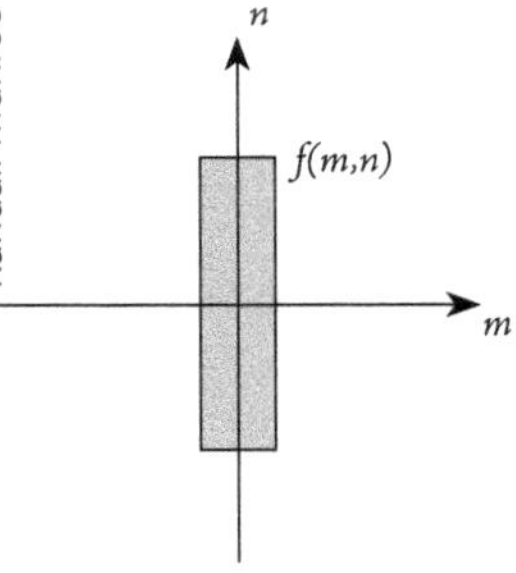

Figure 8.3

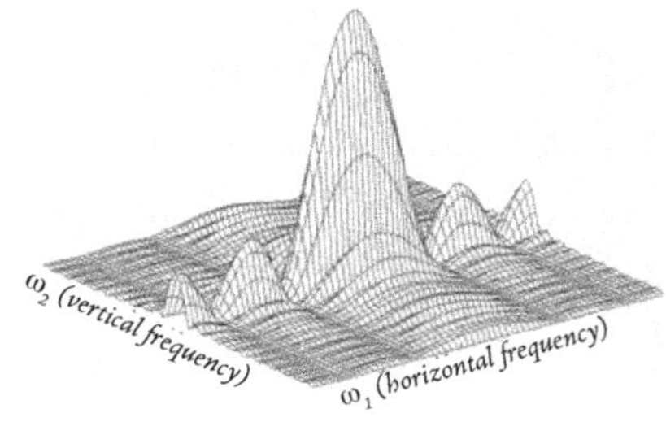

Figure 8.4

Figure 8.2: Visualizing the Fourier Transform [94] Visualising an unfamiliar concept is not always easy but sometimes a picture can convey an idea where words struggle to impart the full meaning.

Figure 8.3: Rectangular Function [95] Figures 8.3, 8.4 and 8.8 illustrate how a 3D shape can be progressively generated through the mathematics of Fourier Transforms. It is not necessary to understand the mathematics but merely to appreciate that it can be done. *[Reprinted with permission of The MathWorks, Inc.]*

Figure 8.4: Magnitude Image of a Rectangular Function This is a mesh plot of the magnitude of the Fourier transform of the rectangular function shown in the preceding figure. The mesh plot of the magnitude is a common way to visualize the Fourier transform.[95] *[Reprinted with permission of The MathWorks, Inc.]*

Alan Aspect's Two-Particle Experiment results suggest that matter has holographic properties yet they are not quite the same as a hologram which is a non-material projection of a material entity (for example the rose described previously). Instead, Bohm's model is a projection of a physical object from a non-material entity in the underlying (hidden) frequency domain. This is the reverse of a hologram and maybe should be called an 'infogram' or 'mattergram' as it is a projection of the information from the Implicate Order. What we perceive as a 'solid' entity, such as a subatomic particle, like a quark, was described by David Bohm as a *'spherical standing wave'* which is connected to its frequency domain 'shadow' by some process that seems to be analogous to a Fourier transform. Each subatomic particle in our space-time domain has a counterpart that is like its 'shadow' in the frequency domain. These non-material entities combine with their neighbouring subatomic particles to form atoms, molecules and ultimately life as we know it. In Bohm's words: *"Put another way, electrons and all other particles are no more substantive or permanent than the form a geyser of water takes as it gushes out of a fountain. They are sustained by a constant influx from the implicate order, and when a particle appears to be destroyed, it is not lost. It has merely enfolded back into the deeper order from which it sprang."* [96] As this pattern is built up to the size of the Universe, Bohm says, it can be seen that we live in a (digital) 'holographic' universe. This relationship is illustrated in Figure 8.5.

In this 2D representation, the material dimensions 1 to 4 are shown above the dotted material/non-material dimensions boundary in the space-time domain and the non-material dimensions are below the boundary in the frequency

93. Mathworks.com *at* http://www.mathworks.com/help/toolbox/images/f21-17064.html

94. Visualising the Fourier Transform. by Randall Munroe at http://xkcd.com/26/

95. Reprinted with permission of The MathWorks, Inc.: http://www.mathworks.com/help/toolbox/images/f21-17064.html

96. Essay on the Life & Ideas of David Bohm (1917–1992) *at* http://www.spaceandmotion.com/Physics-David-Bohm-Holographic-Universe.htm

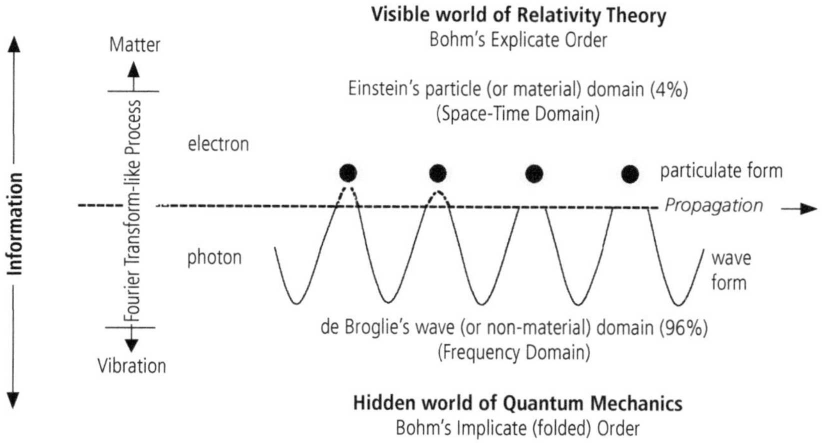

Figure 8.5: The Duality of Light showing the Explicate and Implicate Orders

domain. Strangely, this relationship appears to have been perfectly described in 1637 by Descartes who wrote: *"I am not just lodged in my body like a pilot in his ship, but I am intimately united with it, and so confused and intermingled with it that I and my body compose, as it were, a single whole."* [97]

As unusual as Bohm's conclusions are, he was not the first to suggest this idea. Entomologist and philosopher, Bernard d'Abrera points out: *"Others, following the work of Paul Dirac (1902–84), suggest the theoretical existence of an 'anti-universe' composed of Dirac's 'anti-particles' (=positrons, antiprotons etc.)."* [98] Dirac was a British physicist and Professor of Mathematics at Cambridge who won the Nobel Prize in 1933 (with Schrodinger) for his work on the development of Quantum Mechanics.[99] Dirac's anti-universe appears to have been different to Bohm's concept but the ideas were not completely dissimilar.

Because direct measurements can only be made in the space/time domain, when we look at the vacuum of space between galaxies, space appears relatively empty apart from some interstellar dust but in Bohm's Implicate Order there is a whole other dimension of dynamic flux which is forever flowing and changing. It could be construed that the long imagined *'ether'* that was hypothesized to link everything together does actually exist but in the non-material and currently unmeasurable frequency domain. All things in the Universe arise from it and eventually all things within our universe will dissolve back into it. When the Universe formed, the Implicate Order would appear to have formed first, composed of digital information, out of which precipitated the material Explicate Order. The two orders now appear to be locked together, back-to-back as it were, even though they are very different and reside in completely different domains. Trying to separate them in our four dimensional space-time domain would be like trying to detach an object's dimension of width from its height and depth; within the current space-time continuum they cannot be disconnected.

Inverse Law of Quantum Expression

Stephen Hawking and Leonard Mlodinow in their book: *The Grand Design*, describe how in 1999 a team of Austrian physicists fired tiny clusters of carbon atoms, called buckyballs that contained sixty carbon atoms, through a double slit and found that the buckyballs recreated the classical interference patterns as light waves do.[100] To help explain the analogy Hawking and Mlodinow compare

97. Descartes, R.: *Discourse on method*, (1637) trans. Wollaston, A.: Penguin, Harmondsworth (1960) *in* Ward, K.: *God as the ultimate informational principle*, *in* Davies, P. and Gregersen, N.H. (eds.): *Information and the Nature of Reality: from Physics to Metaphysics*. Cambridge University Press (2010) p.297.

98. d'Abrera, B.: A History and Philosophy of the Science of Time. Hill House Publishers, London (2005) p.6.

99. *The World Book Encyclopaedia*. Chicago (1984).

100. Hawking, S. & Mlodinow, L.: (2010) p.63.

the buckyballs to full-sized footballs, which of course *do not* behave in a quantum fashion. They subsequently comment: "*...in general, the larger the object the less apparent and robust are the quantum effects.*" [101] The logical question then arises: at what size do particles stop behaving in a quantum fashion and start behaving like everyday objects with which we are familiar, such as marbles and footballs? Hawking and Mlodinow do not comment on this question, however it is clear that Quantum Mechanical effects apply to both electrons and buckyballs but not to full-sized footballs. From this it can be inferred that there is an inverse Law of quantum expression, where the larger the object, the less likely it is going to behave in a quantum manner. This explains why Newtonian mechanics is a good explanation for the behaviour of macroscopic objects, such as footballs and planets but not electrons or buckyballs.

Black Holes

And now it is time to move from the ultra-small world of Quantum Mechanics to the extra-large study of galaxies and black holes. Gonzalez and Richards wrote: "*Black holes are fearsome objects, distorting space, time and common sense, so densely packed that not even light can escape their horizons...Like a giant dragon sleeping in the heart of a mountain, a massive black hole probably dwells in the centre of our galaxy, though at the moment it is fairly inactive. Like our Milky Way, most large galaxies in the nearby universe do not have active nuclei, but recent research with the Hubble Space Telescope is revealing that almost all probably have a giant black hole lurking there.*" [102] Dolphin described their formation as: "*Stars are usually stable. They burn steadily aeon after aeon, scarcely wavering even a few per cent in energy output. Stellar evolution is usually slow and stable but not always. Sometimes stars go through violent, sudden and catastrophic changes in very short time periods...There are two very different fates which befall those stars that undergo sudden catastrophic changes. The difference in fates is determined by the mass of the star in question. The critical mass is roughly* [three][103] *times greater than the mass of our sun. Stars smaller in mass than the Sun, when they collapse inwardly upon themselves, reach a new stable state remaining in the sky as visible neutron stars. On the other hand, stars somewhat larger than our sun can collapse into tiny invisible spheres or 'black holes' less than a distance of ten kilometres (six miles) across...In the case of stars larger than the Sun the collapse of the core is not halted when the density that sets off these nuclear forces is reached. Only a temporary slowing occurs as matter streams inwardly, torrentially falling in upon itself to smaller dimensions and greater densities. In less than one tenth of a second a particle of matter caught in such a vortex is crushed out of existence! No expanding shock wave, no resulting supernova gives outward, visible evidence to the collapse of these larger stars. They remain as invisible 'black holes' that emit no light, although they can emit powerful gravitational radiation...densities rapidly approach infinity...Yet careful mathematical* [calculations] *seem to indicate that black holes are stable. That is, all the matter in such perfect spheres does indeed collapse into a point rather than contorting and squirting out matter in snake-like fashion. The science of black holes almost seems magic, for black holes are apparently holes in space itself — silent, invisible doorways into other universes through which matter cannot pass unmodified by the force of the collapse.*" [104]

Although black holes are still not properly understood, scientists have suspected their existence for more than two hundred years. Jim Al-Khalili writes: "*Back then they were known as invisible stars and their existence followed logically*

101. *Ibid.* p.68.

102. Gonzalez, G. & Richards, J.W.: *The Privileged Planet.* Regnery Publishing, Washington DC (2004) p.162.

103. Al-Khalili, J.: (1999) p.85.

104. Dolphin, L.: *Lord of Time and Space.* Good News Publishers, Westchester, Illinois (1974) pp.15–18.

and reasonably from a combination of Newton's law of gravity and his particle theory of light. Until fairly recently it was thought that the first person to predict the existence of black holes was the world famous French mathematician and astronomer Pierre Laplace in 1795. It is now clear that he was beaten to it by an English geologist named John Michell, who was Rector of Thornhill Church in Yorkshire. Michell is considered to be the father of the field of seismology...His ideas on the formation of black holes in space were presented to the Royal Society of London in 1783. Both Michell and Laplace had based their quite similar arguments on the idea of escape velocity." [105] Escape velocity is the speed an object requires to escape the gravitational pull of a larger body such as a rocket leaving Earth to travel to the moon. Al-Khalili continues: *"The escape velocity of the surface of the Earth is eleven kilometres per second (or forty thousand kilometres per hour)...The escape velocity on the Sun is 620 kilometres per second. This is a number which Michell had worked out, based on the size and density of the Sun. He also knew with some accuracy another figure: the speed of light, which had been measured a century earlier, and which is 500 times bigger than the Sun's escape velocity. Michell therefore calculated that a star 500 times bigger than the Sun, but with the same density, would have an escape velocity equal to the speed of light. Michell was following the conventional wisdom of the time; that of Isaac Newton, and believed that light was composed of particles. It therefore follows that light should be affected by gravity like any other object. But a star with the same density as the sun, but more than 500 times as large, would have an escape velocity that exceeded the speed of light, and so the particles of light would not be fast enough to escape its gravitational pull. Such a star must therefore look black to the outside world. In fact, it would be invisible!"* [106] History has vindicated Michell and Laplace. Al-Khalili continues: *"...although we have never come face to face with a black hole, we have a rough idea what it would be like to fall into one (not very nice). The reason for this confidence is our trust in Einstein's general theory of relativity. For if general relativity is correct, and we have no reason to doubt it so far, then it suggests that black holes not only exist in our Universe but are an inevitable consequence of Einstein's version of gravity. One of the world's leading experts on general relativity, Kip Thorne, goes so far as to state that 'the laws of modern physics virtually demand that black holes exist.'* " [106]

So physicists are now convinced that black holes do exist but since we cannot (and do not want to) get near them, what are they and what do they do? Al-Khalili takes up the story once more: *"As soon as Einstein completed his general theory of relativity in 1915 he began trying to solve his field equations. These equations were the complicated (yet mathematically beautiful) embodiment of his ideas on the connection between matter, space and time. But being able to write down the equations is only half the battle. They then have to be applied to particular situations and scenarios which involves much more than simply 'putting' numbers into a formula, but involves page after page of tedious and complicated algebra...The first exact solution of the equations of general relativity was obtained by a German astronomer named Karl Schwarzschild. He completed his calculations while on his death-bed, having contracted an incurable skin disease during the First World War, and only a few months after Einstein had himself published his work. The Schwarzschild solution, as it is now known, described the properties of space and time due to the gravitational field around any spherical concentration of mass. It was only later realised that Schwarzschild's result contained a description of a black hole in space. In fact, it was not until 1967 that the American physicist John Wheeler first coined the phrase 'black hole' which has since captured the public's imagination so spectacularly."* [107]

105. Al-Khalili, J.: (1999) p.82.
106. *Ibid.* pp.83–4.
107. *Ibid.* pp.86–7.

To understand the characteristics of a black hole, imagine what will happen to a star that is ten times the size of our Sun as it begins to collapse. The law of gravity as Newton understood it, says that the force exerted is proportional to the square of the radius, therefore as this radius halves, the force increases four times. In reality, the radius of the collapsing star reduces more than one thousand times so the force increases one million times. Thus the forces are immense. Al-Khalili continues: *"If it were possible for all the mass of the star to be squeezed to a pinpoint of zero size (it now has a radius of zero) then the force of gravity would be infinite. General relativity says something dramatically different. If a star were to collapse down to some critical size, such that its escape velocity equals the speed of light, then the gravitational force on its surface would be infinite! By this I mean the force that would be required to stop it from collapsing further would be infinite. The radius of this critical size is called the Schwarzschild radius and marks the boundary of a black hole. Now we see that the collapse must continue beyond this radius."* [108]

"Within the Schwarzschild radius, nothing — not just light — can escape. Imagine a sphere that has a radius equal to the Schwarzschild radius and which surrounds the collapsed star. Such an imaginary spherical surface is known as the event horizon and is an artificial boundary in space which marks the point of no return. Outside the horizon gravity is strong but finite and it is possible for objects to escape its pull. But once within the horizon, an object would need to travel faster than light to escape, and this is not allowed. Thus the event horizon is a rather unpleasant concept in that it allows one-way traffic only." [108]

"Once a collapsing star has contracted through its event horizon, nothing can stop it from continuing to collapse further until its entire mass is crushed down to a single point. This is called the singularity *and it is a very strange entity indeed. It is so strange in fact, that the laws of physics that work — as far as we know — perfectly well everywhere else, describing the behaviour of the tiniest subatomic particle to the properties of the whole Universe, break down at a black hole's singularity."* [108] Al-Khalili summarises it concisely when he writes: *"A black hole is therefore very simple in its structure. It has a centre (the singularity) and a surface (the event horizon). All else is gravity."* [109] One way to imagine this is to go back to our trampoline example in Chapter 6. Al-Khalili explains: *"A black hole corresponds to the case when a very heavy, yet point-sized, object causes the rubber (space) to be curved and stretched down into an infinitely deep cone-shaped hole.* [See Figure 8.6] *The event horizon here*

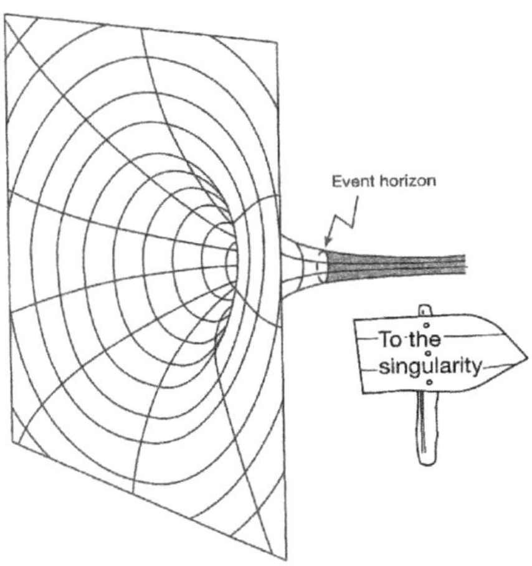

Figure 8.6: A Black Hole drawn in Two-dimensional (2D) Space.[110]
If you imagine that the Universe was flat, instead of three-dimensional (3D), this is how a black hole would appear.
[Reproduced with permission of Al-Khalili, J.S.: Black Holes, Worm Holes & Time Machines. Taylor & Francis Group, NY (1999) via Copyright Clearance Center.]

108. Al-Khalili, J.: (1999) pp.88–9.
109. *Ibid.* p.91.
110. *Ibid.* p.92.

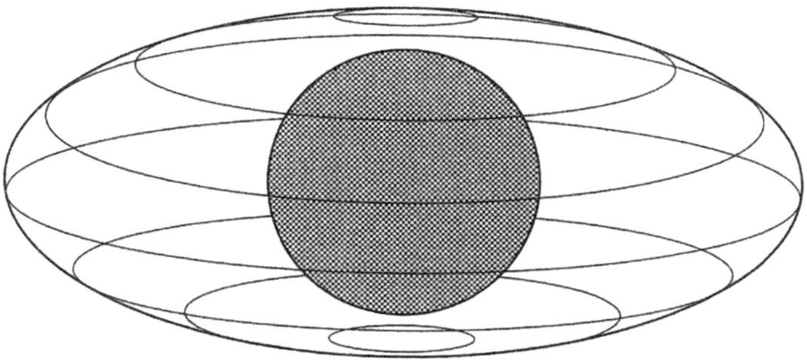

Figure 8.7: A Spinning Black Hole with its Ergosphere surrounding the Event Horizon.[111] *[Reproduced with permission of Al-Khalili, J.S.: Black Holes, Worm Holes & Time Machines. Taylor & Francis Group, NY (1999) via Copyright Clearance Center.]*

corresponds to a circle somewhere inside the rim of this bottomless pit beyond which there would be no escape." [110]

In reality, like the stars from which they develop, black holes spin so they have a 3D shape and the extreme gravity distorts the space around them. Al-Khalili writes: *"This region where space is dragged around a spinning black hole is called the ergosphere. It means that a spinning black hole will have two horizons: an inner, spherical, one which is the original event horizon and from which nothing can escape and an outer, bulged-out at the equator one which marks the surface of the ergosphere.* [See Figure 8.7] *Within the ergosphere, the dragging is so strong that nothing can stand still. However, an object that falls into the ergosphere can still escape again, as long as it does not stray within the event horizon."* [112]

We have already discovered that inside black holes (beyond the event horizon) the usual laws of physics, which apply elsewhere in the Universe, fail to apply and this is true of time too! Al-Khalili explains: *"Inside a black hole space and time are so warped that the distance from the event horizon to the singularity is not a distance in space in the normal sense (in the sense that it can be measured in kilometres or some other appropriate unit of length). Instead it becomes a time direction. Basically the radial distance to the centre of the hole is interchanged with the time axis!... Physicists liken this to the unavoidable way we move in time towards the future. What is more, since you can get no further once you have reached the singularity, this point must mark the end of time itself! This is where black hole singularities differ from the Big Bang which is a singularity that marks the beginning of time. They are more like the Big Crunch singularity (the one that marks the end of space and time if there were enough matter in the Universe to cause it to collapse in on itself)."* [113] From this description it can be seen that time (in Einstein's General Relativity) began with the Big Bang and finishes with the Big Crunch (if there is enough mass in the Universe). Thus time only exists within our physical universe. Since time also ends at the singularity within a black hole, it implies that if you were to travel through the event horizon in a black hole you would be leaving our physical universe and indeed once the event horizon was crossed, there would be no physical way of returning to our material universe.

Dolphin expands on this subject by writing: *"Just as stars can collapse in the ways we have described, Einstein's Theory of Relativity shows also that the universe, if closed...as it is thought to be, will expand to some maximum size and then eventually contract and suffer the same fate as those stars which become black holes. Present day astronomers talk not only about the supposed origin of the universe as a burst of light but also of its termination in a gigantic collapse inward upon itself in a manner similar to black holes...What happens to matter which chances to fall into a black hole in space? Or what is the fate of our universe after its collapse? Astronomers are not sure. The*

111. Al-Khalili, J.: (1999) p.96.
112. *Ibid.* pp.95.
113. *Ibid.* pp.97–8.

properties of matter change drastically even at the atomic nuclear level during the final stages of gravitational collapse. But speculation is running high in the conservative courts of science that black holes connect our universe with others." [114] The problem is that we cannot leave our universe to investigate this possibility so how could we possibly know? The only way is through mathematical modelling which brings us to another interesting phenomenon: *white holes.*

White holes

Al-Khalili writes: *"There are certain solutions of the equations of general relativity which allow for the existence not only of black holes but of objects called white holes too. But it turns out that this would be possible only if the universe had certain properties, called initial conditions, at the time of the Big Bang. These highly speculative objects are the opposite of black holes which rather than sucking matter in, would spew matter and energy out into the universe. Their singularities would be different too and would mark the beginning rather than the end of time."* [115]

Although the mathematics of General Relativity Theory predicts the existence of white holes, they have not been observed in our physical universe. One reason is that they cannot exist in the presence of matter (which causes them to disappear). White holes may have existed at the formation of the Universe when there was no matter but only light energy which must have come into our universe at its formation or sometime shortly thereafter; however as physical matter decoupled from this initial radiation, the matter would have caused the white holes to disappear. Al-Khalili writes: *"...a white hole...is the opposite of a black hole from which matter will emerge rather than fall into. White holes must therefore be surrounded by the opposite of an event horizon, something known as an anti-horizon, which would allow one-way traffic out and never in. Unfortunately, anti-horizons are very unstable and get converted to normal horizons in a matter of seconds after forming."* [116] As a result, unlike black holes which can form at the collapse of a dying star, white holes have no naturalistic or physical way of being formed. There are other problems with white holes too, such as how were the necessary initial conditions set up at the time of the Big Bang (which would have required information) and where did the matter and energy come from if there was nothing there before the formation of the Universe? White holes also appear to violate the Second Law of Thermodynamics as they create more order not less but as they communicate with a space or void outside our current universe, there is no longer a closed system, thus the Second Law may no longer apply. Further, since the General Theory of Relativity is symmetric with respect to time, theoretically black holes would turn into white holes if time could run backwards![117] Stephen Hawking from his study of Quantum Mechanics described Hawking radiation, the behaviour of which suggests that black holes and white holes may be different forms of the same thing.[118] Black holes and white holes can be described mathematically as linking together to form a portal which could possibly link between different universes which is called an Einstein-Rosen bridge or a Schwarzschild wormhole. Indiana University theoretical physicist, Nikodem Poplawski has recently suggested that the formation of a black hole in our universe results in the formation of a white hole somewhere else which leads to the birth of a new universe.[119]

Everything we know about black holes and white holes shows that you cannot physically travel through one and come out alive at the other end, as all

114. Dolphin, L.: (1974) pp.18–19.

115. Al-Khalili, J.: (1999) pp.106–7.

116. *Ibid.* p.199.

117. Wheeler, J. Craig: *Cosmic Catastrophes: Exploding Stars, Black Holes, and Mapping the Universe.* Cambridge University Press (2007) pp.197–8.

118. Hawking, S.W.: Black Holes and Thermodynamics. *Physical Review D* 13:191–7 (1976).

119. Poplawski, N.J.: Radial motion into an Einstein–Rosen bridge. *Physics Letters B* 687:110–3 (2010). http://newsinfo.iu.edu/news/page/normal/13995.html

physical matter would be crushed at the event horizon but it may be possible that black holes and white holes form portals between our Universe and the void from which our universe formed (Voids One and Zero respectively in Figure 6.4). It seems unlikely that white holes and worm holes are useful in our present age but as far as we know, there would be no bar on information traversing them and they may have been significant in the formation of the Universe.

The Information Plane

Returning to the subatomic level, it is time to tackle the business of 'empty' space. David Bohm, while studying the behaviour of superheated gases known as plasma, came to the conclusion that all particles were linked to all other particles. If information in the form of digitally-specified vibrations underlies all we see in our space-time domain then there should be a vibration shadow which underlies every object and every 'landscape' as in Figure 8.8 below. This can be thought of as a hidden information layer or '*Information Plane*'. As we have seen previously, when we think of deep space, we think of utter emptiness, void of everything but Quantum Theory predicts, and experiments verify, that so-called empty space (the vacuum) contains an enormous residual background energy known as Zero-Point Energy or ZPE. This energy derives its name from the fact that at temperatures of absolute zero (-273° Celsius), elementary particles continue to exhibit energetic behaviour when Relativity Theory suggests there should be none.[120,121]

Australian physicist Barry Setterfield explains: "*...there is an intrinsic energy in every cubic centimetre of the vacuum, this energy may manifest as virtual particle pairs like electron/positron pairs that flit in and out of existence.*" [122] These virtual particles which "*flit in and out of existence*" are all that we can discern of the information plane. Thus, although the void of outer space is empty of *tangible* entities it still has *intangible* entities that we cannot always identify. If the electron/positron pairs project from the Implicate Order and separate momentarily they can manifest as matter briefly before recombining and annihilating each other again.

In contrast to the vacuum, physical matter such as an electron/photon appears to perpetually oscillate between the Information Plane, in Bohm's non-material, Implicate (frequency) Order and the physically explicit, Explicate (space-

120. Andrews, E.: (2009) p.151.

121. Calphysics Institute: *Zero Point Energy and Zero Point Field at* http://www.calphysics.org/zpe.html

122. Setterfield, B.: Recent Lightspeed Publicity. 10/2/2002 *at* http://ldolphin.org/recentlight.html

123. For the mathematicians and engineers: The Fourier transform of the impulse response of a linear filter gives the frequency response of the filter. See *Designing and Implementing 2-D Linear Filters for Image Data* for more information about linear filtering, filter design, and frequency responses *at* http://www.mathworks.com/help/toolbox/images/f21-17064.html.

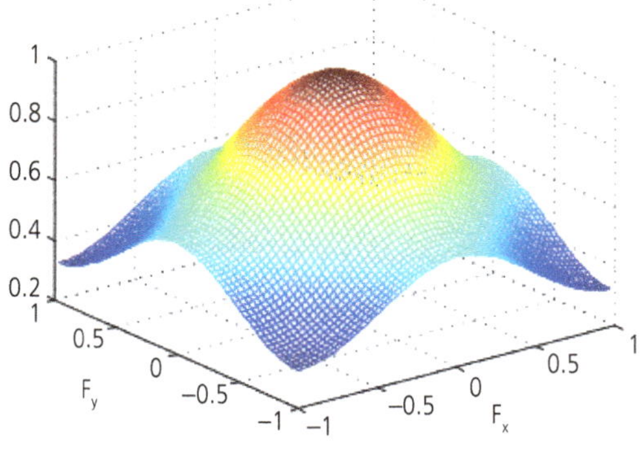

Figure 8.8: Frequency Response of a Gaussian Filter.[123]
This is an example of an information terrain in three dimensions created from digital information. If we could imagine this projection in five, six, seven or more dimensions then we would be able to start to appreciate the way in which our universe might be constructed if M-Theory is proven to be true as Hawking and Mlodinow suggest.[65] *[Reprinted with permission of The MathWorks, Inc.:* http://www.mathworks.com/help/toolbox/images/f21-17064.html]

time) Domain. Remember that the photon/electron is neither exclusively in one or the other order but *both simultaneously*; it is just that we cannot *detect* it in both simultaneously which is consistent with the Heisenberg Uncertainty Principle. If it is detectable in the Explicate Order as a particle then we would not be able to detect it in the Implicate Order *at the same time* and vice versa.

The Information Plane's ability to connect all points in the Universe is hard to imagine due to our limiting four-dimensional perceptions but imagine the way that all the water molecules in an ocean interconnect by being adjacent to each other. To illustrate this, consider the tides in the Pacific Ocean and instead of focusing on the individual water molecules, view the behaviour of the ocean as one entity. When the tide goes out on Australia's east coast on the south-western rim of the Pacific Ocean, it comes in on the Californian coast on the north-eastern edge of the ocean, half the world away *at the same time* but *always in the opposite direction!* That does not mean that the water molecules which went out on the Australian tide are the same molecules that came in on the Californian tide but the effect was the same even though they are 12,000 kilometres (7,500 miles) apart. This is because the ocean behaves as a single entity, even though it is made up of many discrete water molecules. This concept of interconnectivity, although it is not identical to the Information Plane, acts as a model for Quantum Mechanics and could be a clue to understanding Alain Aspect's two-particle experiment, where the particles appear to travel faster than the speed of light (just like the water molecules within the tide). It also demonstrates the principle of non-locality, which is characteristic of Quantum Mechanics. However, an alternative explanation might be that although we are limited by the speed of light in the Explicate Order, it may be that there is no such speed restriction in the Implicate Order, where instantaneous events may be the norm.

Although this concept of an information plane in the frequency domain, running invisibly through/ under/behind the entire Universe like a shadow, may seem far-fetched and frankly 'impossible', Aspect's Two-Particle Experiment suggests that the concept of an information plane or something equivalent, far from being impossible, is necessary. We need to remember that Quantum Mechanics is non-local. Edgar Andrews explains: *"According to quantum mechanics, a subatomic particle like an electron is not so much in two places at once as in every place at once (at least potentially)! Instead of thinking of the electron as a tiny billiard ball having a precise location in space, quantum mechanics says we have to think of it as being associated with a 'wave function' or probability wave that spreads out over the whole of space and time. This wave tells us the likelihood or probability that the electron will be found in a particular place at a particular time."* [124] Now if you are struggling with these concepts you are not alone. Andrews continues: *"The philosophical impact of quantum theory was immense because it overthrew the long-cherished concept of determinism in nature. Determinism is the idea, first propounded by the French mathematician Pierre-Simon Laplace (1749–1827), that if you knew everything about every particle in the universe you could predict 'what happens next' and so determine the future course of all physical events."*

"Albert Einstein agreed with Laplace; he didn't like the 'spooky' implications of QM one little bit. He could not deny that the theory was highly successful in describing and predicting what goes on in the sub-microscopic underworld — indeed, he himself made notable contributions to the development of the theory. But he always believed that QM was not, and could not be, the ultimate reality — because it relies on ideas like probability waves and indeterminacy that are seriously counter intuitive (that simply

124. Andrews, E.: (2009) p.33.

don't 'make sense')...Einstein maintained that QM could only be a provisional and approximate description of the fabric of the universe. Maybe it's all we can know, he argued, but there must be an underlying reality in which particles do have well-defined positions and speeds and do follow deterministic laws. He believed that quantum mechanics 'would eventually go beyond a mathematical description of probabilities and find an explanation in the form of deeper causal mechanisms.'" [125] The Information Plane in its de Broglie/Bohmian waveform (implicate) state would appear to satisfy Einstein's specification but when we do fully understand it, it may not be like anything with which we are familiar. Richard Feynman wrote: *"Things on a very small scale [like electrons] behave like nothing you have any direct experience about. They do not behave like waves, they do not behave like particles, they do not behave like clouds, or billiard balls, or weights or springs, or like anything you have ever seen."* [126] Instead imagine an immense frenetic ocean of jostling Higgs bosons or something equivalent which although it behaves as one entity is still composed of discrete parts and even this analogy is probably a pale imitation of the real thing.[120]

Multiverses

We have met the concept of the Multiverse previously when looking at the probability that we could be living in the infinitesimally rare universe that has all the parameters that, by chance, provided precisely the requirements necessary for life to exist. The possibility of black and white holes linking universes has also been discussed although Professor Al-Khalili comprehensively demonstrates why black holes and wormholes are not a viable means for human beings to travel between different universes (or even different places within our own universe), at least within our physical or explicate domain. Despite Dawkins and Ward describing five models for possible multiverses,[127,128] they all would require laws to govern them and bring them into existence. As described in Chapter 2, laws contain intrinsic information, which specifies them and that information had to come from somewhere. Professor Keith Ward of Oxford University writes: *"But something must cause the forces and constants to vary in a systematic way in order to cover the whole range of different possibilities."* [129] That is, something or someone had to specify the laws (information) in the first instance. So who or what specified the information that describes the laws that control the universes within the Multiverse?

Andre Linde of Stamford University suggested that we are living in an eternal superspace that contains a (theoretical) *"inflation field"* which, through quantum fluctuation of the superspace, caused the birth of multiple soap-like 'bubbles', each of which contains a different universe and ours is just one of them. Each bubble is finite in that it had a beginning and will have an end but the surrounding superspace is infinite.[130] The Multiverse option (the second in Al-Khalili's list of alternatives to explain Quantum Mechanics) is also the model favoured by Hawking and Mlodinow in their recent book: *The Grand Design*. They conclude: *"Bodies such as stars or black holes cannot just appear out of nothing. But a whole universe can...Because there is a law like gravity, the universe can and will create itself from nothing...We've seen that there must be a law like gravity and... that for a theory of gravity to predict finite quantities, the theory must have what is called supersymmetry between the forces of nature and the matter on which they act. M-theory is the most general supersymmetric theory of gravity. For these reasons*

125. *Ibid.* p.37.

126. Feynman, R.: Lectures from a series given at Caltech in the early 60s - intended for freshmen, to stimulate their interest in physics. p.116 *at* http://www.librarything.com/work/23020

127. Dawkins, R.: *The God Delusion*. Bantam, London (2006) p.144.

128. Ward, K.: *Why There Almost Certainly Is A God.* Lion Hudson, Oxford (2008) pp.67–70.

129. *Ibid.* p.72.

130. Strobel, L. & Vogel, J.: (2004) p.53.

M-theory is the only candidate for a complete theory of the universe. If it is finite - and this has yet to be proved - it will be a model of the universe that creates itself...If the theory is confirmed by observation, it will be the successful conclusion of a search going back more than 3000 years. We will have found the grand design." [131] The problem is that Hawking and Mlodinow have failed to undertake an information audit. It may be possible in their theory for whole universes to appear out of nothing but it is impossible for information to do so. Expecting the quantum fluctuations to create galaxies without information, is like Miller and Urey in 1953 trying to produce life without the information to specify the DNA sequences. Robin Collins, a physicist, mathematician and Professor of Philosophy lists some of the mechanisms (think 'information') that would be required to produce a functional Multiverse:

- *"A mechanism to supply the energy needed for the bubble universes. That would be the inflation field that Linde has hypothesised.*
- *A mechanism to form the bubbles. This would be Einstein's equation of general relativity.*
- *A mechanism to convert the energy of the inflation field to the normal mass/energy that we find in our universe.*
- *A mechanism to vary the constants of physics so that by random chance it would produce some universes, like ours, that have the right fine-tuning to sustain life.*
- *The right background laws in place: the principle of quantisation, the Pauli-exclusion principle, a universally attractive force between all masses - such as gravity. If just one of these components was missing or different, it's highly improbable that any life-permitting universes could be produced."*

Collins continues: *"And keep in mind,...you would need to make trillions upon trillions upon trillions upon trillions of universes in order to increase the odds that the cosmological constant would come out right at least once, since it's finely tuned to an incomprehensible degree."* [132] Each new universe without the above laws would amount to nothing very interesting. Even universes with laws but without the fine-tuning described by Robin Collins would be untenable for life.

Alternatively, the 'Multiverse' may be spaces *within* our universe, in different dimensions, called: *Hyperspaces*. Instead of Multiverses, they could be thought of as 'Multi-dimensions', where we exist in dimensions yet to be described and understood. This is a deduction of String Theory but it is speculative at present due to a lack of information to confirm or deny it.[57]

The Physical Information Generator

Whichever mechanism turns out to be correct, it is clear that a vast amount of precisely-specified information is involved. Therefore it will be more productive to focus on that which is known, the existence of the information, rather than that which is unknown, the exact mechanism by which it acts. Whether you subscribe to one precisely defined universe or a multiverse with an infinite number of constituent universes, this immense quantity of information had to be generated somehow. Andrew Gosler of Oxford University concludes: *"Hence we can argue that since, within this universe, everything that exists, either realised or potential, is a*

131. Hawking, S. & Mlodinow, L.: (2010) pp.180–1.

132. Collins, R.: interview *in* Strobel, L.: (2004) p.144.

133. Gosler, A.: Surprise and the Value of Life. *in* Berry, R.J.: (Ed.) *Real Scientists Real Faith.* Monarch Books, Oxford (2009) p.183.

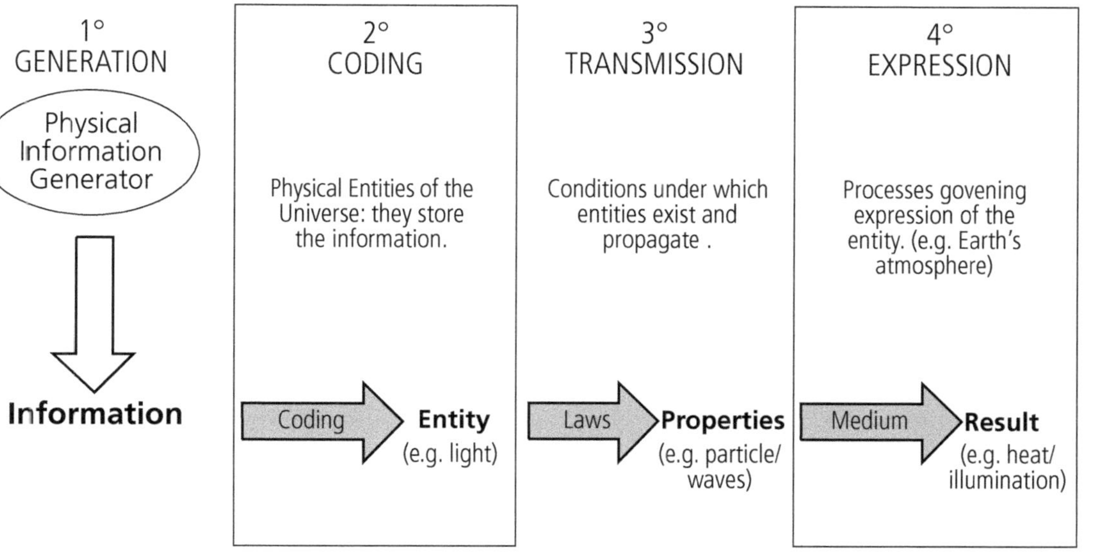

Figure 8.9: The Physical Information Generator
Arrows show the coding information flows. From this diagram, it is immediately apparent that the behaviour of physical information is very similar to the biological information we discovered in Chapter 3. Notice that once again it is a one-way process. The illumination provided by the light that has been guided by the laws of the Universe cannot change the information that determined the composition or behaviour of any of these entities. An information generator is required.

manifestation of information, we might regard information as the signal of existence, as opposed to the noise of chaos." [133] This information cannot have appeared from nowhere, so it must have been generated by some mechanism and for simplicity I shall call it the 'Physical Information Generator'. (See Figure 8.9)

Characteristics of Information

To summarise our progress to date, we have now discovered a number of characteristics of information such as that:

- information is intangible and yet it is real;
- information is neither matter nor energy;
- information is transmissible but;
- information requires a vehicle to be transferred in the space-time domain;
- information is independent of time;
- information can be both quantitative and qualitative;
- meaningful information needs to be actively generated;
- information can be stored;
- information is digital;
- information comes at a cost, both to generate and maintain it;
- the existence of information does not necessarily explain its causation.

Summarising this knowledge that has been gleaned so far allows two basic principles of information to be defined:

1. **The law of information primacy** which states that: ***information is the fundamental entity of the Universe and specifies all matter and energy***. The corollary of this is that: all outcomes are information dependent as they require prior information input or control.

2. ***Information is independent of time.***

The Physical Information Hypothesis

Thus we live in a finite universe which had a beginning and must have an end. Any explanation of our Universe must take these parameters into account. Edgar Andrews summarises it as: *"This beginning can be modelled in various ways such as a 'hot big bang' singularity, but all such models point to an origin in which space, time, matter and energy came into being ex nihilo (out of nothing)...Current scientific cosmologies thus imply the existence of a non-physical realm (we'll call it 'eternity') that transcends space and time and within which the physical universe was created."*[134] We previously met this place as 'Void Zero' in Chapter 6. (See Figure 6.4) We do not have much information about 'Void Zero' but from the preceding discussion we do know that information is required to specify all matter in the Universe. This is the *Physical Information Hypothesis*. As such, the information had to exist prior to the matter, that is, at or before the formation of the Universe. This is the Hypothesis of Information Primacy for Physical Entities. Scientists have wondered at the simplicity and predictability of the laws of the Universe. In 1967, Eugene Wigner (1902-1995), Nobel laureate in Physics in 1963 wrote of the *"unreasonable effectiveness of mathematics in natural science."* [135] Professor Lenny Susskind of Stanford University goes much further when he states: *"It is not only that it becomes easier to describe* [the Universe] *with mathematics, as you go deeper and deeper into reality, mathematics becomes the only way to describe reality."* [136] Max Tegmark, cosmologist of the Massachusetts Institute of Technology sums it up by saying: *"To me, math is a window on our Universe...the master key to understanding what is out there...Equations...describe exactly what is going on outside the window, in our Universe...I think our Universe isn't just* described *by math, I think it is math. I think our entire Universe is a giant mathematical structure that we are part of. And that's the reason why the more we study physics, the more mathematical regularities we keep discovering...I think the Universe is a mathematical object...Space and time exist inside of it."* [136] Therefore, if the Universe is specified by information, which is digital, and mathematics is the domain of digits, then it would be unreasonable *not* to expect mathematics to describe not only the natural sciences, but the whole Universe (wherever digital information exists, which is almost everywhere you look). It would be more appropriate to conclude that if mathematics cannot be found to describe an aspect of science, then our concept of that aspect of science may be faulty and may need to be revised. Perhaps Lord Kelvin was right! (See Chapter 2)

The other unexpected outcome from examining information and the duality of light is its double existence in both our physical space-time domain as a particle and the frequency domain as a vibration. Physicists who have been frustrated by their inability to understand the way fundamental entities make up the Universe, which includes the Higgs boson, have begun to look for ways to expand the Standard Model of particle physics that was developed in the 1970s. *Cosmos* notes: *"Physicists need something beyond this framework, something that can explain why the Higgs…has the mass it does. Extra dimensions beyond the four that people experience as space and time have been proposed, as has a new kind of symmetry hidden in the universe...'supersymmetry'* [is] *one of the most popular extensions of the Standard model. According to supersymmetry, every particle has a shadowy twin that differs in a quantum property called spin,..."* [137] So this duality does not *only* appear to apply to light; the de Broglie/Bohm Model predicts that a frequency domain

134. Andrews, E.: (2009) p.93.

135. Wigner, E.: The Unreasonable Effectiveness of Mathematics in the Natural Sciences. *in Symmetries and Reflections.* Indiana University Press (1967) pp.222–237 *in* Gonzalez, G. & Richards, J.W.: (2004) p.212.

136. Shariatmadari, H.: *What is Reality?* Horizon, BBC Productions, UK (2011).

137. Powell, D.: Atom Smashers *Cosmos* 43:77 (2012).

underlies the entire Universe in a similar fashion to that illustrated in Figure 8.5. We therefore have not only a duality of light but a universal duality, which includes life, with matching facets of a physical explicate existence and an non-material implicate reality as well, which will be further investigated in subsequent chapters.

9

The Information Model

The whole history of science has been the gradual realisation that events did not happen in an arbitrary manner, but that they reflect a certain underlying order, which may or may not be divinely inspired.[1]

Stephen Hawking

Order and the Universe

Albert Einstein struggled to understand the surprising degree of organisation within the Universe. He wrote: "*You find it strange that I consider the comprehensibility of the world (to the extent that we are authorised to speak of such a comprehensibility) as a miracle or as an eternal mystery. Well,* a priori, *one should expect a chaotic world, which cannot be grasped by the mind in any way…the kind of order created by Newton's theory of gravitation, for example, is wholly different. Even if man proposes the axioms of the theory, the success of such a project presupposes a high degree of ordering of the objective world, and this could not be expected* a priori. *That is the 'miracle' which is being constantly reinforced as our knowledge expands.*" [2] Michio Kaku amplified this in an interview in 2005: "*It turns out that it is no easy matter to create a universe where matter, space and time are stable. For it all to work, the fundamental forces must be unified. And that unification can only take place in higher dimensions…The universe is gorgeous, and it did not have to be that way. It could have been random, lifeless and ugly, but instead, it is full of rich complexity and diversity.*" [3] Edgar Andrews adds further perspective to these comments when he writes: "*Disordered states of matter arise spontaneously; ordered states do not.*" [4] The inescapable conclusion is that the Universe is exquisitely organised, a fact that requires explanation. However, to discover how and why this is so, we will need to look at the fundamentals specifying the construction of the Universe.

We have already established that the Universe is highly ordered, it had a beginning and it has an expanding edge and this is what we call our *physical* universe but what is outside that expanding edge? What does the Standard Model predict was there before the initial singularity which physicists call the 'hot Big Bang'? Edgar Andrews explains: "*The answer is truly nothing. Everything that exists materially is contained within the expanding volume of space itself. There isn't even a vacuum or a void outside of space - just nothing at all. And if the big bang theory is correct, and the universe (and space itself) began as an infinitely small, infinitely hot pinhead, then there was at first an awful lot of nothing and not much something*

1. Hawking, S.: *A Brief History of Time: From Big Bang to Black Holes*. Random House (1988) p.127.

2. Einstein, A.: Letters to Solovine: 1906-1955. Philosophical Library, NY (1987) p.131 *in* Lennox, J.C.: *God's Undertaker. Has Science Buried God?* Lion Books, Oxford (2007) p.59.

3. Finkel, E.: Fish out of Water. *Cosmos* 2:78–9 (2005).

4. Andrews, E.: *Who Made God? Searching for a Theory of Everything*. EP Books, England (2009) p.22.

(in fact, none)." [5] The operative word here though is *"materially"*; the answer to this conundrum lies in the fact that a significant amount of our existence (such as information) is *non-material*. We have no physical way of detecting what came before the initial singularity or what is beyond the outer boundary of our expanding universe but what if some source of information were outside our material universe which caused it, not ex nihilo but by injecting information, in a similar manner to the way that an electromagnetic wave brings information in the form of a TV signal into your television and creates a 'world' within your TV or DNA brings information to create a new human being? These examples are not exactly analogous but they suffice to demonstrate the effect that the injection of extrinsic information into a system can achieve. This information injection model avoids the problems of creating order out of an uncontrolled explosion and being outside of our material world, this information source would not be subject to the same limitations as previous models, such as the creation of an entire material universe out of nothing, which is required for the Big Bang model to work. However, if the Big Bang was a 'controlled' event, rather than an uncontrolled one, then there must have been an information generator to create this controlling effect but any information generator that is wholly within our universe or that is not non-material will have physical constraints, which would make the model untenable, such as explaining how the Universe originated. This is discussed in greater detail at the end of this chapter.

Entropy and Time

In Chapter 8 we saw that the Second Law of Thermodynamics, which was alluded to in 1851 by the English physicist William Thomson (later Lord Kelvin) and formally expressed by German physicist and mathematician Rudolf Clausius (1822–1888) in 1865,[6] states that in any closed system, entropy, which is the tendency of any system to move toward complete disorder, inexorably leads to the winding-down of all physical systems. Thus order constantly decreases in our physical universe too, as it runs in parallel with time. Although time in pure mathematics can theoretically run either forwards or backwards, in our physical universe it can only run forwards to greater disorder due to energy dissipation. To make entropy run backwards (that is to create more order) we need to put in energy but not *just* energy. Uncontrolled energy in the form of an explosion is liable to *increase* entropy. Edgar Andrews writes: *"To reduce entropy the energy supplied must be specifically controlled and directed."* [7]

This control of energy which is necessary to reduce entropy implies that there is information required as well as the energy. Just as we have friction (energy) losses in our material universe, there are continual information losses too. Thus it can be seen that entropy and information, although they are different, run very similar races through the history of the Universe and with time, both result in greater disorder within the material universe. Edgar Andrews summarises it as: *"Put briefly, the only way science can explain the phenomenon of time, with its one-directional 'arrow', is by assuming that the universe began in a highly ordered (low entropy) state. Physicist Brian Green puts it thus: 'Conditions at the birth of the universe are critical to directing time's arrow. The future is indeed the direction of increasing entropy. The arrow of time — the fact that things start like this and finish like that but never start like that and end like this — began its flight in the highly ordered, low-entropy state of the universe at its inception'...Scientifically speaking, then,*

5. *Ibid.* pp.105–6.

6. Gregersen, N.H.: God, matter, and information: towards a Stoicizing Logos Christianity *in* Davies, P. and Gregersen, N.H. (Eds.): *Information and the Nature of Reality: from Physics to Metaphysics*. Cambridge University Press (2010) p.327.

7. Andrews, E.: (2009) p.115.

time is simply the measure (and experience) of change, as the universe is transformed progressively from an initial highly ordered state of low entropy to some final condition of maximum entropy. In this final condition…entropy would stop increasing and time would cease to flow. The question then arises: what or who provided the energy input that established this low-entropy condition in the early universe?" [8]

It is clear that both energy *and information* had to have been input into the early Universe to achieve this state of low entropy. In Chapter 6 we encountered the inflation field, which is the Big Bang Theory's answer but Andrews notes: *"In the inflationary scenario, of course, it was the negative gravitational field (called the 'inflation field') that donated this energy. But where did the inflation field get its energy? From 'a statistical fluctuation from primordial chaos', we are told."* [9] If this is the correct explanation then we are all very fortunate that the statistics stacked up at just the right moment in history to create the Universe but it does not explain the informational component that specified the low entropy state noted above. The other problem, as described by Jim Al-Khalili in Chapter 8, is that the 'borrowed' energy must be repaid promptly (in less than a second), so it cannot explain the vast amounts of energy required to create and sustain our physical universe for the estimated 13.7 billion years that it has existed.

From the foregoing information, the pre-eminent conclusion is that time began at the formation of the Universe, when there was minimal entropy (maximal order) and will end with high entropy (maximal disorder) at the end of the (physical) Universe. This is where the *'Big Crunch'* comes in; where the Universe expands until it runs out of outward motion, before it is dragged back in by gravity as the Universe implodes, only to explode again (we are told) as the next 'Big Bang' and so it cycles forever, sometimes called the *'Big Bounce'*. Andrews points out that this perpetual bouncing causes a problem in that: *"…entropy would have been increasing forever. But sooner or later the entropy of the universe must reach a maximum, which will happen once complete disorder is established. If entropy had been increasing for an infinite time, then maximum entropy would have been achieved and time would have ceased to flow. When nothing changes, time stands still."* [10] There is also the small matter of the provision of information!

Al-Khalili describes five 'arrows of time' that are associated with increasing entropy. He writes: *"Another way of defining increasing entropy is through loss of information. By saving a file on computer you are creating order and decreasing entropy locally. The reverse happens when you erase a file. You are losing information and entropy increases."* [11] At the beginning of the Universe, all energy and matter were condensed into one place so the entropy was very low. If as we have seen above, it takes energy and information to achieve this low state of entropy the question then follows: how did the Universe get to be in such a low entropy state in the first place? If the Universe expands through entropy and contracts again in the Big Crunch then you cannot use entropy to explain the contraction also. Al-Khalili writes: *"An obvious difference between the expanding and contracting phases is that in the former there would have been some initial conditions that set the universe expanding in the first place."* [12] Once more we are forced to contemplate some information or instructions which preceded the formation of the Universe, to define the 'initial conditions' that caused it to come into existence. When we look at the conundrum of inflation at the start of the Universe, as per General Relativity Theory, the addition of extra information does not make inflation impossible, it may just mean that it requires pre-specified information (the *"initial conditions"*), to cause it to happen.

8. Andrews, E.: (2009) p.116.

9. *Ibid.* p.117.

10. *Ibid.* pp.119–120.

11. Al-Khalili, J.: *Black Holes, Worm Holes & Time Machines.* Taylor & Francis Gp, New York (1999) p.127.

12. *Ibid.* p.135.

Information as a Fundamental Quantity

John Lennox wrote: "*In an editorial in the* New Scientist [in 1999], *Paul Davies writes: 'The increasing application of the information concept to nature has prompted a curious conjecture. Normally we think of the world as composed of simple, clod-like, material particles, and information as a derived phenomenon attached to special, organised states of matter. But maybe it is the other way around: perhaps the universe is really a frolic of primal information, and material objects a complex secondary manifestation.'* [13] *Davies says that this idea was first proposed in 1989 by the well-known physicist John Archibald Wheeler, who said: 'Tomorrow, we will have learned to understand all of physics in the language of information.'* " [14]

Lennox continues: "*And in the* New Scientist, *under the intriguing title, 'In the beginning was the bit', there is an account by Hans Christian von Baeyer of the work of physicist Anton Zeilinger, of the University of Vienna. Zeilinger advances the thesis that, in order to understand quantum mechanics, one has to start by associating information (in terms of bits) with so-called elementary systems in quantum mechanics which, like the spin of an electron, 'carry' one bit of information (there are only two possible outcomes from measuring spin - 'up' or 'down'). Zeilinger argues that his basic principle gains credibility by leading directly to 3 pillars of quantum theory - quantisation itself, uncertainty and quantum entanglement. This proposal that information be regarded as a fundamental quantity has profound implications for our understanding of the universe.*" [15] Note that these 'bits' of information are far more fundamental or basic than the 'infobits' which we discussed in Chapter 1. Professor of Quantum Mechanical-Engineering at MIT, Seth Lloyd refers to these as 'quantum bits' or 'qubits'.[16] At this stage the qubit is a useful concept but impossible to use to calculate information content due to the vast complexity of our universe. Yet if Quantum Mechanics can be measured in qubits of information (that is, zeros and ones), then it is digital, meaning that our universe is digital from its most basic particles to its most complex structures. Table 9.1 summarises some of these significant entities.

Table 9.1: Examples of Fundamental Entities in the Universe
Infobits are apt for measuring macroscopic entities while Plank scale information is more appropriately measured in qubits.

Fundamental Entity	Type of Information	Transmission (vehicle)	Expression (result/effect)
Light (electro-magnetic radiation)	Intrinsic	Particle/wave (photon)	Universal power source
Information	Intrinsic	Mathematics	Laws of the Universe (physics)
Genome (DNA)	Extrinsic	Biochemistry	Life (biology)
Chance	None (noise)	Variable	Randomness/disorder

The Beginning of Time

As we live with time all our lives and it is inextricably linked to everything that we do, it is hard to think of it as having had a 'beginning' but most physicists believe that time did have a beginning. Jim Al-Khalili writes: "*...time itself is thought to have started at the Big Bang and is part of the fabric of the Universe. The Big Bang cannot even be considered as the 'first event' since that would require it to have happened within time...Physicists are now trying to understand why the Big Bang happened in the first place. What caused it? Unfortunately, cause and effect are notions that require time, and since the Big Bang marked the beginning of time we cannot say that something 'prior' to it caused it. ...One explanation which physicists are fond of using to describe how the Universe came into existence is that the rules of*

13. Davies, P.: Bit before it? *New Scientist* 161(2171):3 (1999) *in* Lennox, J.C.: (2007) p.167.

14. Lennox, J.C.: (2007) p.167.

15. von Baeyer, H.C.: In the Beginning was the Bit. *New Scientist* 169(2278): 26 (2001) *in* Lennox, J.C.: (2007) p.167.

16. Davies, P. & Gregersen, N.H.: (2010) p.3

quantum mechanics would have allowed the Big Bang to happen on the understanding that the Universe should quickly 'pop back out' of existence again. For reasons we do not fully understand yet, what may have happened next was that the Universe quickly underwent a brief period of extremely rapid expansion after which it became a permanent fixture, still expanding but at its current, more sedate, rate." [17] We met this phenomenon called 'quantum fluctuation' in the previous chapter, where a particle and its antiparticle pair can momentarily form out of the empty vacuum of the Universe and then almost instantly mutually annihilate themselves again. This explanation for the origin of the Universe has significant problems. Foremost, it ignores the information which Davies, Wheeler, von Baeyer and Zeilinger have just described as necessary to specify the fundamental functioning (including the initial conditions) for our universe. Further, since time began with the formation of space, light and matter when the Universe began, that information must have preceded the formation of the Universe.

Information and time

Einstein's General Theory of Relativity says that time varies with mass, acceleration and gravity.[18] To have time, you must have mass. If there is no mass then the entity does not experience time and is 'timeless'. Therefore as information has no mass it is not constrained by time and is outside time or transcends time. This can be observed by looking at ancient writings from the Middle East. The hieroglyphs in the pyramids in Egypt can be read thousands of years later; one only needs to know the code to understand the message recorded by the author. The same applies to the Dead Sea Scrolls, although the fragility of the parchment on which they were recorded helps to illuminate the difference between the 'hardware' (stone versus parchment on which it is recorded), the 'software' (the language in which it is written) and the information (the recorded message). Whereas the Egyptian hieroglyphs, carved into stone inside the pyramids have weathered very little over the millennia, the Dead Sea Scrolls have pieces of information missing, because the hardware (the papyrus scroll) on which the software (the ancient Hebrew language) was recorded, has disintegrated. In the case of the pyramids, the hardware (the stone) was particularly robust and has hardly changed, so that the software, the system of hieroglyphs, has been well preserved and the information has not physically disappeared from view. From these examples, it can be seen that the information is an entity that is different to either the hardware or the software but not completely independent of it in our physical space/time domain. The information is dependent on the software to record its meaning *and* on the hardware for its physical storage. Without both hardware and software, the information still exists but it has no *physical* presence, which is why when listening to a lecture, the information conveyed by the speaker must be written down before the student forgets it! Similarly, when typing a new document on a computer, the information must be saved to the physical storage within the computer; otherwise the information is lost. Information can therefore be viewed as an invisible entity that is made visible by the different media in which it is recorded.

When describing the origin of life in *Scientific American* in 1953, George Wald wrote: *"Time is in fact the hero of the plot…given so much time the 'impossible' becomes possible, the possible probable and the probable virtually certain. One has only to wait: time itself performs miracles."* [19] The problem with this statement is that a

17. Al-Khalili, J.: (1999) p.117.

18. Missler, C.: *Cosmic Codes.* Koinonia House, Coeur d'Alene, Idaho (2004) p.369.

19. Wald, G.: The origin of life. *Scientific American* 191:53 (Aug. 1954) *in Journal of Creation* 23(3):126 (2009).

string of amino acids will not make a functional protein without first possessing a string of DNA containing information, even with billions of years at their disposal. Without information it is still a string of amino acids. *Thus the true hero is invisible, intangible information!*

When Egyptologists (palaeontologists) first discovered hieroglyphs, they did not at first understand their significance as a language (the software), even though they had a strong suspicion that there was a message recorded within the drawings that they had found. In this case, the hardware (the stone) and the software (the writing system) were intact but the information was not interpretable. In 1799 Napoleonic forces found the Rosetta Stone which recorded the same passage in three different languages; the ancient Egyptian writings were then able to be decoded and the information became meaningful. In this case, even though the information was intact, the code needed to be interpreted before the information could be understood. Note that no matter how long it took to decipher the message recorded in the hieroglyphs, the information did not change; once it has been formulated, the information transcends time. To escape time therefore, one has to escape our material universe to a place outside of time (for example beyond the expanding edge of our universe).[20] There may, however, be other places to escape time as we shall see; but first we need to take a detour to learn about 'Chaos'.

Chaos Theory

Chaos Theory, in the mathematical sense, describes behaviours that repeat themselves but the patterns are so complex that the events appear to occur in a random way (such as weather patterns, economic cycles and biological ecosystems) and yet the functioning of these systems can be described by the information supplied at the outset. It is formally referred to as: *'Deterministic Chaos'* or 'Chaos Theory' for short. The effect can be very small, such as the vibrations of molecules or very large, such as magnetic field changes in celestial bodies.[21] It was not until the advent of computers that the speed of number computation became sufficient for Chaos Theory to be fully recognised and developed as a legitimate branch of science. The term was coined by James Yorke in the mid 1970's but mathematically, 'chaotic' behaviour had been observed for a century before that.

Complex systems can appear chaotic because the effects are often non-linear, so we cannot readily perceive all the variables controlled by the different laws of the Universe that determine their outcomes. The reason that Chaos Theory exists, however, is that there is unrecognised information embedded within the system being studied. A magic show to a child is surprising because the nature of the illusion is not understood. But once the child understands the trick, the illusion is not 'magic' any more. In the same way, complex behaviours in the Universe can be understood once sufficient knowledge of their deterministic laws is known. It is probable that there are still complex systems in the World that are determined by laws that are not yet fully understood, meaning that their behaviour appears 'chaotic' at present. This is the same problem that the Egyptologists had before the discovery of the Rosetta Stone; the information existed, it just needed to be deciphered.

Chaos Theory has been applied to the growth of populations of living things and to describe the branching patterns of blood vessels and the bronchi in the lungs. These are further examples of the information embedded in biological

20. Sarfati, J.: *By Design.* Creation Book Publishers (2008) p.229.

21. Oracle Education Foundation ThinkQuest *In A World of Order…Chaos Reigns!* (1997). http://library.thinkquest.org/3120/

systems. Another characteristic of a chaotic system, which is also called a *'dynamic system'*, is that it is sensitive to its initial conditions, which then determine its subsequent behaviour.[22] As the relationships are non-linear, small changes in the inputs can result in large changes in the outputs. So if phenomena such as the turbulence in fluids, as seen in the vortex at the tip of an aeroplane wing, were not seen as deterministic one hundred years ago, (Birchoff first described turbulence in 1927), where might our understanding of complex systems be in another hundred years? Chaos Theory is yet another demonstration of hidden, invisible information within our universe.

The Digital Universe

In 1623, as Galileo Galilei (1564–1642) began to understand the order within the Universe better, he wrote: *"The great book of nature is written in mathematical characters."* [23] Since that time, subsequent discoveries have continued to reveal the uncanny degree of order governing the Universe and most surprising to many was that it is finite. To summarise the parameters of the Universe discussed previously, it was originally thought to be infinite in space and time but progressive scientific discoveries have confirmed that it had a beginning and it will have an end. We also now know that the Universe has an edge or outer boundary and although very large, the number of stars and galaxies in the Universe is finite. So space within our universe is not infinitely expansive and surprisingly, nor is it infinitely divisible, as discussed in Chapter 8. Remember that space cannot be divided smaller than 10^{-33} centimetres, after which it still exists but it loses its physical properties and enters the quantum realm of non-locality. Similarly, Plank time of 10^{-43} seconds is the smallest unit of time. Thus, there are boundaries on physical space/time, both great and small. Time, according to the Standard Model, began with the formation of the Universe at the 'Big Bang' and most scientists would consider that there will be an end to the Universe (and hence time) when all energy is evenly distributed throughout the whole Universe in the ultimate heat death, although this will be many years into the future. If the Universe were infinitely old, then it would be uniform in temperature, which it is not and therefore it cannot be infinitely old.[24]

The Universe is therefore made up of many tiny blocks of space and time and can be thought of as modular. In physics, a small part or portion is often called a quantum (an amount), such as a 'quantum of light' or a small, defined packet of light which physicists call a 'photon'. This modular arrangement of the components within our universe is sometimes referred to as 'quantisation', meaning it is made up of a discrete number of units which are theoretically countable (even if it is completely impractical to do so). But if they can be counted, then they could be expressed as numbers or digits and measured, meaning that our universe is digital in its construction. A recent article quotes Caltech cosmologist, Sean Carroll who said: "Each state of space… [is] a matrix of numbers, one number for each point in space." [25] If information and biology are digital, and the Universe is also defined by information and is countable, then it must follow that the Universe is also digital. A digital universe is an unexpected finding and once more points to an underlying order which was recognised by Galileo in his statement in 1623 (not to mention Einstein and Whitten). It will become apparent as we study the Cosmos that many entities in our universe are quantised.[26] However, if the Universe is defined by information and is digital, the implication is that our universe was pre-specified and not a chance event. But if the formation of our universe was not a random

22. Hasselblatt, B. & Katok, A. *A First Course in Dynamics: With a Panorama of Recent Developments*. Cambridge University Press (2003).

23. Galileo Galilei: *Il Saggiatore* [The Assayer] 1623 *in* Fowler, T.B. & Kuebler, D.: (2007) p.127.

24. Missler, C.: *The Beyond Collection*. (DVD) Version 2.0. Koinonia House, Coeur d'Alene, Idaho (2010).

25. Wertheim, M.: Are we just numbers in the matrix? *Cosmos* 52 (2013) p.65.

26. Missler, C.: *Cosmic Codes*. Koinonia House, Coeur d'Alene, Idaho (2004) p.335.

or chance event, then there is the implication that there was information and hence intention behind its formation. This begs the controversial question, was it designed?

Design

In information terms, 'design' is the intentional arrangement of information. Sometimes the purpose of a system is obscure to the observer who stumbles upon it for the first time and the information (and hence the design) goes unrecognised but usually the purpose can be discerned by examining the system; an aeroplane or an education system is obviously designed but an economy is less obvious. After the purpose is discerned, often the information can be recognised and finally the design can be identified.[27] A more obscure example might be the head of a stone axe. To the untrained observer, the irregularly shaped piece of rock may not appear remarkable, as the original handle has now disintegrated. But once the rock is discovered to have a cutting function, the pattern of the chips that have been flaked off the rock can be discerned and the design becomes apparent. From this example, however, it can be seen that to recognise design, our brains need to match recognisable patterns (which are pieces of information that have been previously identified).

When we view a skyscraper, we all acknowledge that it has been designed. Positively identifying design is not always that easy though. If something was designed in the past, and it can positively be identified as such, it is referred to as an artefact, for example, a cuneiform tablet. In this case, the writing on the cuneiform tablet identifies it as something manufactured rather than being an object made by random processes. So what happens if no one was there to witness the event, but it is still suspected that the object was designed? How can we tell? Does the Information Model help us here? The answer is that it most emphatically does; the hallmark of any design is the introduction of information. The difficult thing is to discern if there was information input or not. Even with very complex events or systems, there can be great difficulty in assessing the presence of information input. A typical example might be the weather. Weather patterns reoccur quite regularly but they are never quite the same, so are they random or is some form of pre-specified information controlling their behaviour? The advantage of viewing the question from an information standpoint is that if meaningful information input is identified, whether intrinsic or extrinsic, whether the introduction of the information was observed or not, there is an element of 'design' involved; whereas if there is no information input, then the only conclusion is that the events occurred by 'chance'. Time and direct observation can be removed from the equation.

The next point to note is that design requires a 'designer' and chance requires the absence of a 'designer'. As computer programs require elements arranged in a specified way, which we recognise as the information in the software, we know they are not there by random chance. They are the product of a 'program designer'. With computer programs it is easy to discern the design (information) within the digital system. But do all things that are digital, imply design? If events occurred by chance then a limited degree of order may occur randomly but it cannot be expected. When throwing a standard die, to throw three sixes in a row can occur by chance but to throw one hundred sixes in a row would make one doubt that it was a standard die. Certainly the multilayered, digital precision

27. Behe, M.J.: *The Edge of Evolution*. Simon & Schuster, New York (2007) pp.239–240.

that is observed in our universe appears beyond the realms of blind chance. If we discover discrete pieces of information arranged in an organised way, then this is the essence of digital construction and digital constructions require meaningful information input. So if the Universe is digital, does the same logic apply? It would appear that there is no other conclusion possible as the only choices are shown in Figure 9.1 below:

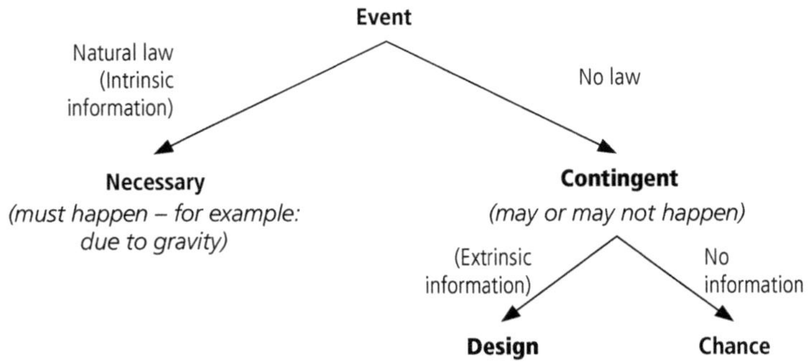

Figure 9.1: Information Outcomes in Our Universe Note that Natural Law results from previously specified or intrinsic information.[28]

Paul Davies, former Professor of Theoretical Physics at Cambridge University wrote: *"It is hard to resist the impression that the present structure of the universe, apparently so sensitive to minor alterations in numbers, has been rather carefully thought out... The seemingly miraculous concurrence of these numerical values must remain the most compelling evidence for cosmic design."* [29] Davies statement is reiterated by Owen Gingerich, Harvard Professor of Astronomy and senior astronomer at the Smithsonian Astrophysical Observatory who wrote: *"A commonsense and satisfying interpretation of our world suggests the designing hand of a super intelligence."* [30]

In 2007, Paul Davies wrote an article exploring the possible origins of the laws of the Universe and the possible options. He begins by considering the Multiverse option: *"This 'multiverse' explanation is just a little too glib. For example, what about the multiplicity of other universes? Not only is it extravagant to invoke all that cosmic real estate, there has to be some sort of universe-generating mechanism to create the varied patches in the cosmic quilt.*

Several mathematical theories exist that describe the creation of universes. In fact, one of these, called eternal inflation, is the currently favoured explanation for the Big Bang. And a by-product of the inflation theory is the prediction of countless other Big Bangs, in vastly distant regions of space and time. However, the need to assume — without any explanation whatsoever — a universe-generating mechanism, an inflating space, and a host of other advanced physical ideas, leaves the multiverse theory falling far short of a complete explanation for why the universe is as it is. So while a multiverse is a distinct improvement on Intelligent Design [in Professor Davies' paradigm], *it is by no means a panacea."*

He continues: *" I suspect we can do better than this. Previous attempts to explain why the universe is fit for life all suffer from a fundamental shortcoming: they appeal to something outside the universe: an unexplained god, physical laws that exist reasonlessly, or a vast ensemble of other universes.*

It would surely be better to try and explain as much as possible about the universe by appealing to processes within it. But here we hit a major snag. Scientists

28. Gonzalez, G. & Richards, J.W.: *The Privileged Planet*. Regnery Publishing, Washington DC (2004) p.297.

29. Davies, P.: *God and the New Physics*. Simon and Schuster. NY (1983) p.189 *in* Strobel, L.: *The Case for a Creator*. Zondervan, Grand Rapids (2004) p.125.

30. Gingerich, O.: Dare a Scientist Believe in Design? *In Evidence of Purpose*. Templeton, J.M. (Ed.) Continuum, NY (1994) p.25 *in* Strobel, L.: (2004) p.128.

have always assumed that the basic laws of physics are immutable mathematical relationships, imprinted on the universe at the moment of the Big Bang. It is a viewpoint often called Platonism, after the famous Greek philosopher. According to Plato, mathematics isn't an invention of the human mind; rather, it exists independently in a non-material realm of perfect, idealised forms that lie outside the physical universe. Mathematics is discovered, not created.

Many theoretical physicists follow in the Platonic tradition, and envisage the laws of physics as infinitely precise eternal mathematical relationships that are simply 'there', transcending physical reality. Note the curious asymmetry involved here: the universe depends for its properties on the laws, but the laws are in no way affected by the universe. I believe we will never achieve a satisfactory and complete scientific account for why the universe is as it is so long as we cling to Platonism — to externally-imposed, immutable mathematical laws." [31] Note that this is Professor Davies "*belief*" within his preferred paradigm but not a shred of evidence is proffered to support it.

Davies continues: "*But what's the alternative? In recent years, a radically new view of physical laws has emerged, prompted by the growth of the science of computation. At rock bottom, a law of physics is simply an algorithm that takes input data and returns output data. The motion of the planets round the Sun is an easy example — it can be calculated using Newton's laws of motion and gravitation. Knowing the positions and motions of the planets today, we can work out where they will be, say, this time next year. So today's information about the planets is 'processed' by the laws, and next year's information is delivered as output. Looking at it this way, the laws of physics are akin to computer software. And the hardware? Well, the universe itself, of course.*

Regarding the universe as a gigantic computer on which the laws of physics are 'run' prompts us to ask a curious question. Any computer is fundamentally limited in its performance by two factors: speed and resources. So is there an analogous limit to the power of the Great Cosmic Computer?

The answer is yes. The universe may be vast, but it is finite in both age and size. The reason for the latter concerns the finite speed of light. Since the Big Bang that gave birth to the cosmos 13.7 billion years ago, light can have travelled at most 13.7 billion light-years. So there is a 'horizon' in space beyond which we cannot see, however good our instruments may be. As nothing can travel faster than light, the horizon represents a basic limit to communication. Expressed simply, two regions of the universe beyond each other's horizon cannot combine the results of their 'computation'.

It is fairly straightforward to work out the theoretical maximum number of bits of information contained within a volume of space encompassed by the horizon. For example, our volume contains about 10^{80} atoms, each capable of representing a few bits of information. Taking into account all the known particles, including photons, neutrinos and gravitons, the total information content of the universe is no more than about 10^{120} — that is, one followed by 120 zeros.

What is the significance of the finite informational capacity of the universe? In the Platonic view, there is no significance, because Mother Nature computes in the Platonic heaven of infinitely precise mathematical relationships and infinite resources. But according to the software view of the laws, it is meaningless to invoke any mathematical procedure or relationship that exceeds the theoretical information limit."[31] This argument assumes that the Universe is purely material and that there is nothing else that cannot be seen or counted. However, this assumption contradicts the findings in Chapter 8, where it was established that only 4 percent of the

31. Davies, P.: Life, the universe and everything. *Cosmos* 14 (2007) *at* http://www.cosmosmagazine.com/ .

Universe was able to be accounted for to date and Professor Davies has no idea what the other 96 percent contains – see Figure 8.1.

Davies resumes once more: "*For example, a well-known law of physics states that electric charge cannot be created or destroyed. The orthodox interpretation of this statement is that the charge carried by, say, an electron, cannot vary one iota: the charge is fixed to infinite precision. But the software view of the laws denies that any physical quantity can be specified with a precision better than one part in 10^{120}. If the laws of physics are the cosmic program, then there will be irreducible sloppiness, or wiggle room, in their operation.*" [31] Once again, this makes the unsubstantiated assumption that the laws of the Universe are a product of the Universe rather than the accepted view that they are the other way around. As we have seen in previous chapters, the Information Model of the Universe works very nicely with precisely defined laws that are the 'software' of the Universe.

"*The number 10^{120} is so huge that for almost all practical purposes it might as well be infinite. Electric charge, for example, cannot be measured to better than one part in a trillion. So we are unlikely to notice any 'flaws' of nature arising from the inherent lack of precision in the cosmic computation.*" Though Professor Davies' inability to measure the entities to that degree, is not confirmation of his theory.

"*However the information limit on the universe, though enormous today, was smaller in the past. That is because the universe was younger and light had not travelled as far. Most cosmologists believe the basic structure of the universe was forged in the first split second after the Big Bang, at roughly 10^{-32} seconds. At that time, the universe contained maybe only 10^{20} bits of information, implying a serious uncertainty or imprecision in the results of the operation of the laws of physics.* [Once again, this is guesswork; not proven fact.] *So we can think of the universe starting out with very ill-defined or approximate laws.* [He supposes!] *Over time, the laws sharpened and focussed into the form we observe today. What we want to explain is why, from this higgledy-piggledy start, the laws zeroed in on such an uncannily bio-friendly form.*" [31] Ah, and there's the rub. If "*the laws zeroed in on*" anything then there must have been information to specify that end-point, which requires a credible mechanism to generate it!

Davies therefore asks: "*Could the universe have fine-tuned itself in order to bring about life and consciousness? At first sight, the idea seems preposterous. How could the universe in the first split second, when its laws were still malleable, 'know about' the emergence of life billions of years later?*" [31] If it did, it would have required the prior input of information.

Davies replies: "*A possible answer to this question lies with quantum physics. At the atomic level, the orderly process of cause and effect — seen in macroscopic systems like planets in their orbits or billiard balls on a table — dissolves into ghostly patterns of vibrating energy. This indeterminism is captured by the famous uncertainty principle of Werner Heisenberg: it is impossible to know all the variables describing a quantum particle at the same time.*"

"*The way quantum uncertainty is usually confronted is as follows: a system such as an atom can be prepared to be in a certain definite state, such as in a particular position or moving at a particular velocity. At a later time the experimenter measures some property of the atom, such as its position. In general, the outcome of the measurement cannot be reliably predicted from the initial state: there will be a spectrum of possible outcomes, each with a certain probability.*"

"*Compare this with billiard balls. If you know where one is and you roll it in a certain direction at a certain speed, you'll be able to predict where it will be at a later time. However, with atoms, even if you know precisely what state an atom is in at one moment, you generally cannot predict what will be found at a later moment when the measurement takes place."*

"That much is agreed, and very well confirmed by experiment. But a problem arises when the quantum system is the entire universe. The initial state of the universe is not something any experimenter can prepare, so we need to think about quantum uncertainty differently when it comes to cosmology. We make observations of the universe today, and can use them to infer something about the past. The laws of physics are symmetric in time, so quantum uncertainty works both forwards into the future and backwards into the past. Therefore, any given observation of the universe made today is consistent with a large number of possible histories, stretching all the way back to the Big Bang. However, we can reject any histories that are incompatible with the emergence of life and observers, otherwise we would not be here to make the observations in the first place. So the very fact that an observation gets made today in some sense helps shape the reality of the past — even the far past."

"Bizarre it may seem, but linking the present with the past via quantum measurement is part and parcel of standard quantum mechanics, and can even be demonstrated experimentally, although so far only over very short time scales. This backwards-in-time effect cannot be used to send information into the past, or to change the past, but it does constrain the past to conform to the present."

"So can the 'retro-causation' aspect of quantum physics explain why the laws of physics are fine-tuned for life? Not in the usual formulation, no. Although quantum mechanics requires the presence of many alternative pasts, every allowed history develops over time in conformity with the same laws of physics. The differences come about purely from inherent quantum uncertainty, not from any variations in the laws of physics as such. What we would like to explain is why the laws themselves are bio-friendly, thus permitting at least some quantum histories containing observers."

"To do so, we would need to find some way of applying the general principle linking future to past through quantum observations to the laws of physics themselves. Until now, such an application would have been meaningless, because the laws were regarded as fixed and infinitely precise. But treating the laws as cosmic software, with an inherent flexibility, neatly lends itself to the task."

"Observations made throughout the entire duration of the universe can contribute to fashioning the form of the laws in the first split second after the Big Bang, when they were still significantly malleable." [31] Yet there is not one iota of evidence that they ever were *"malleable"*. In fact the opposite is true! Fine-tuning laws requires information input but where is the mechanism in Professor Davies' model to generate the information (*"fine-tuning"*) that is required? Davies continues: *"Thus the potential for future life acts like an attractor, drawing the emerging laws towards a bio-friendly region of the available parameter space. In this way, life, mind and cosmos form a self-consistent explanatory loop."* [31] This could only happen if there was pre-specified information *"drawing"* the process, which is not what I think Professor Davies had in mind!

"For four hundred years, science has been based on the implicit belief that the laws of nature are themselves supernatural, and thus off limits to scientific inquiry. The time has come [in Professor Davies' opinion] *to challenge this fundamental assumption and seek a natural physical mechanism that enables the universe to generate its own bio-friendly laws."*

"In the eternal quest to explain life, the universe and everything, it could be that life explains the universe even as the universe explains life. Which pretty much covers everything." [31] However, there are major problems with this solution to explain away the formation of the Laws of the Universe and the existence of life:

1. It assumes that the cause lies within our universe but gives no scientific reason;
2. It gives no explanation as to what these laws might consist of;
3. It assumes that the universal laws are imprecise which is not the observation of science;
4. It assumes they can form and fine-tune themselves for life but gives no mechanism;
5. It assumes that *"the potential for future life acts like an attractor"* which requires pre-specified information but does not explain the source of that information.

Thus, while it is laudable to seek a mechanism to explain the origin of the Laws of the Universe and why they are fine-tuned to support life, there is not a shred of scientific evidence to support the notion that there is *"a natural physical mechanism"* or that there is a materialistic explanation within our universe to account for it. In reality, it has been seen that the laws had to predate the Universe, or at the very least they must have been a product of the process that formed the Universe and are specified by information. Thus the Laws of the Universe are the *product* of information not the generator of it. Once again this appears to be wishful thinking emanating from Tier 4 of the Explanations Table (Table 1.5), rather than observation and scientific rigor from Tier 1.

Gödel's Second Incompleteness Theorem

In 1931, Kurt Gödel, a mathematician, published his incompleteness theorems. Lennox writes: *"Gödel in his Second Incompleteness Theorem...proved that one of the statements that cannot be proved in a sufficiently strong formal system is the consistency of the system itself. In other words, if arithmetic is consistent then that fact is one of the things that cannot be proved in the system...In informal terms, as the British-born American physicist and mathematician Freeman Dyson put it, 'Gödel proved that in mathematics the whole is always greater than the sum of the parts.' "* [32] The reason is that there is added information in a complex structure and in analysing any complex structure, the information specified in the relationships between the components of the system must also be taken into account. If a system is deconstructed, such as the library at Alexandria, that information is likely to be lost to our physical universe, unless it is carefully identified and stored. The next question however is: 'Is it lost altogether?'

A Self-Generating Universe

As Paul Davies argues above, that the Universe generated itself from its own defining laws, which are the natural laws of the Universe, Stephen Hawking and Leonard Mlodinow also assert in their book: *The Grand Design*. So do they fare any better than Davies in demonstrating that the Universe could have self-generated and hence be self-explanatory? Unfortunately for Hawking

32. Lennox, J.C.: (2007) p.52.

and Mlodinow, Information Theory demonstrates that this is not possible, as the very laws which Hawking and Mlodinow appeal to for self-organisation must first be specified by information! Information defines the Universe and everything is subject to it, as Paul Davies intimated in 1999.[13] Wherever you look in the Universe there is information embedded within it, determining its activity. In a system which increases in complexity, there must be an introduction of information which must come from somewhere. Bernd-Olaf Küppers explains it another way: *"...Information conveying meaning cannot be compressed without change in, or even loss of, its meaning...This is the case for any meaningful information...Viewed technically, this means that no algorithms – that is, computer programs – exist that can extrapolate arbitrarily chosen parts of the message and thus generate the rest of the message. But if there are no meaning-generating algorithms, then no information can arise de novo. Therefore, to understand a piece of information of a certain complexity, one always requires background information that is at least of the same complexity. This is the sort-after answer to the question of how much information is needed to understand some other information. Ultimately, it implies that there are no 'informational perpetual motion machines' that can generate meaningful information out of nothing."* [33] Thus the formation of the Universe with its complex structure of laws, stars and galaxies required information input (not to mention DNA for the formation of life, as discussed in earlier chapters). This information needed to be generated at some point but from Küppers statement it can be seen that the infant Universe could not have generated *"meaningful information out of nothing."* As described in Chapter 2, the information is either: intrinsic, environmental or extrinsic. However, when we consider the formation of the Universe, the information cannot be intrinsic to something that is not there yet. Neither can it be environmental since the environment within the Universe does not exist before its formation, so it must have been extrinsic. In other words, the information to specify the formation of the Universe must have existed prior to its formation. Further, from Küppers statement, that information source must have been at least as information-rich as the information that is found in the Universe today!

The Hypothesis of Universal Information Primacy

This brings us to the Universal Law of Information Primacy, which combines the Biological and Physical Laws of Coding. It states: **'No process or product within the Universe can alter the information that specified it.'** Restated, this means that information must have preceded the formation of the Universe and with the 'Big Bang', or any alternate process that constructed the Cosmos, came information, which specified not only the energy and mass that formed the Universe but also the way in which it occurred. Therefore, at the beginning of the Universe, a primary information generator must have produced the information that was subsequently introduced into our universe, along with energy/mass, which caused it to form. This is the Hypothesis of Universal Information Primacy or, in short, the Universal Information Hypothesis. Further, we previously observed that we live in a digital universe, thus it would be reasonable to suppose that all *origins* have been written in the language of mathematics. Once the Universe formed, there may have been a myriad of potential histories due to quantum behaviour, as described by Davies above but indeterminism was not part of its formation.

33. Küppers, B.-O.: *Information and communication in living matter. in* Davies, P. & Gregersen, N.H.: (2010) pp.180–1.

The Universal Information Generator

If the Law of Universal Information Primacy is true, then a Universal Information Generator (UIG) is a logical deduction from information analysis. The problem is that there is very little more that can be deduced. The UIG must have been outside our universe to generate the information to create the Universe, so it lies outside time and has no mass and thus is non-material. Therefore, we cannot use material methods to detect or communicate with such a UIG. As the UIG is outside time, it must be 'timeless' so must have had no beginning and will have no end, although this cannot be verified experimentally, as the UIG is outside our physical domain. The options for the identity of the UIG are limited and not new. They are summarised in Figure 9.2. This subject will be discussed in greater depth in Chapter 12.

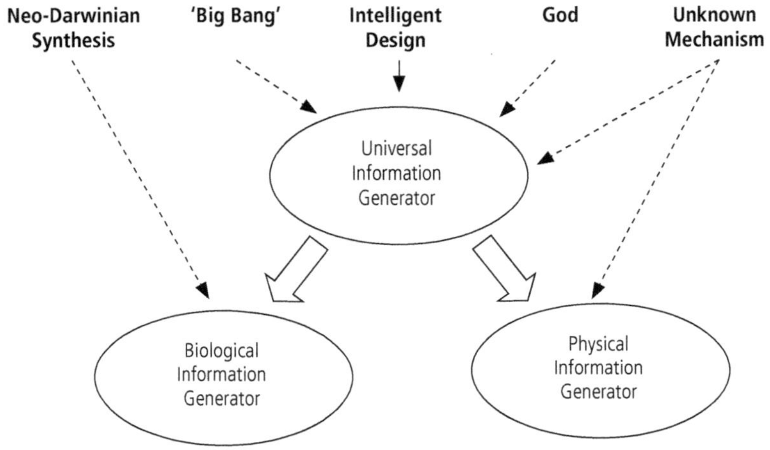

Figure: 9.2: The Universal Information Generator
It can be seen from this diagram that the number of possible alternatives for the causation of the Universe is small, as seen in the Levels of the Explanation Table 1.5. Chance has not been included as an option as it has already been shown to be inadequate to explain the extraordinary volume of information contained within the Universe. The 'Big Bang' has been included but it does not provide a mechanism for inducting information, which is a prerequisite for the formation of the Universe. Likewise, the Neo-Darwinian Synthesis also has a problem with information (as seen in Chapter 3) but has been included here to demonstrate the limitation of its scope in universal terms.

The Law of Conservation of Information

We previously asked the question as to whether information that is lost to our material world is lost altogether? It has already been shown that information is non-material, and that in our material world, information requires a vehicle or medium for storage and transmission; we observe this every day. Take for example, this page you are reading, which is probably printed in book form but may be recorded on a DVD, have been downloaded from the Internet via a copper wire through your telephone line, have arrived via optic fibre cable or beamed to your TV by an electromagnetic wave. Someone may even be reading the text to you, and it reached your brain via soundwaves from their voice via your ears. Whatever the method, there is a transmission vehicle required. If the medium fails, however, the message (information) is lost to our material state but is it irretrievably lost altogether? A clue to the answer comes from our investigation of the reproduction of living organisms. Professor-Emeritus Holmes Rolston III writes: "*What is conserved is not the matter, not the organism, not the somatic self, not even the genes, but a message that can only be conserved if – and only if – it is distributed, disseminated.*" [34] Edgar Andrews suggests that Relativity Theory mandates that all information through time is preserved. He writes: "*The fact that we are confined to 'now' and can visit neither yesterday nor tomorrow, doesn't mean*

34. Rolston, H.III: Care on Earth: generating informed concern. *in* Davies, P. & Gregersen, N.H.: (2010) pp.219–220.

that yesterday has ceased to exist or tomorrow doesn't yet exist. It is, in fact, one of the inevitable conclusions of relativity theory that the whole *of space-time must have a real and continuing existence - regardless of our perception of time as being divided into past, present and future. If you doubt my word, physicist Brian Greene sets out detailed arguments to prove this and concludes: 'Just as we envision all of space as* really *being out there, as really existing, we should also envision all of time as* really *being out there, as* really *existing too'(his emphasis)."* [35]

This statement by Brian Greene means that all of space-time is conserved in some way but what about information? Seth Lloyd, Professor of Quantum-Mechanical Engineering and Director of the W.M. Keck Centre for Extreme Quantum Information Theory at the Massachusetts Institute of Technology explained in a recent BBC Horizon programme: *"The fact that information is conserved in ordinary physics is at the very basis of physical law."* [36] Yet it has not always been seen this way. After attending a presentation in San Francisco by Stephen Hawking in 1981, theoretical physicist, Professor Lenny Susskind of Stanford University became disturbed by what he had learnt about black holes: *"I just knew or felt deep in my gut that Stephen had to be wrong. That lecture set me on a mission...* [which] *was to reconcile the two competing and conflicting points of view about black holes: that they eat information and evaporate but information is not allowed to be lost."* [36] The consensus after four decades of tussling with this conundrum is that: *"information can never be destroyed."* [36]

So, if physics mandates that this information, that can never be destroyed, along with all space and time that ever existed, is really *'out there',* where is *'there'?* For the answer to this question we have to go back to Bohm's Explicate and Implicate Orders Model from the previous chapter. After intensively studying the Universe for well over a century, no such place has been found within our physical universe to date, so perhaps the *'there'* is not to be found in our physical space/time domain. If it was then insurance companies would not pay out so much money when houses burn down because what was lost should be able to be restored. So Susskind has hypothesised that it is located beyond or on the outer edge of the physical universe, as no closer physical repository has been identified to date.[36] However, if the record of the space-time continuum cannot be found in the Explicate Order then maybe it exists in the Implicate Order in the frequency domain. This would mean that a frequency version of our physical existence that underlies our space-time domain was persisting and (under the right conditions) might be induced to regenerate the physical entity if required. The startling implication, if this is true, is that the frequency domain would be the primary domain and our familiar space-time realm is a dependent secondary domain — a projection or manifestation of an unrecognised, non-material digital realm. This was the conclusion Bohm came to from his study of Quantum Mechanics.

This is consistent with the Law of Information Primacy, which states that information precedes everything else in the Universe. Therefore if the physical information carrier is lost in our material domain it cannot automatically be redeemed in our four dimensional physical state but there may be another way. For example *Tyrannosaurus rex* cannot be regenerated as it has become extinct and its information code (both DNA and cytoplasmic) has been lost from our physical domain.[37] However, it is entirely consistent with the characteristics of information that a form of non-material, vibrational code could be stored in another domain or dimension and that the information may be conserved and recoverable if it could be retrieved from that other place. This is similar to an internet banking

35. Andrews, E.: *Who Made God? Searching for a Theory of Everything*. EP Books, UK (2009) p.110.

36. Shariatmadari, H.: *What is Reality?* (TV) Horizon, BBC Productions, UK (2011).

37. Some DNA has survived in the marrow of *T. rex* bones but a complete genome has not survived.

transaction, where you can print out the transaction that you have just made, to keep in a hardcopy format at home but the bank stores an electronic record of the same transaction on its computers remotely. If you were to lose your printout of the transaction, it is possible to go to the bank and request another copy, even if this reprinting of the information is not straight forward or easy. If the storage of a non-material copy of the physical information was subsequently shown to be true, then the Law of Conservation of Information would be confirmed; unfortunately as we are confined to the first four dimensions, we do not know of any way to access that information at present. Although this notion of information in another dimension (which cannot currently be accessed) sounds far-fetched, it follows logically from the Information Model.

As previously discussed, String Theory describes ten dimensions and M-theory, eleven dimensions, only four of which are measurable (length, width, height and time) with six (or seven in M-theory) which are 'curled' and unmeasurable from within our material world. Remember that 96 percent of mass/energy in the Universe is thought by physicists to be hidden as 'dark matter' and 'dark energy'. If the information produced by the biological and physical information generators came from another dimension (for example, a fifth, sixth or seventh dimension), then that may also identify the location of at least some of the 96 percent of the Universe that currently cannot be found. Although the *T. rex* became extinct on Earth and its information was lost from our physical domain, the information may be preserved in one of Hawking's higher dimensions along with the 96 percent of dark energy/dark matter, estimated by physicists to exist somewhere else in the Universe. As will be seen in Chapter 10, thought and consciousness also appear to be a projection from a non-physical domain and hence outside the constraints of our space-time domain, so they should be preserved too. Thus, if Mozart had written one last symphony, and the only physical copy of it was burnt accidentally after he died, the information is lost from our space-time domain but the information, which formed his consciousness in the sixth dimension, may still be preserved. (See Figure 12.1) This would seem to be confirmed by Brian Greene's comment that: "*all of time* [is] really...*out there, as* really *existing too.*" [38]

Lennox writes: "*More recently, William Dembski has argued for a non-deterministic law of conservation of information in the sense that, although natural processes involving only chance and necessity can effectively transmit complex specified information, they cannot generate it...For, if information is conserved in some sense, then we might logically expect that any origin of life simulations that claim to get information 'for free' by purely natural processes must somehow, in spite of their claim, be smuggling that information in from outside.*" [39] Thus a Law of Conservation of Information means that information cannot be inexplicably produced (nor lost) and further, that it should be possible to audit it with the same rigor as a company's balance sheet.

Summary of the Information Laws

In Chapter 8, two basic laws of information were discussed, and now two more can be added to the list. The Third Law of Information states: within a closed physical system, all information is bound by the Law of Entropy (as it depends upon a physical vector to carry or store it). The Fourth Law states that: information is conserved; therefore it can be audited.

38. Greene, B.: *The fabric of the cosmos.* Alfred Knopf, NY (2004) pp.132–9 *in* Andrews, E.: (2009) p.110.

39. Lennox, J.C.: (2007) p.153.

In summary, the Laws of Information are:

1. Information is the fundamental entity that pre-specified the entire Universe.
2. Information is independent of time.
3. Within a closed physical system, all information is bound by the Law of Entropy.
4. Information is conserved and may be audited.

The Information Model

Our investigation of information has led us from the smallest cell to the complexity of the human brain and from the non-material, subatomic realm of quantum mechanics to beyond the edge of the known Universe. Underpinning everything is information, which appears to be distributed in a digital, holographic manner. This has allowed the description of new laws in biology and information and opens vast new areas for future scientific research. It also gives a new information paradigm to help us understand our universe. As information is intangible and non-material it is not able to be perceived directly by our senses in the first four dimensions, unless it has a materialistic vector or carrier that allows us to identify it. However, it is to be hoped that science will in the future discover better ways to detect and access information that lies within the frequency domain.

For all that has been discovered above, there is still much to be done. In particular, there are problems with explaining the origins of the information that we can detect. Walter Bradley writes: "*...There do not seem to be any free lunches in nature when one is trying to explain the origin of information...No cosmological information (i.e., finely tuned cosmological constants) results from mutation or natural selection.*" [40] And he continues: "*Arno Penzias, Nobel laureate in physics and the director of Bell Laboratories until its recent fragmentation, makes this observation about the enigmatic character of the universe: 'Astronomy leads us to a unique event, a universe which was created out of nothing, and delicately balanced to provide exactly the conditions required to support life. In the absence of an absurdly-improbable accident, the observations of modern science seem to suggest an underlying, one might say, supernatural plan.'* " [41] This conclusion will be discussed further in Chapter 12.

Thus, structured information preceded the formation of the Universe and controlled the roll-out of the galaxies, the formation of the Earth, the appearance of life and everything else; furthermore, it is digital. Information is a fundamental quantity of the Universe, and it is multidimensional, being expressed in both the physical space-time domain as well as the non-material frequency domain. Information is all of these and yet, and this is the paradox, it has no mass, cannot be seen, heard or touched; it is invisible to our physical senses, which is to say that it is intangible. Therefore, information is not readily apparent and it goes unnoticed unless we intentionally search for it. This paradox is one of the mysteries of the Universe: that information, one of the fundamental entities of the Universe, is concealed in this way and we often miss it unless we diligently seek it out, yet it is the foundation of the Information Model. However, now that we have discovered and defined the paradox and described this new information paradigm, in Chapter 10 we will begin to explore the implications of the Information Model and the ways that it alters our view of the Universe.

40. Bradley, W.L.: 'Nature. Designed or Designoid' *in* Dembski, W.A. (Ed): *Mere Creation. Science, Faith and Intelligent Design.* InterVarsity Press (1998) pp.48–9.

41. *Ibid.* p.40.

10

Exploring the Information Model

Everything we call real is made of things that cannot be regarded as real.[1]

Niels Bohr

Having completed our survey of the Universe and having assessed our findings in the light of information, we have been able to describe an Information Model about which we can now ask questions such as: are there any implications that stem from this Information Model? We are also ready to ask some of the most difficult questions and formulate some conclusions based on the evidence already collated, once again using the tool of information to aid our interpretation. As we approach these topics it is worth noting Bernd-Olaf Küppers' lament that: *"Phenomena associated with meaning, as expressed in the semantic dimension of information, appear to evade completely all attempts to explain them on a naturalistic basis, and thus also to escape scientific description."* [2] So, as the traditional approach has failed, it will be necessary to use a different method of analysis and in this chapter a number of topics will be assessed using our yard-stick of information instead of naturalism. The first topic is: 'Mind versus Matter'.

Mind versus Matter

Bertrand Russell (1872–1970) contemplated the question of mind versus matter writing: *"Is the world divided into mind and matter, and, if so, what is mind, and what is matter? Is mind subject to matter, or is it possessed of independent powers? Has the universe any unity or purpose?"* [3] The debate regarding materialism and science has been referred to earlier in this book and has been going on for a long time. Andrew Briggs writes: *"Thomas Huxley coined the word 'agnostic' in 1869 at a party held to mark the formation of the short-lived Metaphysical Society."* [4] Huxley, whom we met in Chapter 1 as a colleague of Charles Darwin, defined the term agnostic as: *"One who holds that the existence of anything beyond material phenomena cannot be known."* [5] But is it valid to assume that reality must involve physical characteristics exclusively?

The human mind has fascinated generations of enquirers and still steadfastly resists scientific understanding. As far as we know, William Shakespeare had little knowledge of science, yet from his direct observation of people he wrote: *"What a piece of work is man...how noble in reason! How infinite in faculty."* [6] Shakespeare is famous for his poetic flair but he was a keen witness

1. Bohr, N.: *at* Jones, A.Z.: About.com Education: Physics – Niels Bohr Quotes http://physics.about.com/od/nielsbohr/tp/Niels-Bohr-Quotes.htm.

2. Küppers, B.-O.: *Information and communication in living matter, in* Davies, P. & Gregersen, N.H.: *Information and the Nature of Reality: from Physics to Metaphysics.* Cambridge Univ. Press (2010) p.177.

3. Russell, B.: *History of Western Philosophy*. Routledge, London (2000) p.13 *in* Lennox, J.C.: *God's Undertaker. Has Science Buried God?* Lion Books, Oxford (2007) p.41.

4. Briggs, A.: And Information Became Physical. *in* Berry, R.J. (Ed.): *Real Scientists Real Faith.* Monarch Books, Oxford (2009) p.80.

5. *Ibid.* p.77.

6. Shakespeare, W.: The Tragedy of Hamlet, Prince of Denmark. *The Complete Works of Shakespeare*. Octopus Books, London (1982).

to human behaviour and he appears to be implying here that man has a greater depth to his existence than the mere physical form that we observe in our space-time domain. Eastern mysticism has for many centuries endeavoured to explore the mind and 'inner-self' through meditation and detachment from the physical world[7] and treated the mind as more than an extension of the body, like a hand or a leg. Niels Bohr (1885–1962), particle physicist and Nobel laureate was referring to Quantum Mechanics when he wrote: *"Everything we call real is made of things that cannot be regarded as real."*[1] but he may just as well have been referring to the human mind. It is from this enigmatic point that we will delve into subjects which science has shied away from in the past, in an effort to better understand the human mind, the thinking process and consciousness itself. However, as noted above, the traditional approach has not produced any effective models, so we will need to look for a different approach and for that we need some new tools.

Fourier Transforms and Holograms

In Chapter 4 it was observed that eyesight, hearing and possibly brain function use Fourier Transforms to process information. The Fourier Transform is a mathematical tool for resolving or converting any time-domain function into its frequency spectrum. It is used by mathematicians and engineers to analyse wave patterns so the above observations suggest that our senses take in information from the environment in the physical space-time domain but then convert that input into waveforms to process the information. This does not seem so strange when it is realised that sound is a wave, light is a wave and an EEG measures our 'brainwaves'. This also opens up possibilities in Bohm's Implicate Order as discussed in Chapter 8. If electrons and buckyballs can traverse the interface between the Explicate (space-time domain) and the Implicate (the frequency domain) then maybe our brains can process information that resides in both domains too.

Holograms are also Fourier transforms that distribute information across an entire piece of photographic sheet. A standard photograph is a *space-time* representation of the image of the object photographed by a standard camera but a hologram is a *frequency* representation of the image of the object, which has been photographed using laser light. If you view the hologram in natural light, it looks like a series of blurry smudges and is unintelligible because the hologram is a three-dimensional frequency representation of the original image that is now projected into two dimensions in the space-time domain. It can only be interpreted using the same frequency of light used to create it. If this is done, it produces the strange phenomenon that although the holographic print only takes up the same amount of space as your reflection in a mirror, it has three dimensional characteristics such that the viewer is able to peer around objects within the hologram that might otherwise obscure items behind them. Thus, entities within the hologram that were initially hidden, can now be seen. In Bohm's hypothesis, the Explicate Order of matter is a four dimensional Fourier projection of information that is stored in the hidden Implicate Order in a different domain.[8] The implication is that all the matter in the Universe in the entire space-time domain is a four dimensional projection of digital information in the (currently) hidden frequency domain.[9] Thus, just as holograms are a 2D representation of 3D objects, in Bohm's model, 4D objects are a projection of 5D or 6D entities; or maybe up to 11D if M-theory is correct.

7. Dolphin, L.: *Lord of Time and Space.* Good News Publishers, Westchester, Illinois (1974) p.41.

8. Missler, C.: *Cosmic Codes – Metacodes.* (CD) Koinonia House, Coeur d'Alene, Idaho (2004).

9. Missler, C.: *Cosmic Codes.* Koinonia House, Coeur d'Alene, Idaho (2004) pp.339–41.

Living in a Multidimensional Universe

As already discussed in Chapter 8, Stephen Hawking and Leonard Mlodinow believe that ten dimensional String Theory and its eleven dimensional brother, M-theory, will turn out to be the basis of a future grand unifying theory of the Universe. Their hypothesis is highly controversial and not all mathematicians and physicists agree with it but if Hawking and Mlodinow are right, String Theory leads to some startling implications. As previously discussed, proponents of String Theory have suggested that the Universe exists in at least ten but maybe eleven or even twenty-six dimensions[10] but in this discussion I will take Hawking and Mlodinow's interpretation of eleven dimensions, although the argument is just as valid with ten, twenty-six or any other number of dimensions above six. If this is so, it would seem logical that there is something in the Universe that exists in at least eleven dimensions and that the formation of the Universe involved these eleven dimensions in some way, even though we can only appreciate and measure four dimensions at present. It is possible for human beings to exist in an eleven-dimensional universe, yet not appreciate them all but it is hard to understand how an eleven-dimensional universe could have been created by an entity that can appreciate and exist in less than the eleven dimensions. Therefore any explanation for the formation of the Universe must involve an eleven dimensional process or entity as a minimum. However, as it appears that there was nothing before our universe was formed, if an eleven-dimensional universe came into existence, then it is reasonable to assume that there is another dimension outside the eleven dimensions, where the universe-forming-entity or process resided prior to the formation of our eleven-dimensional universe. A simple way to illustrate this is to imagine a cartoonist who creates a character, such as Charlie Brown. The cartoonist can write themselves into their comic strip if they choose to; but they existed in at least an additional two dimensions outside of the two-dimensional cartoon strip prior to Charlie Brown and Snoopy coming into existence (as we live in a four dimensional material universe).

It is now time to return to the missing matter and energy that physicists predict should be present in the Universe. A *multi-dimensional* universe (as opposed to multiple universes or a 'Multiverse') may be able to help explain this problem. In April 2008, *Cosmos* addressed this difficult issue of: "*...solving The Mystery of the Missing Mass. In an embarrassing development, over the past decade scientists have come to realise that they've been overlooking about 96 per cent of the universe. The visible 'ordinary' matter constitutes a mere four per cent, according to current thinking. Observations and theory say the rest is either dark energy (about 74 per cent) or cold dark matter (some 22 per cent). Since cold dark matter emits neither heat nor light, no one has yet been able to directly detect it. However, if ordinary matter were the only thing holding clusters of galaxies together, then astronomers say they would have long ago flown apart. There isn't anywhere near enough matter to produce the necessary gravitational pull. Ergo, the pressing need for dark matter to close the gravitational deficit. One widely accepted hypothesis is that dark matter particles could have been produced in copious quantities in the Big Bang, and that enough of them survived until today to constitute the invisible cosmological soup*".[11] Is it possible that information in the Implicate Order is one of the missing pieces that might explain these puzzles? If the Universe exists in more than the four space-time dimensions, as some cosmologists suggest, then human beings and indeed, all life may also, which could also provide a possible solution to the conundrum of consciousness

10. Finkel, E.: Fish out of Water. *Cosmos* 2:77 (2005).

11. Calamai, P.: Cosmic Roulette. *Cosmos* 20:57 (2008).

and memory which have also steadfastly eluded investigators. We will return to consciousness and memory later but first we need to address the entity that is life.

What is life?

Edgar Andrews observes that: *"Although proteins and nucleic acids are essential for life, they do not constitute life…the creation of a living cell (or a house for that matter) requires ingredients* plus *organisation."* [12] The textbooks say that life is self-sustaining and self-reproducing and it is commonly referred to as the 'Spark of life' but what is it? Richard Dawkins writes: *"The difference between life and non-life is a matter not of substance but of* information. *Living things contain prodigious quantities of information. Most of the information is digitally coded in DNA, and there is also a substantial quantity coded in other ways…there are three further ways in which information about the past is archived in such a way that it can be used to improve future chances of survival. These are the immune system, the nervous system, and culture."* [13] The immune system learns from experiences within an individual's lifetime to prevent further infection (such as the chicken pox virus) but it only applies to that individual's lifetime. *"The third memory"* Dawkins writes: *"is the one we ordinarily think of when we use the word: the memory that resides in the nervous system. By mechanisms that we don't yet fully understand, our brains retain a store of past experiences."* [14] Intellectual memory is analysed later in this chapter. Dawkins goes on to describe the fourth memory as: *"your brain includes collective memories inherited non-genetically from past generations, handed down by word of mouth, or in books or, nowadays, on the Internet."* [15] But none of these *"four memories"* really explains life; all this information is essential for human life as we know it but there is more to life than this. My impression as I studied biology at school and university was that the genome explained life and while to a certain extent that is true, I have now realised that a string of DNA from a *Tyrannosaurus rex* bone does not make a living dinosaur. American biologist, Professor Stuart Kauffman wrote: *"autocatalysis and molecular reproduction are necessary for life, but not yet sufficient. Life possesses deeper, still more mysterious properties".*[16]

Holmes Rolston III observes: *"Living things are self-maintaining systems. They grow; they are irritable in response to stimuli. They reproduce. They resist dying… They gain and maintain internal order against the disordering tendencies of external nature. They keep recomposing themselves, whereas inanimate things run down, erode, and decompose. Life is a local countercurrent to entropy. Organisms pump out disorder. This self has to be maintained against entropy, or, if you like, against 'ageing'. The constellation of these characteristics is nowhere found outside living things,…A crucial line is crossed when abiotic formations get transformed into loci of information."* [17] Thus life is denoted by processes that intrinsically oppose the Law of Entropy, whereas inanimate objects are powerless to defy it.

Richard Dawkins makes the observation that: *"DNA doesn't care about being 'dead', it is just coded information."* [18] It is true that a dead body has a genome but it must be stated that neither the DNA nor the information that it stores is 'alive'. It is only the fully competent cell that can live. There must be something more. Simply possessing physically stored information in the form of a genome is insufficient. Again, the genome is essential for life and is a marker of life but it does not define it *per se*. These are some of the problems associated with defining life, so the question of: 'what is life?' may be easier to assess if we could first decide what defines the absence or loss of life, which is death.

12. Andrews, E.: *Who Made God? Searching for a Theory of Everything*. EP Books, Darlington, UK (2009) p.203

13. Dawkins, R.: *The Greatest Show on Earth*. Bantam Press, London (2009) pp.405–6.

14. *Ibid.* p.407.

15. *Ibid.* p.408.

16. Kauffman, S.: *Investigations*. Oxford University Press, Oxford (2000) p.47 *in* Fowler, T.B. & Kuebler, D.: *The Evolution Controversy*. Baker Academic, Grand Rapids (2007) p.317 footnote 77.

17. Rolston, H.: *Care on Earth: generating informed concern*, pp.205–243 *in* Davies, P. & Gregersen, N.H.: (2010) p.212.

18. Dawkins, R.: (2009) p.303.

What is Death?

Life requires matter, energy and information but we have seen that on their own they do not create life. Matter, energy and information can be used to create a computer which functions and can produce real outputs. It is also possible to visualise the construction of a factory that would allow our computer to make more copies of itself, which is equivalent to 'reproduction' but does that confer the status of 'life' on the computer? There is something else, call it a 'life factor' that distinguishes a computer from a living organism. It also distinguishes a living organism from a dead one. When an organism dies, five minutes after life becomes 'extinct' the body and DNA are physically the same in the space-time domain but 'life' is missing.

So, what is death? The obvious answer is the absence or unravelling of life but what does that mean? There are different types of death. Cell death is when the cell no longer functions and the chemical processes in the cytoplasm of the cell irretrievably break down. If a person has a cardiac arrhythmia and their heart suddenly stops beating, the person loses consciousness within seven seconds due to the loss of an oxygen supply to their brain, as the brain cannot store oxygen. The oxygen is delivered by the haemoglobin transport system within the red blood cells in the bloodstream, which in turn is dependent on the beating heart to pump the blood. If the heart is not started again within ten minutes (under standard conditions), then 'brain death' occurs.[19] However if the person is defibrillated within the first few minutes and their heart beat returns to normal, it is possible for the patient to make a full recovery. In severe head injuries or near drowning, sometimes the victim's physical body can be resuscitated and it survives on a ventilator but the person is declared 'brain dead'; they have lost their consciousness, their personality and all the essential qualities which we think of as 'human' are missing but their physical body still functions perfectly (sometimes for many years if they remain on the ventilator) and their organs (such as their kidneys) are often able to be successfully transplanted into someone else and survive for decades. However if the life support system is switched off but the body continues to breathe and the heart continues to pump yet they remain unconscious and 'not there', are they alive or dead? Their physical body is 'living' (at least the cells are) and the brain stem (lower brain) which controls breathing is functioning but as human beings they are, for all intents and purposes, 'dead' as they cannot function like a 'living' person. There are therefore at least two types of death: cell death and death of the consciousness. Cell death can be seen when the blood supply to a limb is lost, such as in a smoker or diabetic who develops gangrene of a toe from blockage of their arteries to that foot. Their brain is still completely normal, but some of their peripheral cells have died, their toe turns black and needs to be amputated before the gangrene progresses to the rest of the body. In contrast, when we lose the information that defines our consciousness we become 'brain dead', even though as we have just seen above, the physical body can be 'alive'. Thus there is a stream of consciousness that is distinguishable from our physical body, which appears to epitomise human life but that is massless and is non-material. We know that this consciousness is dependent on the physical network of nerves in the brain for it to function, in the same way that a computer program is dependent on its computer circuitry (hardware) to operate but we have seen that there is more to life than this. So could these thoughts and consciousness also reside in the Implicate Order in an information repository,

19. If the temperature is very low then brain death can be delayed as in people trapped under ice or in cryosurgery, where the body's temperature is temporally lowered to lower respiration to preserve brain tissue.

perhaps as a waveform? (Remember that we observed that EEGs are waveforms.) A mechanism is not yet known.

Cell death occurs when the life-sustaining processes within the cell cease. This may be due to lack of energy, oxygen starvation, nutrient deficiency or irreparable mechanical disruption within the nano-machinery or external cell membrane. When someone dies, cell death is sequential, depending upon the oxygen requirements of each tissue; for example, brain death is first, which occurs within minutes but the cells within the cornea (the clear covering of the eye) can survive for a number of hours.

Thus, when brain death occurs, some vital ingredient is lost from the living body but what is it? One clue is that consciousness (which is just as difficult to characterise as 'life') is lost as well. Much research has gone into trying to determine if life and consciousness are emergent properties of the functioning body and brain, such as the magnetic field that arises when a current flows in a completed electrical circuit; or are life and consciousness extrinsic properties, which are joined before birth and lost at death, such as the software programme in a computer? Are they due to functioning biochemical, cellular activity alone or is there an extra entity (perhaps in another domain) which is present but 'curled', as in String Theory, such that we cannot measure it? In information terms, are life and consciousness due simply to 'dynamic activity', such as the electro-chemical impulses that travel through the physical components of the neural network of the brain or the result of extrinsic information within our space-time domain, such as the picture signals beamed to your television or even from a different dimension?

Weighing up the preceding information, cell death appears to involve biochemical breakdown that results in the loss of cellular, material activity, whereas the death of consciousness appear to involve the loss of an additional non-material entity, as a human being who is 'brain dead' on a ventilator is physically identical, yet mentally different to their former healthy self. Thus the 'life' that we observe in an amoeba (or a single cell in a multi-celled organism) appears to be a dynamic (material) process, as is seen in the flow of electrons in an electrical circuit and is hence an emergent property that is dependent upon the physical integrity of the cell and the intrinsic information in its DNA, whereas the phenomenon of consciousness that is seen in more complex organisms, such as mammals and human beings requires the input of additional information. This information or entity has not been found in any of the first four dimensions, so it would appear to be in another place; let us call it the 'fifth dimension' to give it a name for discussion purposes. If that is the case, then it can now be seen that at organismal death, information in another dimension might become separated from its physical hardware platform in dimensions one to four. The physical body of the deceased person retains its genomic information in the space-time domain, which is in the first four dimensions, but it has lost the 'spark' of life, which is in addition to the molecular activity within the formerly living cells. In teasing apart the above mechanisms of death, it is important to remember that every complex system is dependent upon the competency of all of its critical subsystems. Failure in critical cellular machinery may cause cell death but this is principally a hardware problem. In contrast to the emergent quality of the cellular activity that results in 'simple life', the elements that produce 'consciousness' are an information or software issue not primarily a hardware matter and we must not muddle them up.

Thus, as a cell must already be living when a new life materialises at conception, the 'spark of life' is handed down from one generation to the next.

If we were to attempt to build a cell from scratch, as previously mentioned, the friction forces within the cell would prevent its animation. In contrast, if there is an extra dimension housing consciousness, then, when the gametes are formed from the living, parental sex-cells, they would bring the extra dimension that has the potential for consciousness with them, in the same way that our third dimension is inextricably linked to our first two dimensions. In our three spatial dimensions, you cannot physically make yourself into a two-dimensional being, although you can create a two-dimensional image, such as a photograph; however, that is not alive. When the sperm and oocytes unite, they form a new competent life in our space-time domain, with the fifth dimension that has the prospect of consciousness attached. Before the union of the sex gametes, each gamete has half the potential for new life, but is incompetent in itself. Similarly, new human life is dependent upon a physical platform in the first, second, third and fourth dimensions being partnered with a hidden information/software entity in the fifth dimension that forms consciousness. If the physical body in the first four dimensions dies, then the dynamic activity, which is the 'spark of life' that runs on the physical platform composed of living cells is lost, while the entity that is in the fifth dimension, which is consciousness, disappears from view, as we are physically confined to our space-time domain. But since this information defining consciousness is not in our physical Universe, it may persist in its non-material location as the information is massless and hence timeless. Separated from our physical world containing entropy, our consciousness presumably does not age either.

Therefore, five minutes after death, the physical substance of the cadaver with its constituent DNA information are the same as they were five minutes earlier when the person was alive and any stored chemical memory, such as immunological memory is presumably unchanged if it was possible to access it. We know this because doctors can recover viable donor organs such as hearts, kidneys and corneas from people who have died (such as in a car accident) and the donor organs can function perfectly well in the recipient's body for many years and the lifespan of the donor organs does not appear to be reduced just because the original owner has died and 'life' for the original owner has ceased. Thus we have established that there are two types of death - cell death and death of the consciousness. Cell death appears fairly straightforward but death of the consciousness still has unanswered questions. To improve our understanding of this elusive subject, we will now look more closely at the boundary between life and death.

Viruses and the Boundaries of Life

Viruses are very simple in comparison to bacteria and other forms of life. As a common example, Influenza A has only enough genetic (RNA) coding to make eleven proteins [20] and you will recall that two hundred and fifty-six proteins was considered the minimal set for the simplest cell in Chapter 3, so Influenza A needs to co-opt or commandeer a host cell's metabolic machinery to reproduce. But since they are so simple, the question arises: Are viruses living? It could be argued that viruses are living organisms because they reproduce and have genetic material and are autonomous but many scientists consider that they are non-living because they cannot reproduce without invading a host cell and hijacking the cytoplasmic machinery. Another way of saying this is that viruses are not independent, self-

20. Chanda, S. et al.: Human host factors required for influenza virus replication. *Nature* 463: 813–7 (2010).

contained units. This also means that they could not have been precursors to the first cell as they require a fully competent cell in order to reproduce. At this point, it needs to be pointed out that the nomenclature in science and medicine is not always consistent. Although many consider that viruses are not truly alive, we do talk about 'killing' viruses, which implies that they must have been alive in the first place! So what do physicians mean by this? In medicine, if a virus is dead then it is not infectious and can no longer reproduce. Thus viruses can be considered to be not completely alive and yet they must be so in some respect in order that they can be killed; hence the debate. This reproductive definition of life and death is used for simple cells as well as complex life, however it does not work for the death of consciousness, so we need to address awareness or consciousness from a different perspective.

Consciousness and Awareness

Consciousness or awareness of our surroundings is such an elusive concept that most of us rarely contemplate it but we use it all our waking hours and it needs an explanation. Yet science is currently at a loss to even know where to start. Oliver Sacks, British-born neurologist and author, who is Clinical Professor at the Albert Einstein College of Medicine in New York, lamented in a recent essay that despite all our advances in neuroscience, " ' *...neurological activity and psychic activity still seem utterly different,' even 'incommensurable' in kind."* [21] However, can a fresh approach using the Information Model help to elucidate this problem? Andrew Gosler of Oxford wrote: *"Information exists most powerfully in the mental realm; it is surely the very stuff of consciousness and of relationship, whether mathematical, ecological or personal."* [22] Consciousness is like an operating system on a computer; it is a platform on which other useful programs, such as thoughts and feelings run. Without this software package there is no useful output or productivity from the computer. Conversely, the hardware system by itself yields nothing either; it is the integrated hardware and software package combined into one system that is productive. As such, consciousness appears to be a facilitating software programme that allows multiple applications packages to function – once again, layer upon layer of information!

Further, Christof Koch, Professor of Computation and Neural Systems at the California Institute of Technology in Los Angeles points out that most of what we do is not under our conscious control. Nor are we *aware* that we are doing it. "'*When you walk, you don't think lift leg, move leg forward, put leg down. You just walk!' The same is true for talking. When you speak, you do not have to think about grammar and syntax and vocabulary, you just open your mouth and the words come out. 'If you had to consciously think through that stuff,' says Koch, 'you'd never get anywhere.'* " [23] Speech is controlled by a complex interplay between our cerebral cortex, or Neocortex, where our conscious thoughts occur and the cerebellum, which automates physical routines, such as how to make the sound of a letter with our tongue and voice box without thinking consciously about it. This is an over-simplification as there are other specialised centres for speech, such as Brocca's and Wernike's areas to initiate and execute the act of speaking but it gives an idea of the processes involved.[24] Speech relies on the competency of both the subconscious and the conscious parts of our brain to function harmoniously in the same way that to type this paragraph on my computer I needed both a fully competent operating system and a word-processing package, yet neither of these

21. Wertheim, M.: Alas Poor Zombie. *Cosmos* 11:90 (2006).

22. Gosler, A.G.: Surprise and the Value of Life, *in* Berry, R.J. (Ed.): (2009) p.187.

23. Wertheim, M.: Alas Poor Zombie. *Cosmos* 11:86–8 (2006).

24. Greenfield, S.: *Brain Story.* BBC video series (2001).

software tools were responsible for the words that you are reading. Remember that the electrochemical messages in the brain and the software packages in the computer are simply vehicles or conduits that carry or facilitate your conscious processing of the information. However, if you think that this level of co-ordination sounds complicated, it is just the tip of the iceberg. Most of our bodily functions, such as our heart rate, blood pressure and digestion are completely automated and are controlled by yet another mechanism called the *autonomic nervous systems*. Our autonomic nervous system controls functions which operate in the background, completely independently of our consciousness. If you think of the autonomic nervous system as the '*automatic* nervous system' you won't go far wrong. So why does consciousness exist at all?

Margaret Wertheim of Los Angeles writing in *Cosmos* continues. "*In principle, Koch says, there is no reason why consciousness is necessary to life. With enough 'input sensors and output effectors' it is conceivable that 'a zombie could pretty much do anything'. But since every zombie behaviour must be hardwired, the more situations it must respond to, the more complex its internal mechanism becomes. Instead,…'a much more powerful and flexible system' [is]that [which] we call 'consciousness'. The main function of this innovation, Koch proposes, is to enable organisms to deal rapidly with unexpected events and to plan for the future. As Koch likes to say, consciousness puts us 'online', allowing us to override our instinctual 'offline' programming.*" [25] Without consciousness we have to rely on pre-programmed information as computers, insects or robots do, which removes any form of spontaneity, which we treasure as a hallmark of being human. Koch's analogy with zombies starkly reveals the significance of consciousness. Table 10.1 compares these two similar, yet very different programming types.

So consciousness is a fundamental quality that makes us human but where does it come from? There does not appear to be a 'consciousness centre' within human beings, but consciousness appears to be the result of global brain activity, which suggests that consciousness is 'distributed' throughout the brain.[24] Mervyn Eadie, Professor of Neurology in Brisbane, Australia writes: "*The seat of consciousness, if it has a single localised seat, is not yet unequivocally determined and the phenomenon may really reflect a continuing dynamic interplay between upper brain stem and neo-cortical regions.*" [26] Thus consciousness exhibits holographic properties as previously discussed.

In these discussions we will begin by considering the *human* brain (in contrast to other organisms), as there is no doubt about the relationship between the human brain and consciousness. The cognitive abilities of other living creatures will be discussed later in the chapter. In addition, the human brain, which is estimated to contain thirty billion neurones,[27] stands out from every other living entity that has ever been known to have lived, in the way that it can look beyond mere survival. Chimps and dolphins are not interested in creating

25. Wertheim, M.: Alas Poor Zombie. *Cosmos* 11:91 (2006).

26. Eadie, M.: A Molecular Basis for Spiritual Experience? *Luke's Journal* 11(2):6 (2006).

27. Wertheim, M.: Alas Poor Zombie. *Cosmos* 11:89 (2006)

	Computers	Human Brain
Hardware	Computer components	Nerve cells and supporting tissues
Software	Machine Language	Biochemistry
	Operating system	Consciousness
	Applications packages	Speech
	User input	Thoughts and feelings

Table 10.1: Hardware and Software Comparisons between Computers and Human Beings Note that all software, whether for computers or living organisms, is ultimately information.

works of art or wearing clothing.[28] The human brain is three times the size of the chimp's brain, although brain size alone does not determine intelligence. Women's brains are almost universally smaller than men's but there is no substantiated data demonstrating that there is any difference in the level of intelligence between men and women. Instead intelligence appears to depend upon the total number of connections between the individual nerve cells. However, as chimps' brains are only one third the size of human brains, it is likely that chimps have fewer neuronal connections also. By way of a typical example, it has been shown that chimps can plan over a short period of time, such as hours or days, but only humans can plan for years, decades or centuries into the future.

So is consciousness purely an emergent property of electrical brain activity running on biological hardware or is there more to it? Alan Gijsbers, Professor of Medicine in Melbourne, Australia wrote: "*George Engel's biopsychosocial model of understanding humans, challenges the simple biomechanical approach to patients. Engel describes humans as a series of emergent systems from sub-atomic particles, through cells, organs/organ systems to the person…Engel uses the concept of emergence, whereby, as we move to the next level, new properties emerge which could not be predicted from understanding the state of the lower level. A useful slogan for emergence is, 'We cannot experience wetness by looking at the isolated H_2O molecule' because surface tension comes about by the interaction of many water molecules together.*" [29] But how might this emergent property which causes consciousness work? Professor Gijsbers continues: "*Neurological diseases illustrate the tight link between the mind and brain, and we recognise that defects in brain function (Alzheimer's, strokes, tumours etc) have a profound effect on mind function. But mind functions can also affect brain function…The trouble started with Descartes. As a mathematician he tried to extrapolate the principles of Euclidean mathematics into a full philosophical system. He tried to work out what were the most basic axioms, and from there deduce the full system via pure reason. Such an approach is called foundationalism. Of course no one's system of knowledge develops like this. It develops more like a web of knowledge which is continually changing as new knowledge challenges old knowledge. Descartes subjected all he knew to radical doubt; what was left in the end? That he existed and that he thought. He therefore developed his famous axiom, cogito ergo sum, I think therefore I am. Even that is challenged as illogical these days. Further, his thinking world was separate from the physical world. The world of the mind is the world of the soul is the immaterial world of the spirit. This was of a completely separate substance than the physical substance of the material world. Thus he entrenched dualism into our thinking. This did allow physicians to study the body as a mere machine, but it also forged the separation between the world of the mind, soul spirit and the world of the body or the environment. We are still grappling with Descartes' legacy - both its benefits and drawbacks.*[29]

Hawking and Mlodinow add: "*Descartes, for instance, in order to preserve the idea of free will, asserted that the human mind was something different from the physical world and did not follow its laws. In his view a person consists of two ingredients, body and soul. Bodies are nothing but ordinary machines, but the soul is not subject to scientific law.*" [30] Hawking and Mlodinow who profess a belief in the concept of scientific determinism[31] continue: "*It is hard to imagine how free will can operate if our behaviour is determined by physical law, so it seems that we are no more than biological machines and that free will is just an illusion.*" [32] However, just because something is difficult to imagine, does not make it impossible. Einstein's theories could not be conceived of before 1900 but now no physicist doubts them. It

28. That is not to say that they do not play with things from time to time but they do not intentionally wear articles of adornment or create memorials to the past.

29. Gijsbers, A.: The Dialogue Between Neuroscience and Theology. *Luke's Journal* 11(2):8–9 (2006).

30. Hawking, S. & Mlodinow, L.: *The Grand Design*. Bantam Press, London (2010) pp.30–1.

31. *Ibid.* p.34.

32. *Ibid.* p.32.

appears that Hawking and Mlodinow are forced to accept their conclusion because of their presupposition that *"physical law"* is the ultimate entity in the Universe. It is not surprising therefore that they continue by writing: *"While conceding that human behaviour is indeed determined by the laws of nature, it also seems reasonable to conclude that the outcome is determined in such a complicated way and with so many variables as to make it impossible in practice to predict…Because it is so impractical to use the underlying physical laws to predict human behaviour, we adopt what is called an effective theory. In physics, an effective theory is a framework created to model certain observed phenomena without describing in detail all the underlying processes."* [33] However, having stated that they believe that free will is an illusion, on the next page they write: *"In the case of people, since we cannot solve equations that determine our behaviour, we use the effective theory that people have free will. The study of our will, and of the behaviour that arises from it, is the science of psychology."* [33] By assuming that scientific determinism is the only possible answer, Hawking and Mlodinow have forced this contradiction upon themselves. Alan Gijsbers expresses an alternative view: *"If on other grounds we believe in an unseen world acting (somehow) on the seen world, then our failure to see how that should be, does not rule out that the unseen world can act on the seen world. If we do not understand how the mind affects the brain, it does not mean the mind does not affect the brain. We know that thoughts lead to action. Atomic bombs and jet aeroplanes were thoughts and plans before they were reality…Neuroscience has not obviated the need to consider higher-level phenomena, even if we don't fully understand the relations between the layers. Mind activity, while dependent on brain activity, cannot be reduced to brain activity. Even reductive scientists would agree to such a statement."* [29]

So, since the purely reductive model has proven inadequate, is Gijsbers' suggestion of parallel *"seen"* and *"unseen"* worlds any more productive? No generally accepted model exists for consciousness, although scientists such as Stanford neurophysiologist, Carl Pribram have made suggestions. Michael Talbot proposed the integration of Pribram's ideas with Bohm's theory of Quantum Mechanics which was discussed in Chapter 8. Talbot explains: *"…the most mind-boggling aspect of Pribram's holographic model of the brain is what happens when it is put together with Bohm's theory. For if the…world is but a secondary reality and what is 'there' is actually a holographic blur of frequencies,…As the regions of the East have long upheld,…although we may think we are physical beings moving through a physical world, this too is an illusion."* [34] This description may initially seem incongruous because we are not used to thinking of ourselves as existing in more than four dimensions. We are familiar with our material state which we consider is the 'real' world and which Bohm termed the Explicate Order but as Gijsbers points out, it does not rule out an information mediated frequency domain, which Bohm termed the Implicate Order. Each mirrors the other, like the image of a landscape reflected in a lake. If our existence in the Universe is as a multi-dimensional entity, as String Theory predicts, then we have both a material, physical presence and an illusory (but still real) waveform identity in another domain but they are not completely separate. In the same way that our depth cannot be separated from our height or our width, while we are still living, our waveform dimension is married indissolubly to our familiar physical dimensions. The separation can only occur after death with the loss of the physical body.

Finally, in the same way that an appreciation of death assisted in understanding life, so the study of unconsciousness can be helpful in gaining

33. *Ibid.* p.33.

34. Talbot, M.: *The universe as a hologram: Does objective reality exist…or is the universe a phantasm?* Harper Collins, NY (1991).

a better understanding of consciousness. In medicine we not infrequently see patients who have experienced a period of unconsciousness after head injuries. Although the mechanism is not understood, it is presumed that consciousness is lost due to damage to a significant number of neurones, which reduces the active network to less than the threshold required for consciousness. When the unconsciousness is prolonged it is called coma. In coma, the person is still alive but not conscious, as is also seen in sleep and anaesthesia. Thus, the state of consciousness is independent of the state of being alive. As discussed above, Engels sees consciousness as an emergent property of our physical state but a detailed credible mechanism has so far evaded detection and not for want of trying. However, if consciousness is viewed as an interaction between an information-specified non-material faculty, which is independent of the space-time domain (that is a software component), that requires a sufficiently large and competent neural platform, which is the brain's nerve cell network on which to operate (the hardware component), then a credible mechanism begins to emerge. Since consciousness has not been satisfactorily explained from the physical domain alone, we will investigate this compound model of consciousness instead and allocate the non-physical component, as it has not been found in the space-time domain, to a place that we shall call the 'fifth dimension' so that it can be identified for discussion purposes. However, it is now time to step up to the next level and examine *self*-consciousness.

Self-consciousness

No one doubts that self-consciousness, the state of self-awareness, is a trait possessed by all normal human beings.[35] However, are chimpanzees, which are generally considered to be the next most intelligent species after man, self-conscious? The evidence would suggest that they are not. They do not wear clothes or apparel for modesty (whereas almost every human people group does). They do not appear to reflect on their own existence or death; no chimps have been observed to practice funereal rituals or create memorials to the dead as humans do. Although primates can be observed to be thinking and able to perform simple problem solving exercises, there is no evidence that they possess self-consciousness.

Robin Dunbar, Professor of Evolutionary Anthropology and Director of the Institute of Cognitive and Evolutionary Anthropology at the University of Oxford writes: "*Despite all the fuss being made about cultural traditions in chimpanzees and orangutans (and occasionally among whales, too), only one species has built the Great Wall of China, put men on the Moon, thought up Pythagoras's theorem, conceived the existence of life after death, written the plays of Shakespeare and composed the music of Johan Sebastian Bach. In short, it is in human's capacity for culture, our capacity to live in a world constructed by ideas, that we really differ from the other apes… There is a growing consensus among both developmental and comparative psychologists that the phenomenon known as 'theory of mind' – the capacity to imagine another individuals 'mind states' — is a critical difference between humans and other animals. It is widely accepted now that most higher animals have knowledge of their own mind states. In terms of philosophy of mind, this is what is usually referred to as first-order intentionality. 'Intentionality' is a general term covering those mind states that involve beliefs. They are evidenced by the use of words like believe*

35. When we were studying psychiatry in medical school we visited a centre for the care of profoundly mentally-impaired people who appeared to lack self-consciousness. These people had impaired mental functioning, which suggests that a fully competent hardware platform is necessary for the process of self-consciousness to operate.

[that something is the case], know, suppose, intend, want, desire, wish, *etc. They imply that we have a certain level of awareness about the contents of our minds, that we are aware that we believe, in some intuitive sense, that something is the case."*

"This state of self-knowledge is the world inhabited by very young children. It is not necessarily the same thing as self-consciousness, *but is certainly a form of consciousness — and probably one shared with a wide range of birds and mammals, at least. However, at around four years of age, children develop the capacity to reflect on the minds of others as well. This capacity is referred to as 'theory of mind' (literally intended to mean that they have a theory about minds, a theory about other people's capacities to have beliefs). This may even be the key to self-consciousness in some formal sense: they can believe that they themselves believe that something is the case. The key distinction in this respect lies in two particular aspects of young children's behaviour. Once they have acquired the capacity for theory of mind, they are able to engage in fictive play (to imagine that a dolly can drink tea from a teacup, even though the cup is empty) and they can lie with conviction. From the age of about four, they begin to appreciate that other people have minds like their own, and that sometimes these minds see the world differently and so can have different beliefs about the world. As a result they can begin to exploit those minds by feeding them false information in order to deliberately mislead them. And this raises the ante considerably, because the skill and effectiveness with which they can lie is greatly magnified. The acquisition of the theory of mind takes their mental world into a new and higher plane."*

"With theory of mind in the mental toolkit of young children, they become good psychologists — they understand the mind behind the behaviour, and begin to understand how false information can mislead that mind. They are in what philosophers of mind would call second-order intentionality — they believe that someone else believes [that something is the case]. Before this, they have always assumed that everyone else shares the same knowledge of the world as they do. But now, with theory of mind, they realise that others can hold false beliefs (or beliefs that they suppose to be false). There is no strong evidence to suggest that any species of animal other than humans has theory of mind. The only case where it might be so (and the evidence is weak and open to dispute) is among chimpanzees and the other great apes. In my view, the evidence is suggestive enough to imply that chimps are at about the same level as children aged four to five — that is, the point at which children are acquiring theory of mind, but are still not wholly proficient at it in the way that, for example, six-year-olds are."

"This is enough, I think, to suggest that the great apes may indeed be a cut above the average monkey, dolphin, lion, and rat. But in the grand scale of things, they are still a long way short of what adult humans can do. If we considered intentionality as a natural reflexive scale that extends to infinity — I believe *that you* suppose *that we* want *to ensure that everyone else* understands *that we would* like *them all to* believe *that we know [what we are talking about] — then we can compare how well humans of different ages do on understanding these kinds of tasks."*

"We find that, on the scale, babies have first-order intentionality, five-year-olds have second-order and normal adult humans scale in at fifth-order. We can cope competently with thoughts as convoluted as 'I believe *that you* suppose *that we* want *to ensure that everyone else* understands *that we would* like *them...', but after that we get lost. It seems that this is not just a memory problem (that is, remembering long strings of words in sentences) but one of keeping track of the various mind states. Even so, by comparison with what human toddlers and chimpanzees can do, even fifth-order intentionality is in a different and very grand league. It's the difference between running a computer program on your Nintendo Gameboy and running it on a powerful modern*

PC. It makes it possible to do certain things that simply cannot be done on the smaller computer. Let me illustrate this with a strictly literary example."

"Consider Shakespeare sitting down to write Othello. To make the play work, he had to have at least three core characters: Othello himself, Iago and the ill-fated Desdemona. This involved Shakespeare persuading his audience (when they eventually got to see the play) that Iago intended *that Othello* believe *that Desdemona* was in love *with someone else. That involved three separate mind states on the stage. But to make the story really convincing, he added Cassio, the apparent object of Desdemona's desires. If Desdemona merely fantasised about Cassio, Othello would surely have been much less bothered by it all. It might have led to a bit of leg-pulling in the garden, but why should Othello have been so tormented by the intelligence that Iago offered him, if he believed that Cassio was not involved? The answer is that Othello is led to believe that Cassio reciprocates Desdemona's interest. It is this that racks up the intensity of Othello's angst and causes him to do what he eventually did. So, to make the story really sell, Shakespeare has to handle four mind states on the stage: Iago* intends *that Othello* believes *that Desdemona* loves Cassio and Cassio loves *her."*

"So just to put the bones of the story together on the stage, Shakespeare had to handle four sets of minds, and that meant working at fourth order intentionality. But this is not the end of the story, because Shakespeare had to persuade the audience to believe all this stuff. If they were not taken in by it, the play would be dead in the water. So Shakespeare had to factor the minds of the audience (or, at least, the virtual mind of a nominal member of the audience) into his calculations. And last, but not least, he had to be doing the imagining of all this himself. So when he sat down with his quill pen poised over a sheet of foolscap one wet Monday morning in Elizabethan London, he had to be able to work – minimally – at sixth-order intentionality: he had to intend *that the audience* believed *that Iago* wanted *Othello* to suppose *that Desdemona* loved *Cassio and that Cassio, in return,* loved *her (and hence that they planned to run away together). That's no mean feat, because he is already working at one level of intentionality above what the average adult human can cope with. Notice that he is also pushing his audience to its limits — fifth-order intentionality. It is probably precisely because Shakespeare could work successfully at this level, and so challenge his audience to their limits, that he came to be such a successful playwright."*

"However, the real issue for our present concerns is that only a human could have done this. With their cognitive limits set firmly at second-order intentionality (at best!), even the proverbial chimpanzee sitting down at its typewriter could never have produced the script for Othello*...So the lesson for us is that the flights of fancy that we engage in when dabbling in literature, even when just telling stories around the campfire, are far beyond the cognitive capabilities of any other species of animal currently alive...Only adult humans could ever intentionally produce literature of the kind that we associate with human culture. It is possible, of course to produce stories with third- or even fourth-order intentionality (the cognitive equivalent of eight-and eleven-year-old children), but they would inevitably lack the sophistication of the stories told by the average adult, never mind those of a Shakespeare or a Molière."*[36]

Thus, putting aside Dunbar's speculation about second-order intentionality in chimpanzees, for which he admits there is no convincing evidence, it appears that human beings are uniquely self-conscious, that is, aware of their own self-existence, have an appreciation that they are alive and therefore one day must die and the personal implications thereof. This is why human beings uniquely have funerary rights, burying the dead and practicing some form of religious ceremonies, unlike any other living creature.

36. Dunbar, R.: Why are Humans not Just great Apes? *in What makes us human?* (ed.) Pasternak, C.: Oneworld Publications, Oxford (2007) pp.37–48 *in Cosmos* 20:47–9 (2008).

So if a chimpanzee has awareness or consciousness and a human being has *self*-consciousness, how might this work? There is little evidence. The fact that birds, mammals and human beings have consciousness but that only the latter possess self-consciousness, suggests that these are discrete entities and it does not simply reflect the different ends of a spectrum, as there is a quantum leap of gigantic proportions between the cognitive abilities of chimpanzees and adult human beings, as Dunbar demonstrates above. If they are indeed discrete and consciousness domiciles in the fifth dimension, then I would suggest that self-consciousness resides in a 'sixth' dimension (Although it is equally possible to argue that they are two distinct types of information in the fifth dimension).

So what happens if there is separation of the non-material fifth and sixth dimensions from dimensions one to four, which presumably happens after death, as the physical body in the first four dimensions decays and disappears? Remember that we saw in Chapter 8 that non-material entities without mass are outside of time and hence timeless and not subject to decay. The answer to this must surely depend on how the fifth and sixth dimensions are connected to the physical body in the first four dimensions. At this stage, it is completely unknown and as the fifth and sixth dimensions are curled, this is impossible to physically investigate. If we did understand the connection between the first six dimensions we might also comprehend how the physical and non-material components behave during near-death experiences, where people 'come back to life' after a cardiac arrest when the person is successfully resuscitated. The presumption is that in 'near-death' experiences, the non-material consciousness does not completely separate from the physical body. Table 10.2 explores some of the possibilities.

If understanding a multidimensional universe is challenging, and most of us have never thought about this before, let me give you an analogy that may bring this subject closer to home. If you have a computer and you use the Internet, then you can visit many sites. You may have friends on 'Facebook' with whom you correspond and you may have an electronic banking facility where you can access your bank account from the privacy of your own home. The information on Facebook is stored on one or more host computers somewhere in the World but the information, as far as you are concerned, is somewhere 'out there'. Programmers sometimes refer to the information as being stored *"in the cloud"*. This information is fluid and can flow from one computer to another but it is still stored in a physically discrete place ultimately. Similarly, your bank has computers that store your bank balance, which you may be able to access online. The bank and the account are real but they could be anywhere in the World. You only need a suitable interface to be able to connect to them, which might be an automatic teller machine (ATM) or your home computer. In the same way, could it be that if we are multi-dimensional beings that we are able to work simultaneously in different spheres of influence, some in the space-time domain and others in the

37. Davila Ross, M., Orwen, M.J. & Zimmermann, E.: Reconstructing the Evolution of Laughter in Great Apes and Humans. *Current Biology* 19(13): 1106–11 (2009).

38. Morelle, R.: *Scientists tickle animals to find laughter clues.* BBC News 30 December 2011 *at* http://www.bbc.co.uk/news/science-environment-15880045

Table 10.2: Characteristics of the Domains of Life
A study published in 2009 suggested that laughter is a universal language which is shared by all humans, primates and probably mammals. This suggests that mammals experience feelings or emotions too.[37,38]

Dimensions	Active Mind States	Examples
1-6	Life with self-awareness: self-consciousness and conscience.	Adult awake human beings.
1-5	Life without self-awareness: simple thought and emotions.	Complex animals (birds and mammals); human infants.
1–4	Life without consciousness.	Plants; simple animals (e.g. bacteria & insects); complex animals that are sleeping or in a coma.
1-4	Nil	A dead person (cadaver) or dead animal or plant

frequency domain? The problem is that we do not know exactly how the interface might work yet; however, as discussed previously, the evidence suggests that the mechanism employs a Fourier transform-like mechanism to facilitate this process. So to reiterate, you can access information in many different locations in the World via the Internet if you have a computer and the knowledge to use it. You do not have to physically travel to a library or other data repository to access the stored information and some higher dimensions may operate in a similar way. It is to be hoped that in the future it will be possible to intentionally access these dimensions in the frequency domain from our space-time domain but that will require a better understanding of the interface between the two. One can only say that with our present knowledge of science, this sounds like a farfetched objective, so is there any evidence from which we might develop a better understanding of the interface? Not surprisingly, these areas are not well understood, so we need to examine other conditions that appear to share some properties in common with the interface, such as sleep, dreaming, thoughts and memory.

Sleep and Dreaming

Sleep is a temporary absence of consciousness, so it would appear that if consciousness operates in, from or with a fifth dimension, then the parameters that define either the non-material domain or our physical space-time domain (or the interface between the two) must vary in some way when sleep cycles occur. We know that some natural hormones, such as rising melatonin levels in the evening, precede sleep and rising cortisol levels early in the morning precede waking, so there would seem to be an interplay between the space-time domain and the fifth dimension to permit sleep and waking but this relationship is not readily apparent yet. Research published in the Journal *Neuroscience* in 2009 by Professor Marshall Devor of the Hebrew University in Jerusalem suggests: *"...that a small group of neurons near the base of the brain, in the Mesopontine tegmentum, has executive control over the alert status of the entire cerebral and spinal cord."* [39] This mechanism appears to be analogous to the 'hibernation' mode on your computer, which is usually controlled by a button that is part of the hardware, which causes the software on which you are working to be placed in a 'safe mode' for a temporary shutdown. Once again, hardware and software are woven together for normal functioning.

Studies of brainwave patterns show that there are three different types of sleep: deep sleep, light sleep and dreaming sleep. A typical, healthy, adult human being has an average of three hours deep sleep, three hours light sleep and two hours dreaming sleep each night but our periods of slumber are divided into alternating segments of the various types of sleep. Each phase of sleep lasts between twenty minutes and one and a half hours. Deep sleep is when the body appears to partially shut down, to physically rest, to rejuvenate its cells in the space-time domain. Dreaming sleep, which is also called rapid eye movement (REM) sleep, allows us to replay past events where we can revue, explore possibilities and problem-solve; this rejuvenates the mind in its non-material domain. Light sleep appears to transition between the other two types of sleep and it acts like the neutral gear in a motor car's gear box, allowing smooth transitions from wakefulness to sleep and then in and out of deep and REM sleep; waking directly from REM sleep risks sleep paralysis, which is discussed in sleep disorders below. As mentioned previously, Electroencephalograms (EEGs) are sensitive

39. Abulafia, R., Zalkind, V. & Devor, M.: Cerebral activity during the anesthesia-like state induced by mesopontine microinjection of pentobarbital. *Journal of Neuroscience* 29:7053–64 (2009).

machines which measure our brainwave activity in the network of neurones that forms our brain. Prof Roy Beran, neurologist and Professor of Medicine at Griffith University, Brisbane, Australia writes: *"sleep is a neurological state of being, with defined EEG patterns which reflect the stages of sleep emanating from the brain."* [40] EEGs measure the activity within our brains in the same way that an electrician can measure the strength of an electrical current running through your computer's circuitry but just as the electrician cannot use his voltmeter to identify the contents of the electronic documents on which you were working with your computer, the EEG can identify the broad type of brain activity that is present but not what you are thinking. That would require a form of thought analyser and the Information Model suggests that thoughts reside in the fifth or sixth dimension which is beyond the measuring capability of an EEG machine in our space-time domain.

There are many intriguing questions about sleep. An obvious one is: what causes the switching between the different types of sleep; from deep sleep to light sleep or dreaming sleep? It is interesting that in dreaming sleep, the mind appears to be disconnected from the physical body and one can have the sensation of trying to run away in a dream but our legs will not physically move until waking, when we find ourselves thrashing in the bed, as the physical brain and non-material mental processes are suddenly re-engaged. Thus, whilst dreaming in the frequency domain in the fifth dimension, our physical body in the first four dimensions of the space-time domain appears disconnected from our cerebral activity except our eyes which are considered by neurologists to be a direct extension of the brain. 'Daydreaming' seems to be an intermediate state between consciousness and REM sleep. In daydreaming we become intellectually disconnected from our physical world but our physical status is unaffected. Deep sleep appears more akin to coma, where there is minimal activity in the fifth dimension; in this state a person is still alive but it is much harder to rouse an individual from deep sleep, as the level of consciousness is in a state of temporary shut-down.

Sleep is essential to our health and so we suffer if we do not get enough of it. Sleep both restores our physical strength and our mental well-being and systematic sleep deprivation is a form of torture. Sleep deprivation also impairs the optimal functioning of the immune system and sleep deprivation studies in animals have led to increased rates of illness and disease.[41] However, illness associated with sleep abnormalities may present in a number of different ways. Professor Beran notes: *"Altered behaviour during sleep, such as REM behaviour disorder...may be increased in Parkinson's disease."* [42] Rapid Eye Movement Sleep Behaviour Disorder (RBD) is associated with olfactory (nasal) dysfunction and Parkinson's disease (a movement disorder) which is associated with a lack of the hormone serotonin, so the serotonin system may be part of the system that connects between the physical brain in the space-time domain and dreaming in the frequency domain.[43] Restless Legs Syndrome, where sufferers cannot prevent ceaseless movements of their legs when they try to sleep, is often helped by medications that restore dopamine levels in the brain. This is also a treatment for Parkinson's disease, so it may be that the neuromuscular connection/disconnection between the mind and the peripheral nervous system, via the neurones in our brain, is incomplete in Restless Legs Syndrome and Parkinsonism. Thus, the usual separation between the two in REM sleep to facilitate the rehearsal of co-ordinated routines, such as a tennis or golf swing, is inadequate

40. Beran, R.: The Neurology of Sleep: Part 1. *Medical Observer* St Leonards, NSW Australia (5 Feb. 2010) p.23.

41. *Sleep and Disease Risk.* Division of Sleep Medicine at Harvard Medical School. (18 Dec., 2007). http://healthysleep.med.harvard.edu/healthy/matters/consequences/sleep-and-disease-risk

42. Beran, R.: The Neurology of Sleep: Pt 1. *Medical Observer*, St Leonards, NSW Australia (5 Feb., 2010) p.24.

43. Stiasny-Kolster, K.: Combination of 'idiopathic' REM Sleep Behavior Disorder and olfactory dysfunction as possible indicator for α-synucleinopathy demonstrated by dopamine transporter FP-CIT-SPECT. *Brain* 128(1): 126–37 (2005).

and the muscles become partially activated in the same way that a broken cog wheel may sometimes connect with its motor and sometimes it does not. It is also interesting to note that the manipulation of serotonin with antidepressant medications can assist patients with depression, which will be discussed further in the section on mental illness.

Sleep Disorders

There are a number of recognised sleep disorders. An uncommon but significant one is narcolepsy, which is a condition associated with excessive sleep. Rob Moodie, a physician and Professor at the University of Melbourne's Nossal Institute for Global Health writes that narcolepsy: *"...is a chronic disorder where the brain can't effectively control the sleep-wake cycle. It is characterised by excessive daytime sleepiness (EDS) with frequent, brief and uncontrollable bouts of sleep – regardless of the amount of sleep taken at night. This excessive sleepiness can be accompanied by cataplexy: an episodic loss of muscle function ranging from slight weakness to complete body collapse, usually triggered by strong emotions such as fear, anger or laughter. The patient may also experience hallucinations and sleep paralysis: an inability to move just before falling asleep or upon waking up. There is a wide spectrum of severity and the disorder can inconvenience people's personal, social and professional lives. If untreated, it can compromise exercise, work and, of course, driving. This under-diagnosed disease affects one to two people per 2000. It usually begins in the teenage years but may take 10 to 15 years before it is adequately diagnosed. Its cause is unconfirmed, but researchers have discovered that abnormalities of genes on chromosome six are strongly, but not always, associated with narcolepsy. There is no definitive cure but the symptoms can, in most cases, be well managed using drugs combined with behavioural strategies. These include regular naps, and improving night-time sleep with a regular schedule, avoiding alcohol and caffeine, and increasing exercise."* [44]

Thus, narcolepsy has many features in common with REM sleep, however, the patient has lost control of their consciousness and can drift off to sleep suddenly and inappropriately. Neurologist, Professor Roy Beran writes: *"Narcolepsy with its associated tetrad of symptoms includes excessive sleepiness (with the rapid onset of REM sleep); sleep paralysis, in which the patient may be conscious but cannot move until stimulated by something as innocuous as a light touch; cataplexy with sudden loss of muscle tone, which results in collapse that may be provoked by emotional activity such as laughter; and hypnagogic (drifting off to sleep) or hypnopompic (awakening) hallucinations, namely, vivid and realistic dreams associated with REM sleep, although the patient is sufficiently awake to recall the experience."* [45] Thus narcolepsy also appears to blur the boundary between consciousness and sleep. Professor Beran also writes that there has been a: *"...recent finding that narcolepsy is the result of a deficiency of hypocretin, also known as orexin transmission. Orexin enhances wakefulness and inhibits REM sleep."* [45] From this finding it would seem that narcolepsy results from interference with the physical switching mechanism that engages REM sleep, while at the same time disengaging the mind from the physical body. This suggests that there is a physical mechanism in the brain that engages with the non-material process of consciousness, however to date there is no model to explain how this occurs. It is interesting to note that narcolepsy can be triggered by *"strong emotions"*, which are non-material entities that precipitate the physical responses! It is certainly a fertile area for further

44. Moodie, R.: Nodding off. *Cosmos* 20:35 (2008).

45. Beran, R.: The Neurology of Sleep: Part 2 of Clinical Review Update. *Medical Observer*, St Leonards, NSW Australia (12 Feb., 2010) p.27.

research but it is now time to move on to our next cluster of topics that have previously defied explanation, which includes thoughts, feelings and emotions.

Thoughts, Feelings and Emotions

Rene Descartes (1596–1650) was the pre-eminent scientist and philosopher of his generation; he is also one of the outstanding minds in history, who radically changed the thinking of the world. As a mathematician, he developed mathematical rules so that geometric shapes can be expressed as algebraic equations, called the Cartesian Co-ordinate System and he has been acknowledged as the father of analytical geometry. He was also a key figure in the scientific revolution that was gathering pace in Europe at that time. As a philosopher, he inspired the development of current Western philosophy and is referred to as the father of modern philosophy. Descartes attempted to deduce a fundamental set of principles, which he was certain to be true, through his system of Methodological Scepticism or 'radical doubt'. Initially the only thing which Descartes could not doubt was that he *thought*.[29]

Thought may be defined as a process that occurs in a living brain, which has the ability to generate or manage information. More fundamentally, thought can be understood as an information manipulation, packaging and distribution mechanism. Thought is very poorly understood at present. As a result, the potential of thought is probably underestimated. A recent example from the BSI-Toyota Collaborative Centre in Japan will help illustrate this. *Australian Medicine* in 2009 reported: *"The BSI-Toyota Collaboration Centre (BTCC) in Japan has developed a system, using a new signal processing brain-machine Interface (BMI) technology, which will allow elderly or handicapped people using signals from their brains to control a wheelchair in as little as 125 milliseconds".*

"In the system, commands for left-right turns and forward-backwards motion of the wheelchair are processed every 125 milliseconds by analysing brainwaves and displaying the data on a screen in real time, giving neuro-feedback to the driver for efficient operation. The system can adjust itself to the characteristics of individual drivers, and can improve the efficiency with which it senses the driver's commands, so that drivers can get the system to learn their commands (forward/right/left) quickly and efficiently. BTCC says that it has achieved an accuracy rate of 95%...So far its research has centred on brainwaves related to imaginary hand and foot control but, through further measurement and analysis, it is anticipated that the system may be applied to other types of brainwaves generated by various mental states and emotions." [46] Thus our non-material thoughts are able to physically change our immediate environment. The big question is: where and how does this happen? When we think, our thoughts appear to be generated in the fifth dimension but then relayed to the first four dimensions for execution. If that is so, then our thoughts are still within our brains but happening in a non-physical manner. This distinction between hardware, software and information means that thoughts could be conserved in another dimension after we die;[8] it depends upon the extent of the relationship between the space-time and frequency domains. Certainly the physical connection is severed at brain death but it is currently impossible to say what may happen to the 'software' and information, which are our thoughts and personalities.

To date, no specific location has been identified for the physical *storage*

46. Brain-controlled wheelchair. *Australian Medicine* Australian Medical Association. (Sept 21 2009) p.20.

of thoughts and feelings within our brains, however we do know that certain functions are processed in different parts of the brain. We have already seen that sleep is initiated in a localised part of the base of the brain while vision is interpreted at the back of the brain and speech interpretation is usually processed in specific parts of the left hemisphere. In most right-handed people, linguistic ability is located in the left hand side of the brain. Susan Greenfield, Professor of Synaptic Pharmacology at Oxford University states: *"Inner thoughts are all in words, which come from the linguistic left"* while problem-solving appears to be associated more with the creative right brain. Professor Greenfield believes that consciousness is caused by the *"biggest ripple"*, which is happening in our brains at any one particular time.[24] If this is the case, the next question is: does that ripple just reside in the space-time domain or somewhere else as well?

If life comes from the dynamic functioning of dimensions one to four and conscious thought derives from the fifth dimension, this would also provide an explanation for the placebo effect, where approximately one third of subjects in controlled trials will respond positively to the treatment (or whatever intervention is being tested) even if they are simply given an imitation sugar tablet. Many people would say that the placebo effect is "all in your mind" but we know from medical research, that it is a confounding factor that must be taken into account when doing controlled trials. Any medication being tested must be in the physical domain, so it would seem impossible that a placebo tablet (the imitation sugar pill) could be impacting the non-material mind (because the tablet is inactive) but we will see in Chapter 12 that the converse can be true, where the non-material mind can alter the physical brain to which it is attached, simply by the process of active thinking, resulting in, for example, new nerve cell development. So if the person in the trial believed that the inactive tablet was real, their mind could still stimulate changes in the physical body to emulate the anticipated findings, thus confounding the results.

Other lessons have been learned from localised brain damage, such as that sustained from a shrapnel injury to a Vietnam veteran's right frontal lobe, which left him lacking insight and unable to plan. This has allowed scientists to map the various functions of different regions of the brain.[24] From this type of research, it is clear that wherever thought resides, a fully competent physical brain architecture is necessary for normal brain functioning but that in and of itself, the physical brain is insufficient to explain thoughts and the process of thinking.[47] This discussion raises the question of what do the different groups of living organisms appreciate when they exercise their mental ability, which is summarised in Table 10.3.[48,49,50]

From this categorization in Table 10.3 it can be seen that the simpler organisms such as viruses, bacteria and plants do not display features suggestive of a non-material dimension. Insects are difficult to place as it is still uncertain as to their capacity for sleep whereas crustaceans, fish, amphibians and reptiles all display some non-material characteristics that would appear to be associated with a fifth dimension. Birds, mammals and primates demonstrate the additional characteristic of conscious reasoning, while human beings alone demonstrate self-consciousness. Professor of Biological Anthropology and Neuroscience at Berkeley, Terrence Deacon notes that there are: *"...hundreds of thousands of species with brains, and tens of thousands with complex behavioural, perceptual, and learning abilities. Only one of these has ever wondered about its place in the world,"* [51]

47. Occasionally selective brain injuries such as strokes may allow normal functioning but this is the exception rather than the rule.

48. Thuesen Bleeg, M.: Cat naps. *Cosmos* 34:46–7.

49. Cirelli, C. & Bushey D.: Sleep and Wakefulness in Drosophila melanogaster. Molecular and Biophysical Mechanisms of Arousal, Alertness, and Attention. *Annals of the New York Academy of Sciences* 1129: 323–9 (2008).

50. Fanjul-Moles, M.L. et al: Circadian Modulation of Crustacean Hyperglycaemic Hormone in Crayfish Eyestalk and Retina. *Chronobiology International* 27(1):34–51 (2010).

51. Deacon, T.W.: *The Symbolic Species: The Co-Evolution of Language and the Brain.* Norton, NY (1997) p.21 *in* Rolston, H.III: Care on Earth: generating informed concern. *in* Davies, P. & Gregersen, N.H.: (2010) p.226

Dimension	1–3	4th	5th						6th
Characteristic / Genus	Space	Time	Life	Thought & memory	Deep sleep	Hibernation	REM sleep/ dreams	Conscious reasoning	Self-awareness
Rocks	√	√	×	×	×	×	×	×	×
Viruses	√	√	≈	×	×	×	×	×	×
Bacteria	√	√	√	×	×	×	×	×	×
Plants	√	√	√	×	×	×	×	×	×
Insects	√	√	√	≈	≈	×	×	×	×
Crustaceans	√	√	√	√	≈	×	×	×	×
Fish	√	√	√	√	√	×	×	×	×
Amphibians	√	√	√	√	√	√	×	×	×
Reptiles	√	√	√	√	√	√	×	×	×
Birds	√	√	√	√	√	×	√	√	×
Mammals	√	√	√	√	√	√ (+/-)	√	√	×
Primates	√	√	√	√	√	×	√	√	×
Humans	√	√	√	√	√	×	√	√	√

Table 10.3: Categorisation of Mental attributes according to Genus
As hyper-dimensions are not directly measurable, we must infer their characteristics from the discontinuities that may be discerned within our space-time domain, in the same way that Riemann did in 1854. By comparing the characteristics between plants, animals and humans, some differences can be deduced. Insects and crustaceans demonstrate some features of deep sleep compared with mammals but they differ in other respects. [48,49,50]

And this is not the only distinguishing characteristic of human beings either. Rob Brooks, Professor of Evolutionary Biology at the University of New South Wales in Sydney describes other ways that the significant differences between *Homo sapiens* and the rest of the animal kingdom are expressed, such as: *"...the belief in fairness and reciprocity, a capacity for empathy and impulse control, and a willingness to work cooperatively in ways that even our smartest primate kin cannot match."* [52] These observations are in keeping with the conclusion that self-awareness operates at a different level to conscious reasoning, thoughts and memory and it is therefore not unreasonable to surmise that it either resides in a different domain or is in a unique class when compared with similar characteristics in all other species. Human beings alone appear to possess the ability to imagine how the world might be, rather than just responding to the way that it is.

David Bohm believed that thoughts and behaviour could not arise from our physical domain alone but that: *"Every action starts from an intention in the implicate order."* [53] This would be logical if our thoughts were primarily a function in the fifth dimension. There are important implications that follow from these deductions as Bohm describes: *"In a brain that operates holographically, the remembered image of a thing can have as much impact on the senses as the thing itself...* [As a result] *imagination and reality are ultimately indistinguishable, and it should therefore come as no surprise to us that images in the mind can ultimately manifest as realities in the physical body."* [53] This explains psychosomatic illness but more importantly, forensic experts and lawyers are very aware of this phenomenon when it comes to verifying the evidence that may be admitted to a court of law, as witnesses to a crime can easily confuse thoughts that occurred sometime after an event with the event itself and for this reason barristers are sometimes reminded not to 'lead the witness' when they ask questions so that the original memory is not contaminated with false information; instead, the witness is encouraged to

52. Brooks, R. Playing fair keeps us grounded. *Sydney Morning Herald* 24–25th Sep, 2011 News Review p.18.

53. Haselhurst, G.: *Essay on Life & Ideas of David Bohm (1917–1992)* p.9. http://www.spaceandmotion.com/Physics-David-Bohm-Holographic-Universe.htm. See: *Holographic Universe* http://www.essays.cc/free_essays/e4/dkt106.shtml

give their testimony 'in their own words'. Rehearsing alternative scenarios can also induce the visualisation by the witness of situations that may never have happened and usually, only notes made at the time of the incident (or artefacts pertinent to the event) are acceptable as physical evidence in a court of law.

Personality and emotions are just as poorly understood as thoughts and consciousness. Professor Michael Gazzaniga of Dartmouth College, New Hampshire, USA has discovered that personality does not change by splitting the corpus callosum (the nerves that form the bridge between one half of the human brain and the other) when this procedure was performed for the treatment of epilepsy.[24] This is consistent with a non-physical, holographic mechanism for brain function that involves more than the space-time domain. In neurology, emotions have been associated with the hippocampus for many decades and they can be manipulated by administering hormones, such as adrenaline or noradrenaline or drugs such as amphetamines and hallucinogens. However, emotions can also be influenced by psychological effects, such as intentionally thinking peaceful or angry thoughts. Once again this suggests that there is an interface within the brain that connects the physical platform that is the brain substance with thoughts, emotions and feelings which behave rather like the software and information within a computer but the evidence to date does not help us to narrow down our search for the location or mechanism by which that interface might operate, so our next area of study will be the thought disorders.

Thought Disorders

In medicine, it is sometimes easier to work out the normal functioning of an organ by studying what happens when the process goes awry rather than just by studying healthy individuals alone. However, there are no physical markers, such as blood tests or radiological studies, to define the difference between normal thought and formal thought disorders, such as schizophrenia and manic-depressive states (now referred to as 'bipolar' disorder). This is one of the reasons why psychiatry is such a challenging discipline. However, if thought involves the fifth dimension, then thought disorders such as schizophrenia, bipolar disorder, personality disorders and hysteria are likely to involve the fifth dimension as well. In contrast, epilepsy and delirium have physical markers and are the result of material problems in the space-time domain in dimensions one to four. This division equates respectively to the disciplines of psychiatry, which studies all the conditions described as 'mental illness' (think of them as being similar to software problems) and neurology, which manages physical ailments of the brain and nervous system (which are akin to hardware problems). Further, it is known that certain psycho-active substances such as marijuana and psychotropic drugs, such as LSD can alter thoughts and perceptions and thus would appear to be acting on the material platform in the space-time domain. From this it may be inferred that schizophrenia and bipolar disorder are associated with the interface between the material hardware and the non-material thought domain while personality disorders and hysteria relate more to software problems.

In addition, manipulating neurotransmitters, for example serotonin and noradrenaline (norepinephrine), can sometimes help people suffering depression, while antipsychotic medications can often assist patients with mental illness too. However, as we discovered with consciousness and sleep, the relationship must be complex because we know that there is often an overlap between these medications and the induction of sleep. Using medications for the treatment of these illnesses is equivalent to mending malfunctioning hardware in your computer, such as fixing a faulty circuit board, when it was not performing the way it was supposed to. In contrast, psychotherapy and counselling would appear to be equivalent to repairing the software programme or correcting the data running on the software within your computer and is often used for personality disorders and depression.

Interestingly, qualitative electroencephalograms (QEEG's) have been introduced to research some psychiatric conditions. In a QEEG the standard EEG is split into its different frequencies by computers, which has revealed differing patterns for the different psychiatric conditions. *"Brain waves are electrical signals generated when neurons fire, and like instruments in an orchestra, different neurons have different frequencies at which they discharge."* [54] Using this technique, psychiatrists have been able to distinguish between normal and abnormal patterns of brain activity. Neuropsychiatrist, Leslie Prichep, Associate Director of the Brain Research Laboratories at New York University, based at New York's Bellevue Hospital, has spent more than thirty years researching QEEG's. She said: *"The profile of a schizophrenic patient looks very different from a depressed patient. They each indicate abnormalities, but there are distinctive patterns."* [54] Although it is not yet available within the wider community, the emergence of this technology is allowing psychiatrists to objectively diagnose conditions in a way that has never been possible before, when the only tool available was the analysis of the symptoms presented by the patient.[54] Once again the evidence suggests that there is both a physical, hardware component, which is the network of neurones as well as the non-physical software component that conveys the semantic thought processes that run on it. This combination produces the electrical signals measured by the QEEG. At this juncture there is no definitive evidence to demonstrate the location of the semantic component that carries the meaning (the patient's thoughts), as the EEG signal only measures the electrical discharges, which is *evidence* of the transmission of 'moieties' of syntactic information but the QEEG waveform does not measure the *meaning* of the information which steadfastly remains opaque to scientific enquiry. It is analogous to counting postal trucks travelling along a highway. The presence of the moving postal trucks is evidence of the movement of information (letters and parcels within the postal service) but it cannot tell you the information contained within the letters in each of the trucks. QEEGs are the equivalent of being able to discern the difference between letters and parcels within the trucks but they still do not tell us the contents of the letters or parcels. Once again, it is interesting to note that these thought processes present as a waveform, as evidenced by the QEEG, which is a characteristic of the non-material, Implicate Order. Thus, there are three possibilities for mental illness: most obviously, hardware problems are seen in the disintegration of the nerve cells in senile dementia or a lack of the neurotransmitter serotonin in depression. Secondly, there is the possibility of maladaptive processing of the information in the non-material domain, such as in personality disorders (although we have no way of objectively assessing this at present) or there could be faults at the interface between the two domains (which may represent schizophrenia and bipolar disorder). The disease processes involved in mental illness are still very poorly understood and it may take decades to tease out the mechanisms involved but this intersection between the material and non-material domains appears to be critical in the human brain, where the hardware and the software either coordinate

54. Barry, C.: Mind maps. *Cosmos* 20:68 (2008).

harmoniously to produce normal function or behave aberrantly, to cause disease and dysfunction. But now it is time to look at the cumulative effects of thoughts, which can lead to memory.

Memory

Memory is the ability to record, store and recall thoughts and as we noted earlier, Richard Dawkins identified memory as one of the four information repositories in living things. In contrast to thoughts, which appear to be dynamic information, memory seems to be an information warehouse. Memories can generally be recalled at will, which is an amazing ability when we consider the vast amount of information encountered in a single day, let alone a lifetime and the immense number of memories generated. Missler writes: "*The brilliant physicist and mathematician, John von Neumann, once calculated that over the course of the average human lifetime, the brain stores something on the order of* $2.8x10^{20}$ *bits of information.*" [55] When memory is first generated, it appears to be in the form of an 'electrical' signal which can be imagined as a 'brainwave' and this brain activity can be detected by an EEG but thought is associated with a moving electro-chemical signal in a neurone and yet that is no more the thought than the moving messenger-RNA strand on the ribosome, is life. There is some extra quantity or quality that is associated with the moving electrical signals in a functioning brain which generates and communicates the thought. Memory is a stored thought and appears to partly involve the pathway taken by the electro-chemical signal through the network of nerve cells but what directs the switching of the signal at each neurone? We know that the transient 'electrical' signals are converted to some other form of long-term memory with the assistance of the mammillary bodies, which are located on the base of the brain, because people who have severe damage to their mammillary bodies, such as after a severe head injury, can no longer lay down long-term memory. They can remember old memories and remember new events for up to two minutes but cannot generate new long-term memories. One possible explanation could be that the mammillary bodies initiate the process or identify which thoughts are to be laid down in long-term memory. If this were the case, then an error in the switching mechanism which decides whether to record a memory, such as a bird singing outside the window (or not), could explain the rare cases where *all memories* are remembered and the person cannot forget anything. Superficially, this ability to remember everything sounds like a great asset but people who cannot forget anything often see it more like torture, as they drown in vast amounts of useless information, which they can neither filter nor forget.

The anatomy and physiology of brain function (the hardware) has been described in immense detail in medical text books but brain function with regards to thought and consciousness (the software and the way that information runs on the material brain substance) is still very poorly understood. When we attempt to understand brain function there are many aspects that need to be considered, which includes creative thought, personality, logical analysis and memory but we have been hampered to date by what we can measure; the creative output of the brain is orders of magnitude larger than that which our current understanding predicts from the input and processing that we detect within the physical neuronal network. The closest analogy we have is a computer network, such as the Internet, which is vaguely similar to the way that our nerves are connected and yet they do not come close to the extraordinary power of the human brain. Although the

55. Missler, C.: (2004) p.330.

Internet has extensive memory capabilities, it cannot generate original thought or perform critical analysis in the way that all normal human beings are capable of doing. Therefore, brain function appears to be more than just the *communication* between individual nerve cells; there is also a very powerful production process that can generate new original information. Phillip Johnson wrote: "*What makes the mind from the brain is not the neurons but the information that coordinates the neurons and uses them (and perhaps other entities as yet unknown) to produce the phenomena of our mental life.*" [56]

So what information do we have with which to understand our brain and our memory better? In the 1920s, Canadian neurosurgeon, Wilder Penfield (1891–1976) "*electrically stimulated the brains of epilepsy patients and found he could cause them to move their arms or legs, turn their heads or eyes, talk, or swallow. Invariably the patient would respond by saying, 'I didn't do that. You did.'* [57] *According to Penfield, 'the patient thinks of himself as having an existence separate from his body.' No matter how much Penfield probed the cerebral cortex, he said, 'There is no place...where electrical stimulation will cause a patient to believe or decide.'*" [58] These findings were consistent with the previous discoveries that specific brain functions, such as limb movements (from the motor cortex), speech generation (in Brocca's and Wernike's areas) and vision (in the occipital cortex) had specific locations. In contrast, thought showed no such localisation. At about the same time that Penfield was researching, "*In a series of landmark experiments in the 1920s, brain scientist Karl Lashley found that no matter what portion of a rat's brain he removed he was unable to eradicate its memory of how to perform complex tasks it had learned prior to surgery.*" [59] Memory exhibited the same "*whole in every part*" phenomenon which we previously encountered when investigating holograms in Chapter 8. Why should this be? A clue came in the 1960s. "*Carl Pribram, a neurophysiologist at Stanford University and author of the classic neuropsychological textbook* Languages of the Brain,... [also] *discovered that removal of parts of the brain didn't eradicate memories.*" [60]

At this point we must distinguish between the mechanism for the recording of new memories, which does require specific parts of the brain, such as the mammillary bodies versus the retrieval of old memories, which is independent of any particular part of the brain. Pribram "*came to believe that memories were* not *localised at specific brain sites but were somehow spread out or distributed throughout the brain as a whole.*" [60] Later still, in the 1970s, Paul Pietsch, a biologist at Indiana University, "*set out to disapprove Pribram's theories. In a series of over 700 operations on salamanders, he discovered that their learned behaviour was not affected by repositioning, reversing, or even shuffling the brain. After recovering from the operation, their behaviour returned to normal.*" [61] Missler wrote: "*When Pribram discovered holography, he was ecstatic. If it was possible for every portion of a piece of holographic film to contain all the information necessary to create a whole image, then it seemed equally possible for every part of the brain to contain all of the information necessary to recall an entire memory!*" [60]Although this seems like a lot of information in a very small space, Michael Talbot put this into perspective when he wrote: "*...it has been demonstrated that one cubic centimetre of film can hold as many as 10 billion bits of information... This is equivalent to 5 complete sets of the Encyclopaedia Britannica.*" [34] Thus, vast amounts of information can be compressed very efficiently when appropriate storage media and retrieval mechanisms are available. "*Pribram believes memories are encoded not in neurons, or small groupings of neurons, but in patterns of nerve impulses that criss-cross the entire brain in the same way that patterns of laser*

56. Johnson, P.E.: *Wedge of Truth*. InterVarsity Press, Downers Grove, Illinois (2000) p.123.

57. Penfield, W.: *The Mystery of the Mind: a Critical Study of Consciousness and the Human Brain*. Princeton University Press, Princeton, NJ (1975) pp.76–78 *in* Strobel, L.: *The Case for a Creator*. Zondervan, Grand Rapids, Michigan (2004) p.258.

58. Moreland, J.P.: Interview *in* Strobel, L.: (2004) p.258.

59. Haselhurst, G.: *Essay on Life & Ideas of David Bohm (1917–1992)* pp.7–8. http://www.spaceandmotion.com/Physics-David-Bohm-Holographic-Universe.htm

60. Missler, C.: (2004) pp.329–30.

61. Missler, C.: (2004) p.329 – See Paul Pietsch, *Shufflebrain: the Quest for the Hologramic Mind*, Houghton Mifflin, Boston, MA (1981).

light interference criss-crosses the entire area of a piece of film containing a holographic image." [59] In other words, Pribram viewed memory as a holographic image, which is stored on the medium of the collective neuronal network in our physical brain structure. This could be an explanation for short-term memory which lasts up to two minutes but it does not explain how we can remember long-term memories years later. Even people who have recovered from a prolonged coma due to a head injury can still recall memories that were laid down before the accident and if memory was simply an electrical impulse criss-crossing the brain in a particular pattern, how does the brain 'remember' what that pattern is? Pribram's model may explain the mechanism for dynamic, short-term memory in the space-time domain but there must be some extra facility, either in the space-time domain or the frequency domain, to fully explain long-term memory.

Another property of holograms is that more than one hologram can be stored on a single photographic sheet when differing frequencies and angles of projection of the laser beams are employed, which makes the storage capacity of holography immense.[59] In the same way, brainwaves of differing frequencies, following divergent pathways in the same neuronal network, may elicit a myriad of different memories and thought processes. This could be one explanation for the phenomenal storage ability of our brains. Evidence in support of a disseminated mode of storage can be seen in people who have suffered a cerebral vascular accident (sometimes referred to as a 'CVA' or 'stroke'). They do not lose whole chunks of memory but rather their powers of deductive reasoning and thinking become less acute, as if their global 'brain power' had been dulled or reduced. This would be similar to taking a computer with multiple parallel processors and removing one of the processors. This elegantly explains the findings in a study from the Sahlgrenska Academy, Gothenburg University, Sweden, which found that patients with demyelination (damage to the long nerve fibres) were more likely to be apathetic, which suggests a decrease in the power of the entire network rather than the selective loss of specific abilities or memories.

It is also interesting to note that the laying down of memories is enhanced by emotions. Both positive and negative emotions are effective; but the stronger the emotion, the stronger the memory. Emotions are traditionally associated with a different part of the brain, called the hippocampus, which suggests that the process of recording long-term memory is a multistage procedure that is influenced by several hardware components. Certain stimuli, like a smell, can cause you to retrieve a memory and strong emotions can increase the longevity of the memory. If you want to memorise a fact though, you rehearse it and go over it again and again. In doing so, you appear to etch that memory pathway into your brain more deeply, so that it can be recalled more easily.

In contrast to memory, which is constantly being updated and may be forgotten, personality is consistent throughout life and minor skill losses with ageing do not change personality. Even moderate strokes that affect motor skills and speech centres with the loss of global thinking abilities do not usually change basic personality.[62] So whereas motor skills, such as limb movements and speech have relatively localised areas of activation and a comparatively small brain injury can significantly compromise their functioning, it is much harder to damage the personality, suggesting that its mechanism of action is more diffusely spread or disseminated and a holographic model, would fit these observations regarding personality as well.

A holographic mechanism for memory offers possible explanations for

62. Severe strokes may change personality due to the severity of the damage to the physical hardware. The exception with mild strokes is where an antisocial tendency is consciously being repressed by the individual because they know that it is socially unacceptable. A small stroke in the voluntary centres in the frontal cortex may then unmask the antisocial behaviour if the restraining effect of the frontal lobes is removed.

other perplexing brain functions as well. In Chapter 4 we discovered that much of our sensory input and in particular vision, hearing and our sense of smell are processed in the brain as waveforms. In Chapter 8 we found that Fourier transforms are mathematical processes that not only analyse waveforms but also describe holograms, so frequency information is the ideal input format for holographic systems. Michael Talbot wrote: *"Encoding and decoding frequencies is precisely what a hologram does best. Just as a hologram functions as a sort of lens, a translating device able to convert an apparently meaningless blur of frequencies into a coherent image, Pribram believes the brain also comprises a lens and uses holographic principles to mathematically convert the frequencies it receives through the senses into the inner world of our perceptions."* [34]

Having examined memory, thought, hearing, holograms and Fourier transforms, it is now time to return to the scenario posed in Chapter 2, regarding the information in the music score that was lost. If thought is in the fifth dimension with consciousness, then presumably memory (which is the storage of previous thoughts) is also located in the same domain. If the composer had been alive to write the score down again from memory, then presumably it would have been recalled from the fifth dimension. So the answer to the symphonic conundrum is that although the physical music score was burnt in the house fire, its 'memory', via the composer's brain function, should be stored in the fifth dimension. The only problem is that we cannot visit there from our space-time domain to retrieve it once the composer has died. Much of this explanation may sound unusual because these deeper ramifications of information have not been explored before and there is very little data to confirm or deny what is going on. To assist, Table 10.4 summarises some of what we can deduce.

63. The Confessions of St Augustine. Bishop of Hippo. Book XI, Ch XIV *in* Cyber Library at http://www.leaderu.com/cyber/books/augconfessions/bk11.html

64. Harrison, E.: Time Warp. *Cosmos* 20:64 (2008).

65. Of course Einstein's Theory of Relativity says it does vary with mass, acceleration and gravity but on Earth these factors do not vary enough for us to notice. These exceptions will be addressed in Chapter 12.

Intangible entities	Life	Thought	Consciousness	Self-Consciousness
Carrier	Living cell	Neuronal network	Living brain (holographic)	Living complex brain (Implicate Order)
Storage	Living cell	Memory	Living brain	Human brain
Mechanism	Dynamic information in a functioning cell	Dynamic information in a functioning brain	Dynamic thoughts in a functioning brain	Dynamic consciousness in a functioning human brain
Generation	Requires an existing living cell	Requires an active brain	Requires a functioning complex brain	Requires a functioning human brain
Information Type	Dynamic information in a cell	Dynamic information in a brain	Dynamic thoughts within a brain	Integrated, creative consciousness

Table 10.4: Characteristics of Intangible Entities
It is important to note that to generate a thought, it is insufficient to merely send an electrical signal through a neuronal network. A fully functioning brain is the minimum requirement to generate thought; and yet a thought is merely a package of information.

Perception of Time

Time is another entity that we all use every day but struggle to understand. The problem is summed up by St Augustine when he wrote: *"What, then, is time? If no one ask of me, I know; if I wish to explain to him who asks, I know not."* [63] C.S. Lewis summarised the common notion that it is a defined standard that is unchangeable: *"The future is something which everyone reaches at a rate of 60 minutes an hour, whatever he does and whoever he is."* [64,65] But our *perception* of time is very different. We say: 'Time flies when we are having fun' or 'A watched pot never boils' even though we know it is not true but only a perception. In an article on: 'Time' in *Cosmos*, science writer, Erica Harrison wrote: *"In July 1962, the French geologist, Michael Siffre...descended 130 metres underground into a glaciated cave in the Southern Alps where he stayed for 60 days. The 23-year-old wanted to know what would happen to his sense of time when cut off from external cues such as clocks, caffeine and cycles of light and dark. Siffre used a battery-powered lamp to take notes, but the cost of lighting meant he spent most of his time in the dark. His only contact with the outside world was via a telephone he used to let his team know when he lay down to sleep, when he woke, and how much time he thought had passed. Gradually, Siffre lost touch with time...On September 14, a rope ladder was lowered into the cave and a confused Siffre was hauled to the surface. His diary entry from hours earlier was dated August 20 - somehow he'd lost 25 days. But while his consciousness experienced a timewarp, Siffre's body had been running like clockwork: each day in the cave lasted 24.5 hours, of which he spent 16 hours awake. Since then, numerous other experiments have confirmed that the human body is governed by a reliable internal clock...In all of these experiments, after a period of adaptation, the body falls into a precise rhythm of days lasting 24.5 to 26 hours, depending on the individual. This rhythm is set by a biological timepiece scientists call the circadian clock – from the Latin circa ("about") and diem ("a day"). "It's incredibly precise, within minutes per day."* [66]

So we all have a very accurate internal biological clock, which has been located in the suprachiasmatic nucleus (SCN) near the base of the brain in the hypothalamus, one for each cerebral hemisphere. The central time processing function that it performs is resistant to change as anyone who has suffered jetlag will tell you. In contrast, our *perception* of time, which makes time appear to 'speed up' or 'slow down' depending upon our surroundings, is called the 'interval timing system'. Scientists do not understood exactly how it works but the conventional theory proposes a *pacemaker accumulator,* which takes note of an event when it is triggered, just as one would start and stop a stopwatch for a race. This is unlike our circadian timing system which ticks away regardless of external or internal events.[67] The principal way that the body controls our pacemaker accumulator is through the hormone called 'dopamine' but a number of other factors can influence our dopamine levels such as: fear causing adrenaline release or illness such as Parkinsonism. The more common factors that can influence our perception of time are summarised in Table 10.5.

Causes	Accelerators	Decelerators
Illnesses	Schizophrenia (increased dopamine)	Parkinson's disease (decreased dopamine)
Temperature	Fever	Hypothermia
Emotions	Fear (adrenaline)	Boredom
Mood	Elation/mania	Depression
Medication	Stimulants (caffeine, nicotine)	Depressants (alcohol, Valium)
Other substances	Cocaine, amphetamines	Marijuana

Table 10.5: Time Perception Modulators
Various factors can cause our perception of time to appear to speed up or slow down.

When we are busy or engaged with an activity, we underestimate the time that elapses during the event (we say 'time flies') but when we remember it later, the incident seems to have taken longer than the event actually took. John Wearden, Professor of Psychology at Keele University in Staffordshire, England relates: *"The explanation...is that 'in the moment' judgements depend on attention to the clock, while retrospective judgements depend on how much information we processed in a period."* [68] Neuroscientist, Dr David Eagleman of the Baylor College of Medicine in Houston, Texas continues: *"Novel events may also play a part. 'When you're a child, you lay down rich memories for all your experiences; when you're older, you've seen it all before and lay down fewer memories...Therefore, when a child looks*

66. Harrison, E.: Time Warp. *Cosmos* 20:62 (2008).

67. For a more detailed explanation see: Harrison, E.: Time Warp. *Cosmos* 20:60–5 (2008).

68. Harrison, E.: Time Warp. *Cosmos* 20:64-5 (2008).

back at the end of the summer, it seems to have lasted for ever; adults think it zoomed by.'...Frightening events are associated with richer and denser memories. And the more memory you have of an event, the longer you believe it took...a surge of adrenaline that modulates dopamine levels and speeds up the inner clock..." would therefore account for the sensation that time goes more slowly in a crisis situation.[68] Associate-Professor Catalin Buhusi, who researches the interval timing system at the Medical University of South Carolina, Charleston, USA notes: "*...the pacemaker accumulator model is a 'powerful theoretical tool,' which is good at explaining how people perform when asked to judge the length of a short interval - the duration of a tone or flashing light, for example. But it runs into serious problems at the physiological level. The idea of a neural 'accumulator' that counts ticks indefinitely is especially problematic. As Buhusi explains, 'There's no way for something like this to be coded in the brain.'* "[69] There are therefore two very different timekeeping mechanisms in our brains: the circadian clock, which is located in the hypothalamus, acts as our timekeeper and would appear to be part of the Explicate Order, whereas the pacemaker accumulator, which is associated with events and memory, neatly fits with the characteristics of the Implicate Order, as was the case with memories.

Language

The single word 'language' encompasses a large body of information and is another characteristic that is universal to human beings. Language is the ability to express or communicate information.[70] Language involves:

1. formulation of the content (the message),
2. design of the vehicle,
3. synthesis of the transmission,
4. delivery of the information,
5. reception of the information,
6. interpretation of the information and finally,
7. understanding the message.

Since in all complex systems, as we saw in Chapter 2, the integrity of the whole depends upon the competence of each and every component subsystem, there are a lot of places where effective language (which is successful information transmission) can break down. This is why communication errors are a rich source of comedy in literature and the arts.

There are estimated to be 6,300 languages in the world.[71] The basic syntax of all languages is the same. Only the grammar and characters vary significantly. Philip Johnson writes: "*As with observed biological evolution, language evolution involves no increase in complex information or invention of new capabilities.*"[72] Stephen Pinker expressed the same idea slightly differently when he wrote: "*There are Stone Age societies, but no Stone Age language.*"[73] Does this mean that there is only one method of language construction? Or is language such a recent development that variations have not had time to develop? Or have the languages borrowed from each other so extensively that there is no appreciable difference between them? The jury is still out.

Information can be expressed in words, which is the traditional way that we think about language but information can also be conveyed through music to add emotion and colour. We know from ancient written texts such as the book

69. Harrison, E.: Time Warp. *Cosmos* 20:63 (2008).

70. Gitt, W.: *In the Beginning was Information.* (3rd Ed.) CLV Neuhausen, Germany (2003) p.206.

71. Sarfati, J.: *Refuting Evolution 2.* Answers in Genesis, Acacia Ridge, Qld, Australia (2002) pp.192–3.

72. Johnson, P.E.: (2000) p.137.

73. Pinker, S.: *The Language instinct.* Harper, NY (1994) p.27 *in* Johnson, P.E.: (2000) p.137.

of Psalms in the Old Testament of the Bible that not only prose and poetry but songs and music have been written for thousands of years too but until the 12th-Century, when staves were devised, music could not be recorded and this innovation was only achieved in European culture, which suggests that it is much harder to devise a language to record music than to compose the music itself. However, it can clearly be seen that music and song are valid languages that convey mood and feelings.

Professor Robert Pennock of Michigan State University compared the process that caused the facility for language to arise with the process which led to the appearance of life, declaring both to be: *"a mystery"*.[74] Thus the facility for language is yet another information dependant ability that is ubiquitous amongst human beings, which has been utilised for thousands of years, yet we struggle to explain the fundamentals of its appearance and development. It obviously depends on intelligence and it can be learnt but this does not tell us how it originated. What we can say is that it is associated with information and is non-material but now it is time to move on to intelligence, knowledge, understanding and wisdom.

Intelligence and Learning

What is intelligence? Superior intelligence is another distinctive of human beings; but can it be learned or is it innate? The human brain has 100 billion cells and 100 trillion connections which gives it unparalleled ability to learn, ability to organise and reason and ability to configure information.[71] Thus learning is the ability to acquire or manipulate information. Greater intelligence appears to be associated with the number of neuronal connections, that is, the richness of the network; whereas Alzheimer's dementia is associated with the degradation or thinning of the number of connections in the brain's neural network. One of the discoveries of the last decade has been that learning involves the generation of new connections. So to some extent we can increase our intelligence by growing new connections through learning but no matter how much you train a goldfish it will never reach the intelligence of a chimpanzee and a chimp is never going to match an adult human being, so although there is some environmental effect, intelligence appears to be mostly innate or pre-specified through information inherited through the genome. Twin studies in human beings conducted by Professor Thomas Bouchard, Jr., a psychologist at the University of Minnesota, suggest that about 70 percent is inherited and 30 percent is learnt.[75] Thus, intelligence represents the creative processing power of a living neural network and this is primarily specified by the information in the organism's DNA but it can be enhanced through active teaching and learning.

Knowledge, Understanding and Wisdom

Knowledge is the possession of a range of information and the ability to apply it in an optimal way. The Oxford English Dictionary defines wisdom as: *"The possession of experience and knowledge together with the power of applying them critically or practically; sagacity, prudence, commonsense."* Sagacity means: *"mentally penetrating, gifted with discernment, practically wise, acute minded* [or] *shrewd."* [76]

Understanding is defined as: *"to comprehend...grasp mentally, perceive the significance or explanation or cause or nature of, know how to deal with; having insight;...power of apprehension, power of abstract thought."* [76] Regius Professor-

74. Johnson, P.E.: (2000) p.138.

75. Bouchard, T.J. Jr et al: Sources of human psychological differences: the Minnesota Study of Twins Reared Apart. *Science* 250(4978): 223–8 (1990).

76. *Shorter Oxford English Dictionary*. Oxford University Press, London (1986).

Emeritus of Divinity at Oxford University, Keith Ward writes: "*...understanding, and intelligent action for the sake of realising some envisaged but not yet existent goal, are properties, not of physically measurable entities, but of a distinctive sort of reality that is not material.*" [77] Understanding is a derived effect, which requires the achievement of a critical threshold of connectivity. Bernd-Olaf Küppers explains: "*Information...is always related to an entity that receives and evaluates the information. This in turn means that evaluation presupposes some other information that underlies the process of registration and processing of the incoming information. But how much information is needed in order to understand, in the foregoing sense, an item of incoming information? This question expresses the quantitative version of the hermeneutic thesis, according to which a person can only understand some piece of information when [he] has already understood some other information.*" [78] Susan Greenfield, Professor of Pharmacology at Oxford summarised this concept when she said: "*Recognition turns seeing into understanding.*" [24] So what is the difference in terms of information between learning & understanding? Learning places information into memory but understanding places information into context and hence gives it meaning.

Higher level understanding is a uniquely human ability. Paul Davies describes this unexpected occurrence in *The Goldilocks Enigma*: "*We human beings have been made privy to the deepest workings of the universe. Other animals observe the same natural phenomena as we do, but alone among the creatures on the planet, Homo sapiens can also explain them. How has this come about? Somehow the universe has engineered not just its own awareness, but its own comprehension. Mindless, blundering atoms have conspired to make, not just life, not just mind, but understanding. The evolving cosmos has spawned beings who are able not merely to watch the show, but to unravel the plot. What is it that enables something so small and delicate and adapted to terrestrial life as the human brain to engage with the totality of the cosmos and the silent mathematical tune to which it dances?*" [79] Andrew Briggs refines Davies' statement when he writes: "*...to say that we understand something is an imprecise statement; the best that we can usually do is to understand an observation to a given level of explanation.*" [80] From the above it can be seen that there are levels of mental functioning possessed by different creatures on Earth. Larger mammals can obviously think, learn and gain limited levels of knowledge and understanding, especially in learning to react to human beings. The degree however, will be hotly debated. A simple example is the pod of killer whales in Twofold Bay (see Chapter 5) which learnt to co-operate with human beings for their own benefit. This information points to the existence of a hierarchy amongst the fundamental entities within the Universe and it can clearly be seen that new information is required to progress up the hierarchy. See Figure 10.1.

Each entity is discrete and uniquely different from the ones before it but irrevocably dependent upon them for its existence. The paradox is that the defining difference between each entity is information; although differing in content each time, the information is non-material and yet both the information and each resulting entity is undeniably 'real'! Keith Ward affirmed this when he wrote: "*...conscious intention and understanding...are irreducible and distinctive forms of reality.*" [81]

The non-material entities within this fundamental hierarchy can be expanded into a more detailed list as shown in Figure 10.2:

77. Ward, K.: *God as the ultimate informational principle, in* Davies, P. & Gregersen, N.H. (Eds.): (2010) p.287.

78. Küppers, B.-O.: *Information and communication in living matter, in* Davies, P. & Gregersen, N.H.: (2010) p.180.

79. Davies, P.: *The Goldilocks Enigma: Why is the Universe Just Right for Life?* p5 (2007) *in The Good News* 14(6):5 (2009).

80. Briggs, A.: And Information Became Physical. *in* Berry, R.J. (Ed.): (2009) p.82.

81. Ward, K.: *God as the ultimate informational principle, in* Davies, P. & Gregersen, N.H.(Eds.): (2010) p.289.

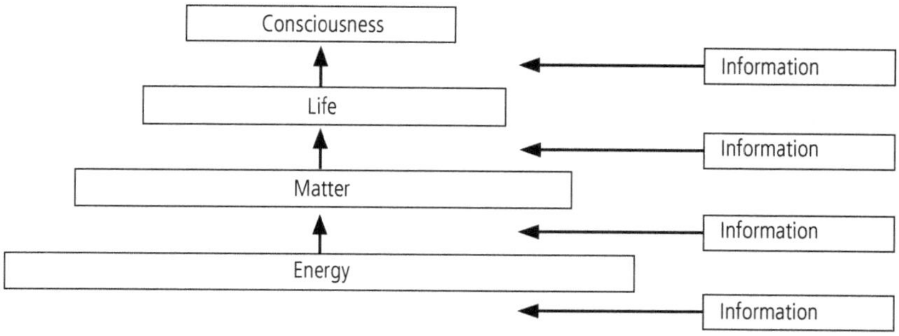

Figure 10.1: Relationships between the Fundamental Entities within the Universe

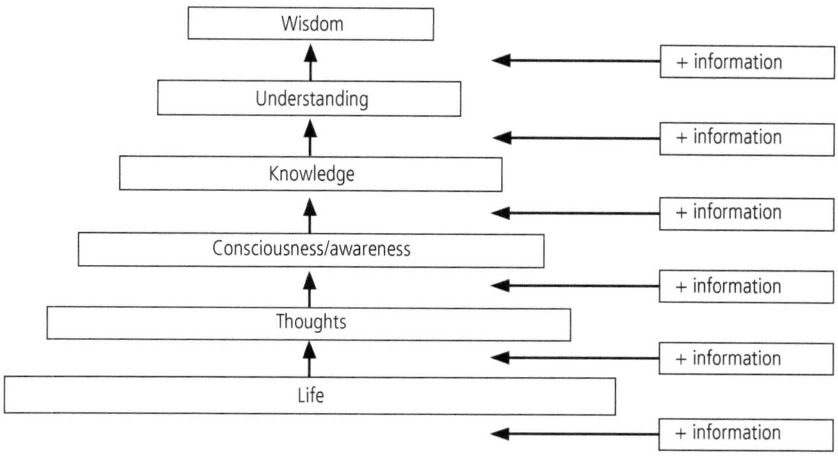

Figure 10.2: Relationships between some Non-material Entities of the Universe

Wisdom is the application of knowledge in an optimal way; with the amount of information that is required to gain it, it can be clearly seen why wisdom is so highly prized.

Music, the Arts and Information

There are many more non-material entities that deserve an explanation, such as what is it that constitutes beauty? Lee Strobel writes: "*In his book* The Analysis of Beauty, *written in the mid-1700s, William Hogarth said the defining feature of beauty or elegance is 'simplicity with variety'.*" [82] A recent article by Bevil Conway from the Department of Neurobiology at Harvard Medical School and Alexander Rehding from the School of Music at Harvard University entitled "Neuroaesthetics and the Trouble with Beauty" in *PLoS Biology* concluded that a materialist definition, let alone an explanation was very difficult.[83] I would assert that beauty equates to the harmonious integration of information. This can be readily seen when considering for example: sound (music), vision (art), touch (texture), smell (aroma) and taste (flavour). This effect on the harmonisation of information is amplified when different senses are combined, such as in a musical or opera. And mathematics is the mechanism which not only constructs the information but also defines the relationships between the individual pieces of information to create that harmony. When our senses perceive a mathematical relationship with balance ('not too much and not too little but just right') between the pieces of information that are delivered to us, we experience the condition that we identify as harmony. A simple analogy is the blending of two or more musical notes to produce a 'harmony' that is not possible with a single note alone. The performance of a beautiful piece of music is an extension of this principle. And

82. Strobel, L.: *The Case for a Creator*. Zondervan, Grand Rapids, Michigan (2004) p.147.

83. Conway, B.R., Rehding, A.: Neuroaesthetics and the Trouble with Beauty. *PLoS Biol* 11(3):e1001504 (2013).

the extraordinary finding noted by cognitive scientist, Josh McDermott of the University of Minnesota, who wrote in an article in *New Scientist* in 2008, is that: *"Music is universal, a significant feature of every known culture and yet* [it] *does not serve an obvious uncontroversial function."* [84] Charles Darwin had previously come to the same conclusion writing: *"As neither the enjoyment nor the capacity of producing musical notes are faculties of the least direct use to man…they must be ranked amongst the most mysterious with which he is endowed."* [85] Similarly, McDermott concludes that the origins of music remain an *"enduring puzzle"* to the evolutionary paradigm. Richard Dawkins' explanation is that: *"…without evolved brains spirituality and music would be impossible."* [86] However this does not help us to understand their existence. Darwin proposed that life involves *"survival of the fittest"* but is that all there is to life? Music is instantly recognisable by everyone. If you visit a culture whose language or customs are unknown to you, music can instantly make a connection. It is a universal language which we can all appreciate. But why is this? Jonathan Rhys Meyers who plays the character of Lewis in the movie *August Rush*, commenting on the film said: *"…music, it's just international. Even little two-year-old kids…you notice that when you put music on, they immediately start dancing… everybody has it…Music goes right inside you, whereas touch is always external."* [87] In the same movie, Robin Williams who plays the character, Wizard, asks August Rush (Freddie Highmore): *"Do you know what music is? …Harmonic connection between all human beings"* is his answer.[88] August Rush's closing line summarises one of the major themes of the movie when he says: *"Music is all around us… all you have to do is listen."* [88] Although this is not a scientific analysis of the topic, it puts into words what we experience when we listen to beautiful music. Little children do not have to be taught to appreciate music, they sense it automatically; as if it was hard-wired. Music has a capacity for impacting our emotions more easily than most other media. It is for this reason that most television advertising and most movies have musical sound tracks to augment their visual effect. My headmaster, Alastair Mackerras, who was an Oxford-trained mathematician and whose brother Charles was conductor of the London Philharmonic Orchestra, maintained that there was a close relationship between mathematics, music and learning and I suggest that the impression that *"music is all around us"* is partly because our universe is constructed from information that is digital and that beauty and harmony are mathematically defined.

Music is another information-intangible. It is real and yet we cannot physically hold it. Unless it can be recorded while the music is being played, it is gone. And yet, as Jonathan Rhys Meyers states, it can instantly touch our inner being in ways that few other things can. We know that music contains information and that it is digital. It is propagated by a sound wave but its universal effect of arousing emotions suggests that human beings have a facility not just for sound but its harmonious blending, which we call music. Music is like crystallised emotion. And it is not just *Homo sapiens* that responds to music. It has been known for some time that whales make music, which they learn and recognise.[89] Further, since music is an harmonic collection of frequencies, it is amenable to being manipulated by Fourier transforms. Therefore, from our previous discoveries regarding the way our brains process frequency information, it will be realised that the human brain is ideally suited for appreciating music and we should therefore not be surprised that the *"music goes right inside you"*.

The other point to note is that the beauty that results from an harmoniously played piece of music is an emergent property that has no physical

84. McDermott, J.: The evolution of music. *Nature* 453(7193): 287–8 (2008) *in Creation Magazine* 30(4):11 (2008).

85. Darwin, C. R.: *The descent of man, and selection in relation to sex.* John Murray, London. Vol. 2 (1871) http://darwin-online.org.uk/.

86. Dawkins, R.: (2009) p.402.

87. Sheridan, K. (Dir.): *August Rush* (DVD) Special Features interview with Jonathan Rhys Meyers. Warner Bros, USA (2007)

88. Sheridan, K. (Dir.): *August Rush* (DVD) Warner Bros, USA (2007).

89. The World Book Encyclopaedia. World Book Inc., Chicago (1984). *The World Book Encyclopaedia*. World Book Inc., Chicago (1984).

state. The notes can be identified and graphed on an oscilloscope but the beautiful harmonies and their effect on our thoughts and feelings are intangible and need to be studied using a different instrument, as an oscilloscope is not capable of measuring either beauty or harmony. In this instance, only a complex brain will suffice! It is also interesting to note that of all living species, only human beings are known to manufacture musical instruments.[90]

Thus we have seen that there are many non-material entities within the Universe including: life, consciousness, self-consciousness, thoughts, memory, intelligence, knowledge, understanding, wisdom, language, harmony, beauty, sleep and of course information. In Figure 8.1 we began to plot various physical and non-physical entities within the Universe. We are now in a position to elaborate upon and expand Figure 8.1, as shown in Figure 10.3:

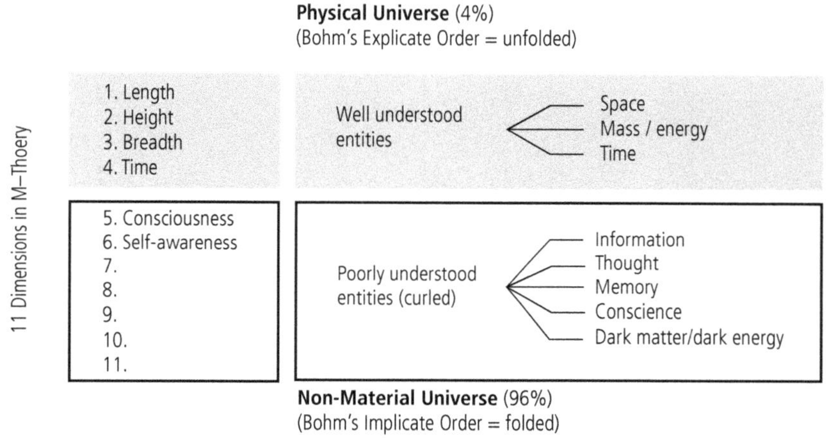

Figure 10.3: An expanded Illustration of both the Physical and Non-Material Universe
As the non-material Universe is curled (according to Relativity Theory) and hence unable to be probed from the physical Universe directly, most of the dimensions are unknown and hence have been left blank. Dimensions 5 and 6 are discussed in greater detail in Chapter 12.

Conclusion

In an interview in 2009, Richard Dawkins identified four entities with which evolution still struggles: the origin of life, the origin of sex, the origin of consciousness and the rise of morality.[91] The Information Model can answer all of these questions (see Chapter 12 for a discussion of morality), whereas Neo-Darwinian Evolution cannot. Just as DNA is a physical molecule which acts as a vehicle for genomic information and *"Human languages are the vehicles of semantic information,"* [92] the human body appears to be the physical vehicle which carries the entity of human life, while the complex brain is the conduit (hardware) for thought, consciousness and morality. Holmes Rolston III notes that: *"A crucial line is crossed when abiotic formations get transformed into loci of information."* [93] With increasing levels of organisation of information, Rolston notes: *"At some critical level, complexity can become aware and, at a still higher threshold, self-aware."*[94] However, thorough research in these information-containing fields will require a relaxation of the dogma that materialism is the only valid paradigm and that nothing else is worthy of study or even allowed to be investigated by serious scientists. Through an appreciation of information and the acknowledgement of information-bearing, non-material entities though, much of the Universe that remains a mystery to the materialist paradigm becomes understandable.

90. Sacks, O.: *Musicophilia, Tales of Music and the Brain (revised and expanded)* Vintage Books, NY, (2008) *in* Demme, G.: Does music have evolutionary origins? *Journal of Creation* 25(1): 40-3 (2011).

91. Boyle, A.: The not-so angry evolutionist. http://cosmiclog.msnbc.msn.com 14 Oct 2009 *in* Sarfati, J.: (2010) p.27.

92. Ward, K.: *God as the ultimate informational principle, in* Davies, P. & Gregersen, N.H. (Eds.): (2010) p.296.

93. Rolston, H.: *Care on Earth: generating informed concern, in* Davies, P. & Gregersen, N.H.: (2010) p.212.

94. *Ibid.* p.224.

Although this requirement will challenge a lot of scientists, Thomas Nagle, Professor in the Department of Philosophy and the School of Law at New York University writes: "*Major scientific advances often require the creation of new concepts, postulating unobservable elements of reality that are needed to explain how natural regularities that initially appear accidental are in fact necessary. The evidence for the existence of such things is precisely that if they existed, they would explain what is otherwise incomprehensible...we should expect theoretical progress in [the mind-body problem] to require a major conceptual revolution at least as radical as relativity theory, the introduction of electromagnetic fields into physics — or the original scientific revolution itself...We ourselves are large-scale, complex instances of something both objectively physical from outside and subjectively mental from inside. Perhaps the basis for this identity pervades the world.*" [95]

For the past 10 chapters we have been using information analysis as a probe to explore our universe, which has revealed the Information Model; it is now time to critically evaluate this new paradigm, which we will do in Chapter 11.

95. Nagle, T.: Mind and Cosmos. *Why the Materialist Neo-Darwinian Conception of Nature is Almost Certainly False.* Oxford Univ. Press (2012) p.42.

11

Testing the Information Model

The most incomprehensible thing about the universe is that it is comprehensible.[1]

ALBERT EINSTEIN

The definition of science used by Fowler and Kuebler in Chapter 1 required that good science must be testable. Further, Alistair and Joanna McGrath add: *"Every intellectual tool that we possess needs to be calibrated - in other words, to be examined to identify the conditions under which it is reliable."* [2] Thus, best practice requires that the Information Model be tested against an objective standard, to *"calibrate"* it, after which it can be measured against competing theories to see if it provides more or better answers than the existing models of the Universe.

As previously discussed, the usefulness of a scientific theory may be gauged by its ability to predict laws which lead to further useful discoveries. This is sometimes referred to as 'explanatory power'. Kepler used Copernican principles to formulate his laws. Newton used Kepler's laws to write his Laws of Gravity. Einstein extended Newton's Laws which were a reliable starting point from which to develop the Theory of Relativity. Physics, chemistry and cosmology have a plethora of laws and each new discovery validates the previous ones on which it is based. In contrast, the geocentric Aristotelian Theory of the Solar System, especially with Ptolemy's epicycles, led to no new advances but only confusion and European science languished in the academic doldrums for over one thousand years. Looking at the inorganic sciences, it would be fair to say that since Copernicus put everyone back on the right track, the physical sciences deserve an "A" for their achievements, especially with regard to the discovery of new laws that have helped us to understand the Universe. Looking at living things, however, how would we assess our biological report card with regards to new laws? The result is really rather mediocre. Pasteur's Law, which rescued the French wine industry, and from which was developed the process of pasteurisation, which is used in many countries to ensure that milk is safe for human consumption, is the only universal law of consequence. But if Neo-Darwinian Theory is correct, then life arose from inorganic life somewhere in the past history of the Earth or Universe, which contradicts Pasteur's Law that spontaneous generation of life cannot occur. Looking at the Darwinian paradigm over the last one hundred and fifty years, there are no universal laws and no subsequent discoveries leading to further laws that came about, either directly or indirectly, from the process of random mutation coupled with natural selection describing the process of Evolution. Microevolution

1. Lennox, J.C.: *God's Undertaker. Has Science Buried God?* Lion Books, Oxford (2007) p.58.

2. McGrath, A & J.C.: *The Dawkins Delusion.* SPCK, London (2007) p.15.

(variation within a species) has been observed, such as in the British peppered moths but this had already been described before Charles Darwin wrote *The Origin of Species* and no revelations have come from Darwin's theory. Mendelian inheritance, DNA structure and the Genome Project have occurred independently of the Theory of Evolution and it has even been suggested that Mendel published his work in 1866 as a response to Darwin's publication of *The Origin of Species*, to show that change was not gradual (pea plants were tall or short not a blend of the parental plants).[3]

So how does the Information Model compare?

A. Are there any predictions?
B. Is it testable? And,
C. Does it provide more or better answers than the existing explanations?

The answer is yes to all three. The three biological laws and the four information laws flow directly from it. The First Biological Law, which is Pasteur's Law, has stood for over a century and is entirely consistent with the Information Model. Each of the proposed biological and information laws will be addressed in turn.

Testing the Information Model

In Chapter 1, we listed ten essential characteristics of any genuine scientific theory. So, how does the Information Model measure up against the list of the ten essential characteristics, and if so, how? [4]

1. **Compactness: relatively simple with tight boundaries.**
   Yes. The Information Model can be summarised as: *Information defines and underlies the entire Universe. It cannot generate itself but requires an information generator; it is intangible but the Universe cannot exist or function without it.*

2. **Simplicity: simple and elegant.**
   Yes. It is one of the simplest of theories: digital information describes all laws and all structures in the Universe without exception.

3. **Falsifiability: capable of falsification.**
   Yes. If it could be shown that information is not the basis of the Universe or the natural laws of the Universe or if information is self-generating, such as would be the case if life could spontaneously appear, then the theory would be falsified. This is dealt with in greater detail below.

4. **Verifiability: verifiable by experiment and observation.**
   Although the Information Model does not directly propose formulae, which can be mathematically verified as Einstein's Theory of Relativity did, all the fundamental measurements in the Universe are digital and hence it is to be expected that mathematical relationships will be common within the Universe and that elegant mathematically defined laws will describe the Universe which indeed they do.

3. D'Abrera, B.: Personal communication. (2011).

4. Fowler, T.B. & Kuebler, D.: *The Evolution Controversy*. Baker Academic, Grand Rapids (2007) pp.129–31.

5. **Retrodiction: ability to explain known facts and observations.**
Yes. As we have seen in the previous ten chapters, all the examples cited are neatly explained by the Information Model without exception. The recent discovery that the redshifts in light from distant stars is quantised is similarly consistent with the mathematical basis of the Information Model.

6. **Prediction: ability to direct researchers to yet unknown facts.**
Yes. All entities in the Universe will ultimately be shown to have a digital information foundation, which will inevitably provide an enhanced explanation when compared with competing theories or hypotheses where there is no information basis, such as Macroevolution.

7. **Exploration: suggest new areas of fruitful research.**
The Information Model predicts that a digital information foundation underlies the entire Universe, so evidence of this should be sought in all areas of science and especially areas which are perplexing at present, such as thought, consciousness and life itself. At first approach, this suggestion may appear incongruous but before Newton described the Laws of Gravity, few (if any) scientists were expecting to find such precise mathematical relationships between the planets and the Sun within the Solar System and as scientists had not been directed to actively look for these mathematical links, the laws remained invisible; but the mathematical relationships were still there.

8. **Repeatability: able to be repeated or observed by independent researchers.**
As digital information is responsible for the construction of the entire Universe, it should be present in every experiment and every activity, if we choose to look for it carefully. As mathematical relationships are precise, there should be no trouble in repeating any experiments that are built upon a digitally-defined, information foundation.

9. **Clarity: general agreement about what the theory says and means.**
Due to its simplicity, the Information Model is hard to misconstrue; it says what it means and it means what it says - digital information is the explanation.

10. **Intuitiveness: transparent and intrinsically satisfying.**
Intuitiveness is a relativistic term and as such, it is appreciated in the mind of the beholder but in my opinion, once the observer becomes attuned to seeking out the informational foundation of our universe, they will perceive the Information Model as both transparent and intrinsically satisfying.

Critiquing the Biological and Information Laws

It has already been stated that the veracity of an hypothesis can be measured by its usefulness and testability. From the preceding chapters, it has been seen that the Information Model has yielded many useful new insights, such as an explanation for the wave-particle Duality of Light, an improved understanding of thought, consciousness, sleep and self-consciousness and a better concept of the processes involved in life and death, to name just a few. So how might the Information Model be tested? In science, the aim is to find an exception and hence to demonstrate that a premise is not true in all circumstances; in which case we need to ask how might the Information Model be falsified? Thus, are there any situations where the four information laws, or the three biological laws, which were derived from the Information Model in Chapters 7 and 9, might be proven to be incorrect? We will deal with each in turn.

First Law of Biology: Law of Biogenesis

As we saw in Chapter 1, Louis Pasteur showed that only *'life begets life'*. This law is refutable by building a new life form from scratch. Charles Darwin described a scenario, where in *"some warm little pond"* the first life might have spontaneously appeared and it appears to be largely taken for granted that this must have happened *'because we are here'* but it has never been demonstrated nor observed. In a much publicised case in May 2010, Craig Venter published an account in *Science* where he artificially copied the RNA of a bacterium and then inserted it into a living cell after removing the bacterial DNA. Three years earlier, Fowler and Kuebler wrote: *"Craig Venter and colleagues are attempting to create a new life-form from an existing bacterium. This, however, would not satisfy the criterion of creating a new life form from scratch because the bacterium from which they are starting supplies most of the design work."* [5] Apart from the design work however, by inserting his artificially manufactured copy of the original bacterial genome into an extant living cell, rather than building the cytoplasm from its constituent raw materials, Venter has inherited all the cytoplasmic information without which a living cell cannot survive. To build the cell completely from scratch, would require the artificial construction of each and every nano-machine and catalyst in the cytoplasm but if an organism really could be built entirely from basic building blocks then this would refute Pasteur's law. Darwin's hope of new life beginning in *'some warm little pond'* and its usual justification through the argument that 'it must be true because we are here and there is no other *acceptable* explanation' is *not* science and does not refute this law.

Second Law of Biology: Information Primacy

The Law of Information Primacy states that information coding precedes structure and function. If information is required to specify the DNA required for life, then new information must be generated in some manner although as we saw in Chapter 3, Neo-Darwinian Evolution is incapable of doing this. Therefore when bacteria develop resistance there must be mutation of old information, unless there is some other mechanism to generate new information. Thus, this law could be refuted if it could be shown that multi-resistance in bacteria is the product of *new* information rather than simply the mutation of existing DNA. A logical prediction of this theory is that bacterial resistance to antibiotics, such as Multi-Resistant *Staphylococcus aureus* (MRSA) will diminish and eventually disappear in the community if antibiotics could be completely removed from their immediate environment for a significant period of time, as MRSA would then

5. Fowler, T.B. & Kuebler, D.: (2007) p.259 footnote.

have to compete with the more efficient wild-type strains which lack the broken machinery of the MRSA strains. This is hard to test as antibiotics are used in the community too and the people who have acquired MRSA often have significant problems and make frequent trips back to hospital but it is instructive to note that MRSA and Vancomycin Resistance Enterococcus (VRE) first appeared in the big tertiary-referral hospitals where much of the very potent antibiotics are used, not in community medicine or general practice where, although the total volume of antibiotic usage is greater, the strongest antibiotics such as gentamicin, ticarcillin and vancomycin, which most powerfully select out the multi-resistant strains, are rarely used. At present, the best management for MRSA, when all known antibiotics have failed, is to send the patients out into the community and scrupulously avoid antibiotics all together. This is supported by a study published in the *British Medical Journal* in 2002, where antibiotic resistance was found to persist in children for two months after taking antibiotics, after which the incidence of bacterial resistance decreased over the next year, if further antibiotics were avoided.[6] Another confounding factor is that multi-resistant bacterial strains are emerging from intensive animal husbandry practices, where antibiotics are sometimes used in significant quantities to accelerate the growth of the stock and these multi-resistant bacterial strains are passing down the food chain before they are eliminated by natural selection through competition with wild-type strains. The answer is still to avoid antibiotics completely and let the wild strains out-compete the mutated and hence weakened, multi-resistant strains.

Third Law of Biology: Genetic Preservation

The Law of Genetic Preservation states that random variation with natural selection reverts to the mean unless there is environmental change, in which case the variation is limited by the pre-existing genetic information within the organism's DNA. This permits Microevolution but denies Macroevolution. This law would therefore be disproved if, by observing a defined population such as the Galapagos finches, it could be shown that over a number of generations that the beak size continued to grow until a new species (such as a tern) emerged. As discussed in Chapter 5, Peter and Rosemary Grant of Princeton University studied the Galapagos finches over a number of breeding seasons, however, the Grants showed that beak sizes oscillated about or toward the mean rather than away from it.[7,8]

First Law of Information: Information is the fundamental entity in the Universe.

This law states that information pre-specifies all matter and energy, which as far as we know, were the raw materials from which the Universe was constructed. To refute the law, an entity ultimately needs to be demonstrated to be independent of information; that is, to have another type of construction at its core. Since information is digital, this might take the form of proving that an entity has some other form of construction which is not digital.

Second Law of Information: Information is independent of time.

This is a corollary of Einstein's General Law of Relativity which infers: "*That which has no mass has no time*".[9] To disprove the law requires the refutation of the

6. Nasrin, D. et al.: Effect of βlactam antibiotic use in children on pneumococcal resistance to penicillin: prospective cohort study. *BMJ* 324:28 (2002).

7. Wells, J.: *Icons of Evolution*. Regnery Publishing, Washington, DC (2002) pp.165-169.

8. Johnson, P.: *Wedge of Truth*. InterVarsity Press, Downers Grove, Illinois (2000) p.48.

9. Missler, C.: *Cosmic Codes*. Koinonia House, Coeur d'Alene, Idaho (2004) p.369.

General Law of Relativity or an alternative possibility would be to show that an information store could vary due to time alone. For example it would be disproven if it could be shown that a message which had been buried in a bottle was different when dug up two years later.

Third Law of Information: Information is bound by entropy within a closed system.

This law states that within a closed physical system, all information systems are bound by the Law of Entropy. This is a corollary of the Second Law of Thermodynamics and means that information within the space-time domain tends to greater disorder unless information and energy are introduced into the system. The reason for this is that information within our space-time domain requires a physical carrier, which is invariably governed by the Law of Entropy. As the material vehicle which carries the information becomes degraded, the information also becomes scattered. This is in contrast to the non-material information within the vehicle, referred to in the Second Law of Information above, which is outside the space-time domain and does not spontaneously degrade. Once again, this means that information within a system cannot spontaneously organise itself but some external source is required for the organisation of information. To disprove this law, would require the demonstration that physical information systems were self-perpetuating and did not require extraneous information or energy input for their maintenance. An example might be to demonstrate that a computer can run indefinitely without malfunctioning and without maintenance.

Fourth Law of Information: Information conservation

The Law of Conservation of Information states that information from a primary information generator is conserved and can be audited. This hypothesis is much harder to prove or disprove as it requires an ability to access the frequency domain (Bohm's Implicate Order) which we lack the technology to do at this time. At first glance this law would appear to contradict the previous one, however, the Third Law of Information refers to closed *physical* systems and the Fourth Law refers to the entire realm of information. Although information can become scattered in our space-time domain if its physical vehicle is damaged by the effects of time, its non-material information component lies outside the physical domain and is unaffected.

Objections to the Information Hypothesis

Long before reaching this point in the book, there will have been loud cries from some quarters of *objection!* However, until the information content is described in the myriad of ways that we have seen, both in life and in the Universe, it is not easy to appreciate how all the elements fit together but now that we have a good understanding of the basic principles and implications of the Information Model, it is time to address the most obvious objections.

1. **Information is only a concept**
   In the same way that Charles Darwin, and Richard Dawkins have maintained that biological entities only have the *appearance* of design,

is the information in biological and universal systems real or do they only appear to have information? Jonathan Wells writes: "*Dawkins defines biology as: 'the study of complicated things that give the appearance of having been designed for a purpose.'* "[10] Likewise, does the Universe really contain information and is it a real entity or does it, as David Hume and Richard Dawkins have suggested, only have the *appearance* of being so? This subject was discussed in Chapter 1, where it was shown that there is a huge body of evidence which refutes Dawkins' assertions. The simplest demonstration is a new computer which has not had its software installed. It is not "just a concept" that turns this new computer, which is a useless machine without its software, into a powerful tool; it is sophisticated and costly information!

2. **Non-material concepts are not "science".**
   Thought, consciousness and memory have never been demonstrated to possess mass or physical dimensions and yet they are all part of mainstream science and medicine. No physician or neuroscientist in the modern era would suggest that the study of any of these was not "scientific". Non-material entities, which include information, do not lie outside science; they are just harder to study due to their lack of physical attributes. The 'Information Age' is not just part of our imagination.

3. **There is no evidence.**
   Presenting *physical* evidence of a non-material entity (such as information) is obviously an impossibility but denying that a murder occurred when a body or murder weapon is not obvious does not prove that a murder never occurred. If a person is missing in suspicious circumstances, the police keep working until they solve the case. If no satisfactory resolution is reached initially, the case is kept open until a satisfactory explanation is found. The police do not accept that a lack of physical evidence in suspicious circumstances is proof that a murder never occurred. The problem with information is that its true character is non-material and it requires a paradigm shift in thinking to take advantage of this brave new world. However, the results are worth all the effort!

4. **But if there had to have been something before the Big Bang to create the Universe, it cannot be studied and therefore cannot be science. In other words, it is not a legitimate scientific pursuit.**
   Science has always pursued what it cannot yet prove. Scientific knowledge grows through inquisitive people asking questions to which they do not fully know the answer, in an attempt to understand their situation better. If science cannot study why we are here, then someone needs to look at the definition and change the parameters so that science can adequately examine our own origins. The alternative is that science must consult philosophy to discover how the Universe came to be, which includes the origins of science too. The philosophers have no problem with this but more than a few scientists may be upset. As per Figure 1.4, an idea awaiting evidence may remain an hypothesis until

10. Wells, J.: *Icons of Evolution*. Regnery Publishing, Washington, DC (2002) p.204.

evidence is forthcoming but denying that non-material entities are a legitimate area of study is anti-intellectual and unscientific.

5. **How could Darwin be wrong when "everybody knows" he was right. (In other words, how could the majority of the scientific community have been mistaken for so long?)**
   No human being is infallible and even Einstein admitted that he had made significant mistakes. A good scientist will be able to recognise when new information means he or she needs to amend a point of view which was previously ill-informed. This is the history of science. Although Darwin wrote *The Origin of Species* more than one hundred and fifty years ago, Copernicus overturned more than fifteen hundred years of geocentrism, so time alone is not a valid argument. I simply ask you to look at the body of information that is before you and make a judgement based on the *evidence* that is there.

6. **If there are at least two more dimensions, why haven't they been detected previously?**
   String Theory states that the extra dimensions are curled in 10^{-33} centimetres and are too small to be detected with our existing technology but Michio Kaku, Professor of Physics at the City University of New York, anticipates that the large Hadron collider at the European particle physics laboratory (CERN) in Switzerland, may provide evidence. In a 2005 interview he stated: *"the Large Hadron Collider...won't be able to detect the strings themselves; but it will be able to sense their vibrations, potentially in the form of 'sparticles'. (Sparticles are theoretical ghostly partners of the fundamental particles — leptons, photons and quarks - that were produced in the Big Bang). 'It won't clinch it [string theory]. But string theory is the only theory that gives you the sparticles.' "* [11]
   The Large Hadron Collider was switched on in 2010 but the evidence is not yet forthcoming, either for or against.

7. **Much of the Information Model is neither testable nor falsifiable. So, how can it be a valid theory?**
   If you still hold this view then I suspect you have skipped Chapters 1 to 10 and started with this paragraph. I know that no amount of information will convince some readers but I believe that there is a significant body of evidence which demonstrates that the ideas embodied in the Information Model are true and that they are both testable and falsifiable.

Conclusion

Fowler and Kuebler asked: *"Should nonnaturalistic explanations be admitted into science? This question is a complete red herring. The important issue is not what we define science to be, but whether purely naturalistic explanations are adequate to explain all observed phenomena. If science is restricted to naturalistic explanations, as traditionally it has been, then the issue is whether there are limits to what science can do."* [12] If your response to this chapter were to be that the Information Model is "not science", all the non-material entities that are presently inexplicable using the

11. Finkel, E.: Fish out of Water. *Cosmos* 2:79 (2005).

12. Fowler, T.B. & Kuebler, D.: (2007) p.356.

current traditions of science, such as the existence of life, coma, sleep, thought, memory, the Dual Theory of Light and the mutual contradiction between the Theories of Relativity and Quantum Mechanics still need to be researched and understood. If the current definition of science cannot accommodate or explain these entities then there needs to be a new definition for science or a new discipline that is able to investigate these conditions. As Fowler and Kuebler conclude: "*...legitimate scientists have nothing to fear from* [investigating non-naturalistic phenomena] *- that is, nothing to fear if truth is the ultimate goal... as opposed to declaring natural causes sufficient by fiat.*" [13] Instead, the significant questions should include: does the new model make predictions which will lead to future discoveries and can it answer questions that other hypotheses cannot? We need to decide what the purpose of science is and use it, like any other tool, for the purpose for which it was designed rather than using it as a weapon to suppress viewpoints with which we disagree on philosophical grounds.

13. Fowler, T.B. & Kuebler, D.: (2007) p.357.

12

Implications of the Information Model

The entire universe is a cryptogram set by the Almighty.[1]

ISAAC NEWTON

We have come to the end of our information quest and it is time to reflect upon what we have learnt and the implications of these discoveries. We have travelled from the philosophy of the ancient Greeks, through information technology to biology, chemistry, cosmology, physics, and even a little geology. Theodosius Dobzhansky, the noted Neo-Darwinist wrote in 1973: *"...nothing in biology makes sense, except in the light of evolution"* which has been referred to as Dobzhansky's maxim.[2] However, although the acquisition of antibiotic resistance in bacteria illustrated Microevolution within defined boundaries, which was demonstrated in Chapter 3, it is simply wishful thinking to apply Dobzhansky's proposition to everything, as it is completely untrue of Macroevolution. In 2000, Jonathan Wells restated Dobzhansky's, maxim as: *"Nothing in biology makes sense, except in the light of evidence"* [3] which is true but not as useful as the interpretation by Jesper Hoffmeyer, Professor-Emeritus in Molecular Biology at the University of Copenhagen who noted in 2010 that it: *"...should nowadays be reframed as 'nothing in biology makes sense, except in the light of information'* " [4] because the evidence is that all science, and indeed, the entire Universe, is based on information.

In Chapter 1 we began with the emergence of the Information Age but the far-reaching effects of information and the accelerating pace of change are threatening to turn it into a revolution. Revolutions can be messy, chaotic affairs, like the French Revolution or they can be orderly and productive, like the Newtonian Revolution. The Information Model is likely to continue this process of change but hopefully in a methodical, constructive way, as the information paradigm is just a *reappraisal* of the facts, which have not changed; only our perspective has changed. However, sometimes a change in perspective opens our eyes to different possibilities. To take advantage of this opportunity however, we need to open our minds to the scientific evidence that is in front of our noses. One of the biggest problems is that information is non-material and science has become used to dealing with material entities. Materialism argues that non-material entities cannot exist, as science cannot readily conduct experiments to verify them. But, as we have seen, information is intangible, and there is no doubt that information not only exists but that it can influence material outcomes as well.[5] Ernan McMullin, Professor-Emeritus of Philosophy at Notre Dame University in the USA comments that Aristotle described just such a non-material model over

1. Missler, C.: *Cosmic Codes.* Koinonia House, Coeur d'Alene, Idaho (2004) p.333.

2. Wells, J.: *Icons of Evolution.* Regnery Publishing, Washington, DC (2002) p.247.

3. *Ibid.* p.248.

4. Hoffmeyer, J.: *Semiotic freedom: an emerging force, in* Davies, P. & Gregersen, N.H. (Eds.): *Information and the Nature of Reality: from Physics to Metaphysics.* Cambridge University Press (2010) p.185.

5. Johnson, P.E.: *Wedge of Truth.* InterVarsity Press, Downers Grove, Illinois (2000) p.118.

2,300 years ago: *"The world for Aristotle was made up of substances...This led him to introduce first, or 'prime', matter, something that lacks all properties of its own and functions only to ensure the continuity, and hence the reality, of substantial change...In the case of ordinary (non-substantial) change, the 'matter' ('second' matter, as it came to be called) is just the subject of that change, whatever its properties happen to be (a leaf, for instance)."* [6] Aristotle's 'prime' matter corresponds to the information that appears to reside in the Implicate Order while 'second' matter corresponds with the physical objects that we experience in our material universe, the result of the Fourier transform-like process described in Chapter 8. Aristotelian metaphysics, as it became known, was developed further by Thomas Aquinas in the 13th-Century and will not be dealt with in depth here but it is thought-provoking to note the high degree of congruence between Classical metaphysics and the Information Model which derive from completely different lines of argument.[7,8]

These two opposing models of Materialism and Information cannot both be true so how will the impasse be resolved? Charles Darwin wrote: *"A fair result can be obtained only by fully stating and balancing the facts and arguments on both sides of each question;"* [9] this has been the aim of this book. Every author will be surpassed by discoveries that are based on information which comes after them to which they were not privy and Mendelian genetics and Information Mechanics are two areas that were not understood in Darwin's day. To survive, each reigning paradigm must grow to encompass new discoveries as they occur, as Neo-Darwinian Theory did when Mendelian genetics was fully appreciated, otherwise it will be eclipsed and be replaced, as Lamarckian inheritance was.[10] Thus we are each called upon to extend the boundary of knowledge so that we heed William Bean, Professor of Medicine at the University of Iowa, who counselled in 1966 that: *"...students should be better than their teachers. If you as our followers are not better than we are there will be no progress and the World will either stand still or go backwards."* [11]

Sometimes advances require only minor changes, such as Darwinian Evolutionary Theory accepting Mendelian genetics but sometimes radical changes in direction are required, such as when the concept of an eternal universe was superseded by a finite universe. The idea that our universe had a beginning initially seemed counter-intuitive to the scientists of the day (including Einstein) but with time, the evidence was re-assessed and accepted. Information Mechanics may be viewed in a similar way by those who are unfamiliar with its ideas and it will take some time to re-appraise the evidence in the light of the new paradigm. However, Information promises to be the answer to Aristotle's search for *quintessence*, the *"Fifth Essence"*, which he believed would tie everything else together (which for him was the four basic elements). Rather than linking Aristotle's four elements though, our challenge is to combine the four fundamental forces described in Chapter 1; but the principle is the same. It is from Aristotle's quest that we derive the word *quintessential* which describes information's role precisely. All the information is there; it just requires an understanding of the relationships between the puzzle pieces so that they can be put together in an orderly way.

Precision Thinking

So how do we work out the best way to connect all the pieces of information? In Chapter 1 we looked at discontinuities and how to reconcile Pasteur with Darwin because they both could not be right even though both are highly esteemed

6. McMullin, E.: *From matter to materialism...and (almost) back, in* Davies, P. & Gregersen, N.H. (Eds.): (2010) pp.15–16.

7. Feser, E.: *Aquinas – A Beginner's Guide.* Oneworld publications, Oxford, England (2009).

8. Feser, E.: *The Last Superstition. A Refutation of the New Atheism.* St Augustine Press, South Bend, Indiana (2008).

9. Darwin, C.: *The Origin of Species.* John Murray, London (1859) reprinted by Penguin Classics, London (1985) p.66.

10. Pearce, E.K.V.: *Who Was Adam?* Africa Centre for World Mission, Walkerville, South Africa (1987) p.109.

11. Bean, W.: 9th Lambie-Dew Oration 1966, Univ. of Sydney *in Sydney Univ. Medical Society Senior Year Book* (1982) p.8.

scientists (at least within their own spheres of influence). George Wald, who was awarded the Nobel Prize for Physiology (Medicine) in 1967 for his work on the chemistry of vision, helped shed some light on this situation when he wrote: " '*We tell this story [regarding Pasteur's experiments disproving spontaneous generation] to beginning students of biology as though it represents a triumph of reason over mysticism. In fact it is very nearly the opposite. The reasonable view was to believe in spontaneous generation; the only alternative, to believe in a single, primary act of supernatural creation. There is no third position.*' " [12] But where does this leave science?

Richard Dawkins in his concluding chapter in *The Greatest Show on Earth* wrote: "*This is a book about evidence, and we have no evidence bearing upon the momentous event that was the start of evolution on this planet. It could have been an event of supreme rarity. It only had to happen once, and as far as we know it did happen only once.*" [13] He continues: "*The beginning had to be simple, and evolution by natural selection is still the only process we know whereby simple beginnings can give rise to complex results.*" [14] However, if there is no evidence regarding the beginning of life, then how does he know that the beginning "*had to be simple*"? Dawkins is so convinced that Macroevolution is true (tier 4) that he *assumes* that the origin of life was simple (tier 1) yet he is pinning his hopes on a process for which he admits he has no evidence and no idea of its origin. This is woolly thinking at best and wishful thinking at worst. It is vitally important to leave woolly thinking behind and to be very clear about what is being discussed, so that the different concepts are not muddled up. The Information Model, in which the formation of the Universe is initiated by the introduction of information, provides a much better fit with the evidence that we can detect and has explanatory power regarding the origin of life, which Dawkins admits the Neo-Darwinian Theory lacks.

In the Western World, six generations of scientists have now been taught Darwinian or Neo-Darwinian Theory as truth and many have accepted this position uncritically, so it will take more than a year or two for a new understanding of life based on information generation and information flows to alter established ideas, maybe even several generations. Yet information is digital and a digital framework can deliver precision. Thus the Information Model offers us a precise tool with which we can more accurately benchmark what we observe in our universe. With greater precision, there is a lower level of uncertainty, so the error range will be less and previous errors in interpretation are more likely to be identified and corrected.

If the information in DNA had to be actively generated rather than having evolved by chance, then the DNA code was intentional rather than having been an accident, which implies deliberate intervention in its formation. If information is premeditated and not mishap, then life and by extrapolation, the Universe, is not an accident either. If this is true, then we as human beings are not an accidental collection of proteins but the result of optimisation in the same way that a building which has been designed and built by an architect or an engineer has been optimised.[15] George Wald validated this conclusion when he wrote: "*One only has to contemplate the magnitude of the task to conclude that spontaneous generation of a living organism is impossible*" however he then unexpectedly concluded that: "*Yet we are here - as a result, I believe, of spontaneous generation.*" [16]

Through this curious logic, Wald manages to uphold both Pasteur, who demonstrated that spontaneous generation does *not* occur and Darwin, who believed that it must! Thinking from an informational frame of reference precludes

12. Wald, G.: The Origin of Life, *in Life: Origin and Evolution*, W.H. Freeman, San Francisco (1979) p.24 *in* Sarfati, J.: *By Design*. Creation Book Publishers (2008) p.190.

13. Dawkins, R.: *The Greatest Show on Earth*. Bantam Press, London (2009) p.416.

14. *Ibid.* p.417.

15. See the discussion on fitness landscapes and Figure 3.11.

16. Wald, G, The origin of life, *Scientific American*, 191:46 (1954) *in* Sarfati, J.: (2008) p.189.

this type of woolly thinking, as information (digital code) is precise and specific. Another example of this is the study of similar anatomical structures in different species, which is the discipline of homology (as discussed in Chapter 5). Darwin's tree of descent is based on similarities between species, such as the fluke of the whale, a bats wing and a human being's upper limb.[17] Based on similar anatomy and especially bone structure, they would appear to have common ancestry but when gene mapping (which provides more information) became available, the gene that induces limb development was found in such diverse creatures as the mouse, sea urchins, spiny worms, and velvet worms and is almost identical in each species, even though some of them lack limbs all together! Further, the creatures are from very diverse phyla, placing them on different branches of Darwin's Tree and are hence considered by Darwinians to be very distantly related in their evolutionary history and descent.[18] The Neo-Darwinian Theory cannot explain these findings. As more information allows digital comparisons, greater rigour will be applied to many areas of science, just as the digital precision of Kepler's mathematics changed astronomy in the 17th-Century.

Science is supposed to be the study of precise entities using precise methods by precise minds, so how can scientists using the same data come to diametrically opposed conclusions for so long? The problem is that science is the study of observations, from which we construct theories to fill the gaps in our knowledge (information which is not available at that time). There are many critical points in constructing an accurate scientific theory but one crucial starting point is in deciding which information is relevant, such that it warrants inclusion, as opposed to that which does not. This is where our preconceived bias comes into play and that bias can continue if we persist in inappropriately weighting the value of the evidence we have selected. If the theory turns out to be erroneous but we then prefer the *theory* rather than new (conflicting) observations, we have put the cart before the horse and the resulting predictions may be deceiving. The main problem is that despite all our emphasis on scientific method, we as human beings have a tendency to see what we want to see.

The next important point is that more recently, certain influential sections of the academic community have attempted to artificially restrict science to the study of physical entities, which are more easily demonstrated by experimentation but just because something cannot currently be confirmed experimentally, does not *prove* that it does not exist and should hence be excluded from the study of science. It may just show the limitations of the science of the day. It is clear that non-material entities such as information cannot be treated materially but does that require a new discipline outside science to acknowledge that there are entities which, though they can be logically inferred, they are not yet acceptable to scientists with a materialistic mindset? Instead, I am proposing the term: 'Information Mechanics' be used to describe that subset of science in which non-material entities are investigated; it would allow science to be inclusive of those instances where the materialist, scientific model has failed to provide useful answers. This does not imply that the discipline of Information Mechanics is negating those parts of the existing materialistic, experimental model that are successful. It just points out that the study of information is of equal importance with the study of particle physics; it recognises that a different modality is required to study it. In fact, it is my belief that for science to advance, the study of both the material and the non-material *in parallel* will be unavoidable, as information underpins the entire Universe. However, the opportunities offered by digital

17. Wells, J.: (2002) pp.60 & 76.
18. Wells, J.: (2002) pp.74–5.

information will not be an instant panacea for everything that is currently unclear or unknown, as information is a tool to aid analysis, not an answer in itself. What is patently clear is that integrating Information Mechanics into mainstream science will undoubtedly reveal new insights which may not be apparent without it.

Information as the Fundamental Entity of the Universe

The time has come to properly consider the evidence offered by Information Mechanics and its implications for the sciences. Andrew Gosler, Chair of the Institute of Human Sciences at Oxford University writes: *"...all that exists, exists as information."* [19] In Chapter 8, we examined Stephen Hawking and James Hartle's mathematical model for the formation of the Universe, which involved energy fluctuations in a quantum vacuum, to explain the source of space, mass, energy and time. However, from Chapter 9, where the Universal Information Hypothesis was discussed, it was seen that with the 'Big Bang', at the formation of the Universe, there was also a requirement for the introduction of information, to specify the unfolding of the Cosmos. More information was required to describe everything within the Cosmos, including mass-energy and the fundamental laws of the Universe which carry the information that defines its behaviour. Lennox writes: *'...information...therefore, is more fundamental than mass-energy. Mass-energy belongs to the category of the created.'* [20] Lennox goes on to point out that information is non-material, that is, it has no mass outside its carriers. Thus the material Universe must be based upon the invisible and non-material entity of information.[21] Material entities are the visible and measurable expressions of information and also its mode of residence within the space-time domain. From the preceding chapters, it would appear that the Universe began in a frequency domain from which matter 'precipitated'. Jim Al-Khalili says that: *"mass can be thought of as frozen energy"* [22] but since mass (matter) and energy are simply the vectors of information, matter can be viewed as *'frozen' information* and energy is *'liquid' information*. To carry the theme slightly further, Bohm's Implicate Order is analogous to *'gaseous' information* although it behaves more like *plasma* than a gas. Further, it can be seen that information can change the state of its carrier but not its content. In the same way that light can be both a particle and a wave, Hawking and Hartle's *"virtual particles"* can materialise from the frequency domain, whence they go if they mutually annihilate, as they usually do. This frequency dimension must have had information embedded in it, which would appear to have been associated with the unfolding of the Universe. This reveals our tenth information paradox which is that the most significant determinant of the Universe is the least detectable. Thus, all physics and chemistry can be seen as the mechanisms by which information manages the Universe and are hence encompassed by our newly described discipline of 'Information Mechanics'. It should be noted that although the term: 'Information Science' already exists, it needs to be broadened to look at the wider implications of information, not only the material vectors of information which our senses can verify but the non-material message too, which tends to have been overlooked in the past.

Thus, information can be viewed as the 'gold standard' for the Universe and in particular, science; but not only the gold standard which keeps science accountable but also the benchmark by which all science can be measured.[23] Professor of Quantum Mechanical Engineering at MIT, Seth Lloyd summarises the topic well when he writes: *"Information and energy play complementary roles in*

19. Gosler, A.G.: Surprise and the Value of Life, *in Real Scientists Real Faith*. Berry, R.J. (Ed.) Monarch Books, Oxford (2009) p.184.

20. Lennox, J.C.: *God's Undertaker. Has Science Buried God?* Lion Books, Oxford (2007) p.167.

21. *Ibid.* p.168.

22. Al-Khalili, J.: *Black Holes, Worm Holes & Time Machines.* Taylor & Francis Gp, New York (1999) p.31.

23. The gold standard was a measure developed by the British Government in the 19th-Century when paper money was introduced. Coins used to contain precious metal equivalent to the value imprinted on their face, so that the person who possessed the coin could resort to melting it down to retrieve the value of their investment if the issuer of the coin became unable to redeem the coin for gold or silver. Once paper notes were introduced, this was no longer possible, so the Bank of England retained the equivalent value of gold bullion to match the value of the notes that the bank issued. To guarantee that value, it undertook to buy and sell gold at a fixed price. This was called 'the Gold Standard'. The standard was maintained until 1913 when Britain was no longer able to fund it. The USA used the Gold Standard from 1900 to 1917 and from 1919 until it was finally discarded in 1933 due to lack of financial resources. [*World Book Encyclopaedia*. World Book Inc. Chicago Il. (1984).]

the universe: energy makes physical systems do things. Information tells them what to do." Thus *"...the primary actor in the physical history of the universe is information"* [24] In the introduction to their book: *Information and the Nature of Reality*, Paul Davies and Niels Gregersen amplify Lloyd's statement when they write: *"Davies suggests that instead of taking mathematics to be primary, followed by physics and then information, the picture should be inverted in our explanatory scheme, so that we find the conceptual hierarchy: information → laws of physics → matter."* [25] This concept has been previously discussed in greater detail at the beginning of Chapter 8. Walter Bradley, Professor of Mechanical Engineering at Texas A & M University concludes: *"Not only has science failed to provide naturalistic explanations for the mathematical form of nature, the coincidence of cosmological constants, and the emergence of living things, but also these facets of nature all demonstrate the essential element of design, namely, information...Constraint of systems far from equilibrium can produce self-organisation that takes the form of spatial ordering and/or cycling over time. However, neither biological information (e.g., aperiodic specificity in sequencing of biopolymers) nor cosmological information is created in such processes. The limited ordering and information produced in such systems is consistent with the information level of the constraints imposed on the system."* [26] I am therefore proposing that the current materialist paradigm be replaced by an information paradigm which includes the former but extends it. The materialistic paradigm has been very useful for centuries but in the same way that the study of mathematics was incomplete without the concept of zero, the study of the Universe is incomplete without the entity of information and its attendant non-material characteristics.

The Copernican Principle

So with this new perspective on the Universe afforded by information analysis, where does this leave man's place? When Copernicus described the Sun as the centre of the Solar System, there began a long chain of discoveries about the Earth's positioning within the Universe. The realisation that the Earth was not at the centre of the Solar System, let alone at the centre of the galaxy, led some people to question the significance of the Earth and the importance of humankind's existence in 'the grand scheme of things'. If human beings, and hence the Earth on which they reside, were a special creation of God, wouldn't He have placed them in the most prestigious place, at the centre of the Universe? Instead, Earth appears, as Carl Sagan, described it as: *"a lonely speck in the great enveloping cosmic dark."* [27] Gonzalez and Richards have comprehensively discussed this in *The Privileged Planet*, but can humankind's place in the Cosmos be described in terms of information? It has been noted in Chapters 3 to 6, that there is a vast amount of information in our universe and the complexity hierarchy in Chapter 2, shows how it is connected, from the smallest known subatomic entity (currently the muon or quark) to the Universe as a whole. But there is general consensus amongst scientists that the most complex entity in the Universe is the human brain. So in terms of the maximum information content within a single entity, the pinnacle of information processing and complexity within the Universe is the human brain and hence human beings. Although many have tried to find other intelligent life, none has been found to date; so based on current information, far from being an insignificant species on an insignificant planet, the Earth and especially humankind represent the informational pinnacle of the Universe. New information in the future may change this conclusion but today, based on the

24. Lloyd, S.: *Programming the Universe*. Knopf, NY (2006) p.40 *in* Gregersen, N.H.: God, matter, and information: towards a Stoicizing Logos Christianity *in* Davies, P. & Gregersen, N.H. (Eds.): (2010) p.334.

25. Davies, P. & Gregersen, N.H. (Eds.): (2010) p.3.

26. Bradley, W.L.: 'Nature. Designed or Designoid' *in* Dembski, W.A. (Ed): *Mere Creation. Science, Faith and Intelligent Design*. InterVarsity Press (1998) p.48.

27. Sagan, C.: *Pale Blue Dot*. Ballantine Books, NY (1994) p.7 *in* Gonzalez, G. & Richards, J.W.: *The Privileged Planet*. Regnery Publishing, Washington DC (2004) p.247.

evidence available, this is *fact*. Sagan's assessment regarding the significance of the World, when he described it as a *"pale blue dot"* is based on its 'geographical' ('cosmographical' sounds funny) significance and if you are a real estate agent, then position is everything. But Gonzalez and Richards have shown that being in the centre of things in universal terms (that is, at the centre of the galaxy) will probably fetch you up in the middle of a giant black hole! If we take an information perspective rather than a cosmographical one, however, humankind sits at the summit of the information mountain and the inhabitants of Earth, as Gonzalez and Richards concluded, have an optimal location from which to survey and gain an understanding of the Universe. By this line of thinking, humankind occupies, not one of the least, but one of the most significant (or valuable) places in the Universe as far as human kind is concerned.

Time

Time is an enigma. Remember, when we examined the concept of time in Chapter 10, our perception of time passing, meant that at school the winter term seemed to drag along whereas the summer holidays fairly whizzed past; but now we are going to leave our *perception* of time to one side to look at what physics tells us about the *entity* of time. Jim Al-Khalili writes: *"Let's get one thing straight. Whatever you have read or heard, nobody understands what time really is."* [28] Like information, time is another intangible entity and yet it is a physical property with dimensions in the space-time domain and it varies with gravitation, mass and acceleration.[29] We often have problems with time, as the human brain is familiar with linear relationships and that is the way we analyse the data that we observe around us, however we live in a non-linear universe. We intuitively sense that the flow of time is constant *for us*, however, as the gravitation, mass and acceleration affecting the Earth almost certainly have not been constant since the formation of the Universe, as strange as it may appear, time almost certainly has not been constant on the Earth since its formation either.

The ancients noted the cyclical nature of the days and seasons. In ancient Egypt they invented the sundial five thousand years ago to mark the passing hours, while in Britain the early inhabitants built Stonehenge to mark the passing seasons. Although the ancient Babylonians defined sixty seconds to make a minute and sixty minutes to make an hour, it was difficult to measure the seconds and minutes until: *"...the sixteenth century when Galileo discovered that a pendulum of a given length will always take the same time to complete one full swing. But it was not until the mid-seventeenth century that the first pendulum clock was built."* [30] The pendulum held sway as the gold standard in timekeeping until the latter 20th-Century when it was discovered that quartz crystals vibrate regularly and reliably when an electrical current is passed through them and so the quartz digital watch was born. Even more accurate are the atomic clocks which: *"...rely on the fact that certain atoms, when pumped with energy, emit light at a precise frequency that is unique to that type of atom. The most famous of these are caesium clocks which now set the world standard for time."* [31]

But this ultra-brief history of timekeeping describes the 'how' of time measurement but does not address the entity of time itself. Al-Khalili writes: *"Until Isaac Newton completed his work on the laws of motion, time was considered to be the domain of philosophy rather than science.*[32] *However, Newton described mathematically how objects move under the influence of forces, and since all movement*

28. Al-Khalili, J.: (1999) p.111.
29. Missler, C.: (2004) p.369.
30. Al-Khalili, J.: (1999) p.112.
31. *Ibid.* p.113.
32. Before Newton, Philosophy was considered synonymous with Science and vice versa.

and change requires the notion of time for it to make sense, he used what is known as a realistic view of time…Newtonian time is absolute and relentless. He described it as a medium which exists entirely on its own outside space and independent of all processes that occur within space. In this view, time is said to flow at a constant rate as though there were an imaginary cosmic clock that marks off the seconds, hours and years regardless of our, often, subjective feelings about its passage. According to Newton time is absolute, true and mathematical. We have no influence over its rate of flow and cannot make it speed up or slow down." [33] Al-Khalili continues: *"According to the mechanistic view that Newton's laws of motion give us of the Universe, it is possible, in principle at least, to know the position and velocity of every particle in the Universe. Given that each particle will follow a well-defined trajectory and be under the influence of forces that, again in principle, can be well defined, it is possible to work out their positions at any future time and hence to know the state of the Universe in the future. The future is therefore mapped out and preordained."*

He continues: *"Such a reductionist view of the world might seem to leave no room for human free will. Since we too are made up of atoms we are subject to the same laws of physics as any other object; then presumably what we consider to be free will is no more than mechanical processes in the brain obeying Newton's laws like everything else."*

"In practice of course we are not even able to calculate the future positions of a few balls on a pool table after they are scattered by the cue ball, let alone the future positions of all the particles in the Universe. But, according to this 'deterministic' view it should at least be possible in principle *to do so, provided we had a powerful enough computer."* [34]

Despite the ability of Newtonian physics to theoretically describe the future of our Universe, this is not what most physicists believe today. Al-Khalili explains: *"Today physicists no longer adhere to this idea of a deterministic universe. That way of thinking was overthrown when the theory of quantum mechanics was developed in the mid-1920s showing that, at its most fundamental level, nature is random and unpredictable. Despite this, many physicists believe that the future is already out there, not because of the Newtonian picture of a clockwork universe, but because it follows on from the way the theory of relativity unifies time with space. This idea that the future already exists goes beyond the Newtonian view, which only claims that the future can be predicted."* [35]

And so we come to our third type of time which is the Einsteinian version, as described by Relativity Theory, which says that time is relative or can be altered by speed and gravity. If we had two identical clocks and kept one on Earth and could send the other on a trip at half the speed of light to our nearest neighbouring galaxy, the Andromeda Nebula, the clock in the rocket ship when it returned would register a significantly different time. Jim Al-Khalili explains: *"The closer to the speed of light that a clock moves, the slower it will tick. If it were to travel at the speed of light relative to us then we would see its time stand completely still."* [36] This phenomenon is referred to as *"time dilation"*. The other effect of travelling close to the speed of light is that distances become compressed so that the distances appear to be physically less too. The Special Theory of Relativity forbids anything to go faster than the speed of light but if it could, theoretically, time would run backwards which would play havoc with the arrow of time. So logically, anything faster than the speed of light (and yes, physicists have hypothesised such a particle called a 'tachyon') would be outside our physical universe as we presently understand it.[37] *"A rather baffling consequence of this is that for light itself time stands still. If you could attach a clock to a light beam it would not tick at all. We say that to*

33. *Ibid.* p.114.
34. *Ibid.* p.115.
35. *Ibid.* p.116.
36. *Ibid.* p.152.
37. *Ibid.* p.159.

a photon, time does not go by at all (maximum time dilation) and the whole Universe has zero size (maximum length contraction)!!" [38] Al-Khalili summarises Einsteinian time as: *"...an interval of time depends on your perspective in relativity. Less time will elapse for someone who accelerates to very high speeds or spends some time in a strong gravitational field."* [39]

Information and Time

It has previously been noted that information is outside of time and does not change with time. The physical medium in which information is stored (such as papyrus) may well age in the space-time domain (dimensions one to four) but the information itself does not. Therefore the information itself, strange as it may seem, must be located somewhere else, possibly in dimensions five or six (or higher). This means that information does not tell you how old it is, such as when it was generated. Thus, information itself does not help with determining the age of the Universe or the age of life on Earth.

Our growing knowledge of information and its multilayered arrangement within the Universe has improved our understanding of why the Universe is the way it is but there are still questions that cannot be answered from an analysis of information, such as: 'why does the Universe exist?' This question is outside the scope of this book. If information derives from, or is stored outside our present frame of reference, to access that information the observer needs to be able to access the place where the information is located. To understand this better, we need to return to the way we conduct internet banking. If you have the appropriate hardware (a computer) and software (to access the internet) and you know your bank account details with the appropriate passwords (access information) then you can find out your bank balance (retrieve the information) but that does not tell you why the bank exists. To find that piece of information you need to visit the bank and ask the people there or have someone else inform you of the bank's history (perhaps by reading about it). For this reason, information regarding time prior to the formation of the Universe and information regarding why human beings are on the Earth needs to be sought from a different place, since this information is outside our physical space-time domain. The problem is that we do not know how to consciously access the frequency domain to retrieve the information.

Dolphin expanded on these issues when he wrote: *"...chronological time is measured & recorded in some sense by our physical bodies not by our spirits. To die physically is to leave this* [universe] *and* [our Universe's] *time reference."* [40] The non-material (information) self is therefore outside the space-time domain and losing mass at death, it loses chronological time but continues to exist in a 'timeless' domain. Our physical bodies are therefore vehicles that are only necessary for our non-physical entities to function within the physical space-time domain. Jim Al Khalili extended this concept regarding time to the entire history of the Universe when he wrote: *"If Einstein's general theory of relativity is correct, and we are confident that it is, then the Big Bang not only marked the birth of the Universe but the beginning of time itself. Asking questions about what happened* before *the Big Bang necessitates having time to imbed the word 'before' in. Since there simply was no time before the Big Bang, the question doesn't make sense."* [41] Thus time is a peculiarity of our physical universe but nowhere else.

38. *Ibid.* p.161.

39. *Ibid.* p.174.

40. Dolphin, L.: *Lord of Time and Space.* Good News Publishers, Westchester, Illinois (1974) p.63.

41. Al-Khalili, J.: (1999) p.76.

The Mind

The mind, as opposed to the physical brain, is a difficult entity to understand. No physical basis has been found for it. One of the few ways to objectively detect the workings of the mind can be seen in the electrical activity on an EEG tracing, which I would suggest is a secondary result of the mind in action rather than the mind itself. It is analogous to the noise that an internal combustion engine makes as a car travels along a highway. The sound of the car's engine is the result of the motor running; it does not explain how the engine works. Yet the human mind, which is the expression of our sense of self-consciousness, is a real phenomenon which needs to be understood and explained. Edgar Andrews writes: *"All living things respond to stimuli; this is a basic sign of life. Higher animals also display 'intelligence' which broadly defined, reveals itself in such things as complex behaviour, the ability to learn, and capabilities for observation, emotion, communication and action. But as far as we know, only man possesses 'mind' — the capacity for self-knowledge and abstract thought...As evolutionist Theodosius Dobzhansky wrote: 'The pressing problems [for evolutionary theory] today centre on the mechanisms of evolution and the biological uniqueness of man.' Although he wrote this in 1958 it remains true, and nothing about man is more unique than his mental powers — abilities that represent an enormous gulf between Homo sapiens and every other species."* [42] But describing the problem is the easy part, explaining how it works is the difficult bit. Andrews explains: *"Put simply, the chief antagonists are dualism in the blue corner and 'monism' (as a materialistic psychological theory) in the red corner. Monism denies that there is any real distinction between mind and matter and claims that mind is simply the brain at work."* [43] In contrast, a dualist model, where the non-material mind is discrete from the physical brain was developed by Descartes who, Andrews continues: *"started the snowball of modern debate that continues to roll and grow - creating a gargantuan literature."* [44]

Andrews cites the atheist philosopher Bertrand Russell who presented the consequences of the monist view in 1903: *"...even more purposeless, more void of meaning, is the world which science presents for our belief...That man is the product of causes which had no prevision of the end they were achieving; that his origin, his growth, his hopes and fears, his loves and his beliefs, are but the outcome of accidental collections of atoms;...all these things, if not quite beyond dispute, are yet so nearly certain, that no philosophy which rejects them can hope to stand. Only...on the firm foundation of unyielding despair, can the soul's habitation henceforth be safely built."* [45] Andrew's comments: *"The core problem with all modernistic explanations is, of course, that mind becomes meaningless. To modernise Russell's terminology, our thoughts are preordained by self-maintaining electro-chemical processes, which are themselves dictated by our genes. Reasoning has no validity and all our theories and explanations — no matter how energetically promoted or zealously defended — have no more substance than the bubbles in a glass of champagne. Avowed rationalist though he was, J.B.S. Haldane confessed the monist's dilemma when he wrote: 'If my mental processes are determined wholly by the motions of atoms in my brain, I have no reason to suppose that my beliefs are true. They may be sound chemically, but that does not make them sound logically. And hence I have no reason for supposing my brain to be composed of atoms.'...A concrete example of monism is provided by evolutionary psychology, which teaches that our brains are genetically preconditioned to produce certain forms of behaviour. The antisocial person and the criminal are no longer responsible for their*

42. Andrews, E.: *Who Made God? Searching for a Theory of Everything*. EP Books, Darlington, UK (2009) pp.249–50.

43. *Ibid.* pp.250–1.

44. *Ibid.* p.250.

45. *The Collected Papers of Bertrand Russell*, Vol. 12: 'Contemplation and Action', 1902–1914 Routledge, London (1985) *in* Andrews, E.: (2009) pp.251–2.

actions but are the victims of their genetic inheritance. Indeed, whatever our lifestyle, character and inclinations, we cannot help ourselves because we are simply living out our genetic predestination."[46]

So which theory is true? What evidence do we have? The surprising discovery of the last decade is that thoughts can change (and hence shape) our brains, rather than the other way around! When we learn a new task we actually grow new brain cells from resident stem cells and/or re-wire the existing neurones.[47] This process takes about six weeks. Thus, active thought that results in learning, actually changes our physical brain structure; so it is not the physical structure of our brains that determines our thoughts but completely the opposite. However, our genetic inheritance does affect the way we think, to the extent that it determines the physical components of our brain hardware. As a result, we process information differently to sea urchins and frogs (although at the most basic level, they still have nerve cells and synapses) but the latest research suggests that in the end it is our thought processes which determine the ultimate way that the hardware is networked and hence the final outcomes. These findings vindicate dualism but contradict monism because if thought was purely a by-product of our physical brains, there would be no way that thoughts could retro-wire the physical brain structure.

Therefore if a non-material basis for thought is true, how does the mind (which would appear to inhabit the frequency domain) connect to the body in the space-time domain? We do not know but a Fourier transform-like projection (as discussed in Chapter 8) could be a distinct possibility. Arthur Peacocke, biochemist and Oxford University theologian wrote: *"Our intentions and purposes seem to transcend our bodies, yet in fact are closely related to brain events and can be implemented in the world only through our bodies."* [48] At this stage we need to consider whether, as Western science suggests, that the physical space-time domain is the primary entity or, as David Bohm suggested, and as Aquinas previously asserted in 1266,[49] that the non-material frequency domain is the primary entity and that the physical space-time domain is a secondary manifestation. This becomes important when we consider what happens to self-awareness in the frequency domain when the physical body dies. However, mind and body are not the only factors to be considered. As our mind needs a functioning, material platform to express its presence in dimensions one to four, the actuation of consciousness and self-awareness also depend upon the presence of life. If life fails, then so does the expression or projection of consciousness into the physical body within the material domain. (See Figure 12.1.)

If String Theory is correct and multiple dimensions do exist as Hawking and Mlodinow believe, then it is not illogical that consciousness and self-awareness might reside as digital entities (Aristotle called them 'essences') in their respective fifth and sixth dimensions. Extending the internet banking analogy described previously: to access our bank account details, the computer (our brain), is the physical hardware platform required to run the information programs. Once the physical hardware is set up correctly it can be switched on (at conception) and the pre-installed machine language, which is analogous to life, is pre-programmed to run automatically. The next requirement is an operating system, which is analogous to consciousness in the fifth dimension. Finally we can choose which applications software we wish to run on our computer. If our computer is connected via the internet to our bank (equivalent to the mind-brain interface) we can access the bank's software, which is analogous to our self-awareness in

46. Andrews, E.: (2009) pp.252–3.

47. Brooks, R.: Professor of Evolutionary Biology University of NSW, Sydney. *Sydney Morning Herald*: News Review (Sep 24–25, 2011) p.18.

48. Peacocke, A.: The sciences of complexity: a new theological resource? *in* Davies, P. & Gregersen, N.H. (Eds.): (2010) p.263.

49. D'Abrera, B.: Personal communication (2012).

the sixth dimension to retrieve our bank account information stored there (our memories and thoughts). Thus, retrieving the information (the bank balance), requires a multi-level information system: the hardware, the machine language, the operating system, the internet connection, our applications software, the bank's software and the digital repository with the information. They can all exist in their own right but they do not exhibit their full capability until they are combined in a specific configuration to create an operational information system. It is interesting to note that machine language in computers is specific to each individual piece of hardware, and that, in general, applications software packages are specific to their own type of operating system software for which they were written. Microsoft applications do not usually run on an Apple operating system and vice versa, unless they are specifically designed to do so. As the brain architecture of the different species vary, it raises the question as to whether brain operating-systems and software are also brain specific.

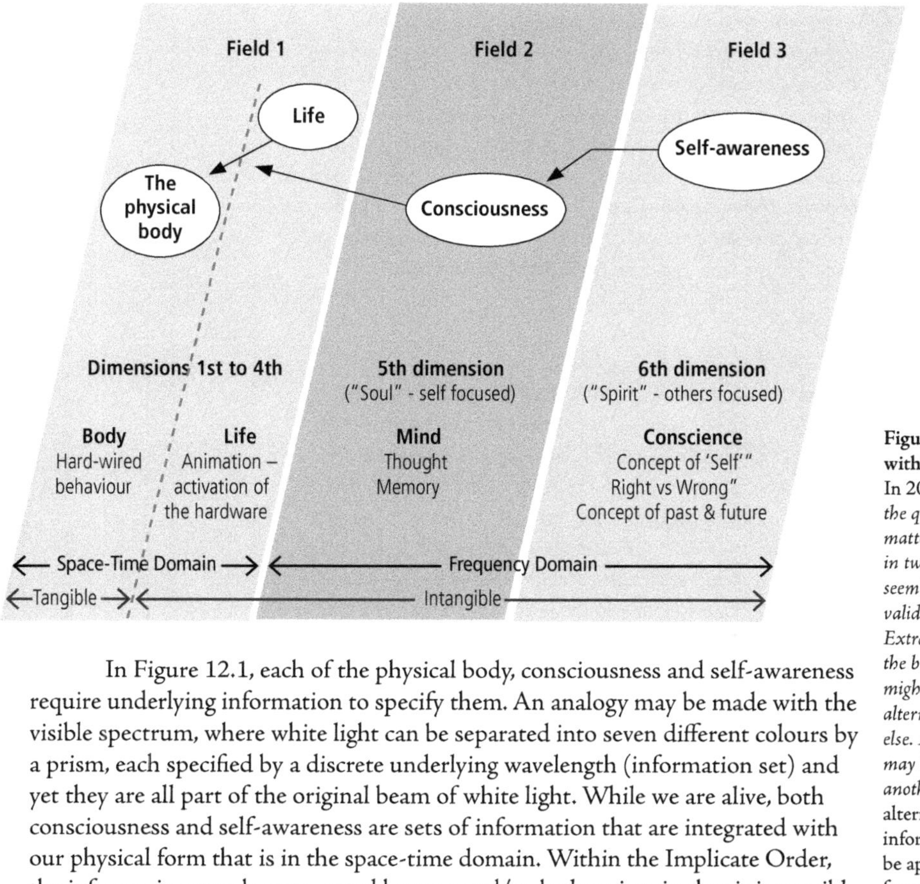

Figure 12.1: Information Sets within the Domain of Life
In 2005, *Cosmos* wrote: *"At the quantum level, particles of matter such as electrons can be in two places at once. In fact, this seemingly absurd statement is validated by modern electronics. Extrapolated, this means that the building blocks of our world might actually be leading an alternative existence somewhere else. As you sit reading this, there may be another you reading in another parallel world."* [50] An alternative interpretation of this information is that you might be appreciating this material from more than one location simultaneously, such as a sixth dimension (See also Figure 10.3).

In Figure 12.1, each of the physical body, consciousness and self-awareness require underlying information to specify them. An analogy may be made with the visible spectrum, where white light can be separated into seven different colours by a prism, each specified by a discrete underlying wavelength (information set) and yet they are all part of the original beam of white light. While we are alive, both consciousness and self-awareness are sets of information that are integrated with our physical form that is in the space-time domain. Within the Implicate Order, the information may be segregated by type and/or by location, in that it is possible that there may be different regions or sub-domains of the Implicate Order (such as Fields 2 and 3 above). At this stage, more information is required to know how this data is arranged but I have illustrated them as spatially segregated (differing fields) for clarity.

So what happens to our self-awareness in the sixth dimension when we die? If the electrical circuits in a computer burnout and the hardware 'dies', the

50. Finkel, E.: Fish out of Water. *Cosmos* 2: 77 (2005).

machine language stops and hence the operating system software malfunctions also. The applications software and bank balance still exist in another place (in the bank's computers across the internet) but they become inaccessible by this route. My connection to the bank's software can be reinstated if my existing computer hardware can be repaired (such as cardiac resuscitation), but, if not, a new computer must be purchased (equivalent to a new physical body in the first four dimensions), with functioning machine language software (life) and a new connection must then be established to the bank's software over the internet (consciousness in the fifth dimension). In living things, the first law of biology states that only life begets life. So once the human body dies, a new body must be manufactured from another living cell if material life is to be restored in the first four dimensions.

But, from whence does a new life acquire a new consciousness and self-awareness? The logical answer would be: the same way that a new body acquires its new existence - through the fusion of a human ovum and a human sperm. The new physical body in the first four dimensions develops from the DNA information stored in the genome which is activated at fertilisation once it has the full complement of chromosomes. Animation is inherited from the already existing living cell (the mother's ovum) with its nascent mind-brain interface (material/non-material linkage) already connected, to permit access to the new 'consciousness' which will develop as the child grows. Just as the third dimension is attached to the first and second, the fifth and sixth dimensions are also inseparable in living human beings in the space-time domain. This is an integrated event that cannot be separated if a fully competent living human being is to result.

Self-Awareness

If the mind and consciousness are difficult to grapple with, then the concept of self-awareness which I will refer to as 'conscience' (sometimes referred to as 'the soul' or 'spirit') is even harder. Dr J.P. Moreland, who has degrees in chemistry, theology and philosophy, observes: *"...computers have artificial intelligence, not intelligence. And there's a huge difference...A computer has...no awareness, no first-person point of view, no insights into problems...A computer can engage in behavior if it's wired properly, but you've got to remember that consciousness is* not *the same as behavior. Consciousness,...* [is] *what* causes *behaviour in really conscious beings. But what causes behaviour in a computer is electrical circuitry...* [in contrast] *Animals are not simply machines. They have consciousness and points of view. But the animal soul is much simpler than the human soul. For example, the human soul is capable of free moral action...Augustine said animals have thoughts, but they don't think about their thinking. And while we have beliefs about our beliefs, animals don't."* [51] Conscience involves the concept of self which confers distinctive qualities, such as the ability to know the difference between 'right' and 'wrong', which is a uniquely human trait. It is not the knowledge that I shall be punished or that there will be negative consequences for certain actions that I know will attract disapproval. It is doing the 'right thing' because you know it is the 'right thing' to do, such as being considerate to a stranger because you perceive that they have a need with which you can assist, *for no personal gain to yourself.* Helping a little old lady to cross a busy street would be a typical example. This is called *altruism*. The question that arises from these observations is from where does this ability come? Andrew Gosler of the University of Oxford posed the question: *"If all 'things' (material and*

51. Moreland, J.P.: Interview *in* Strobel, L.: *The Case for a Creator.* Zondervan, Grand Rapids, Michigan (2004) pp.262–3.

spiritual) exist as information, while some also exist, or are 'realized', as matter-energy, then might we consider the soul as the information 'image' of the living being, including the consciousness of that being as well as the information associated with physical form, function and process?" [52] (See Figure 12.1) In the past, the terms: consciousness, self-consciousness, mind, will, conscience, soul and spirit have often been used interchangeably by different authors. To clarify the nomenclature here, I will be equating *consciousness* with the self-focused 'soul' and *self-awareness* with the altruistic, others-focussed 'conscience' or 'spirit'.

Sources of Information

We live in a precisely defined universe. The degree of precision can be gauged by the error tolerance in the replication of the DNA code or the extreme degree to which the charge on an electron is balanced (not to mention the combined effect of the thirty-nine other variables listed in Chapter 6). This precision reflects the quality of the information generator(s). Walter Bradley expressed it as: *"It has been widely recognised for some time that nature assumes a form that is elegantly described by a relatively small number of simple, mathematical relationships...None of the various proposals...to explain the complexity of the universe address this issue."* [53] We have seen that through the processes of science, the presence of information can be detected within the Universe which is one obvious source for our knowledge about information but it can also be revealed by ancient written religious texts. Lennox writes: *"Francis Bacon talked of God's Two Books — the Book of nature and the Bible. Reason, rationality and evidence apply to both."* [54] The question is: Is the information accurate and can it be relied upon? The Information Model poses other questions too, such as: Was information generated only at the 'Big Bang' or later as well? And, is it still being generated?

However, the ultimate question is not the nature of the information within the Universe, although there is an immense amount to be learnt by studying the information that specifies the four fundamental forces alone. The more significant questions are why there should be a universe for the model to describe[55] and where did that information come from? If information can create itself or there is some naturalistic mechanism that can generate all information, then looking for an external source to reveal the origin of the Universe is probably unnecessary (although it is not impossible that both an external creator and an internal information generator could exist). An exclusive naturalistic information generator would also imply that the Bible and the other ancient religious books were erroneous because there would be no one from whom that revelation could have come but if there is no naturalistic information generator to account for all the information we observe in our universe, then we are forced to conclude that there is an external information creator and in the same way that a benevolent parent instructs a child, it is most likely that the creator of the Universe would have revealed something about themselves and some of the history of our cosmos in some way. So as Lennox implied above, divine revelation in a book like the Bible becomes a *logical* expectation.

Before leaving this section on information sources, we need to revisit Chapter 2, where we looked at the multiple layers of information within a number of different systems including living systems. As information is a discrete entity and must be generated in its own right, Table 2.4 can now be expanded to include the sources of the information as shown in Table 12.1 following:

52. Gosler, A.G.: Surprise and the Value of Life. *in Real Scientists Real Faith.* Berry, R.J. (Ed.) (2009) p.187.

53. Bradley, W.L.: 'Nature. Designed or Designoid' *in* Dembski, W.A. (Ed): (1998) p.38.

54. Lennox, J.C.: (2007) p.43.

55. *Ibid.* p.63.

	Chronicle	Music	Life
	Author	*Composer*	*Information generator*
	Narrator	Conductor	Epigenome
+ Information →	Storyline	Symphony	Life
+ Information →	Book	Score	Cell
+ Information →	Chapter	Movement	Chromosome
+ Information →	Page	Sheet	Gene
+ Information →	Sentence	Phrase	DNA
+ Information →	Word	Chord	Triplet (codon)
Information →	Letter	Note	Nucleotide

Table 12.1: Sources of Information

Whoever you are, you cannot escape the fact that there was an information generator, the only question is who or what was responsible? Materialists will be searching for materialistic information generators and theists will be invoking the power of God. It depends on the evidence you deem to be accurate and correct.

Truth

It is now time to return to our question from Chapter 1 which derived from Linus Pauling's definition of science. You will remember that Pauling defined science as: *"the search for the truth"*. So to understand Pauling's definition of science we need to be able to answer the question: *"What is truth?"* We are now in a better position to answer this question. If we do not like the idea of a divine creator who might have revealed 'ultimate truth' in a book like the Bible and choose to ignore classical metaphysics, such as that described by Thomas Aquinas, who also described a methodology for discerning morality and truth,[56] is there any other way to define it? Well, the Information Model offers a third avenue. Truth has a positive predictive value and hence it helps us to anticipate the future. By having an accurate weather forecast we will know whether it is safe to put to sea in a small boat or whether it is an advantageous time to plant a crop. But can we define truth in purely objective scientific terms? Alistair McGrath and Joanna Collicutt McGrath write: *"...truth is determined...by what the scientific, empirical evidence indicates."* [57] Although this is valid, the problem with this definition is that if there is no accepted *empirical evidence* then does that make something untrue? Richard Dawkins would say that there is no empirical evidence for God whereas John Lennox would say that the Universe exhibits a considerable amount of evidence that God does exist. So the ultimate definition of truth must depend on something that can be defined independently of the selection of the scientific empirical evidence which, as has been seen, has some degree of user-definability. Thus an alternative definition for truth could be: 'an accurate description of the information content within an entity.' The practical problem with this definition is determining how precisely the description in question matches the information content within the entity that is being considered when there is still so much to learn about our universe. It does define an ideal, however, against which everyday situations can be assessed.

This leads to the next question which is: is truth absolute or is it only relative as post-modernism suggests? Our interpretation of the truth is

56. Feser, E.: *The Last Superstition. A Refutation of the New Atheism.* St Augustine Press (2008) pp.223–6.

57. McGrath, A &J.C.: *The Dawkins Delusion? Athiest fundamentalism and the denial of the divine. Society for Promoting Christian Knowledge, London* (2007) p.60.

inevitably affected by observer limitations such as our level of technology and any philosophical bias we bring with our world view. However the ultimate definition of truth should be based upon a fixed standard, not what human beings are capable of discerning today. Thus our perception of the truth should not change due to a lack of ability to investigate and understand the information within our universe; 'ultimate truth' should not change because of our limited perceptions. That the Earth was flat was accepted as 'truth' by much of mediaeval European society but with the advent of telescopes and greater understanding, this notion was realised to be false. The Lamarckian theory of inheritance is another example. 'Truth' in the post-modern period is sometimes described as: 'the world as I perceive it'. However, the thought that: 'I can fly' may seem very real to an individual under the influence of psychotropic drugs such as LSD and the person who is so affected may wholeheartedly believe it. This could be construed as 'personal truth' for them at that moment in time but may not constitute 'universal truth' as the facts relate to the rest of humankind and the universal laws, such as gravity. Thus, personal truth may suggest to such an individual that it would be safe to jump out of a fifth storey window in order to fly like a bird, whereas universal truth would deem it as sheer folly.

Alternatively, since the Universe is digital, truth might be measured by its degree of congruence with the Universe (where that can be determined mathematically). Likewise, untruth could be seen as incongruence with the World or the Universe. But if information is preserved, how does untruth fit with the preservation of information? To understand this, it is necessary to identify why something is untrue. The types of 'untruth' can be seen as:

1. Statement A is true and Statement B is true, but the reason for their association is invalid. (For example, Statement A: 'ducks are birds that fly', because (Statement B), 'an aeroplane has wings'. A is true, and B is true, but the reason is invalid. There is incongruence between statements A and B, even though they are both true. The problem is that the subjects are unlinked.
2. The statement is false because at least one of its elements is untrue, for example: 'The Martian came to Earth.' Multiple probes to Mars have never found any evidence of Martians and neither is there any other evidence for them, thus science suggests that Martians do not exist. So how does information theory, handle imaginary things or things that are yet to be discovered (if you believe that Martians exist but they are much better at hiding than we are at finding?) To answer this question, consider if the information is completely congruent with our knowledge of the Universe. If it is, then it would be considered 'correct' and corresponds to 'truth', whereas corruptions of this information would constitute 'untruth'. And what is a lie? A lie is therefore false or incorrect information that is in some way incongruent with our knowledge of the Universe but in addition it implies intentionality, where there is some degree of intent to mislead by supplying false information or deliberately withholding some information. Thus there is the consideration of 'partial truth' versus 'absolute truth'.

In the end, ultimate truth boils down to your consideration of which evidence is admissible as "science", which depends upon your worldview. John Lennox quotes

Austin Farrar who writes: "*The issue between the atheist and the believer is not whether it makes sense to question ultimate fact, it is rather the question: what fact is ultimate? The atheist's ultimate fact is the universe; the theist's ultimate fact is God.*" [58] The ironic conclusion that derives from this line of reasoning is that if you believe that there is a choice here, then ultimate truth is relative! (It depends upon your worldview.) However, as they are opposite conclusions, we have seen above that Information Theory suggests that only one is true and the other must be untrue. Lennox concludes that the resolution is to ask: "*In which direction does science point - matter before mind, or Mind before matter? The answer to that question will surely have to be determined, as always, by following Socrates' advice, examining the evidence and seeing where it leads, however threatening this may turn out to be to our preconceived notions...The crux of the matter is: are we prepared to follow where the evidence leads - even if it points away from a purely naturalistic interpretation?*" [59] There is a saying which goes: "*The truth won't go away*" and the reason is, that in the end, it is the process or idea that is congruent with the laws of the Universe that ultimately works. Hence this is the solution with which we persist because it delivers the results which help us the most. When we have exhausted the store of everything that does not work, we are left with the ones that do, which we call 'truth'. Another way of saying this is that when the physical manifestation found in the Explicate Order accurately reflects the digital information content in the Implicate Order, there is congruence and 'truth'. An individual may consider from their worldview that truth is relative for them, as they ignore what they do not want to consider but universal truths are not relative and do not change as post-modernism might wish. The reason for this is that the fundamental laws of the Universe do not change.

Schrödinger's Cat

To demonstrate the power that the Information Model can provide to assist in interpreting the Universe (discerning the truth) we will examine 'Schrödinger's cat'. Like everyone else, Austrian physicist, Erwin Schrödinger (1887–1961) struggled to understand Quantum Mechanics and to vent his frustration at the absurdity of the theory, he devised a thought experiment which has been subsequently referred to as 'Schrödinger's Cat'. Remember that in Bohr's interpretation of Quantum Mechanics, the outcome of an event cannot be determined until it has been observed and until that time it is considered to exist in all possible configurations that are open to it. We have previously seen that a photon of light is both a particle and a wave until we observe it when, depending on the experiment that we have used to study it, it is then seen to be one or the other. Until that point it is considered by the Copenhagen interpretation of Quantum Physics to be both a particle and a wave simultaneously (which, as was seen in Chapter 8, is absolutely true although impossible in a purely materialistic paradigm).

Schrödinger described an experiment in which a cat is placed in a box with a sealed vial of cyanide that will kill the cat if the cyanide is released. (Maybe Schrödinger didn't like cats!) The release mechanism involves a hammer controlled by a lever. The lever is controlled by a switch which will be activated if a random radioactive particle from an appropriate source happens to strike the switch. A radioactive source is placed beside the switch within the box and the box is closed so that the activity within the box cannot be observed. If radiation strikes the switch the cat will die but if it does not then the cat lives. Schrödinger commented that until the box is opened so that the system can be observed, under the

58. Lennox, J.C.: *God's Undertaker. Has Science Buried God?* (2nd Ed.) Lion Books, Oxford (2009) p.186.

59. *Ibid.* pp.187&190.

Copenhagen interpretation of Quantum Mechanics, the cat is both dead and alive simultaneously. It is only when the box is opened and the system is 'observed' that the outcome is determined. Not satisfied with this initial conundrum, Schrödinger further questioned: "What does the term 'observed' mean?" Does it require a *human being* to do the observing? If a monkey were to open up the box would that qualify? If a mirror was placed in the box, could the cat observe itself, thus determining the outcome of the experiment? The absurdities are mind-boggling!

From the discussion in Chapter 8, it can now be seen that the answer to the conundrum is that the cat exists in both the space-time domain and the frequency domain simultaneously. The problem is that we only have the ability to reliably observe and measure the space-time domain at present. What we do see (and measure) is the entity in dimensions one to four. The cat still exists in all its other dimensions even though we cannot measure or observe it there. So is the cat dead or alive? The problem is that the cat is not a quantum particle and quantum particles are neither dead nor alive. So the question may be spurious. Remember the Inverse Law of Quantum Expression and that only electrons and buckyballs are small enough to behave as quantum particles. Schrödinger's cat is actually a red herring! (You only *thought* it was a cat!!) In the space-time domain the cat is either dead or alive depending upon whether the switch was activated or not, it does not matter whether we are looking or not. If the experiment is run for long enough then eventually the switch will be activated and the cat will die in the space-time domain. Yet, although we cannot observe it, the expectation is that the information that specifies the frequency modality of the cat persists in the non-material domain, independent of the radioactive switch experiment in the space-time domain. So ironically the cat does end up both dead in the space-time domain yet 'alive' in the frequency domain but I don't think this is what Schrödinger had in mind!

Figure 12.2: Schrödinger's Cat *[Reproduced from Wikimedia Commons.]*

Even though electrons are quantum particles and cats are not, Edgar Andrews writes: "*QM actually does teach that each and every object has a wave function that spreads throughout space.*" [60] However he then points out that this is not as it might initially seem and suggests that Victor Stenger stated the situation accurately when he wrote: "*Quantum mechanics changes smoothly into classical mechanics when the parameters of the system, such as masses, distances, and speeds, approach the classical* [macroscopic-scale] *regime. When that happens, quantum probabilities collapse to either zero or 100%, which then gives us certainty at that level.*" [61] So the moral of the story is: do not mix modalities or you may end up with inappropriate comparisons which confuse rather than enlighten; however it has entertained generations of physics students!

60. Andrews, E.: (2009) p.160.

61. Stenger, V.J.: *God, the failed hypothesis: How science shows that God does not exist.* Prometheus Books, NY (2007) p.125 *in* Andrews, E.: (2009) p.160.

Religion, Faith and Spirituality

Religion, faith and spirituality are not usually considered part of legitimate science. Indeed, the Irish philosopher, Bertrand Russell (1872–1970) wrote: *"Religion is something left over from the infancy of our intelligence. It will fade away as we adopt reason and science as our guidelines."* [62] However, after four hundred years of modern science, religion is taking a long time to die. John Lennox takes the diametrically opposite view maintaining that the logic behind the Biblical worldview led directly to the European renaissance. Lennox stated: *"The driving power behind the rise of science was actually the worldview that created a framework to expect law and order."* [62] In case you think that this topic has nothing to do with science and philosophy, Paul Davies comments: *"People working at the theoretical frontiers of physics can't avoid coming up against the big questions,... The crossover to philosophy and theology are inevitable."* [63] Thus, unless we explore and probe beyond the current boundaries of science we will be severely hampered in our attempts to make new discoveries.

Even so, this has not stopped some groups attempting to silence debate. The US National Academy of Sciences declared that: *"Science is a way of knowing about the natural world. It is limited to explaining the natural world through natural causes. Science can say nothing about the supernatural. Whether God exists or not is a question about which science is neutral."* [64] The problem with this statement is that it does not define 'natural world' or 'natural causes'. Depending what dark matter and dark energy turn out to be, it could be that science (under this definition) would be excluded from studying as much as 96 percent of the Universe; so a new definition of science may be needed fairly soon! If, for argument's sake, a non-material, eternal God did create the Universe but science is not yet able to prove this fact beyond all reasonable doubt, would such a God under the US National Academy of Sciences' definition be 'natural'? After all, He would have had to have existed from before the formation of the Universe! Further, would science be able to legitimately investigate this process which formed the Universe or not?

Faith is often thought of as belief or trust in a transcendent being but faith simply fills in the voids where some of the facts are still unknown or where there is a lack of incontrovertible proof. So faith is involved in any belief system where our knowledge or understanding is incomplete, which applies to many aspects of life and the Universe; it applies to the belief in Darwinian Evolution just as much as to the belief in a transcendent God. Thus, as faith bridges the gaps in our information, we often refer to taking a 'leap of faith' when we jump the holes (or chasms) in our knowledge. John Lennox points out that we need to be able to discern the difference between 'blind faith' and 'evidence-based faith'.[65] When we develop a belief system, it depends upon both the evidence we do have (the facts) and the holes in our information (the gaps). The balance between the evidence and the gaps often determines how successful we are when we make our next 'leap of faith'. Remember our thousand metre long suspension bridge in Chapter 5, where each confirmed piece of evidence (in that case, a fossil) represented one slat. The gaps in our knowledge represent the voids. Thus the success of your next 'leap of faith' (think of it as the distance that your 'leap of faith' needs to carry you so that you do not end up at the bottom of the metaphorical ravine) is directly proportional to the evidence (the number of slats) that you have been able to collect, compared with the number of slats that are yet to be discovered.

62. Lennox, J. & McGrath, A.: *Science and the God Question.* (DVD) Fixed point Foundation, USA (2007).

63. Davies, P.: Interview, *in* Lovett, R.A: Looking to the Skies. *Cosmos* 20: 81 (2008).

64. U.S. National Academy of Sciences *in* Andrews, E.: (2009) p.65.

65. Peter Singer & John Lennox: *Is there a God* (debate) Melbourne Town Hall, Melbourne (20th July 2011).

Hypotheses, theories and laws can serve as towers to which we can attach the end of our 'knowledge bridge'. The higher the tower (the more accurate the hypothesis, theory or law) the further we are likely to be able to see from the known into the unknown when we take that next leap of faith.

The oft touted 'God of the Gaps' represents a misnomer that is only a problem in an exclusively material paradigm. The gaps exist because there is either no information or none that is acceptable to the materialist way of thinking. If we once again envision our knowledge bridge over the thousand metre gorge, with each piece of information representing a slat, then the gaps are the absence of information. If a non-material explanation can provide information to fill the gaps in our knowledge, then that is not invoking the hand of an imaginary God but identifying a piece of (non-material) information from within the *entire* Universe, not just the material part of it. It is an alternative solution, just not a material one but as we have seen, materialism cannot explain either the whole of life or the entirety of the Universe. Both the material and the non-material are complementary and necessary parts of the Universe. For those of us who believe in God, He is not in the gaps in our human knowledge but the exact opposite. As the author, He is behind the information which forms the slats that make up the knowledge bridge; it is just that some of the slats are material and some are non-material. As science has progressed, it has discovered the information that describes more and more of the missing material slats, which augments the utility of our knowledge bridge. I am now asking you to look at the information describing the non-material slats too.

And then there is the question of: 'How can we know if human beings possess such an entity as a 'spirit'?' And if they do, what is it? David Bohm wrote in 1980: *"In the prevailing philosophy of the Orient, the immeasurable (i.e. that which cannot be named, described, or understood through any form of reason) is regarded as the primary reality."* [66] It is interesting that the West has emphasised the material portion of our existence and the Orient has focused on non-material aspects. Both facets are significant and thus each is incomplete if both the material and non-material aspects are not considered simultaneously.

Perhaps it is fitting that the last word should go to Albert Einstein, arguably the greatest scientist of the past century and one of the greatest scientists of all time who stated: *"Everyone who is seriously interested in the pursuit of science becomes convinced that a spirit is manifest in the laws of the universe - a spirit vastly superior to man..."* [67] Einstein is said not to have believed in a personal God but his statement implies a belief in an ultimate source which created the laws of the Universe. You can call this entity the 'Universal Information Generator' or 'God' (or any other name which you consider is more appropriate) but the laws of the Universe certainly did not happen by accident nor materialise out of nothing.

Information Limits

Science is a powerful technique for discovering new information and for testing the veracity of previous discoveries. As an assessment tool, it is good at evaluating those things that are quantifiable but as was seen when we tried to use science to analyse thoughts, there are some things that science cannot measure, test or explain well. For example, why is information here? Science is good for the "what" and "how" but poor on "why"; purpose and meaning need a different assessment

66. Bohm, D.: *Wholeness and the Implicate Order*, Routledge & Kegan Paul, London, Boston (1980) p.48. http://www.spaceandmotion.com/Physics-David-Bohm-Holographic-Universe.htm

67. Einstein, A.: *Letter to Phyllis Wright* 24 January, 1936 *in* Jammer, M.: *Einstein and Religion: Physics and Theology*, Princeton University Press, NJ (1999). http://quoteinvestigator.com/2011/12/16/spirit-manifest/

tool. Understanding our purpose for existence depends upon our conclusion as to what was the causation of the Universe![68] Edgar Andrews writes: *"Science leads us to its boundaries where it introduces us to philosophy"* [69] (See Figure 1.3).

However, in the same way that science is not the universal panacea neither does information explain all things. The richness of the Universe appears to be due to a set of basic, simple laws specified by information, with some variation due to the local environment. A common example would be the shape of a tree; its final form is dependent upon both its intrinsic genetic information and the local forces in its immediate environment, such as rocks that it must grow around. Genetically identical trees when grown next to each other will be very similar, but not perfectly identical in their phenotype (what we see in the first four dimensions), and human identical twins do not have the same fingerprints either. Thus the local environment also has an important role in life. Information is a powerful tool which we can use but it should not be the only method used in research.

Exceeding the Information Limits

Having probed the limits of information, it is appropriate to ask what happens when we exceed the bounds of our information. Miracles, mysteries and magic often stir up controversy so can the Information Model help in investigating the occasions when these are invoked? It is important to be very circumspect when trying to draw conclusions where we lack the information to make logical deductions. Francis Collins, who directed the human genome project, aptly wrote: *"it is crucial that a healthy scepticism be applied when interpreting potentially miraculous events, lest the integrity and rationality of the religious perspective be brought into question. The only thing that will kill the possibility of miracles more quickly than a committed materialism is the claiming of miracle status for everyday events for which natural explanations are readily at hand."* [70] In other words, the information exists to give a logical explanation. John Lennox warns: *"It needs to be made clear at the outset that there is an important distinction to be made between miracles and supernatural events. Miracles (genuine miracles, that is) are supernatural events, but not all supernatural events are miracles in the strict sense. For instance, the origin of the universe and its laws, though a supernatural event, does not come under the rubric of miracle since miracles, strictly speaking, concern events that are exceptions to an already recognised normal course of things and as such they clearly presuppose the existence of such a 'normal course of things'. The creation of the universe together with its regularities which form the 'normal course of things' can scarcely be regarded as an exception to them."* [71]

Thus, it can be seen that every reasonable attempt needs to be made to find a scientific explanation before resorting to any other explanation but sometimes a scientific explanation may not be forthcoming, so what then? Does that mean the event did not occur or as Richard Dawkins has suggested, that science just needs more time to discover the explanation? Science usually chooses to gloss over inexplicable matters (such as the origin of life, the existence of consciousness and the mechanisms of thought) in the hope that an explanation will one day come along. In the meantime we all pretend that science does know when we really do not have a clue! Lennox writes: *"The widely held idea that science has rendered miracles impossible is usually thought to have been given its most powerful expression by the Enlightenment Scottish philosopher, David Hume (1711–1776). Hume was a sceptical naturalist philosopher who wrote in his famous essay* An Enquiry

68. Ward, K.: *Why There Almost Certainly Is a God.* Lion Books, Oxford (2008) p.97.

69. Andrews, E.: (2009) p.209.

70. Collins, F.: *The Language of God.* The Free Press, NY (2006) pp.51–2.

71. Lennox, J.C.: (2009) p.194.

Concerning Human Understanding: *'A miracle is a violation of the laws of nature; and as a firm and unalterable experience has established these laws, the proof against a miracle, from the very nature of the fact, is as entire as any argument from experience as can be imagined.'* "[71] Lennox offers a different explanation to Hume when he comments: "*C.S. Lewis pointed out in his book on Miracles that a miracle can only be recognised if there are rigorous laws in the Universe that determine an expected outcome in the first place. If the outcome is not as the laws of the Universe would predict and there is no other explanation then there is the possibility that a miracle has occurred.*"[72] Thus, if the information within the event is incongruent with the laws of the Universe but there is good evidence that it really happened, that is, it is not a deception or lie, then the bounds of information have been exceeded and a miracle may need to be considered. Note that although both a lie and a miracle represent information that is incongruent with the laws of the Universe, the lie contains information that has been corrupted to create incongruence such that an entity is made to appear to be something that it is not, whereas the miracle implies newly generated information that was incongruent from its formation but a true representation of its form.

In contrast to a miracle, a mystery entails any event where the logical information steps are not self-evident to the observer. It may or may not violate the laws of nature. A 'genius' may appear to 'leap' to a conclusion as they appear to 'skip' steps in deducing the answer to a problem but once the sequence of steps is explained, the 'mystery' disappears and the process becomes incorporated into engineering or science. This is the case with a conjurers 'magic' trick, which becomes a 'sleight of hand' once the technique is understood.[73]

Thus miracles can only occur if there has been a temporary suspension of the laws (and hence the information) that specifies the Universe, whereas mystery and magic relate to our current understanding of the information as it already exists in the Universe. To paraphrase Lennox: *If in the end we believe that a genuine miracle has taken place, it should be because we have been forced by the sheer weight of the direct evidence, not because of our ignorance of nature's laws.*[74] In other words a genuine miracle is one where there is no other possible explanation according to the laws of the Universe (whether we know them yet or not). Thus it is important to know when we are exceeding our information limits so that we do not draw inappropriate conclusions.

When the Paradigm Changes

In 1970, American historian of science, Thomas Kuhn (1922–1996) described paradigm change and its effects. A paradigm is a way of thinking about a particular subject and a paradigm shift is equivalent to changing your frame of reference, meaning that although you are analysing the same information, under the new method of analysis, alternate conclusions or outcomes arise.

Hubert Yockey recognised the influence of paradigms on thought processes when he wrote: "*The history of science shows that a paradigm, once it has achieved the status of acceptance (and is incorporated in textbooks) and regardless of its failures, is declared invalid only when a new paradigm is available to replace it. Nevertheless, in order to make progress in science, it is necessary to clear the decks, so to speak, of failed paradigms. This must be done even if this leaves the decks entirely clear and no paradigms survive. It is a characteristic of the true believer in religion, philosophy and ideology that he must have a set of beliefs, come what may (Hoffer,*

72. Lennox, J. & McGrath, A.:'Lennox on C.S. Lewis' *in Science and the God Question.* DVD (2007).

73. McGrath, A &J.C.: (2007) p.37.

74. Lennox, J.C.: (2009) p.199.

1951). Belief in a primeval soup on the grounds that no other paradigm is available is an example of the logical fallacy of the false alternative. In science it is a virtue to acknowledge ignorance. This has been universally the case in the history of science as Kuhn (1970) has discussed in detail. There is no reason that this should be different in this research on the origin of life." [75]

Robert Higgs, American Professor of Economics writes: "*Science is an odd undertaking: everybody strives to make the next breakthrough, yet when someone does, he is often greeted as if he were carrying the Ebola virus. Too many people have too much invested in the reigning ideas; for those people an acknowledgement of their own idea's bankruptcy is tantamount to an admission that they have wasted their lives. Often, perhaps to avoid cognitive dissonance, they never admit that their ideas were wrong. Most important, as a rule, in science as everywhere, to get along, you must go along."* [76]

Science is a methodology that can be applied to both the theistic and atheistic viewpoints. Where an experiment cannot be repeated (as in the formation of the Universe and the origin of life), the explanation that best fits all the observable facts should win the day. However, the insistence that Naturalism be a pre-requisite for the study of science constitutes a filtering or narrowing of the possible areas of endeavour, which risks missing vital clues and significant information. Although this is very 'safe' in academic terms, it assumes that all that is significant in the Universe is known to be material. However, when we contemplate the 96 percent of the Universe which a significant number of mainstream cosmologists consider is dark matter and dark energy, this assertion regarding the primacy of Naturalism becomes decidedly risky. The problem is that making such 'safe' assumptions is 'unscientific'.[77]

It might be useful at this stage to point out that when we lose an old 'friend' we need to grieve the loss. In 1969, the Swiss-American psychiatrist, Elisabeth Kübler-Ross (1926–2004) described the five stages of grieving which are:[78]

1. Denial — "It cannot be true!"
2. Anger — We are likely to lash out at anyone and everyone in our upset. "Who can I blame?"
3. Bargaining — "I will do anything if it's not true!"
4. Depression — Despondency sets in when we can no longer deny or deflect the evidence.
5. Resolution — By working through the issues we come to a new way of understanding.

For people who have invested their whole lives, careers and reputations in the old paradigm, change will be extremely hard, if not impossible and change is often a generational issue. It took over fifty years for the scientific fraternity to accept that the speed of light was not infinite and the bigger the change, the longer it is likely to take.[79] Similarly, it took from 1915 when Einstein's General Theory of Relativity indicated that the Universe was not eternal until 1965 with the verification of the Cosmic Background Radiation, a period of fifty years, for the scientific community to be completely convinced that the Universe was expanding, had a beginning and was not eternal.

When the paradigm changes, there will be winners and losers and it is not possible to make everyone happy all of the time but as has been seen through the pages of this book, our understanding of the Universe will be increased and the whole of mankind will benefit from the change, even though that change may be

75. Yockey, H.P.: *Information Theory and Molecular Biology*. Cambridge University Press (1992) p.336 *in* Sarfati, J.: *By Design*. Creation Book Publishers (2008) pp.188–9.

76. Higgs, R.: *Peer review, publication in top journals, scientific consensus, and so forth*. George Mason University's History News Network 7th May, 2007 *in* Sarfati, J.: (2008) p.236.

77. Lennox, J.C.: (2007) p.28.

78. Kübler-Ross, E.: *On Death and Dying*. Routledge, London (1969).

79. Missler, C.: (2004) pp.344–5.

painful for some. The Information Model will be a foreign concept to most people but it affects every aspect of life, including our understanding of the Universe and cannot be ignored.

The Information Generators

So finally, what can be said about the information generators? Gödel's Second Incompleteness Theorem states that in mathematics, the information within a system cannot be defined from within that same system, so some external source is required to account for the information within it. To date, Gödel's Second Theorem has not been extended beyond the realm of mathematics but the Laws of Coding and Information Primacy show that the Universe has a digital foundation and uses mathematically defined laws to function, making it a mathematical entity, so Gödel's Second Theorem would appear to be applicable to the Universe too. If this is the case, then the information within the Universe would have to have been specified by an external source. In support of this supposition, no information generator has been demonstrated within the Universe that could account for its formation.[80]

The logical question then follows (and Richard Dawkins asked it in *The God Delusion*): who or what made the Universe? [81] In Chapters 3, 8 and 9 we discovered that a Biological Information Generator, a Physical Information Generator and/or a Universal Information Generator was required. Thus to answer Dawkins' question we first have to ask: 'Are the Biological and Physical Information Generators the same or different entities?' To answer this, we need to note that the biological information in DNA and RNA, which is essential for all life, requires a physical universe in which to live, with physical entities to convey the code. The Physical Information Generator must therefore have been active prior to the Biological Information Generator. It is quite possible however that a single, Universal Information Generator performed both functions. The observation that is immediately apparent is: whatever the process was that supplied enough detailed information to specify the Universe, it must have been unimaginably sophisticated. Ockham's razor [82] would suggest that there was one rather than two information generators but even if there were two information generators, they must be extraordinarily complex and have co-orchestrated their respective parts in the formation of the Universe with an extreme degree of co-ordination. For the Biological Information Generator to manufacture the biological information when the time was propitious for life to be successfully established on the Earth implies that there was some form of communication between the two information generators. In addition, if the Biological Information Generator is not the product of the Physical Information Generator then it is not inconceivable that it was related to the same process that formed or operated the Physical Information Generator. However, it would seem more likely that there was only one Universal Information Generator that performed both roles. If we therefore consider that a single Universal Information Generator ultimately accounts for the whole universe, then we are left with the question Alistair McGrath poses: "*What explains the explainer?*" [83]

Naturalistic and materialist scientists have not been able to satisfactorily explain the generation of information in the Universe directly but there is an implicit assumption that it was generated by the laws of the Universe, bit by bit (literally). The generation of information to form the physical Universe is usually

80. It needs to be emphasised that although Neo-Darwinian Macroevolution has been touted as an information generator, the process of information audit described in Chapter 3 demonstrates that in and of itself, it cannot generate novel information that could build a living genome as Darwin supposed.

81. Dawkins, R.: *The God Delusion.*, Bantam, London (2006) p.109 *in* McGrath, A &J.C.: (2007) p.9 Ref. 18.

82. John Lennox explains Ockham's razor: "*Attributed to the 14th-century English Franciscan thinker William of Ockham, this is the general principle that one should favour a hypothesis that makes the fewest new assumptions among competing hypotheses that are equal in other respects. It does not (falsely) assert that the simplest explanation is always likely to be the correct one.*" in Lennox, J.C.: *Seven Days that divide the world: the beginning according to Genesis and Science*. Zondervan, Grand Rapids (2011) p.89.

83. McGrath, A &J.C.: (2007) p.9.

implied to be intrinsic to the 'Big Bang' and its aftermath, while the generation of biological information is attributed to the repeated cycles of mutation with natural selection, as described by Charles Darwin in 1859. However from the trail of evidence detailed in this book, it has been seen that these explanations are completely inadequate. These scientists must now search for new mechanisms to account for the generation of the information that is so abundant in our universe. There are two choices: if a materialistic origin for information is invoked, then it must come from the space-time domain in dimensions one to four and will have mass, be influenced by time and therefore have had a beginning and will have an end. In contrast, if a non-material origin for information is invoked, then it must come from the frequency domain and be outside time and have had no beginning and will have no end. That is, it is eternal. Keith Ward, Professor of Divinity at Oxford University, writes: *"If there is an eternal thing, then it could not be brought into being by anything else, since it could not be brought into being at all. Since it is not in time, there never was a time when it was not, and it could not possibly be brought into being."* [84] An easier way to view this may be to return to the Kalam Cosmological Argument, about which William Lane Craig stated: *"... the first premise of the Kalam argument...is* not *that* everything *has a cause, but that* whatever begins to exist *has a cause."* [85] If the Universal Information Generator is non-material and outside time, then it is eternal, had no beginning and therefore no cause. If God is that eternal entity, as Christians believe and the Bible claims, then likewise He had no beginning and thus no cause. This concept may be hard to comprehend as we are finite beings operating in a finite space-time domain. It requires us to forget our finite selves and think like a non-material entity. It is comparable to the two-dimensional cartoon character Charlie Brown, trying to think like a six dimensional human animator. It can be done but it is a foreign concept and the mind of Charlie Brown (if he could be imagined to have one) may not understand the reasoning behind the actions of the animator, because in his two-dimensional world, the extra four dimensions are not relevant. Questions such as the reason for Charlie Brown's existence, which is to entertain the readership of the comic strip, of which Charlie Brown has no experience or comprehension whatsoever, are likely to be completely beyond him. You, as the reader of the *Peanuts* comic strip, are completely outside Charlie Brown's two-dimensional universe and hence, also likely to be entirely outside his powers of understanding (assuming he had powers of understanding).

In contrast, theists and creationists when making a decision about the origin of information will invoke the hand of God, a non-material information source and they would be in good company. Sir Isaac Newton, who is arguably one of the greatest thinkers of all time, wrote: *"The entire universe is a cryptogram set by the Almighty."* Newton pondered these questions nearly four hundred years ago and we are still debating them today. Paul Davies observes: *"...the universe, conforms to an orderly scheme, and is not an arbitrary muddle of events."* This thought causes Davies to ponder about: "God or no God — *whether there is some sort of meaning behind it all"*. [86] Richard Swinburne, an Oxford philosopher, takes up this point when he asserts: *"...the intelligibility of the universe itself needs explanation"*. [87] In *A Brief History of Time*, Stephen Hawking observes in Chapter 2: *"The whole history of science has been the gradual realization that events do not happen in an arbitrary manner, but that they reflect a certain underlying order, which may or may not be divinely inspired."* [88] Later in Chapter 8 he wrote: *"In the hot big bang model...there was not enough time in the early universe for heat*

84. Ward, K.: (2008) p.52.

85. Strobel, L.: *The Case for a Creator.* Zondervan, Grand Rapids, Michigan (2004) pp.108-9.

86. Davies, P: *The Goldilocks Enigma: Why is the Universe just Right for Life?* Allen Lane, London (2006) pp.15–16 *in The Good News* 14(6): 7 (2009).

87. McGrath, A & J.C.: (2007) p.12.

88. Hawking, S.W.: *A Brief History of Time.* Random House (1988) Chapter 2 at http://en.wikiquote.org/wiki/Stephen_Hawking.

to have flowed from one region to another. This means that the initial state of the universe would have to have had exactly the same temperature everywhere in order to account for the fact that the microwave back-ground has the same temperature in every direction we look. The initial rate of expansion also would have had to be chosen very precisely for the rate of expansion still to be so close to the critical rate needed to avoid recollapse. This means that the initial state of the universe must have been very carefully chosen indeed if the hot big bang model was correct right back to the beginning of time. It would be very difficult to explain why the universe should have begun in just this way, except as the act of a God who intended to create beings like us." [89] In contrast to Hawking's approach, Dawkins introduces the *probability* of the existence of God and concludes that on the law of probabilities, the God of the Bible probably does not exist but as McGrath remarks: *"We may be highly improbable - yet we are here. The issue then is not whether God is probable, but whether God is actual."*[90]

John Lennox writes: *"One of Richard Dawkins's main* God Delusion *arguments is that, if God created everything, we would have to ask who created God. But the very asking of this question reveals at once that Dawkins has in mind a created God: 'Who* created *God?' Created gods certainly are a delusion…Dawkins's difficulty must be that he cannot believe in something eternal. Why not? Science certainly does not tell us that there is nothing eternal…Furthermore, if Dawkins's question is valid, it can be turned back on him. He believes that the universe created him. Therefore, we are justified in asking him: who created your Creator? The point here is that* all *such questionings must stop with what the questioner believes to be ultimate reality. As we have seen, for the atheist the ultimate reality is the universe, and for the theist the ultimate reality is God. Genesis tells us that God is primary, and the universe derivative. This worldview is the exact opposite of ancient polytheism and contemporary secularism, both of which assume that matter is primary and everything else, including mind is derivative."* [91] Thus, settling these questions of what is primary and what is derived, along with the reality of material and non-material entities, will determine your worldview and hence your interpretation of life and the Universe.

Conclusion

So, we have journeyed to places many of us have never been before; from reading the letter on the page to gaining an understanding of not only what the letter says but the message embedded in the prose and thence to beyond the edge of the Universe. We have seen that real information is not the vehicle carrying the message, but an underlying, intangible entity which conveys meaning and understanding. It has been demonstrated that information is the common currency and objective yardstick by which all things in the Universe can be assessed; as such it becomes the 'gold standard' against which the entire Universe may be benchmarked.

This journey into information has been driven by observed facts with the desire to better explain significant deficiencies in our understanding of the World and the Universe: the Wave/Particle Duality of Light, the phenomena of life, sleep and thought and the mutual contradiction between Einstein's Relativity Theory and Bohr's Quantum Mechanics. Post-modernism's purely materialistic paradigm cannot explain these things satisfactorily (if at all), but Information Mechanics can.

89. Hawking, S.W.: *A Brief History of Time.* Random House (1988) p.127.

90. McGrath, A &J.C.: (2007) p.10. (Emphasis in the original.)

91. Lennox, J.C.: (2011) pp.94–5.

Thus the Universe is information-based. Information is non-material, timeless and can be extremely complex, yet on a conceptual basis, supremely simple. There is much to be done to understand the mechanisms and intricacies of Information Mechanics and much of it may not be directly testable this side of death (that is, it is outside the first four dimensions in our space-time domain) but I have no doubt that future research will shed much light onto this unobtrusive and generally neglected component of the Universe.

The ultimate presupposition is that concept which is above or before all others. It therefore defines the paradigm. So what is our ultimate presupposition? It is that an information generator manufactured all the information in the Universe, which is the prime entity within the Cosmos; information is digital and describes the Universe and everything in it. Further, it appears that much, if not all, the information has a dual expression just as light does; it has a vibrational or waveform-type phase in the frequency domain that is detectable in a matter/energy phase in the space-time domain. From our perspective in the space-time domain, the vibrational or waveform phase seems like an indistinct 'silhouette' but it appears that the frequency domain has pre-eminence over the space-time domain, which may prove to be a Fourier transform-like projection from the underlying frequency phase. There is thus not only a duality of light but a universal duality of all matter and energy, which includes a duality of life too. Although in life we are unable to leave our physical universe, we have seen from Gödel's Second Incompleteness Theorem that the Universal Information Generator must be outside the space-time domain, be non-material, timeless and hence eternal. So, summarising the Information Model, the only way to understand the extraordinary amount of exquisitely complex information in the Universe is to appreciate that information necessarily underlies everything and that the information was actively generated, rather than passively or accidentally acquired. In addition, since the Universe is information-based, it is unavoidably digital, which also helps in understanding why computers and digital technology are so useful in science. An analogy would be that, if you wanted to drive along a railway track, it helps to have wheels like a train, because the system was made that way. You can take almost any other vehicle (except a boat or submarine), along the railway but your progress will be much more difficult, because that system was not designed for other modes of transport.

I realize that Information Mechanics will sound strange and be difficult to accept at first and many scientists may find it too different to what they have been taught, to agree with it. However, as the French Nobel Laureate in Literature in 1947, André Gibe (1869–1951) wrote: "*One doesn't discover new lands without consenting to lose sight of the shore...*" [92] In addition, Jonathan Sarfati notes: "*... Thomas Kuhn...pointed out that, in reality, in periods of 'normal science', scientists do not throw out the ruling paradigm readily, but tolerate a large number of 'anomalies'. It takes many anomalies to build up before there is a scientific revolution.*" [93] However, the aim is not revolution but a better model of the Universe with which to explain the many anomalies that we observe under the current paradigm.

This book began with the aim of revealing Information. It has demonstrated that the ultimate Information Paradox is that as one of the least obvious entities in the Universe, it is ubiquitous. Further, that it has defined the Universe since its inception and we are only now starting to acknowledge it. Once information is recognised and identified it will help us to appreciate our World

92. Gibe, A.: Les Faux-monnayeurs [The Counterfeiters] 1925 in Encyclopaedia Britannica at http://www.britannica.com

93. Sarfati, J.: *The Greatest hoax on Earth? Refuting Dawkins on Evolution.* Creation Book Publishers (2010) p.76.

and the Universe in a way that we never could before and it promises to usher in a new era in our understanding of science. Further, it gives us the ability to separate the wheat from the chaff, as the underlying information reveals reality itself. Thus we live in a material universe which is underpinned by a non-material one; one that is constructed from digitally coded information. Winston Churchill once said: *"Men stumble over the truth from time to time, but most pick themselves up and hurry off as if nothing had happened."* [94] It's time, in this digital age, to cease our stumbling and to seize the new and exciting opportunities provided by the versatility and power of information.

94. Churchill, W.: http://gizmotude.www5.50megs.com/archive_truth.htm

---oOo---

Glossary

ADP adenosine di-phosphate: an intermediate energy storage molecule within living organisms.

Algae single or multi-celled, chlorophyll-containing plants which lack stems, leaves or roots.

Allopatric speciation a population that, in Darwinian Theory, has diverged genetically to form a new species as a result of being geographically separated from its original species group.

Anthropic Principle the observation that the Universe gives the appearance that it has been finely tuned to support life on Earth and in particular, human life.

Apoptosis the auto-destruction process initiated by a cell when there is catastrophic breakdown within the cell.

Archaea one of the 3 domains (super kingdoms) of life. They are single-celled organisms which lack a nuclear membrane (prokaryotic) but differ from bacteria in their cell membranes and organelles.

ATP adenosine tri-phosphate: a high energy storage molecule within all living organisms.

Awareness having knowledge of or an appreciation of one's environment.

ATP-synthase an intra-cellular nano-machine that converts lower-energy ADP to higher-energy ATP.

Bacteriophage a virus that parasitizes bacteria, invading them and using their cytoplasmic machinery to reproduce before being released when the bacterium dies.

Beauty the harmonious integration of information.

Biochemistry the study of the chemical compounds and processes that occur in living cells.

Biology the study of living organisms.

Biological Information Generator a generic term for the process that generated the information that is found in living organisms.

Black hole a region in space where the gravitational force is so great that even light cannot escape; thus it appears black when viewed from Earth.

Catalyst a molecule that accelerates the rate at which a reaction takes place without being changed by the reaction.

Cellular machinery the organelles and nano-machines within the cytoplasm of a cell that perform the functions necessary to allow a cell to function.

Chaos Theory also called 'Deterministic Chaos' describes behaviours that repeat

References:

Concise Oxford Dictionary. (5th Ed.) Oxford University Press, London (1973).

d'Abrera, B.: *Butterflies of the Afrotropical Region*. Hill House Publishers, London (2009).

Fowler, T.B. & Kuebler, D.: *The Evolution Controversy*. Baker Academic, Grand Rapids (2007).

Hawking, S. & Mlodinow, L.: *The Grand Design*. Bantam Press, London (2010).

Johnson, P.E.: *Wedge of Truth*. InterVarsity Press, Downers Grove, Illinois (2000).

Ladiges, P., Evans, B., Saint, R. & Knox, B.: *Biology: an Australian Focus*. (4th Ed.) McGraw Hill, Australia (2010).

McLeod, W.T. & Hanks, P. (Ed.): *The New Collins Concise English Dictionary*, William Collins Sons & Co. Ltd, London & Glasgow (1984).

Meyer, S.C. et al.: *Explore Evolution*. Hill House Publishers, London (2009).

Spinney, L.: Five things humans no longer need. *New Scientist*. 19 May 2008.

themselves but the patterns are so complex that the events appear to occur in a random way (such as weather patterns, economic cycles and biological ecosystems) and yet the functioning of these systems can be described by the information supplied at the outset.

Chlorophyll a light-absorbing, green pigment within plants which aids photosynthesis.

Chloroplast an organelle in plants that contains chlorophyll and aids photosynthesis.

Citric Acid Cycle also known as the Krebs cycle, which is a circular chain of eight biochemical reactions within cells that provides energy for the cell in the form of ATP.

CMBR Cosmic Microwave Background Radiation which is considered to represent the energy left over from the formation of the Universe.

Codon a three letter base sequence in DNA which identifies a unique amino acid.

Co-dependent information a piece of information that cannot have originated in isolation but required the linkage to another piece of information for its generation.

Cognition the mental processes involved in acquiring and processing information including perception, intuition and reasoning.

Cognitive dissonance inconsistency between the information, beliefs or paradigms in which one believes.

Complexity the intricacy or degree of sophistication involved in the way that information is linked; a measure of the number of pieces of information required to precisely define an entity.

Convergent evolution the appearance of similar structures in unrelated species in Darwinian theory.

Cyanobacteria chlorophyll containing prokaryotes that produce oxygen as a by-product of photosynthesis.

Cytoplasm that part of the contents of a cell which is outside the nucleus; it contains the organelles.

Datum (Pl.: data) a piece of information which is known to be true that can be relied upon as a starting point for further reasoning.

Design the intentional arrangement of information.

Determinism the idea that if you knew everything about every particle in the Universe, you could predict what happens next and so determine the future course of all physical events.

DNA deoxyribonucleic acid: a double helix of nucleotides which can encode biological information.

Doppler Effect the change in the frequency of sound or light due to a change in the relative motion of an observer to the wave source.

DVD Digital Video Disc or Digital Versatile Disc — an information storage medium.

Dysteleology the notion of 'poor' or 'unintelligent' design.

EEG an ElectroEncephaloGram is a tracing of the minute wave-form activity of the brain.

Embryogenesis the sequential development of a multicellular adult organism from a single-celled egg.

Emotion an innate mental experience or state of mind that derives from feelings rather than reason.

ENCODE the Encyclopaedia Of DNA Elements (ENCODE) Project, which is a collaboration of eighty research institutions across North America, Asia, Europe and Australia, aims to sequence the DNA of different species.

Energy a fundamental entity within the Universe that enables a body or system to perform work and hence oppose the Law of Entropy.

Entity something that has real existence; it may be material or non-material.

Entropy the natural tendency of all material systems within the Universe to move from order to disorder.

Enzymes proteins which act as catalysts within cells to speed up the rate of chemical reactions.

Epigenome supervisory (meta-) information within the genome of living organisms that carries the instructions for how a cell is to grow and function.

Eukaryote an organism containing cells with membrane-bound organelles such as a nucleus, mitochondria and chloroplasts.

Evidence information that establishes a fact; it is an indicator of the true nature of an entity.

Evo Devo Evolutionary Developmental Biology: the study of how embryogenesis might have evolved within Darwinian Theory.

Extrinsic information information that derives from a coding source outside the entity being studied.

Fact a piece of information that has really occurred or is the case; a datum of experience, as distinct from conclusions. In science it is an observation that has been repeatedly confirmed and for all practical purposes is accepted as 'true'.

Feelings a state of mind that is associated with an emotion or sentiment rather than logic.

Fitness cost the degree to which an organism's previous level of functioning is compromised after undergoing a mutation, such as acquiring antibiotic resistance.

Fitness landscape a diagram that maps the relative advantages and disadvantages of different gene combinations within a population under specified environmental conditions and hence predicts the combinations with better prospects of longer term survival.

Frames of reference pre-specified information that acts as a reference point against which new information is assessed.

Gene a unit of information encoded within a sequence of DNA that specifies a protein.

Genome the complete genetic information of an organism that is specified within its DNA or RNA.

Geodesic path the shortest distance between any two points on a curved surface; it becomes a straight line when the space is flat.

Geologic column the system of geological categorisation and its associated timescale that derives from the successive layering of differing soils and rocks observed within the Earth's crust. It is used to date the fossils that have been found within the various layers.

Gold Standard the monetary system in which the value of a currency was defined by reference to gold.

Golgi apparatus Golgi body or Golgi complex: an intracellular organelle that acts as a packaging plant for the transhipment of manufactured molecules and the export waste products.

Habits learned information shortcuts that allow us to function from day to day without having to rework everything from first principles each time.

Hardware a physical unit that is required for the processing of information.

Hermaphrodite an organism which can be both male and female at different times in its life cycle.

Heterozygous the situation within a gene where one of an individual's allele pairs differs from the other.

Homochirality the condition where all the carbon-based molecules in a specified set have the same 'handedness'. On Earth, all life is 'left-handed'.

Homology the similarity in body plans (such as the bone structure in the forelimbs of various mammals) that, in Darwinian Theory, derives from descent from a common ancestor.

Homozygous the situation within a gene where both of an individual's allele pairs are identical.

Hormones protein molecules within living organisms that act as catalysts.

Hox genes developmental control genes, discovered in 1983, which have a supervisory function.

Hydrophilic molecules with a tendency to dissolve in or mix with water.

Hydrophobic molecules with a tendency to be insoluble in water but share an affinity with lipids.

Hyperspaces extra-dimensional spaces within our universe that are outside of the first four material dimensions.

Hypothesis a supposition or conjecture that accounts for the known facts. It is a provisional idea that serves as a starting-point for further investigation by which it may be verified or disproved.

Innate behavior sub-conscious processing of stored information.

Info-bit a single piece of information that contributes to the specification of a macroscopic entity.

Information integrated data, items of knowledge or news; the instructions required to specify an entity. It is a moiety that conveys meaning.

Information Age the period after 1980 when manipulation of information by computers became sufficiently widespread to materially change the way of life in Western Culture.

Information Mechanics that subset of science which studies non-material entities.

Information Model of the Universe the concept that digital information underlies and defines the entire Cosmos.

Inorganic the absence of specified biological information.

Intangible incapable of being touched or felt, which implies a non-physical existence.

Intelligence the creative generating and processing ability of a living complex brain. It manifests the capacity to acquire and learn new information.

Intrinsic information internally coded information within the structure of an entity such as the laws of the Univers.

Isomers the existence of two or more molecular structures for compounds that share the same chemical formula.

Junk DNA the notion that a significant part of the DNA within living creatures was inactive due to the effect of countless damaging mutations over many generations; it is implied that this damaged DNA remains as it has not been removed by natural selection.

Knowledge intellectual acquaintance with fact or truth; the possession of a range of information and the ability to use it in an optimal way: thus it is applied information.

Language a vehicle or system for the transmission of information.

Law a rule which provides predictions that go beyond the immediate situations upon which it was based; there must be no exceptions to that rule if it is to be classed as a law.

Law of Coding the maxim that no process or product within the Universe can alter the information that specified it.

Learning the ability to acquire or manipulate information.

Life the purposeful activity that stems from the functioning of a viable cell.

Macroevolution the introduction of new information into a living organism to the extent that it creates a new species.

Macromolecules large molecules such as proteins, polymers or compounds of proteins and sugars.

Memes the collective memory of our culture which is learned rather than encoded in our genes.

Messenger-RNA an information transport molecule composed of RNA that transfers data from the DNA to the ribosome for protein manufacture.

Meta-information information about information or information that controls other information.

Metallicity the degree to which the heavier (and especially the metallic) elements feature in the galaxies, stars and planets.

Microevolution change in the genetic information within organisms in an individual species. Changes are limited by the pre-specified information within the genome of the organism.

Micro-RNA single-stranded RNA sequences that control gene expression through the regulation of messenger-RNA in the cytoplasm. They are about twenty-one to twenty-three nucleotides in length and derive from DNA in the nucleus of the cell.

Mitochondria organelles within the cytoplasm of eukaryotic cells that provide energy for the cell.

Materialism the philosophical doctrine which states that matter is the fundamental and only reality and hence that the mind and emotions are merely secondary manifestations of matter.

Matter the physical substance that makes up the tangible entities within our Universe.

Moore's Law that the processing power of new computers doubles every two years.

Multiverse the hypothetical concept of multiple parallel universes.

Mutation a random change in the information that specifies a living organism through the alteration of its DNA or RNA sequence. These chance events usually occur due to copying errors during replication but are sometimes due to the external influence of chemicals or radiation.

Nano-machines various tiny machines within the cytoplasm of all cells that carry out the functions of the cell.

Natural selection the process by which a species adapts to its local environment through the preferential survival of certain members of the species which express characteristics that allow them to survive and reproduce when other individuals die and do not leave offspring. This selection is from amongst the characteristics (information) already contained in the collective genome within the individuals in that species.

Neuro-endocrine system the integration of the nervous and hormonal systems within complex organisms to provide sophisticated responses to a large number of internal and external events.

Newtonian Mechanics the system of three laws of motion defined by Isaac Newton that describe the behavior of macroscopic entities within the Universe.

Non-coding RNA (previously known as 'Junk' DNA) derives from DNA in the nucleus of the cell where the RNA segments are not translated into proteins. This RNA is more easily understood as: 'non-protein-coding RNA.' Also see UTR's.

Non-material entities within the Universe that lack physical qualities in the first four dimensions, such as: information, thought, memory, consciousness and self-awareness.

Nucleotides the basic structural unit of nucleic acids (RNA and DNA). They are composed of nitrogen-containing bases (purine or pyrimidine), a ribose sugar and a phosphate molecule.

Olfactory pertaining to the sense of smell.

Ontology the development of an organism over its life cycle from fertilization until death.

Organelles microscopic apparatus within the cytoplasm of cells that permit a cell to function normally, such as the Golgi Bodies.

Organic chemistry the branch of chemistry concerning the compounds of carbon.

Paradigm an information framework that sets the pattern for the way we interpret ideas and values.

Parthenogenesis the process of hermaphroditic reproduction where the same organism can be both male and female at different times in its life cycle.

Phage see bacteriophage.

Pheromones hormones that are released into the environment which carry information that can influence the behaviour of other individuals in that species. They are a form of chemical communication.

Philosophy the study of the causes and laws of all things, both concrete and abstract (physical or metaphysical); this includes the study of ideas that cannot readily be tested or verified by repeated experimentation, especially that which deals with ultimate reality.

Phylogeny the evolutionary history of a group of organisms, usually depicted as a branching tree.

Physical Information Generator a generic term for the process that generated the non-biological information within our physical universe.

Plasticity the ability that living tissues have to react to or grow in response to stimuli such as the development of new brain cells in response to thought.

Pliotropy the condition where a single mutation affects more than one protein or epigenetic control sequence when they are transcribed, due to overlapping gene sequences on the chromosome.

Polypeptide a molecule formed by the linking of multiple amino acid units via peptide bonds.

Postulate a claim or assumption, usually regarding fundamental conditions or prerequisites that is used as a starting point for a line of reasoning.

Prebiotic a time period before the existence of life on the Earth.

Presupposition a required or necessary prior condition or assumption.

Prokaryotes small single-celled organisms (including some bacteria) that lack internal membranes to define their cytoplasmic organelles and hence they lack a nucleus, mitochondria or chloroplasts.

Protein a molecule composed of strings of amino acid units (called polypeptides) that fold into a specific shape to perform a specific task within a living organism.

Proteomics the study and classification of proteins and their interactions.

Proteasomes large protein molecules in all eukaryotes and Archaea which break down and remove excess or damaged proteins.

Punctuated Equilibrium the stepwise theory of Evolution proposed by Gould & Eldredge in 1972 whereby organisms are seen to stay the same for long periods of time, followed by relatively rapid species change that occurs over comparatively brief time periods.

Q-bit Quantum bit: information that records one characteristic of a fundamental particle, such as its spin, or its speed or its direction.

Quantum an amount or portion of an entity; quanta comprise only whole portions which cannot be broken up into part quantities and hence usually represent the smallest or most basic units within a system.

Quantum entanglement the phenomenon whereby two fundamental particles which have been together can sense the state (spin) of the other particle. This occurs no matter what the distance is between them and the communications between them can exceed the speed of light.

Quantum Theory also called Quantum Mechanics: the branch of physics that studies the behavior of atoms and elementary particles in contradistinction to Newtonian Mechanics which describes the behavior of macroscopic entities.

Random variation the minor differences seen between the individuals within a given species due to the chance alteration or deletion of genetic material from mutations; these deplete the functional genomic information in the affected organisms.

Reasoning the ability to deduce new information from existing information.

Relativity Theory Einstein's Special Theory (1905) and General Theory (1915) describe the behavior of material entities within the Universe relative to one another; Relativity Theory describes the concept of the space-time continuum and the equivalence of mass and energy.

Ribosomes complex protein manufacturing machines composed of RNA and proteins within the cytoplasm of all living cells. They use the information in mRNA to make proteins.

RNA Ribonucleic Acid; a molecule formed from chains of the nucleotides: adenine, cytosine, guanine and uracil that stores and transports information within a cell.

Science the systematic study of the nature and behaviour of the Universe based on observation, experiment, measurement, testing, reasoning and logic with the aim of increasing our knowledge and understanding of the Universe.

Self-awareness having knowledge or an appreciation of oneself and the effect that our presence has in our environment; a sense of self in relationship to the Universe.

Semantic information qualitative information that refers to the meaning encoded within the information.

Shannon Theory Claude Shannon's thesis on the behavior of syntactic information which refers to the quantity or volume of information in a communication.

Singularity a circumstance in physics where the known laws of the Universe breakdown, such as where one variable becomes infinite, as at the 'Big Bang' or beyond a black hole's event horizon.

Software information instructions that permit a process to occur when installed appropriately on suitable hardware.

Species a group of similar organisms that can interbreed to produce fertile off-spring.

String Theory a theory that describes the fundamental particles of the Universe which are hypothesized to be tiny one dimensional strings, 10^{-33} centimeters in length, that vibrate in different ways to form the different fundamental particles that construct our Universe.

Symbiosis mutually beneficial co-operation between two discreet species.

Synapse the small regulatory gap between one nerve cell and the next.

Syntactic information quantitative information which counts the number of bits of information in a message.

System a co-ordinated assemblage of interrelated, interdependent or interacting parts that act together synergistically to form a collective entity for a purpose that is greater than the sum of the individual parts.

Tangible capable to being touched or felt, which implies a physical existence.

Theory a system of ideas or statements that is expressed in general terms as an explanation for a collection of connected phenomena.

Thought the conscious processing of information in a living brain; the ability of biological entities to generate or manage information mentally.

Understanding to comprehend the meaning, significance, cause, nature or explanation of an entity or situation; appreciation of the relationship between differing pieces of information. Understanding places information into context and hence gives it meaning.

Universal Information Generator a generic term for the process that generated the Universe and all the information that is found in it.

Urbilateria the hypothesised universal ancestor of all organisms with two-fold symmetry, the existence of which is a deduction from Darwinian Theory.

UTR's Un-translated Regions of the DNA that were previously known as 'Junk' DNA. These segments of DNA are now known to be translated into RNA (but not proteins) which are sometimes called non-coding RNAs. They are thought to be part of the cell's epigenetic control mechanism.

Values the information behind the societal beliefs and behaviours that has been found to be helpful over a number of generations. Values often relate to longer-term solutions such as respecting your neighbour's property.

Variable Penetrance the propensity of inherited characteristics to be expressed to differing degrees in successive generations which appears to be orchestrated by the epigenome.

Vesicles small fluid-filled sacs that are enclosed by a membrane within the cytoplasm of cells. They distribute macromolecules after manufacture and can be used to export waste products as well.

Vestigial organs organs considered by evolutionary biologists to have passed their evolutionary use-by date but which have not yet been eliminated by natural selection.

Vomeronasal Organs a pair of structures in the nasal lining or the roof of the mouth in rodents and other mammals which secrete chemical signals called pheromones.

White hole an event whereby light may have transiently entered the Universe at its formation; it only allows light in, as a one way process and hence is the opposite of a black hole.

Wisdom the ability to think and act utilizing knowledge, experience, understanding, common sense and insight in an optimal way.

World View the information framework that we use to interpret the world around us; it influences the way we perceive or interpret new information.

Bibliography

Aballay, A.: Nervous System Rallies Immune System Forces against Invading Pathogens. *Howard Hughes Medical Institute News at* http://www.hhmi.org/news/20080919aballay.html (19 Sep. 2008).

Abulafia, R., Zalkind, V. & Devor, M.: Cerebral activity during the anesthesia-like state induced by mesopontine microinjection of pentobarbital. *Journal of Neuroscience* 29:7053–64 (2009).

Ackermann, M. et al.: Self-destructive cooperation mediated by phenotypic noise. *Nature* 454:987–990 (2008).

Alberts, B.: The Cell as a Collection of Protein Machines. *Cell* 92:291–294 (1998).

Alexander, D.R.: *Creation or Evolution: Do we have to choose?* Monarch Books, Oxford (2008).

Al-Khalili, J.: *Black Holes, Worm Holes & Time Machines.* Taylor & Francis Group, New York (1999).

Andrews, E.: *Who Made God? Searching for a Theory of Everything.* EP Books, Darlington, England (2009).

Andriessen, M. et al.: *Physics 1: Preliminary Course.* (2nd Ed.) Jacaranda, Milton, Queensland (2004).

Arkin, A.: Setting the standard in synthetic biology. *Nature Biotechnology* 26(7):771–774 (2008).

Arp, H., Bi, H.G., Chu, Y. & Zhu, X.: Periodicity of Quasar Redshift: *Astronomy & Astrophysics* 239:33–49 (1990).

Atkins, P.: *Creation Revisited.* Harmondsworth, Penguin Books (1994).

Atrain, S.: Unintelligent Design, *in* Brockman, J.: *Intelligent Thought.* (Ed.) Vantage, New York (2006).

Aubisson, P. et al.: *Biology in Context – The Spectrum of Life.* (2nd Ed.) Oxford University Press, Melbourne (2004).

Azbel, M.A.: Exact Conservation Laws and Exclusion Principle of Biological Evolution. *American Physical Society, Annual APS March Meeting* (2003).

Barry, C.: Mind maps. *Cosmos* 20:66–71 (2008).

Bateson, G.: *Steps to an Ecology of Mind: Collected Essays in Anthropology, Psychiatry, Evolution, and Epistemology.* University of Chicago Press, Chicago, Illinois (1972).

Batiz, Z. & Chauhan, B.: *Holographic Principle and Quantum Physics* (2 February, 2008). http://arxiv.org/pdf/physics/0611104.pdf

Batten, D.: The adaptation of bacteria to feeding on nylon waste. *Journal of Creation* 17(3):3–5 (2003).

Behe, M.J.: *Darwin's Black Box: The Biochemical Challenge to Evolution*. Free Press, New York (2006).

Behe, M.J.: *The Edge of Evolution*. Simon & Schuster, New York (2007).

Bejerano, G. & Haussler, D.: Junk DNA now looks like Powerful Regulator. *Science Daily* (24 April, 2007).

Bendewald, J. & Sherwin, F.: *Evolution Shot Full of Holes*. Evidence Press, Madison, Wisconsin (2004).

Beran, R.: The Neurology of Sleep: Part 1. *Medical Observer*. St Leonards, Sydney, Australia, pp.23–26 (5 February, 2010).

Bergman, J. & Howe, G.F.: *Vestigial Organs are Fully Functional*. Creation Research Society, St Joseph, Montana, USA (1990).

Bergman, J.: Does the acquisition of antibiotic and pesticide resistance provide evidence for evolution? *Technical Journal* 17(1):26–32 (2003).

Bergman, J.: Is the human pharynx poorly designed? *Journal of Creation* 22(1): 41–43 (2008).

Bergman, J. & O'Sullivan, N.: Did immune system antibody diversity evolve? *Journal of Creation* 22(2):92–96 (2008).

Berry, R.J. (Ed.): *Real Scientists Real Faith*. Monarch Books, Oxford (2009).

Birney, E.: *Inaugural Francis Crick Lecture of the Royal Society*. Institute of Science in Society (2003).

Blanchard, J.: *Evolution Fact or Fiction*. Evangelical Press. Darlington, England (2002).

Bohm, D.: *Wholeness and the Implicate Order*. Routledge & Kegan Paul, London & Boston (1980).

Bohm, D.: Interview: *On Quantum Physics*. (1987) *in* David Bohm Quotes on Quantum Theory *at* http://www.spaceandmotion.com/Physics-David-Bohm-Holographic-Universe.htm.

Bolte, M. & Hogan, C.J.: Conflict over the age of the Universe. *Nature* 376:399-402 (2002).

Borger, P.: Evidence for the design of life: part 2 - Baranomes. *Journal of Creation* 22(3):68–76 (2008).

Bouchard, T.J. Jr et al.: Sources of human psychological differences: the Minnesota Study of Twins Reared Apart. *Science* 250(4978):223–228 (1990).

Bradley, W.L.: Nature: Designed or Designoid, *in* Dembski, W.A. (Ed.): *Mere Creation. Science, Faith and Intelligent Design*. InterVarsity Press, Downers Grove, Illinois (1998).

Braunwald, E. et al.: *Harrison's Principles of Internal Medicine* (11th Ed.) McGraw-Hill Book Company, New York (1987).

Briggs, A.: And Information Became Physical, *in* Berry, R.J. (Ed.): *Real Scientists Real Faith*. Monarch Books, Oxford, pp.72–84 (2009).

Brooks, M.: 13 Things that don't make sense — The Most Intriguing Scientific Mysteries of our Times. Profile Books, London (2010).

Brown, W.M., George, M. & Wilson, A.C.: Rapid Evolution of Animal Mitochondrial DNA. *Proceedings of the National Academy of Sciences of the USA* 76:1967–1971 (1979).

Brynie, F.H.: *Brain Sense: The Science of the Senses and How We Process the World Around Us*. Amacom Books, New York (2009).

Buchanan, M.: Horizontal and Vertical: The Evolution of Evolution. *New Scientist* 205(2744):34–37 (2010).

Buxton, G., Mulherin, C. & Worthing, M.: *God and Science – in Classroom and Pulpit.* Mosaic Press, Preston, Victoria, Australia (2012).
Cai, D., Verhey, K.J., Meyhöfer, E.: Tracking single Kinesin molecules in the cytoplasm of mammalian cells. *Biophysics Journal* 92(12):4137–4144 (2007).
Cairns-Smith, A.G.: *Genetic Takeover and the Mineral Origins of Life.* Cambridge University Press, New York (1982).
Calamai, P.: Cosmic Roulette. *Cosmos* 20:52–59 (2008).
Campbell, N.A., Reece, J.B. & Meyers, N.: *Biology* (7th Ed.) Australian Version. Pearson Education Australia (2006).
Carroll, S.B.: *Endless Forms Most Beautiful: The New Science of Evo Devo.* WW Norton & Company, New York (2005).
Carroll, S.B., Prud'homme, B. & Gompel, N.: Regulating evolution. *Scientific American* 296:60–67 (2008).
Cavalier-Smith, T.: Obcells as Proto-Organisms: Membrane Heredity, Lithophosphorylation, and the Origin of the Genetic Code, the First Cells, and Photosynthesis. *Journal of Molecular Evolution* 53:555–595 (2001).
Chanda, S. et al.: Human host factors required for influenza virus replication. *Nature* 463:813–817 (2010).
Chevalier, J., Pages, J.M., Eyraud, A. & Mallea, M.: Membrane permeability modifications are involved in antibiotic resistance in Klebsiella pneumoniae, *Biochemical and Biophysical Research Communications* 274(2):496–499 (2000).
Chien, P. et al.: *SEM Observation of Precambrian Sponge Embryos from Southern China, Revealing Ultrastructures including Yolk Granules, Secretion Granules, Cytoskeleton and Nuclei.* Paper presented to the North American Paleontological Convention, University of California, Berkeley (June–July, 2001).
Chimpanzee Sequencing and Analysis Consortium: Initial Sequence of the Chimpanzee Genome and Comparison with the Human Genome. *Nature* 437:69–87 (2005).
Cirelli, C. & Bushey D.: Sleep and Wakefulness in Drosophila melanogaster. Molecular and Biophysical Mechanisms of Arousal, Alertness, and Attention. *Annals of the New York Academy of Sciences* 1129: 323–9 (2008).
Clark, S.: Unknown Earth: How come Earth got all the good stuff? *New Scientist,* 199(2675):29–30 (2008).
Close, F.: *Antimatter.* Oxford University Press, Oxford (2009).
Collins, F.: *The Language of God.* The Free Press, New York (2006).
Concise Oxford Dictionary. (5th Ed.) Oxford University Press, London (1973).
Conway, B.R. & Rehding, A.: Neuroaesthetics and the Trouble with Beauty. *PLoS Biol* 11(3):e1001504 (2013).
Cowen, C.: The Light Fantastic. *Cosmos* 34:58–63 (2010).
Coyne, J.A.: Switching on Evolution. *Nature* 435(7045):1029–1030 (2005).
Crick, F.H.C.: *Ideas on Protein Synthesis.* (1956) Welcome Library, London. http://wellcomelibrary.org/player/b18174139#0/1/-0.0321,0.375, 1.0643,0.7501
Crick, F.H.C.: Ideas on Protein Synthesis. *Symposium of the Society for Experimental Biology* XII, 139–163 (1958) *at Royal Society Publishing* (2009). http://profiles.nlm.nih.gov/SC/B/B/F/T/_/scbbft.pdf .
Crick, F.H.C.: Central Dogma of Molecular Biology. *Nature* 227:561–563 (1970).

Crick, F. & Orgel, L.E.: Directed Panspermia. *Icarus* 19:341–346 (1973).
Crick, F.H.C.: *Life Itself: Its Origin and Nature*. Simon & Schuster, New York (1981).
d'Abrera, B.: *A History and Philosophy of the Science of Time*. Hill House Publishers, London (2005).
d'Abrera, B.: *Butterflies of the Afrotropical Region*. Hill House Publishers, London (2009).
D'Costa, V.M. et al.: Antibiotic Resistance is Ancient. *Nature* 477:457–461 (2011).
Dalsgaard, N.J.: Prion diseases. An overview. *APMIS* 110:3–13 (2002).
Darwin, C.: *The Origin of Species*. John Murray, London (1859) reprinted by Penguin Classics, London (1985).
Darwin, C.: *The Descent of Man*. John Murray, London (1871).
Darwin, C.: *The Life and letters of Charles Darwin*. Vol. 3. John Murray, London (1887).
Davidson, J.D. & Rees-Mogg, W.: The Great Reckoning: How the World will change before the Year 2000. Pan Books, London (1994).
Davies, P.: *God and the New Physics*. Simon and Schuster, New York (1983).
Davies, P.: *The Cosmic Blueprint*. Simon and Schuster, New York (1988).
Davies, P.: *The Cosmic Blueprint: New Discoveries in Nature's Creative Ability to Order the Universe*. Touchstone Books, New York (1989).
Davies, P.: Bit before it? *New Scientist* 161(2171):3 (1999).
Davies, P.: *The Fifth Miracle*. Simon & Schuster, New York (1999).
Davies, P.: Life force. *New Scientist* 163(2204):27–30 (1999).
Davies, P.: Was life on Earth born lucky? *New Scientist* 179(2403):32 (2003).
Davies, P.: *The Origin of Life*. Penguin Books, London (2003).
Davies, P.: *The Goldilocks Enigma: Why is the Universe just Right for Life?* Allen Lane, London (2006).
Davies, P.: Life, the universe and everything. *Cosmos* 14 (2007).
Davies, P. & Gregersen, N.H. (Eds.): *Information and the Nature of Reality. From Physics to Metaphysics*. Cambridge University Press, Cambridge (2010).
Davies, P.: Universe from Bit, *in* Davies, P. & Gregersen, N.H. (Eds.): *Information and the Nature of Reality: from Physics to Metaphysics*, pp.65–91 Cambridge University Press, Cambridge (2010).
Davila Ross, M., Owren, M.J. & Zimmermann, E.: Reconstructing the Evolution of Laughter in Great Apes and Humans. *Current Biology* 19(13):1106–1111 (2009).
Dawkins, R.: *The Selfish Gene*. Oxford University Press, Oxford (1976).
Dawkins, R.: *The Blind Watchmaker*. Longman, London (1986).
Dawkins, R.: *River out of Eden*. Basic Books, New York (1995).
Dawkins, R.: *Climbing Mount Improbable*. Norton, New York (1996).
Dawkins, R.: *The God Delusion*. Bantam, London (2006).
Dawkins, R. & Lennox, J.C.: *The God Delusion Debate*. (DVD) Fixed Point Foundation, USA (2007).
Dawkins, R. & Lennox, J.C.: *The Great Debate, Oxford*. (DVD) Fixed Point Foundation, USA (2008).
Dawkins, R.: *The Greatest Show on Earth*. Bantam Press, London (2009).
de Duve, C.: *Vital Dust: Life as a Cosmic Imperative*. Basic Books, New York (1995).

de Duve, C.: *Life Evolving.* Oxford University Press, New York (2002).

De Valois, R.L. & De Valois, K.K.: Spatial Vision. *Annual Review of Psychology* 31:309–341 (1980).

Deacon, T.W.: *The Symbolic Species: The Co-Evolution of Language and the Brain.* WW Norton, New York (1997).

Dembski, W.A.: *The Design Inference.* Cambridge University Press, Cambridge (1998).

Dembski, W.A. (Ed.): *Mere Creation – Science, Faith & Intelligent Design.* InterVarsity Press, Downers Grove, Illinois (1998).

Dembski, W.A.: (Ed.) *Darwin's Nemesis.* InterVarsity Press, Downers Grove, Illinois (2006).

Dennett, D.C.: *Darwin's Dangerous Idea: Evolution and the Meanings of Life.* Simon & Shuster, New York (1995).

Denton, M.: *Evolution: A Theory in Crisis.* Adler & Adler, Chevy Chase, Maryland (1986).

Dittrich, M., Hayashi, S. & Schulten, K.: ATP hydrolysis in the βTP and βDP catalytic sites of F1-ATPase. *Biophysical Journal* 87:2954–2967 (2004).

Division of Sleep Medicine at Harvard Medical School. *Sleep and Disease Risk.* (18 December, 2007).

http://healthysleep.med.harvard.edu/healthy/matters/consequences/sleep-and-disease-risk

Dobzhansky, T.: *Genetics and the Origin of Species.* (3rd Ed.) Columbia University Press, New York, (1951).

Dolphin, L.: *Lord of Time and Space.* Good News Publishers, Westchester, Illinois (1974).

Drake, S.: *Discoveries and Opinions of Galileo.* Doubleday-Anchor, New York (1957).

Dunbar, R.: Why are Humans not Just great Apes? *in* Pasternak, C. (Ed.): *What makes us human?* Oneworld Publications, Oxford, pp.37–48 (2007).

Eadie, M.: A Molecular Basis for Spiritual Experience? *Luke's Journal* 11(2):4–7 (2006).

Eigen, M.: The Origin of Genetic Information: Viruses as Models. *Gene* 135:37–47 (1993).

Einstein, A., Podolsky, B., & Rosen, N.: Can Quantum Mechanical Description of Physical Reality be Considered Complete? *Physical Review.* 47:777 (1935).

Einstein, A.: *Letters to Solovine: 1906-1955.* Philosophical Library, New York (1987).

Eldredge, N. & Gould, S.J.: Punctuated Equilibria: An Alternative to Phyletic Gradualism, *in* Schopf, T.J.M. (Ed.): *Models of Paleobiology.* Freeman, Cooper & Co., San Francisco, pp.114–115 (1972).

Eldredge, N.: *Reinventing Darwin.* Phoenix, New York (1996).

Fabene, P.F. & Bentivoglio, M.: 1898–1998: Camillo Golgi and 'the Golgi': one hundred years of terminological clones. *Brain Research Bulletin* 47(3):195–198 (1998).

Fanjul-Moles, M.L. et al: Circadian Modulation of Crustacean Hyperglycaemic Hormone in Crayfish Eyestalk and Retina. *Chronobiology International* 27(1):34–51 (2010).

Farley, J.: *The Spontaneous Generation Controversy from Descartes to Oparin.* John Hopkins University Press, Baltimore (1979).

Farmer, M. & Wieland, C.: *Clash over Origins: Creation vs. Evolution* Debate (DVD) Creation Ministries International.

Feser, E.: *The Last Superstition. A Refutation of the New Atheism.* St Augustine Press, South Bend, Indiana (2008).
Feser, E.: *Aquinas – A Beginner's Guide.* Oneworld Publications, Oxford, England (2009).
Finkel, E.: Fish out of Water. *Cosmos* 2:74–79 (2005).
Fowler, T.B. & Kuebler, D.: *The Evolution Controversy.* Baker Academic, Grand Rapids (2007).
Fox, R.: The Conditions of Sexual Evolution, *in* Aries, P. & Bejin, A. (Eds.): *Western Sexuality: Practice & Precept in Past and Present Times.* Basil Blackwell, Oxford (1985).
Freeman, S. & Herron, J.: Evolutionary Analysis. (3rd Ed.) Pearson-Prentice Hall, Upper Saddle River, New Jersey (2004).
Freeth, T., Jones, A.R., Steele, J.M., & Bitsakis, Y.: Calendars with Olympiad display and eclipse prediction on the Antikythera Mechanism. *Nature* 454:614–617 (2008).
Gaensler, B.: Extreme Speed. *Cosmos* 41:68–74 (2011).
Gauger, A., Axe, D. & Luskin, C.: *Science and human Origins.* Discovery Institute Press, Seattle (2012).
Galileo Galilei: *Il Saggiatore* [The Assayer] (1623).
Gee, H.: *In Search of Deep Time: Beyond the Fossil Record to a New History of life.* The Free Press, New York (1990).
Giberson, K.W. & Collins, F.S.: The Language of Science and Faith. SPCK, London. (2011).
Gibson, L.J.: Did Life Begin in an 'RNA World'? *Origins* 20:45–52 (1993).
Gijsbers, A.: The Dialogue between Neuroscience and Theology. *Luke's Journal* 11(2):8–11 (2006).
Gilbert, W.: Origin of life: The RNA world. *Nature* 319:618 (1986).
Gingerich, O.: Dare a Scientist Believe in Design? *in* Templeton, J.M. (Ed.): *Evidence of Purpose.* Continuum, NY, p.25 (1994).
Gitt, W.: *In the Beginning was Information.* (3rd Ed.) CLV Neuhausen, Germany. (2003).
Glatzmaier, G.A. & Roberts, P.H.: Rotation and Magnetism of Earth's Inner Core. *Science* 274 (5294):1887–1891 (1996).
Goldschmidt, R.B.: *The Material Basis of Evolution.* Yale University Press, New Haven (1940).
Gonzalez, G. & Richards, J.W.: *The Privileged Planet.* Regnery Publishing, Washington DC (2004).
Gonzalez, G. & Richards, J.W.: *The Privileged Planet.* (DVD) Illustra Media (2004).
Goodwin, B.: *How the Leopard Changed Its Spots: The Evolution of Complexity.* Charles Scribner's Sons, New York (1994).
Gosler, A.G.: Surprise and the Value of Life. *in* Berry, R.J. (Ed.): *Real Scientists Real Faith.* Monarch Books, Oxford (2009).
Gould, S.J.: Evolution's Erratic Pace. *Natural History* 86(5):12–16 (1977).
Gould, S.J.: *The Panda's Thumb: More Reflections in Natural History.* WW Norton & Co, New York (1980).
Gould, S.J.: The Evolution of life on Earth. *Scientific American* 271:85–91 (1994).
Gould, S.J.: *The Structure of Evolutionary Theory.* Harvard University Press, Cambridge, MA (2002).
Grant, S.: From little things, big things grow. *Univ. Sydney Alumni Magazine* pp.14–17(Autumn 2009).

Grassé, P.: *Evolution of Living Organisms*. Academic Press, p.4 (1973) [English translation (1977)].
Greene, B.: *The fabric of the cosmos*. Alfred Knopf, New York (2004).
Greenfield, S.: *Brain Story*. BBC video series (2001).
Gregersen, N.H.: God, matter, and information: towards a Stoicizing Logos Christology, *in* Davies, P. and Gregersen, N.H. (Eds.): *Information and the Nature of Reality: from Physics to Metaphysics*, pp.319–348 Cambridge University Press, Cambridge (2010).
Haeckel, E.: *The Wonders of Life*. Translated by McCabe, J.: Harper, London (1905).
Hager, A.J., Pollard, J.D. Jr & Szostak, J.W.: Ribozymes: aiming at RNA replication and protein synthesis. *Chemistry & Biology* 3:717–25 (1996).
Haldane, J.B.S.: *Data needed for a blueprint of the first organism*. (Published posthumously) *in* S.W. Fox (Ed.): *The origins of prebiological systems and of their molecular matrices*. Academic Press, New York (1965).
Harrison, E.: Time Warp. *Cosmos* 20:60–65 (2008).
Hart, D.B.: Atheist Delusions: The Christian Revolution and its Fashionable Enemies. Yale University Press, New Haven & London (2009).
Hartnett, J.: Cosmologists Can't Agree and Are Still In Doubt! *Journal of Creation* 16(3) (2002).
Haselhurst, G.: *Essay on Life & Ideas of David Bohm (1917–1992)*. http://www.spaceandmotion.com/Physics-David-Bohm-Holographic-Universe.htm
Hasselblatt, B. & Katok, A.: *A First Course in Dynamics: With a Panorama of Recent Developments*. Cambridge University Press, Cambridge (2003).
Hawking, S.: *A Brief History of Time: From Big Bang to Black Holes*. Random House (1988).
Hawking, S. & Mlodinow, L.: *A Briefer History of Time*. Bantam Press, London (2005).
Hawking, S. & Mlodinow, L.: *The Grand Design*. Bantam Press, London (2010).
Hayes, B.: Experimental Lamarckism. *American Scientist* 87:494–498 (1999).
Heenan, D.: *The Skeptic's Guide to God*. Show and Tell. Sanctuary Cove, Qld, Australia (2008).
Herbert, N.: *Bell's Theorem Proves Physical Reality must be Non-local, at* <http://mail.cruzio.com/~quanta/bell.html>
Ho, S.Y.W., Phillips, M.J., Cooper, A. & Drummond, A.J.: Time Dependency of Molecular Rate Estimates and Systematic Overestimation of Recent Divergence Times. *Molecular Biology and Evolution* 22:1561–1568 (2005).
Hoffmeyer, J.: *Semiotic freedom: an emerging force, in* Davies, P. & Gregersen, N.H. (Eds.): *Information and the Nature of Reality: from Physics to Metaphysics*, pp.185–204 Cambridge University Press (2010).
Hoyle, F.: The Big Bang in Astronomy. *New Scientist* 92:521–527 (1981).
Hoyle, F.: The Universe: Past and Present Reflections. *Engineering and Science* pp.8–12 (1981).
Hoyle, F.: *The Intelligent Universe*. Michael Joseph, London (1983).
Hoyle, F.: *The Mathematics of Evolution*, Weston Publications, Cardiff, University College Cardiff Press, (1987).
Huxley, T.H.: *Autobiography and Selected Essays*. (Ed. Snell, A.L.F.) Riverside College Classics (2009).
Jammer, M.: *Einstein and Religion: Physics and Theology*. Princeton University Press, NJ (1999).

Johnson, P.E.: *Darwin on Trial.* (2nd Ed.) InterVarsity Press, Downers Grove, Illinois (1993).
Johnson, P.E.: *Wedge of Truth.* InterVarsity Press, Downers Grove, Illinois (2000).
Jones, S.: The End of Evolution? *Cosmos* 24:35–39 (2008).
Josephson, B.: Letter to the Editor. *The Independent.* London (12 January, 1997).
Joyce, G.F. & Orgell, L.E.: Prospects for understanding the RNA World, *in* Gesteland, R.F. & Atkins, J.F. (eds.): *The RNA World.* Laboratory Press, Cold Spring Harbour (1993).
Joyce, G.F.: The antiquity of RNA-based evolution. *Nature* 418:214–221 (2002).
Katz, L.A.: The Tangled Webb: Gene Genealogies and the Origin of Eukaryotes. *The American Naturalist* 154:137–145 (1999).
Kauffman, S.: *At Home in the Universe: The Search for Laws of Self-Organisation and Complexity.* Oxford University Press, Oxford (1995).
Kauffman, S.: *Investigations.* Oxford University Press, Oxford (2000).
Keziah: *From a Frog to a Prince.* (VHS video) Answers in Genesis, Acacia Ridge, Queensland (1998).
König, R. et al.: Human host factors required for influenza virus replication *Nature* 463:813–817 (2010).
Koonin, E.V.: How Many Genes Can Make a Cell: The Minimal-Gene-Set Concept. *Annual Review of Genomics and Human Genetics* 1:99–116 (2000).
Koonin, E.V.: The cosmological model of eternal inflation and the transition from chance to biological evolution in the history of life. *Biology Direct* 2:15 (2007).
Kruszelnicki, K.S.: *Great Myth Conceptions.* HarperCollins, Australia (2004).
Kruszelnicki, K.S.: *Game of Knowns.* Macmillan, Sydney (2013).
Kübler-Ross, E.: *On Death and Dying.* Routledge, London (1969).
Kuppers, B.: On the Prior Probability of the Existence of Life, *in* Kruger et al. (ed.): *The Probabilistic Revolution.* MIT Press, Cambridge, MA (1987).
Küppers, B.-O.: *Information and the Origin of Life.* MIT Press, Cambridge, MA (1990).
Küppers, B.-O.: *Information and communication in living matter, in* Davies, P. & Gregersen, N.H.: *Information and the Nature of Reality: from Physics to Metaphysics,* pp.170-184 Cambridge University Press (2010).
Ladiges, P., Evans, B., Saint, R. & Knox, B.: *Biology: an Australian Focus.* (4th Ed.) McGraw Hill, Australia (2010).
Laughlin, R.: *A Different Universe: Reinventing Physics from the Bottom Down.* Basic Books, New York (2005).
Lennox, J.C.: *God's Undertaker. Has Science Buried God?* Lion Books, Oxford (2007).
Lennox, J. & McGrath, A.: *Science and the God Question.* (DVD) Fixed Point Foundation, USA (2007).
Lennox, J.C.: *God's Undertaker. Has Science Buried God?* (2nd Ed.) Lion Books, Oxford (2009).
Lennox, J.C.: *God and Stephen Hawking – Whose Design is it Anyway?* Lion Books, Oxford (2011).
Lennox, J.C.: *Gunning for God. Why the New Atheists are missing the Target.* Lion Books, Oxford (2011).
Lennox, J.C.: *Seven Days that divide the world: the beginning according to Genesis and Science.* Zondervan, Grand Rapids, Michigan (2011).

Li, E., Beard, C. & Jaenisch, R.: Role for DNA methylation in genomic imprinting. *Nature* 366:362–365 (1993).
Lin, H.: piRNA. *Science Daily* (26 October, 2007).
Lindahl, T.: Instability and decay of the primary structure of DNA. *Nature* 362(6422):709–715 (1993).
Linton, A.: Scant search for the maker. *Times Higher Education Supplement*, p.29 (20 April, 2001).
Lloyd, S.: *Programming the Universe*. Knopf, New York (2006).
Lloyd, S.: *The Computational Universe*, *in* Davies, P. & Gregersen, N.H.: *Information and the Nature of Reality: from Physics to Metaphysics*, pp.92–103 Cambridge University Press, Cambridge (2010).
Loewenstein, W.: *The Touchstone of life*. Penguin Books, London (2000).
Lolle, S.J., Victor, J.L., Young, J.M. & Pruitt, R.E.: Genome-wide non-mendelian inheritance of extra-genomic information in *Arabidopsis*. *Nature* 434:505–509 (2005).
Lovett, R.A.: Darwin's Evidence for Evolution. *Cosmos* 24:24–25 (2008).
Lubenow, M.L.: *Bones of Contention. A Creationist Assessment of Human Fossils*. Baker Books, Grand Rapids, Michigan (2007).
Machado, D., Couto, I., Perdigão, J., Rodrigues, L., Portugal, I., et al.: Contribution of Efflux to the Emergence of Isoniazid and Multidrug Resistance in *Mycobacterium tuberculosis*. *PLoS ONE* 7(4): e34538 (2012).
Macnab, R.M.: The Bacterial Flagellum: Reversible Rotary Propeller and Type III Export Apparatus, *Journal of Bacteriology* 181(23):7149–53 (1999).
Marks, R.J.II, Behe, M.J., Dembski, W.A., Gordon, B.L. & Sanford, J.C.: *Biological Information — New Perspectives: Proceedings of a Symposium held May 31 through June 3, 2011 at Cornell University*. World Scientific, New Jersey & London (2013).
Matic, J.S.: *Non-coding RNAs: The architects of eukaryotic complexity*. EMBO Reports 2:986–991, (2001).
Matic, J.: DNA Unravelled, *in The Boston Globe* (24 September, 2007).
May, K.: Born to communicate. *Creation Magazine* 31(2):41(2009).
Maynard Smith, J.: *The Concept of Information in Biology*, *in* Davies, P. & Gregersen, N.H. (Eds.): *Information and the Nature of Reality: from Physics to Metaphysics*, pp.123–145 Cambridge University Press (2010).
Mayr, E.: *Populations, Species, and Evolution: An abridgement of Animal Species and Evolution*. Harvard University Press, Cambridge, MA (1970).
McGrath, A. & J.C.: *The Dawkins Delusion? Atheist fundamentalism and the denial of the divine*. Society for Promoting Christian Knowledge, London (2007).
McGrath, A.: *Why God won't go away*. SPCK, London (2011).
McLeod, W.T. & Hanks, P. (Ed.): *The New Collins Concise English Dictionary*, William Collins Sons & Co. Ltd, London & Glasgow (1984).
McMullin, E.: *From matter to materialism...and (almost) back*, *in* Davies, P. & Gregersen, N.H. (Eds.): *Information and the Nature of Reality: from Physics to Metaphysics*, pp.13–37 Cambridge University Press, Cambridge (2010).
Menton, D.: *The Hearing Ear and the Seeing Eye*. (DVD) Answers in Genesis, Hebron, KY (2003).
Meyer, S.C.: The Explanatory Power of Design – DNA and the Origin of Information, *in* Dembski, W.A. (Ed.): *Mere Creation – Science, Faith & Intelligent Design*. InterVarsity Press, Downers Grove, Illinois (1998).

Meyer, S.C.: *The Return of the God Hypothesis*. Discovery Institute Centre for the Renewal of Science and Culture, Seattle (1998).
Meyer, S.C.: DNA and other Designs. *First Things*. 102:30–38 (2000).
Meyer, S.C. & Allen, W.P.: *Unlocking the Mystery of Life*. (DVD) Illustra Media (2002).
Meyer, S.C.: DNA and the Origin of Life: Information, Specification, and Explanation, *in* Campbell, J.A. & Meyer, S.C.: *Darwinianism, Design and Public Education*. (Eds.) Michigan State University Press, Lansing, Michigan (2003).
Meyer, S.C. The origin of biological information and the higher taxonomic categories. *Proceedings of the Biological Society of Washington* 117(2):213–239 (2004).
Meyer, S.C.: *Signature in the Cell*. Harper One, New York (2009).
Meyer, S.C. et al.: *Explore Evolution*. Hill House Publishers, London (2009).
Meyer, S.C.: *Darwin's Doubt: The Explosive Origin of Animal Life and the Case for Intelligent Design*. Harper One, New York (2013).
Miller, S.L.: A production of amino acids under possible primitive earth conditions. *Science* 117:528–529 (1953).
Miller, S.L.: Production of Some Organic Compounds under Possible Primitive Earth Conditions. *Journal of the American Chemical Society* 77:2351(1955).
Mills, G.C. & Kenyon, D.: The RNA World: A Critique. *Origins & Design* 17:1 (1996).
Minnich, S.A. & Meyer, S.C.: Genetic Analysis of Coordinate Flagellar and Type III Regulatory Circuits in Pathogenic Bacteria. *in* Dembski, W.A.: (Ed.) *Darwin's Nemesis*. InterVarsity Press, Downers Grove, Illinois, pp.214–223 (2006).
Missler, C.: *Quantum Physics: The Boundaries of Reality* (1998) *at* http://www.khouse.org/articles/1998/62/
Missler, C.: *Cosmic Codes*. Koinonia House, Coeur d'Alene, Idaho (2004).
Missler, C.: *Cosmic Codes - Metacodes*. (CD) Koinonia House, Coeur d'Alene, Idaho (2004).
Missler, C. & Meyer, S.: *In the beginning there was...information*. (CD) Interview with Dr Meyer. Koinonia House, Coeur d'Alene, Idaho (2000).
Missler, C.: *The Beyond Collection*. (DVD) Version 2.0. Koinonia House, Coeur d'Alene, Idaho (2010).
Mizuuchi, M. & Mizuuchi, K.: Efficient Mu transposition requires interaction of transposase with a DNA sequence at the Mu operator: implications for regulation. *Cell* 58:399-408 (1989).
Mockler, T.C. & Ecker, J.R.: Applications of DNA tiling arrays for whole-genome analysis. *Genomics* 85:1-15 (2005).
Monod, J.: *Chance and Necessity: An Essay on the Natural Philosophy of Modern Biology*. Vintage, New York (1972).
Morelle, R.: *Scientists tickle animals to find laughter clues*. BBC News (30 December 2011) *at* http://www.bbc.co.uk/news/science-environment-15880045.
Morris, S.C.: *The Crucible of Creation*. Oxford University Press, Oxford (1998).
Morris, S.C.: *Life's Solution*. Cambridge University Press, Cambridge (2003).
Mueller, J.C., Steiger, S., Fidler, A.E. & Kempenaers, B.: Biogenic Trace Amine-Associated Receptors (TAARs) are Encoded in Avian Genomes: Evidence and Possible Implications. *Oxford Journals Life Sciences Journal of Heredity* 99(2):174–176 (2008).

Murphy, N. & Brown, W.S.: *Did my Neurons Make Me Do It?* Oxford University Press, Oxford (2007).

Mushegian, A.R. & Koonin, E.V.: A minimal gene set for cellular life derived by comparison of complete bacterial genomes. *Proceedings of the National Academy of Science of the USA* 93:10268–73 (1996).

Nagle, T.: *Mind and Cosmos. Why the Materialist Neo-Darwinian Conception of Nature is Almost Certainly False.* Oxford University Press (2012).

Nesse, R.M. & Williams, G.C.: Evolution and the Origins of Disease. *Scientific American* 279(5):86–93 (1998).

Ng, S.B.L. & Doig, A.J.: Molecular and Chemical basis of prion-related diseases. *Chemical Society Reviews* 26:425–432 (1997).

Ni, X. et al.: A new Early Oligocene peradectine marsupial (Mammalia) from the Burqin region of Xinjang, China. *Naturwissenschaften* 94(3):237–241 (2007).

Nilsson, D.E. & Pelger, S.: A pessimistic estimate of the time required for an eye to evolve. *Proceedings of the Royal Society, London* B 256:53–58 (1994).

Oberbeck, V.R. & Fogleman, G.: Estimates of the maximum time to originate life. *Origins of Life* 19:549–560 (1989).

Orgel, L.E.: The Origin of Life – How long did it take? *Origins of Life* 28:91–96 (1998).

Palmer, T.: *Controversy – Catastrophism and Evolution.* Kluwer Academic, New York (1999).

Parker, A.: The Eyes have it. *Cosmos* 2:48–49 (2005).

Putnam, N.H., et al.: Sea anemone genome reveals ancestral eumetazoan gene repertoire and genomic organization. *Science* 317(5834):86–94 (2007).

Pauling, L.: *No More War!* Dodd, Mead & Co., New York (1958).

Peacocke, A.: The sciences of complexity: a new theological resource? *in* Davies, P. & Gregersen, N.H. (Eds.): *Information and the Nature of Reality. From Physics to Metaphysics.* Cambridge University Press, Cambridge, pp.249–281 (2010).

Pearce, E.K.V.: *Who Was Adam?* Africa Centre for World Mission, Walkerville, South Africa (1987).

Penfield, W.: *The Mystery of the Mind: a Critical Study of Consciousness and the Human Brain.* Princeton University Press, Princeton, New Jersey (1975).

Pietsch, P.: *Shufflebrain: the Quest for the Hologramic Mind.* Houghton Mifflin, Boston, MA (1981).

Pinker, S.: A Matter of the Soul. *The Weekly Standard,* (2 February, 1998).

Pinker, S.: *The Language instinct.* Harper, New York (1994).

Pinker, S.: *How the Mind Works.* WW Norton, New York (1997).

Pohorille, A. & Deamer, D.: Artificial Cells: prospects for biotechnology. *Trends in Biotechnology* 20:123–128 (2002).

Polkinghorne, J. & Beale, N.: *Questions of Truth – Fifty-one Responses to Questions about God, Science, and Belief.* Westminster John Knox Press, Louisville, Kentucky (2009).

Poplawski, N.J.: Radial motion into an Einstein–Rosen bridge. *Physics Letters B* 687:110–113 (2010). http://newsinfo.iu.edu/news/page/normal/13995.html

Popper, K.: *Scientific Reduction and the Essential Incompleteness of All Science, in* Ayala, F.J. & Dobzhansky, T. (Eds.): *Studies in the Philosophy of Biology: Reduction and Related Problems.* University of California Press, Berkeley (1974).

Porter, K.R., Claude, A. & Fullam, E.F.: A study of tissue culture cells by electron microscopy. *Journal of Experimental Medicine* 81(3):233–246 (1945).
Powell, K.: Enteric Diseases, contagion and miasms in the 19th-Century. *Canberra Doctor* (Feb 2007).
Pratt, D.: David Bohm and the Implicate Order. *Sunrise Magazine*. Theosophical University Press. (February /March 1993).
Prigogine, I. & Stengers, I.: *Order out of Chaos*. University of Michigan, Bantam Books (1984).
Radetsky, P.: Silence, signs and wonder. *Discover Magazine*. 15(8):60–68 (1994).
Raloff, J.: Dirty minds. *Cosmos* 37:74-79 (2011).
Rasmussen, S., Chen, L., Stadler, B.M.R. & Stadler, P.F.: Proto-Organism Kinetics: Evolutionary Dynamics of Lipid Aggregates with Genes and Metabolism. *Origins of Life and Evolution of Biospheres* 34:171–180 (2004).
Razin, A. & Riggs, A.D.: DNA methylation and gene function. *Science* 210:604–10 (1980).
Rees, M.: *Just Six Numbers: The Deep Forces that Shape the Universe*. Weidenfeld & Nicolson, London (1999).
Rennie, J.: 15 Answers to the Creationist Nonsense. *Scientific American* 287(1):78–85 (2002).
Richardson, M.K. et al.: There is no highly conserved embryonic stage in the vertebrates: implications for current theories of evolution and development. *Anatomy and Embryology* 196:91–106 (1997).
Richardson, M.K., Hanken, J., Selwood, L., Wright, G.M., Richards, R.J., Pieau, C., Raynaud, A.: Haeckel, embryos, and evolution. *Science* 280(5366):983, 985–6. (1998).
Rolston, H.III: Care on Earth: generating informed concern, *in* Davies, P. & Gregersen, N.H.: *Information and the Nature of Reality: from Physics to Metaphysics*, pp.205–246 Cambridge University Press (2010).
Ross, H.: *The Creator and the Cosmos*. NavPress, Colorado Springs (1993).
Ross, H.: *Why the Universe is the Way it is*. Baker Books, Grand Rapids, Michigan (2008).
Ross, H.: *More than a Theory – Revealing a Testable Model for Creation*. Baker Books, Grand Rapids, Michigan (2008).
Rouse Ball, W.W.: *A Short Account of the History of Mathematics*. 4th Ed. (1908) *at* http://www.maths.tcd.ie/pub/HistMath/People/Fourier/RouseBall/RB_Fourier.html.
Russell, B.: *The Collected Papers of Bertrand Russell, Vol. 12:'Contemplation and Action', 1902–1914*. Routledge, London (1985).
Russell, B.: *History of Western Philosophy*. Routledge, London (2000).
Sagan, C.: *Pale Blue Dot*. Ballantine Books, New York (1994).
Sagan, L.: On the Origin of Mitosing Cells. *Journal Theoretical Biology* 14(3):255–274 (1967).
Salisbury, F.B.: Doubts about the Modern Synthetic Theory of Evolution. *American Biology Teacher*, 33 (1971).
Sanford, J.C.: *Genetic Entropy and the Mystery of the Genome*. Ivan Press, Lima, New York (2005).
Sarfati, J.: *Refuting Evolution 2*. Answers in Genesis, Acacia Ridge, Queensland, Australia (2002).
Sarfati, J.: *By Design*. Creation Book Publishers, Brisbane, Australia (2008).

Sarfati, J.: *The Greatest Hoax on Earth? Refuting Dawkins on Evolution*. Creation Book Publishers, Atlanta, Georgia (2010).
Schroeder, J.R.: Creation or Evolution: Which is More Believable? *The Good News* 14(6):4–7 (2009).
Schutzenberger, M.P.: The Miracle of Darwinism. *Origins and Design*, 17(2):10–15 (1996).
Seiglie, M.: 10 Ways Darwin got it Wrong. *The Good News* 14(6) United Church of God, Milford, Ohio (2009).
Seiglie, M.: Prove there is a God Even Without the Bible. *The Good News* 17(2):14–15. United Church of God, Milford, Ohio (2012).
Sell, C.: On the Right Scent. *Chemistry in Britain*. 33(3):39–42 (1997).
Setterfield, B.: Recent Lightspeed Publicity. 10/2/2002 *at* http://ldolphin.org/recentlight.html.
Shapiro, J.A.: *A Third Way* (alternative to Darwinism & Creationism). *Boston Review* 22(1):32–33 (1997).
Shapiro, R.: Prebiotic cytosine synthesis: A critical analysis and implications for the origin of life. *Proceedings of the National Academy of Sciences of the USA* 96:4396–4401 (1999).
Shapiro, R.: A replicator was not involved in the origin of life. *Journal of the International Union of Biochemistry and Molecular Biology – Life*. 49:173–175 (2000).
Shapiro, R.: A simpler origin for life. *Scientific American* 296:46–53 (2007).
Shariatmadari, H.: What is Reality? *Horizon*, BBC Productions, UK (2011).
Shorter Oxford English Dictionary. Oxford University Press, London (1986).
Simpson, G.G.: *Tempo and Mode in Evolution*. Columbia University Press, New York (1944).
Simpson, G.G.: *The Meaning of Evolution*. (Rev. ed.) Yale University Press, New Haven (1967).
Skybreak, A.: *The Science of Evolution and the Myth of Creationism: Knowing what's Real and Why it Matters*. Insight Press, Chicago, Illinois (2006).
Smit, A.F.A. & Riggs, A.D.: Tiggers and other DNA transposon fossils in the human genome. *Proceedings of the National Academy of Sciences of the USA* 93:1443–1448 (1996).
Smith, C.: *Codes and Creation*. (CD) Creation Ministries International.
Smith, C.: Codes & Creation (DVD) Creation Ministries International (2008).
Speer, B.S., Shoemaker, N.B. & Salyers A.A.: Bacterial resistance to tetracycline: mechanisms, transfer, and clinical significance. *Clinical Microbiology Review* 5(4):387-399 (1992).
Spetner, L.M.: *Not by Chance: Shattering the Modern Theory of Evolution*. The Judaica Press, Inc., New York (1997).
Spinney, L., The old curiosity shop. *New Scientist* 198(2656):42–45 (2007). Also see: http://www.newscientist.com/article/dn13927-five-things-humans-no-longer-need.html
Stein, B.: *Expelled: No Intelligence Allowed*. (DVD) Condios Entertainment (2009).
Stenger, V.J.: *God, the failed hypothesis: How science shows that God does not exist*. Prometheus Books, New York (2007).
Stiasny-Kolster, K.: Combination of 'idiopathic' REM Sleep Behavior Disorder and olfactory dysfunction as possible indicator for α-synucleinopathy

demonstrated by dopamine transporter FP-CIT-SPECT. *Brain* 128(1):126–137 (2005).

Strobel, L.: *The Case for a Creator.* Zondervan, Grand Rapids, Michigan (2004).

Strobel, L. & Vogel, J.: *The Case for a Creator.* (Student Edition) Zondervan, Grand Rapids (2004).

Stryer, L.: The world's smallest molecular motor. *Biochemistry* (5th Ed.) WH Freeman & Co., New York, Ch.18.4.3 (2002) *at* http://www.ncbi.nlm.nih.gov/books/NBK22388/

Swift, D.: *Evolution under the microscope.* Leighton Academic Press, Stirling, UK (2002).

Szostak, J.W., Bartel, D.P. & Luisi, P.L.: Synthesizing Life. *Nature* 409:387–390 (2001).

Talbot, M.: *The Universe as a Hologram...does objective reality really exist...or is the universe a phantasm?* Harper & Collins, New York (1991). http://www.universenews.net/the-universe/the-universe-as-a-hologram-by-michael-talbot-state-of-earth/

Thaxton, C.B., Bradley, W.L., & Olsen, R.L.: *The Mystery of Life's Origin: Reassessing Current Theories.* Philosophical Library, New York (1984).

The World Book Encyclopaedia. World Book Inc., Chicago (1984).

ThinkQuest: *In A World of Order…Chaos Reigns!* Oracle Education Foundation (1997). http://library.thinkquest.org/3120/

Thompson, D.W.: *On Growth and Form.* Cambridge University Press (1917) at http://archive.org/details/ongrowthform1917thom.

Tipler, F.J.: *The Physics of Immortality – Modern Cosmology, God and the Resurrection of the Dead.* Anchor Books, New York (1994).

Valantine, J.W. & Erwin, D.H.: Interpreting Great Developmental Experiments: The Fossil Record, *in* Raff, R.A. & E.C. (Eds.): *Development as an Evolutionary Process*, pp.71–107. MBL Lectures in Biology Series, Alan R. Liss Inc., New York (1987).

Valentine, J.W.: *On the Origin of Phyla.* University of Chicago Press, Chicago (2004).

van Wyhe, J.: *Darwin: The Story of the Man and his Theories of Evolution.* Andre Deutsch, London (2008).

Vaneechoutte, M.: The Scientific Origin of Life: Considerations on the Evolution of Information, Leading to an Alternative Proposal for Explaining the Origin of the Cell, a Semantically Closed System. *Annals of the New York Academy of Sciences* 901:139–147 (2000).

Venter, C. et al.: Creation of a Bacterial Cell Controlled by a Chemically Synthesized Genome. *Science* 329(5987):52–56 (2010).

Voie, A.: Biological function and the genetic code are interdependent. *Chaos, Solitons and Fractals* 28(4):1000–1004 (2006).

von Baeyer, H.C.: In the Beginning was the Bit. *New Scientist* 169(2278):26 (2001).

Wald, G.: The origin of life. *Scientific American* 191:44–53 (1954).

Wald, G.: The Origin of Life, *in Life: Origin and Evolution.* WH Freeman, San Francisco (1979).

Waldrupp, M.M.: Did Life Really Start out in an RNA World? *Science* 246:1248–49 (1989).

Wallis, M.K. & Wickramasinghe, N.C.: Interstellar transfer of planetary microbiota. *Monthly Notices of the Royal Astronomical Society* 348:52 (2004).

Ward, K.: *Why There Almost Certainly Is a God.* Lion Books, Oxford (2008).
Ward, K.: *God as the ultimate informational principle, in* Davies, P. & Gregersen, N.H. (Eds.): *Information and the Nature of Reality. From Physics to Metaphysics,* pp.282–300 Cambridge University Press, Cambridge (2010).
Watson, J.D. & Crick, F.H.C.: Molecular structure of nucleic acids: A structure for DNA. *Nature* 171:737–738 (1953).
Wells, J.: *Icons of Evolution.* Regnery Publishing, Washington DC (2002).
Wertheim, M.: Alas Poor Zombie. *Cosmos* 11:86–92 (2006).
Wertheim, M.: Are We Just Numbers in the Matrix? *Cosmos* 52:60–65 (2013).
West, G.B., Brown, J.H. & Enquist, B.J.: Life's Universal Scaling Laws. *Physics Today* 57(9):36-42 (2004).
West, J.G. (Ed.): *The Magician's Twin. C.S. Lewis on Science, Scientism, and Society.* Discovery Institute Press, Seattle (2012).
Wheeler, J.C.: *Cosmic Catastrophes: Exploding Stars, Black Holes, and Mapping the Universe.* Cambridge University Press, Cambridge (2007).
Wigner, E.: The Unreasonable Effectiveness of Mathematics in the Natural Sciences, *in Symmetries and Reflections.* Indiana University Press (1967).
Williams, A.: Astonishing DNA complexity demolishes neo-Darwinism. *Journal of Creation* 21(3):111–117 (2007).
Williams, A.: Mutations: evolution's engine becomes evolution's end! *Journal of Creation* 22(2):60–66 (2008).
Williams, A.: How Life Works. *Journal of Creation* 22(2):89–90 (2008).
Woetzel, D.: Evolutionists retreating from the arena of science. *Journal of Creation* 23(3):123–127 (2009).
Wolfenden, R.: *Without enzymes, biological reaction essential to life takes 2.3 billion years.* University of North Carolina website *at* http://www.med.unc.edu/biochem/news/without-enzyme-biological-reaction-essential-to-life-takes-2-3-billion-years-unc-study/
Wolfson, J.S. & Hooper, D.C.: Bacterial resistance to quinolones: mechanisms and clinical importance. *Reviews of Infectious Diseases* 11(Supplement 5):S960–S968 (1989).
Wood, T.C.: Genome Decay in the Mycoplasmas. *Impact* 340 (Oct. 2001); www.icr.org/pubs/imp/imp-340.htm>
Woodward, T.: *Darwin Strikes Back.* Baker Books, Grand Rapids (2006).
Yockey, H.P.: A calculation of the Probability of Spontaneous Biogenesis by Information Theory. *Journal of Theoretical Biology* 67:377–398 (1977).
Yockey, H.P.: *Information Theory and Molecular Biology.* Cambridge University Press, Cambridge (1992).
Zouni, A. et al.: Crystal structure of photosystem II from *Synechococcus* elongatus at 3.8Å resolution *Nature* 409(6821):739–43 (2001).

Illustration Credits

Cover Photo: Great horned Owl. Reproduced under licence from Holly Kuchera/iStock Collection/Thinkstock at http://www.thinkstockphotos.com.au/image/104435332;

15 Figure 1.1: The Historical Ages of Europe illustrating Information Growth. R. Wiles
21 Figure 1.2: Steps in Information Generation, Transmission and Communication. R. Wiles
32 Figure 1.3: Types of Explanation. R. Wiles
35 Figure 1.4: Summary of Scientific Method. R. Wiles
49 Figure 1.5: Summary of the Neo-Darwinian Synthesis and its implied Information flows. R. Wiles
54 Figure 1.6: Layers of Information in the Universe. R. Wiles
66 Figure 2.1: Size Scales in the Universe. R. Wiles/D. Stone after Gonzalez, G. & Richards, J.W.: *The Privileged Planet*. Regnery Publishing, Washington DC (2004) p.216.
67 Figures 2.2: Open Loop System & 2.3: Closed Loop System R. Wiles
71 Figure 2.4: The Generation of Information. R. Wiles
77 Figure 3.1: From Crick, F.H.C.: Ideas on Protein Synthesis. (1956). Wellcome Library, London at http://wellcomelibrary.org/player/b18174139#0/1/-0.0321,0.375,1.0643,0.7501. Reproduced with permission.
78 Figure 3.2: Information Flow in Cells. R. Wiles/D. Stone
79 Figure 3.3: Information Flow in Cells showing the Mutation Pathway. R. Wiles/D. Stone
80 Figure 3.4: Information Flows in Cells showing other Intrinsic Pathways. R. Wiles/D. Stone
82 Figure 3.5: The Basic Biological Information Pathway. R. Wiles/D. Stone
89 Figure 3.6: Categories of Intracellular Information. R. Wiles/D. Stone
90 Figure 3.7: Information Generation in Biological Systems. R. Wiles/D. Stone
93 Figure 3.8: Popper's Conundrum. R. Wiles
94 Figure 3.9: ATP-Synthase Generator. Charles McCown - Copyright Apologetics Press (2005) http://www.apologeticspress.org/apcontent.aspx?category=12&article=1281. Reproduced with permission courtesy of Apologetics Press
95 Figure 3.10: Chemiosmosis and ATP Synthesis - reproduced with permission courtesy of Professor Gary E. Kaiser from his course website: *The Grapes of Staph* at http://student.ccbcmd.edu/~gkaiser/index.html
99 Figure 3.11: Information Flows in Cells illustrating the Epigenome. R. Wiles/D. Stone
101 Figure 3.12: A Fitness Landscape. Reproduced with permission from Fowler, T.B. & Kuebler, D. *The Evolution Controversy*. Baker Academic, Grand Rapids. (2007) p.68
103 Figure 3.13: Minimum Information requirement for the Formation of the First Cell. R. Wiles
113 Figure 4.1: The Citric Acid Cycle. Wikimedia Commons – reproduced under the GNU Free Documentation License. http://en.wikipedia.org/wiki/File:Citric_acid_cycle_with_aconitate_2.svg (Accessed 30th Nov, 2013)

219 Figures 7.1 D'Arcy Thompson's crab 'transformations' and 7.2: D'Arcy Thompson's skull 'transformations' reproduced from: Thompson, D.W.: *On Growth and Form*. Cambridge Univ. Press (1917) p.744 at http://archive.org/details/ongrowthform1917thom. This file is in the public domain. (Accessed 7th July, 2013)

234 Figure 7.3: Material and Non-material Entities Required for Living Organisms. R. Wiles/D. Stone

257 Figure 8.1: Our Information dependent Physical and Non-Material Universes R. Wiles/D. Stone

258 Figure 8.2: Cartoon: 'Visualising the Fourier Transform' by Randall Munroe at http://xkcd.com/26/ (Accessed 29th Nov., 2013).
Figures 8.3 Rectangular Function and 8.4 Magnitude Image of a Rectangular Function are reprinted with permission of The MathWorks, Inc.: Visualising the Fourier Transform at http://www.mathworks.com.au/help/images/fourier-transform.html?s_tid=doc_12b

259 Figure 8.5: The Duality of Light showing the Explicate and Implicate Orders. R. Wiles

262 Figure 8.6: A Black Hole in 2D space. Reprinted from: Al-Khalili, J.S.: *Black Holes, Worm Holes & Time Machines*. (1999) p.92. Reproduced with permission of Taylor & Francis Group LLC - Books via Copyright Clearance Center

263 Figure 8.7: A Spinning Black Hole. Reprinted from: Al-Khalili, J.: *Black Holes, Worm Holes & Time Machines*. (1999) p.96. Reproduced with permission of Taylor & Francis Group LLC - Books via Copyright Clearance Center

265 Figure 8.8: Response of a Gaussian Filter. Reprinted with permission of The MathWorks, Inc.: http://www.mathworks.com/help/toolbox/images/f21-17064.html

269 Figure 8.9: The Physical Information Generator. R. Wiles

280 Figure 9.1 Information Outcomes in Our Universe. R. Wiles

286 Figure 9.2: The Universal Information Generator. R. Wiles

310 Table 10.3: Categorisation of Mental attributes according to Genus. R. Wiles/D. Stone.

321 Figures 10.1 Relationships between the Fundamental Entities within the Universe and 10.2: Relationships between some Non-material Entities of the Universe. R. Wiles/D. Stone

323 Figure 10.3: An expanded Illustration of both the Physical and Non-Material Universe. R. Wiles/D. Stone

345 Figure 12.1: Information Sets within the Domain of Life. R. Wiles/D. Stone

351 Figure 12.2: Schrödinger's Cat. Illustration by D. Hatfield from Wikimedia Commons at http://en.wikipedia.org/wiki/File:Schrodingers_cat.svg (2008). This file is licensed under the Creative Commons Attribution-Share Alike 3.0 Unported license.

Photo of Author – Beth Cole

Index

About the Author

Robert Wiles is a senior lecturer in the Rural Clinical School at the Australian National University and is the rural academic co-ordinator for the Snowy Mountains region to the south of Canberra. After graduating from the Faculty of Medicine, University of Sydney, he spent five years in the United Kingdom pursuing post-graduate studies in obstetrics, anaesthetics, orthopaedics and general practice before returning to Australia. He has been a rural doctor in Cooma for 20 years and has a visiting medical officer appointment in emergency medicine and in-patient care at the Cooma District Hospital where he has previously practised anaesthetics and obstetrics for over a decade.

He has a passion for teaching and learning in Medicine which has extended to exploring and understanding the Universe.

www.ingramcontent.com/pod-product-compliance
Lightning Source LLC
LaVergne TN
LVHW070847210926
848482LV00016B/882